Nissan Maxima Automotive Repair Manual

**by Bob Henderson
and John H Haynes**
Member of the Guild of Motoring Writers

Models covered:
All Nissan Maxima models
1993 through 2001

(12D3 - 72021)

ABCDE
FGHIJ
KLMNO
P

Haynes Publishing Group
Sparkford Nr Yeovil
Somerset BA22 7JJ England

Haynes North America, Inc
861 Lawrence Drive
Newbury Park
California 91320 USA

Acknowledgements

Wiring diagrams originated exclusively for Haynes North America, Inc. by Valley Forge Technical Information Services. Technical writers who contributed to this project include Jay Storer, Jeff Kibler and Eric Godfrey.

© **Haynes North America, Inc. 1999, 2002**
With permission from J.H. Haynes & Co. Ltd.

A book in the Haynes Automotive Repair Manual Series

Printed in the U.S.A.

All rights reserved. No part of this book may be reproduced or transmitted in any form or by any means, electronic or mechanical, including photocopying, recording or by any information storage or retrieval system, without permission in writing from the copyright holder.

ISBN 1 56392 450 1

Library of Congress Control Number 2001097039

While every attempt is made to ensure that the information in this manual is correct, no liability can be accepted by the authors or publishers for loss, damage or injury caused by any errors in, or omissions from, the information given.

Contents

Introductory pages

About this manual	0-5
Introduction to the Nissan Maxima	0-5
Vehicle identification numbers	0-6
Buying parts	0-8
Maintenance techniques, tools and working facilities	0-8
Booster battery (jump) starting	0-16
Jacking and towing	0-17
Automotive chemicals and lubricants	0-18
Conversion factors	0-19
Safety first!	0-20
Troubleshooting	0-21

Chapter 1
Tune-up and routine maintenance — 1-1

Chapter 2   Part A
Single Overhead Cam (SOHC) engine — 2A-1

Chapter 2   Part B
Dual Overhead Cam (DOHC) engine — 2B-1

Chapter 2   Part C
General engine overhaul procedures — 2C-1

Chapter 3
Cooling, heating and air conditioning systems — 3-1

Chapter 4
Fuel and exhaust systems — 4-1

Chapter 5
Engine electrical systems — 5-1

Chapter 6
Emissions and engine control systems — 6-1

Chapter 7   Part A
Manual transaxle — 7A-1

Chapter 7   Part B
Automatic transaxle — 7B-1

Chapter 8
Clutch and driveaxles — 8-1

Chapter 9
Brakes — 9-1

Chapter 10
Suspension and steering systems — 10-1

Chapter 11
Body — 11-1

Chapter 12
Chassis electrical system — 12-1

Wiring diagrams — 12-20

Index — IND-1

Haynes mechanic, author and photographer with 1995 Nissan Maxima

About this manual

Its purpose

The purpose of this manual is to help you get the best value from your vehicle. It can do so in several ways. It can help you decide what work must be done, even if you choose to have it done by a dealer service department or a repair shop; it provides information and procedures for routine maintenance and servicing; and it offers diagnostic and repair procedures to follow when trouble occurs.

We hope you use the manual to tackle the work yourself. For many simpler jobs, doing it yourself may be quicker than arranging an appointment to get the vehicle into a shop and making the trips to leave it and pick it up. More importantly, a lot of money can be saved by avoiding the expense the shop must pass on to you to cover its labor and overhead costs. An added benefit is the sense of satisfaction and accomplishment that you feel after doing the job yourself.

Using the manual

The manual is divided into Chapters. Each Chapter is divided into numbered Sections, which are headed in bold type between horizontal lines. Each Section consists of consecutively numbered paragraphs.

At the beginning of each numbered Section you will be referred to any illustrations which apply to the procedures in that Section. The reference numbers used in illustration captions pinpoint the pertinent Section and the Step within that Section. That is, illustration 3.2 means the illustration refers to Section 3 and Step (or paragraph) 2 within that Section.

Procedures, once described in the text, are not normally repeated. When it's necessary to refer to another Chapter, the reference will be given as Chapter and Section number. Cross references given without use of the word "Chapter" apply to Sections and/or paragraphs in the same Chapter. For example, "see Section 8" means in the same Chapter.

References to the left or right side of the vehicle assume you are sitting in the driver's seat, facing forward.

Even though we have prepared this manual with extreme care, neither the publisher nor the author can accept responsibility for any errors in, or omissions from, the information given.

NOTE
A **Note** provides information necessary to properly complete a procedure or information which will make the procedure easier to understand.

CAUTION
A **Caution** provides a special procedure or special steps which must be taken while completing the procedure where the Caution is found. Not heeding a Caution can result in damage to the assembly being worked on.

WARNING
A **Warning** provides a special procedure or special steps which must be taken while completing the procedure where the Warning is found. Not heeding a Warning can result in personal injury.

Introduction to the Nissan Maxima

The Nissan Maxima is offered as a four-door sedan body style.

The transversely mounted 3.0-liter V6 engines used in these models are equipped with a sequential multi-port electronic fuel injection system. Models from 1993 to 1994 came with a SOHC VG30DE V6, while later models all have a DOHC VQ30DE V6. A limited number of 1993 and 1994 models were equipped with a DOHC VE30DE V6 that is not covered by this manual.

The engine transmits power to the front wheels through either a five-speed manual transaxle or an electronically controlled four-speed automatic transaxle via independent driveaxles.

The Maxima features a steel unibody and independent front suspension with MacPherson strut/coil spring suspension and a stabilizer bar. The rear suspension on 1993 and 1994 models utilized MacPherson struts and parallel links, while 1995 and later models are equipped with a solid axle supported by shock absorber/coil spring assemblies, a lateral link and trailing arms. The rack-and-pinion steering unit is mounted behind the engine with power-assist as standard equipment.

All models are equipped with front disc brakes and power assist. Models in 1993 and 1994 were available with either drum or disc rear brakes, while 1995 and later models all have rear disc brakes. Anti-lock Brake Systems (ABS) are available on all models (optional on 1996 and earlier, standard on 1997 and later).

Vehicle identification numbers

Modifications are a continuing and unpublicized process in vehicle manufacturing. Since spare parts manuals and lists are compiled on a numerical basis, the individual vehicle numbers are essential to correctly identify the component required.

Vehicle Identification Number (VIN)

This very important identification number is located on a plate attached to the dashboard inside the windshield on the driver's side of the vehicle **(see illustration)**. The VIN also appears on the Vehicle Certificate of Title and Registration. It contains information such as where and when the vehicle was manufactured, the model year and the body style.

Two particularly important pieces of information found in the VIN are the engine code and the model year code. Counting from the left, the engine code letter designation is the 4th digit and the model year code designation is the 10th digit.

On the model years covered by this manual the engine codes are:

H..... VG30DE SOHC V6 (1993 to 1994)
E..... VE30DE DOHC V6 (1993 and 1994)
C..... VQ30DE DOHC V6 (1995 and later)

On the models covered by this manual the model year codes are:

P... 1993
R... 1994
S... 1995
T... 1996
V... 1997
W.. 1998
X... 1999
Y... 2000
1... 2001

Vehicle identification plate

This metal plate is riveted to the firewall on the engine compartment side, usually on the right, but sometimes in the center **(see illustration)**. The plate contains codes for paint, trim, and drivetrain information.

Chassis Identification Number

The chassis identification number is stamped in the center of the firewall in the engine compartment **(see illustration)**. Like the VIN it contains valuable information about the manufacturing of the vehicle such as the destination, model variations and transaxle information.

Vehicle Safety Certification label

The Safety Certification label is affixed

The Vehicle Identification Number (VIN) is visible through the driver's side of the windshield

The vehicle identification plate on the firewall includes information about the engine, transaxle, and paint and body codes

Vehicle identification numbers

The Chassis Identification Number is stamped on the engine compartment firewall

The Safety Certification label is affixed to the driver's side door end or post

The Engine Identification Number (arrow) is stamped on the left end of the engine block, near the transaxle

The transaxle identification number (arrow) is stamped near the top of the transaxle, at the juncture of the engine and transaxle

to the left front door pillar or the pillar-side of the door **(see illustration)**. The plate contains the name of the manufacturer, the month and year of production, the Gross Vehicle Weight Rating (GVWR) and the safety certification statement.

Engine identification numbers

The engine code number can be found on a pad on the left end of the cylinder block, near the transaxle **(see illustration)**.

Transaxle identification numbers

The transaxle identification number is stamped on top of the bellhousing **(see illustration)**. In some models, the number may also be on a sticker on the forward side of the transaxle.

Vehicle Emissions Control Information (VECI) label

The emissions control information label is found under the hood, normally on the radiator support or the bottom side of the hood (see Chapter 6). This label contains information on the emissions control equipment installed on the vehicle, as well as tune-up specifications.

Buying parts

Replacement parts are available from many sources, which generally fall into one of two categories - authorized dealer parts departments and independent retail auto parts stores. Our advice concerning these parts is as follows:

Retail auto parts stores: Good auto parts stores will stock frequently needed components which wear out relatively fast, such as clutch components, exhaust systems, brake parts, tune-up parts, etc. These stores often supply new or reconditioned parts on an exchange basis, which can save a considerable amount of money. Discount auto parts stores are often very good places to buy materials and parts needed for general vehicle maintenance such as oil, grease, filters, spark plugs, belts, touch-up paint, bulbs, etc. They also usually sell tools and general accessories, have convenient hours, charge lower prices and can often be found not far from home.

Authorized dealer parts department: This is the best source for parts which are unique to the vehicle and not generally available elsewhere (such as major engine parts, transmission parts, trim pieces, etc.).

Warranty information: If the vehicle is still covered under warranty, be sure that any replacement parts purchased - regardless of the source - do not invalidate the warranty!

To be sure of obtaining the correct parts, have engine and chassis numbers available and, if possible, take the old parts along for positive identification.

Maintenance techniques, tools and working facilities

Maintenance techniques

There are a number of techniques involved in maintenance and repair that will be referred to throughout this manual. Application of these techniques will enable the home mechanic to be more efficient, better organized and capable of performing the various tasks properly, which will ensure that the repair job is thorough and complete.

Fasteners

Fasteners are nuts, bolts, studs and screws used to hold two or more parts together. There are a few things to keep in mind when working with fasteners. Almost all of them use a locking device of some type, either a lockwasher, locknut, locking tab or thread adhesive. All threaded fasteners should be clean and straight, with undamaged threads and undamaged corners on the hex head where the wrench fits. Develop the habit of replacing all damaged nuts and bolts with new ones. Special locknuts with nylon or fiber inserts can only be used once. If they are removed, they lose their locking ability and must be replaced with new ones.

Rusted nuts and bolts should be treated with a penetrating fluid to ease removal and prevent breakage. Some mechanics use turpentine in a spout-type oil can, which works quite well. After applying the rust penetrant, let it work for a few minutes before trying to loosen the nut or bolt. Badly rusted fasteners may have to be chiseled or sawed off or removed with a special nut breaker, available at tool stores.

If a bolt or stud breaks off in an assembly, it can be drilled and removed with a special tool commonly available for this purpose. Most automotive machine shops can perform this task, as well as other repair procedures, such as the repair of threaded holes that have been stripped out.

Flat washers and lockwashers, when removed from an assembly, should always be replaced exactly as removed. Replace any damaged washers with new ones. Never use a lockwasher on any soft metal surface (such as aluminum), thin sheet metal or plastic.

Fastener sizes

For a number of reasons, automobile manufacturers are making wider and wider use of metric fasteners. Therefore, it is important to be able to tell the difference between standard (sometimes called U.S. or SAE) and metric hardware, since they cannot be interchanged.

All bolts, whether standard or metric, are sized according to diameter, thread pitch and

Maintenance techniques, tools and working facilities

length. For example, a standard 1/2 - 13 x 1 bolt is 1/2 inch in diameter, has 13 threads per inch and is 1 inch long. An M12 - 1.75 x 25 metric bolt is 12 mm in diameter, has a thread pitch of 1.75 mm (the distance between threads) and is 25 mm long. The two bolts are nearly identical, and easily confused, but they are not interchangeable.

In addition to the differences in diameter, thread pitch and length, metric and standard bolts can also be distinguished by examining the bolt heads. To begin with, the distance across the flats on a standard bolt head is measured in inches, while the same dimension on a metric bolt is sized in millimeters (the same is true for nuts). As a result, a standard wrench should not be used on a metric bolt and a metric wrench should not be used on a standard bolt. Also, most standard bolts have slashes radiating out from the center of the head to denote the grade or strength of the bolt, which is an indication of the amount of torque that can be applied to it. The greater the number of slashes, the greater the strength of the bolt. Grades 0 through 5 are commonly used on automobiles. Metric bolts have a property class (grade) number, rather than a slash, molded into their heads to indicate bolt strength. In this case, the higher the number, the stronger the bolt. Property class numbers 8.8, 9.8 and 10.9 are commonly used on automobiles.

Strength markings can also be used to distinguish standard hex nuts from metric hex nuts. Many standard nuts have dots stamped into one side, while metric nuts are marked with a number. The greater the number of dots, or the higher the number, the greater the strength of the nut.

Metric studs are also marked on their ends according to property class (grade). Larger studs are numbered (the same as metric bolts), while smaller studs carry a geometric code to denote grade.

It should be noted that many fasteners, especially Grades 0 through 2, have no distinguishing marks on them. When such is the case, the only way to determine whether it is standard or metric is to measure the thread pitch or compare it to a known fastener of the same size.

Standard fasteners are often referred to as SAE, as opposed to metric. However, it should be noted that SAE technically refers to a non-metric fine thread fastener only. Coarse thread non-metric fasteners are referred to as USS sizes.

Bolt strength marking (standard/SAE/USS; bottom - metric)

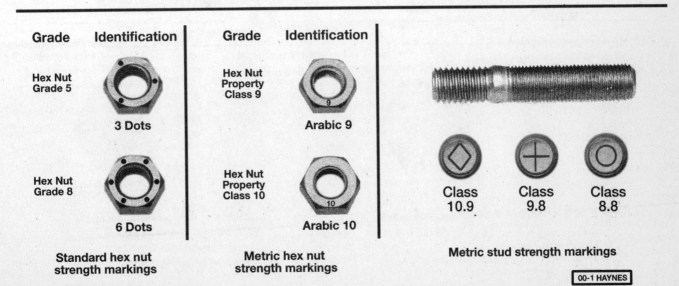

Standard hex nut strength markings

Metric hex nut strength markings

Metric stud strength markings

Maintenance techniques, tools and working facilities

Since fasteners of the same size (both standard and metric) may have different strength ratings, be sure to reinstall any bolts, studs or nuts removed from your vehicle in their original locations. Also, when replacing a fastener with a new one, make sure that the new one has a strength rating equal to or greater than the original.

Tightening sequences and procedures

Most threaded fasteners should be tightened to a specific torque value (torque is the twisting force applied to a threaded component such as a nut or bolt). Overtightening the fastener can weaken it and cause it to break, while undertightening can cause it to eventually come loose. Bolts, screws and studs, depending on the material they are made of and their thread diameters, have specific torque values, many of which are noted in the Specifications at the beginning of each Chapter. Be sure to follow the torque recommendations closely. For fasteners not assigned a specific torque, a general torque value chart is presented here as a guide. These torque values are for dry (unlubricated) fasteners threaded into steel or cast iron (not aluminum). As was previously mentioned, the size and grade of a fastener determine the amount of torque that can safely be applied to it. The figures listed here are approximate for Grade 2 and Grade 3 fasteners. Higher grades can tolerate higher torque values.

Fasteners laid out in a pattern, such as cylinder head bolts, oil pan bolts, differential cover bolts, etc., must be loosened or tightened in sequence to avoid warping the component. This sequence will normally be shown in the appropriate Chapter. If a specific pattern is not given, the following procedures can be used to prevent warping.

Metric thread sizes	Ft-lbs	Nm
M-6	6 to 9	9 to 12
M-8	14 to 21	19 to 28
M-10	28 to 40	38 to 54
M-12	50 to 71	68 to 96
M-14	80 to 140	109 to 154
Pipe thread sizes		
1/8	5 to 8	7 to 10
1/4	12 to 18	17 to 24
3/8	22 to 33	30 to 44
1/2	25 to 35	34 to 47
U.S. thread sizes		
1/4 - 20	6 to 9	9 to 12
5/16 - 18	12 to 18	17 to 24
5/16 - 24	14 to 20	19 to 27
3/8 - 16	22 to 32	30 to 43
3/8 - 24	27 to 38	37 to 51
7/16 - 14	40 to 55	55 to 74
7/16 - 20	40 to 60	55 to 81
1/2 - 13	55 to 80	75 to 108

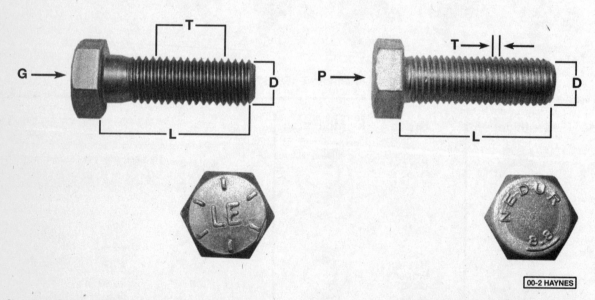

Standard (SAE and USS) bolt dimensions/grade marks
- G Grade marks (bolt strength)
- L Length (in inches)
- T Thread pitch (number of threads per inch)
- D Nominal diameter (in inches)

Metric bolt dimensions/grade marks
- P Property class (bolt strength)
- L Length (in millimeters)
- T Thread pitch (distance between threads in millimeters)
- D Diameter

Maintenance techniques, tools and working facilities

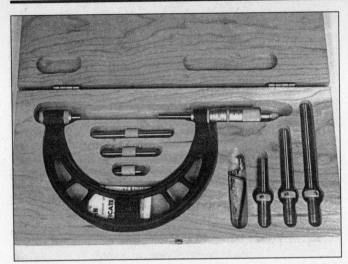

Micrometer set

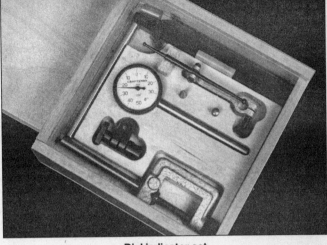

Dial indicator set

Initially, the bolts or nuts should be assembled finger-tight only. Next, they should be tightened one full turn each, in a criss-cross or diagonal pattern. After each one has been tightened one full turn, return to the first one and tighten them all one-half turn, following the same pattern. Finally, tighten each of them one-quarter turn at a time until each fastener has been tightened to the proper torque. To loosen and remove the fasteners, the procedure would be reversed.

Component disassembly

Component disassembly should be done with care and purpose to help ensure that the parts go back together properly. Always keep track of the sequence in which parts are removed. Make note of special characteristics or marks on parts that can be installed more than one way, such as a grooved thrust washer on a shaft. It is a good idea to lay the disassembled parts out on a clean surface in the order that they were removed. It may also be helpful to make sketches or take instant photos of components before removal.

When removing fasteners from a component, keep track of their locations. Sometimes threading a bolt back in a part, or putting the washers and nut back on a stud, can prevent mix-ups later. If nuts and bolts cannot be returned to their original locations, they should be kept in a compartmented box or a series of small boxes. A cupcake or muffin tin is ideal for this purpose, since each cavity can hold the bolts and nuts from a particular area (i.e. oil pan bolts, valve cover bolts, engine mount bolts, etc.). A pan of this type is especially helpful when working on assemblies with very small parts, such as the carburetor, alternator, valve train or interior dash and trim pieces. The cavities can be marked with paint or tape to identify the contents.

Whenever wiring looms, harnesses or connectors are separated, it is a good idea to identify the two halves with numbered pieces of masking tape so they can be easily reconnected.

Gasket sealing surfaces

Throughout any vehicle, gaskets are used to seal the mating surfaces between two parts and keep lubricants, fluids, vacuum or pressure contained in an assembly.

Many times these gaskets are coated with a liquid or paste-type gasket sealing compound before assembly. Age, heat and pressure can sometimes cause the two parts to stick together so tightly that they are very difficult to separate. Often, the assembly can be loosened by striking it with a soft-face hammer near the mating surfaces. A regular hammer can be used if a block of wood is placed between the hammer and the part. Do not hammer on cast parts or parts that could be easily damaged. With any particularly stubborn part, always recheck to make sure that every fastener has been removed.

Avoid using a screwdriver or bar to pry apart an assembly, as they can easily mar the gasket sealing surfaces of the parts, which must remain smooth. If prying is absolutely necessary, use an old broom handle, but keep in mind that extra clean up will be necessary if the wood splinters.

After the parts are separated, the old gasket must be carefully scraped off and the gasket surfaces cleaned. Stubborn gasket material can be soaked with rust penetrant or treated with a special chemical to soften it so it can be easily scraped off. A scraper can be fashioned from a piece of copper tubing by flattening and sharpening one end. Copper is recommended because it is usually softer than the surfaces to be scraped, which reduces the chance of gouging the part. Some gaskets can be removed with a wire brush, but regardless of the method used, the mating surfaces must be left clean and smooth. If for some reason the gasket surface is gouged, then a gasket sealer thick enough to fill scratches will have to be used during reassembly of the components. For most applications, a non-drying (or semi-drying) gasket sealer should be used.

Hose removal tips

Warning: *If the vehicle is equipped with air conditioning, do not disconnect any of the A/C hoses without first having the system depressurized by a dealer service department or a service station.*

Hose removal precautions closely parallel gasket removal precautions. Avoid scratching or gouging the surface that the hose mates against or the connection may leak. This is especially true for radiator hoses. Because of various chemical reactions, the rubber in hoses can bond itself to the metal spigot that the hose fits over. To remove a hose, first loosen the hose clamps that secure it to the spigot. Then, with slip-joint pliers, grab the hose at the clamp and rotate it around the spigot. Work it back and forth until it is completely free, then pull it off. Silicone or other lubricants will ease removal if they can be applied between the hose and the outside of the spigot. Apply the same lubricant to the inside of the hose and the outside of the spigot to simplify installation.

As a last resort (and if the hose is to be replaced with a new one anyway), the rubber can be slit with a knife and the hose peeled from the spigot. If this must be done, be careful that the metal connection is not damaged.

If a hose clamp is broken or damaged, do not reuse it. Wire-type clamps usually weaken with age, so it is a good idea to replace them with screw-type clamps whenever a hose is removed.

Tools

A selection of good tools is a basic requirement for anyone who plans to maintain and repair his or her own vehicle. For the owner who has few tools, the initial investment might seem high, but when compared to the spiraling costs of professional auto maintenance and repair, it is a wise one.

To help the owner decide which tools are needed to perform the tasks detailed in this manual, the following tool lists are offered: *Maintenance and minor repair,*

0-12 Maintenance techniques, tools and working facilities

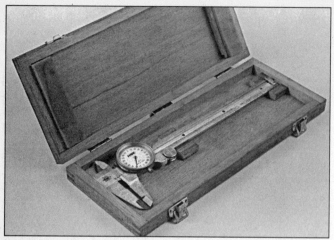
Dial caliper

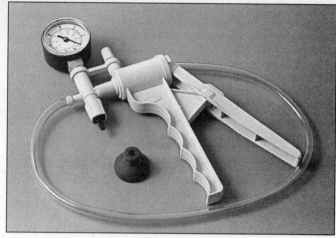
Hand-operated vacuum pump

Timing light

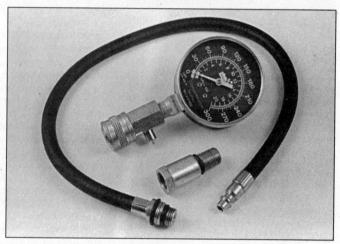
Compression gauge with spark plug hole adapter

Damper/steering wheel puller

General purpose puller

Hydraulic lifter removal tool

Repair/overhaul and *Special*.

The newcomer to practical mechanics should start off with the *maintenance and minor repair* tool kit, which is adequate for the simpler jobs performed on a vehicle. Then, as confidence and experience grow, the owner can tackle more difficult tasks, buying additional tools as they are needed. Eventually the basic kit will be expanded into the *repair and overhaul* tool set. Over a period of time, the experienced do-it-yourselfer will assemble a tool set complete enough for most repair and overhaul procedures and will add tools from the special category when it is felt that the expense is justified by the frequency of use.

Maintenance and minor repair tool kit

The tools in this list should be considered the minimum required for performance of routine maintenance, servicing and minor repair work. We recommend the purchase of combination wrenches (box-end and open-

Maintenance techniques, tools and working facilities

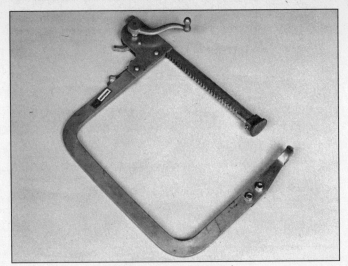

Valve spring compressor

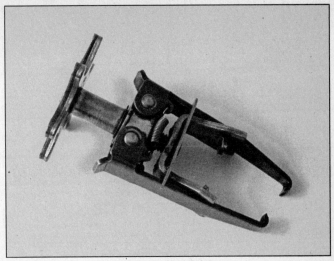

Valve spring compressor

Ridge reamer

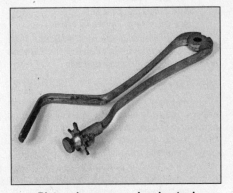

Piston ring groove cleaning tool

Ring removal/installation tool

end combined in one wrench). While more expensive than open end wrenches, they offer the advantages of both types of wrench.

Combination wrench set (1/4-inch to 1 inch or 6 mm to 19 mm)
Adjustable wrench, 8 inch
Spark plug wrench with rubber insert
Spark plug gap adjusting tool
Feeler gauge set
Brake bleeder wrench
Standard screwdriver (5/16-inch x 6 inch)
Phillips screwdriver (No. 2 x 6 inch)
Combination pliers - 6 inch
Hacksaw and assortment of blades
Tire pressure gauge
Grease gun
Oil can
Fine emery cloth
Wire brush
Battery post and cable cleaning tool
Oil filter wrench
Funnel (medium size)
Safety goggles
Jackstands (2)
Drain pan

Note: *If basic tune-ups are going to be part of routine maintenance, it will be necessary to purchase a good quality stroboscopic timing light and combination tachometer/dwell meter. Although they are included in the list of special tools, it is mentioned here because they are absolutely necessary for tuning most vehicles properly.*

Repair and overhaul tool set

These tools are essential for anyone who plans to perform major repairs and are in addition to those in the maintenance and minor repair tool kit. Included is a comprehensive set of sockets which, though expensive, are invaluable because of their versatility, especially when various extensions and drives are available. We recommend the 1/2-inch drive over the 3/8-inch drive. Although the larger drive is bulky and more expensive, it has the capacity of accepting a very wide range of large sockets. Ideally, however, the mechanic should have a 3/8-inch drive set and a 1/2-inch drive set.

Socket set(s)
Reversible ratchet
Extension - 10 inch
Universal joint
Torque wrench (same size drive as sockets)
Ball peen hammer - 8 ounce
Soft-face hammer (plastic/rubber)

Ring compressor

Standard screwdriver (1/4-inch x 6 inch)
Standard screwdriver (stubby - 5/16-inch)
Phillips screwdriver (No. 3 x 8 inch)
Phillips screwdriver (stubby - No. 2)
Pliers - vise grip
Pliers - lineman's
Pliers - needle nose
Pliers - snap-ring (internal and external)
Cold chisel - 1/2-inch

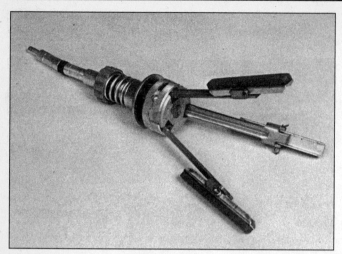

Cylinder hone

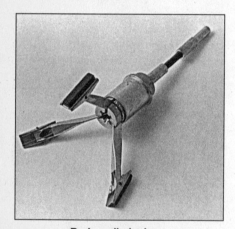

Brake cylinder hone

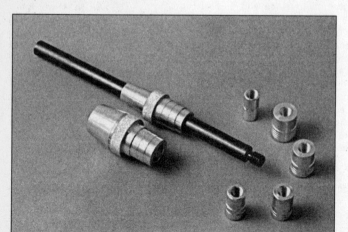

Clutch plate alignment tool

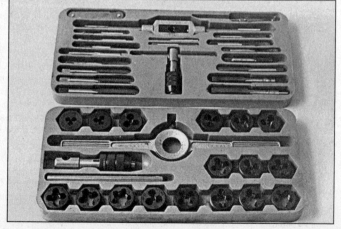

Brake hold-down spring tool

Scribe
Scraper (made from flattened copper tubing)
Centerpunch
Pin punches (1/16, 1/8, 3/16-inch)
Steel rule/straightedge - 12 inch
Allen wrench set (1/8 to 3/8-inch or 4 mm to 10 mm)
A selection of files
Wire brush (large)
Jackstands (second set)
Jack (scissor or hydraulic type)

Note: *Another tool which is often useful is an electric drill with a chuck capacity of 3/8-inch and a set of good quality drill bits.*

Special tools

The tools in this list include those which are not used regularly, are expensive to buy, or which need to be used in accordance with their manufacturer's instructions. Unless these tools will be used frequently, it is not very economical to purchase many of them. A consideration would be to split the cost and use between yourself and a friend or friends. In addition, most of these tools can be obtained from a tool rental shop on a temporary basis.

This list primarily contains only those tools and instruments widely available to the public, and not those special tools produced by the vehicle manufacturer for distribution to dealer service departments. Occasionally, references to the manufacturer's special tools are included in the text of this manual. Generally, an alternative method of doing the job without the special tool is offered. However, sometimes there is no alternative to their use. Where this is the case, and the tool cannot be purchased or borrowed, the work should be turned over to the dealer service department or an automotive repair shop.

Valve spring compressor
Piston ring groove cleaning tool
Piston ring compressor
Piston ring installation tool
Cylinder compression gauge
Cylinder ridge reamer
Cylinder surfacing hone
Cylinder bore gauge
Micrometers and/or dial calipers
Hydraulic lifter removal tool
Balljoint separator
Universal-type puller
Impact screwdriver
Dial indicator set
Stroboscopic timing light (inductive pick-up)
Hand operated vacuum/pressure pump
Tachometer/dwell meter
Universal electrical multimeter
Cable hoist
Brake spring removal and installation tools
Floor jack

Tap and die set

Maintenance techniques, tools and working facilities

Buying tools

For the do-it-yourselfer who is just starting to get involved in vehicle maintenance and repair, there are a number of options available when purchasing tools. If maintenance and minor repair is the extent of the work to be done, the purchase of individual tools is satisfactory. If, on the other hand, extensive work is planned, it would be a good idea to purchase a modest tool set from one of the large retail chain stores. A set can usually be bought at a substantial savings over the individual tool prices, and they often come with a tool box. As additional tools are needed, add-on sets, individual tools and a larger tool box can be purchased to expand the tool selection. Building a tool set gradually allows the cost of the tools to be spread over a longer period of time and gives the mechanic the freedom to choose only those tools that will actually be used.

Tool stores will often be the only source of some of the special tools that are needed, but regardless of where tools are bought, try to avoid cheap ones, especially when buying screwdrivers and sockets, because they won't last very long. The expense involved in replacing cheap tools will eventually be greater than the initial cost of quality tools.

Care and maintenance of tools

Good tools are expensive, so it makes sense to treat them with respect. Keep them clean and in usable condition and store them properly when not in use. Always wipe off any dirt, grease or metal chips before putting them away. Never leave tools lying around in the work area. Upon completion of a job, always check closely under the hood for tools that may have been left there so they won't get lost during a test drive.

Some tools, such as screwdrivers, pliers, wrenches and sockets, can be hung on a panel mounted on the garage or workshop wall, while others should be kept in a tool box or tray. Measuring instruments, gauges, meters, etc. must be carefully stored where they cannot be damaged by weather or impact from other tools.

When tools are used with care and stored properly, they will last a very long time. Even with the best of care, though, tools will wear out if used frequently. When a tool is damaged or worn out, replace it. Subsequent jobs will be safer and more enjoyable if you do.

How to repair damaged threads

Sometimes, the internal threads of a nut or bolt hole can become stripped, usually from overtightening. Stripping threads is an all-too-common occurrence, especially when working with aluminum parts, because aluminum is so soft that it easily strips out.

Usually, external or internal threads are only partially stripped. After they've been cleaned up with a tap or die, they'll still work. Sometimes, however, threads are badly damaged. When this happens, you've got three choices:

1) Drill and tap the hole to the next suitable oversize and install a larger diameter bolt, screw or stud.
2) Drill and tap the hole to accept a threaded plug, then drill and tap the plug to the original screw size. You can also buy a plug already threaded to the original size. Then you simply drill a hole to the specified size, then run the threaded plug into the hole with a bolt and jam nut. Once the plug is fully seated, remove the jam nut and bolt.
3) The third method uses a patented thread repair kit like Heli-Coil or Slimsert. These easy-to-use kits are designed to repair damaged threads in straight-through holes and blind holes. Both are available as kits which can handle a variety of sizes and thread patterns. Drill the hole, then tap it with the special included tap. Install the Heli-Coil and the hole is back to its original diameter and thread pitch.

Regardless of which method you use, be sure to proceed calmly and carefully. A little impatience or carelessness during one of these relatively simple procedures can ruin your whole day's work and cost you a bundle if you wreck an expensive part.

Working facilities

Not to be overlooked when discussing tools is the workshop. If anything more than routine maintenance is to be carried out, some sort of suitable work area is essential.

It is understood, and appreciated, that many home mechanics do not have a good workshop or garage available, and end up removing an engine or doing major repairs outside. It is recommended, however, that the overhaul or repair be completed under the cover of a roof.

A clean, flat workbench or table of comfortable working height is an absolute necessity. The workbench should be equipped with a vise that has a jaw opening of at least four inches.

As mentioned previously, some clean, dry storage space is also required for tools, as well as the lubricants, fluids, cleaning solvents, etc. which soon become necessary.

Sometimes waste oil and fluids, drained from the engine or cooling system during normal maintenance or repairs, present a disposal problem. To avoid pouring them on the ground or into a sewage system, pour the used fluids into large containers, seal them with caps and take them to an authorized disposal site or recycling center. Plastic jugs, such as old antifreeze containers, are ideal for this purpose.

Always keep a supply of old newspapers and clean rags available. Old towels are excellent for mopping up spills. Many mechanics use rolls of paper towels for most work because they are readily available and disposable. To help keep the area under the vehicle clean, a large cardboard box can be cut open and flattened to protect the garage or shop floor.

Whenever working over a painted surface, such as when leaning over a fender to service something under the hood, always cover it with an old blanket or bedspread to protect the finish. Vinyl covered pads, made especially for this purpose, are available at auto parts stores.

Booster battery (jump) starting

Observe the following precautions when using a booster battery to start a vehicle:

a) Before connecting the booster battery, make sure the ignition switch is in the Off position.
b) Turn off the lights, heater and other electrical loads.
c) Your eyes should be shielded. Safety goggles are a good idea.
d) Make sure the booster battery is the same voltage as the dead one in the vehicle.
e) The two vehicles MUST NOT TOUCH each other.
f) Make sure the transmission is in Neutral (manual transaxle) or Park (automatic transaxle).
g) If the booster battery is not a maintenance-free type, remove the vent caps and lay a cloth over the vent holes.

Connect the red jumper cable to the positive (+) terminals of each battery.

Connect one end of the black cable to the negative (-) terminal of the booster battery. The other end of this cable should be connected to a good ground on the engine block **(see illustration)**. Make sure the cable will not come into contact with the fan, drivebelts or other moving parts of the engine.

Start the engine using the booster battery, then, with the engine running at idle speed, disconnect the jumper cables in the reverse order of connection.

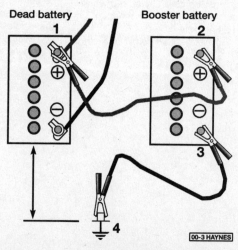

Make the booster battery cable connections in the numerical order shown (note that the negative cable of the booster battery is NOT attached to the negative terminal of the dead battery)

Jacking and towing

Jacking

Warning: *The jack supplied with the vehicle should only be used for changing a tire or placing jackstands under the frame. Never work under the vehicle or start the engine while this jack is being used as the only means of support.*

The vehicle should be on level ground. Place the shift lever in Park, if you have an automatic, or Reverse if you have a manual transaxle. Block the wheel diagonally opposite the wheel being changed. Set the parking brake.

Remove the spare tire and jack from stowage. Remove the wheel cover and trim ring (if so equipped) with the tapered end of the lug nut wrench by inserting and twisting the handle and then prying against the back of the wheel cover. Loosen the wheel lug nuts about 1/4-to-1/2 turn each.

Place the scissors-type jack under the side of the vehicle and adjust the jack height until it fits in the notch in the vertical rocker panel flange nearest the wheel to be changed. There is a front and rear jacking point on each side of the vehicle **(see illustration)**.

Turn the jack handle clockwise until the tire clears the ground. Remove the lug nuts and pull the wheel off. Replace it with the spare.

Install the lug nuts with the beveled edges facing in. Tighten them snugly. Don't attempt to tighten them completely until the vehicle is lowered or it could slip off the jack. Turn the jack handle counterclockwise to lower the vehicle. Remove the jack and tighten the lug nuts in a diagonal pattern.

The factory jack fits under notches (arrows) in the rocker panel flange - there is a front and rear notch on each side of the vehicle

Install the cover (and trim ring, if used) and be sure it's snapped into place all the way around.

Stow the tire, jack and wrench. Unblock the wheels.

Towing

As a general rule, the vehicle should be towed with the front (drive) wheels off the ground. If they can't be raised, place them on a dolly. The ignition key must be in the ACC position, since the steering lock mechanism isn't strong enough to hold the front wheels straight while towing.

Vehicles equipped with an automatic transaxle can be towed from the front only with all four wheels on the ground, provided that speeds don't exceed 30 mph and the distance is not over 40 miles. Before towing, check the transmission fluid level (see Chapter 1). If the level is below the HOT line on the dipstick, add fluid or use a towing dolly.

Caution: *Never tow a vehicle with an automatic transaxle from the rear with the front wheels on the ground.*

When towing a vehicle equipped with a manual transaxle with all four wheels on the ground, be sure to place the shift lever in neutral and release the parking brake.

Equipment specifically designed for towing should be used. It should be attached to the main structural members of the vehicle, not the bumpers or brackets.

Safety is a major consideration when towing and all applicable state and local laws must be obeyed. A safety chain system must be used at all times.

Automotive chemicals and lubricants

A number of automotive chemicals and lubricants are available for use during vehicle maintenance and repair. They include a wide variety of products ranging from cleaning solvents and degreasers to lubricants and protective sprays for rubber, plastic and vinyl.

Cleaners

Carburetor cleaner and choke cleaner is a strong solvent for gum, varnish and carbon. Most carburetor cleaners leave a dry-type lubricant film which will not harden or gum up. Because of this film it is not recommended for use on electrical components.

Brake system cleaner is used to remove grease and brake fluid from the brake system, where clean surfaces are absolutely necessary. It leaves no residue and often eliminates brake squeal caused by contaminants.

Electrical cleaner removes oxidation, corrosion and carbon deposits from electrical contacts, restoring full current flow. It can also be used to clean spark plugs, carburetor jets, voltage regulators and other parts where an oil-free surface is desired.

Demoisturants remove water and moisture from electrical components such as alternators, voltage regulators, electrical connectors and fuse blocks. They are non-conductive, non-corrosive and non-flammable.

Degreasers are heavy-duty solvents used to remove grease from the outside of the engine and from chassis components. They can be sprayed or brushed on and, depending on the type, are rinsed off either with water or solvent.

Lubricants

Motor oil is the lubricant formulated for use in engines. It normally contains a wide variety of additives to prevent corrosion and reduce foaming and wear. Motor oil comes in various weights (viscosity ratings) from 0 to 50. The recommended weight of the oil depends on the season, temperature and the demands on the engine. Light oil is used in cold climates and under light load conditions. Heavy oil is used in hot climates and where high loads are encountered. Multi-viscosity oils are designed to have characteristics of both light and heavy oils and are available in a number of weights from 5W-20 to 20W-50.

Gear oil is designed to be used in differentials, manual transmissions and other areas where high-temperature lubrication is required.

Chassis and wheel bearing grease is a heavy grease used where increased loads and friction are encountered, such as for wheel bearings, balljoints, tie-rod ends and universal joints.

High-temperature wheel bearing grease is designed to withstand the extreme temperatures encountered by wheel bearings in disc brake equipped vehicles. It usually contains molybdenum disulfide (moly), which is a dry-type lubricant.

White grease is a heavy grease for metal-to-metal applications where water is a problem. White grease stays soft under both low and high temperatures (usually from -100 to +190-degrees F), and will not wash off or dilute in the presence of water.

Assembly lube is a special extreme pressure lubricant, usually containing moly, used to lubricate high-load parts (such as main and rod bearings and cam lobes) for initial start-up of a new engine. The assembly lube lubricates the parts without being squeezed out or washed away until the engine oiling system begins to function.

Silicone lubricants are used to protect rubber, plastic, vinyl and nylon parts.

Graphite lubricants are used where oils cannot be used due to contamination problems, such as in locks. The dry graphite will lubricate metal parts while remaining uncontaminated by dirt, water, oil or acids. It is electrically conductive and will not foul electrical contacts in locks such as the ignition switch.

Moly penetrants loosen and lubricate frozen, rusted and corroded fasteners and prevent future rusting or freezing.

Heat-sink grease is a special electrically non-conductive grease that is used for mounting electronic ignition modules where it is essential that heat is transferred away from the module.

Sealants

RTV sealant is one of the most widely used gasket compounds. Made from silicone, RTV is air curing, it seals, bonds, waterproofs, fills surface irregularities, remains flexible, doesn't shrink, is relatively easy to remove, and is used as a supplementary sealer with almost all low and medium temperature gaskets.

Anaerobic sealant is much like RTV in that it can be used either to seal gaskets or to form gaskets by itself. It remains flexible, is solvent resistant and fills surface imperfections. The difference between an anaerobic sealant and an RTV-type sealant is in the curing. RTV cures when exposed to air, while an anaerobic sealant cures only in the absence of air. This means that an anaerobic sealant cures only after the assembly of parts, sealing them together.

Thread and pipe sealant is used for sealing hydraulic and pneumatic fittings and vacuum lines. It is usually made from a Teflon compound, and comes in a spray, a paint-on liquid and as a wrap-around tape.

Chemicals

Anti-seize compound prevents seizing, galling, cold welding, rust and corrosion in fasteners. High-temperature anti-seize, usually made with copper and graphite lubricants, is used for exhaust system and exhaust manifold bolts.

Anaerobic locking compounds are used to keep fasteners from vibrating or working loose and cure only after installation, in the absence of air. Medium strength locking compound is used for small nuts, bolts and screws that may be removed later. High-strength locking compound is for large nuts, bolts and studs which aren't removed on a regular basis.

Oil additives range from viscosity index improvers to chemical treatments that claim to reduce internal engine friction. It should be noted that most oil manufacturers caution against using additives with their oils.

Gas additives perform several functions, depending on their chemical makeup. They usually contain solvents that help dissolve gum and varnish that build up on carburetor, fuel injection and intake parts. They also serve to break down carbon deposits that form on the inside surfaces of the combustion chambers. Some additives contain upper cylinder lubricants for valves and piston rings, and others contain chemicals to remove condensation from the gas tank.

Miscellaneous

Brake fluid is specially formulated hydraulic fluid that can withstand the heat and pressure encountered in brake systems. Care must be taken so this fluid does not come in contact with painted surfaces or plastics. An opened container should always be resealed to prevent contamination by water or dirt.

Weatherstrip adhesive is used to bond weatherstripping around doors, windows and trunk lids. It is sometimes used to attach trim pieces.

Undercoating is a petroleum-based, tar-like substance that is designed to protect metal surfaces on the underside of the vehicle from corrosion. It also acts as a sound-deadening agent by insulating the bottom of the vehicle.

Waxes and polishes are used to help protect painted and plated surfaces from the weather. Different types of paint may require the use of different types of wax and polish. Some polishes utilize a chemical or abrasive cleaner to help remove the top layer of oxidized (dull) paint on older vehicles. In recent years many non-wax polishes that contain a wide variety of chemicals such as polymers and silicones have been introduced. These non-wax polishes are usually easier to apply and last longer than conventional waxes and polishes.

Conversion factors

Length (distance)
Inches (in)	X 25.4	= Millimetres (mm)	X 0.0394	= Inches (in)
Feet (ft)	X 0.305	= Metres (m)	X 3.281	= Feet (ft)
Miles	X 1.609	= Kilometres (km)	X 0.621	= Miles

Volume (capacity)
Cubic inches (cu in; in^3)	X 16.387	= Cubic centimetres (cc; cm^3)	X 0.061	= Cubic inches (cu in; in^3)
Imperial pints (Imp pt)	X 0.568	= Litres (l)	X 1.76	= Imperial pints (Imp pt)
Imperial quarts (Imp qt)	X 1.137	= Litres (l)	X 0.88	= Imperial quarts (Imp qt)
Imperial quarts (Imp qt)	X 1.201	= US quarts (US qt)	X 0.833	= Imperial quarts (Imp qt)
US quarts (US qt)	X 0.946	= Litres (l)	X 1.057	= US quarts (US qt)
Imperial gallons (Imp gal)	X 4.546	= Litres (l)	X 0.22	= Imperial gallons (Imp gal)
Imperial gallons (Imp gal)	X 1.201	= US gallons (US gal)	X 0.833	= Imperial gallons (Imp gal)
US gallons (US gal)	X 3.785	= Litres (l)	X 0.264	= US gallons (US gal)

Mass (weight)
Ounces (oz)	X 28.35	= Grams (g)	X 0.035	= Ounces (oz)
Pounds (lb)	X 0.454	= Kilograms (kg)	X 2.205	= Pounds (lb)

Force
Ounces-force (ozf; oz)	X 0.278	= Newtons (N)	X 3.6	= Ounces-force (ozf; oz)
Pounds-force (lbf; lb)	X 4.448	= Newtons (N)	X 0.225	= Pounds-force (lbf; lb)
Newtons (N)	X 0.1	= Kilograms-force (kgf; kg)	X 9.81	= Newtons (N)

Pressure
Pounds-force per square inch (psi; lbf/in^2; lb/in^2)	X 0.070	= Kilograms-force per square centimetre (kgf/cm^2; kg/cm^2)	X 14.223	= Pounds-force per square inch (psi; lbf/in^2; lb/in^2)
Pounds-force per square inch (psi; lbf/in^2; lb/in^2)	X 0.068	= Atmospheres (atm)	X 14.696	= Pounds-force per square inch (psi; lbf/in^2; lb/in^2)
Pounds-force per square inch (psi; lbf/in^2; lb/in^2)	X 0.069	= Bars	X 14.5	= Pounds-force per square inch (psi; lbf/in^2; lb/in^2)
Pounds-force per square inch (psi; lbf/in^2; lb/in^2)	X 6.895	= Kilopascals (kPa)	X 0.145	= Pounds-force per square inch (psi; lbf/in^2; lb/in^2)
Kilopascals (kPa)	X 0.01	= Kilograms-force per square centimetre (kgf/cm^2; kg/cm^2)	X 98.1	= Kilopascals (kPa)

Torque (moment of force)
Pounds-force inches (lbf in; lb in)	X 1.152	= Kilograms-force centimetre (kgf cm; kg cm)	X 0.868	= Pounds-force inches (lbf in; lb in)
Pounds-force inches (lbf in; lb in)	X 0.113	= Newton metres (Nm)	X 8.85	= Pounds-force inches (lbf in; lb in)
Pounds-force inches (lbf in; lb in)	X 0.083	= Pounds-force feet (lbf ft; lb ft)	X 12	= Pounds-force inches (lbf in; lb in)
Pounds-force feet (lbf ft; lb ft)	X 0.138	= Kilograms-force metres (kgf m; kg m)	X 7.233	= Pounds-force feet (lbf ft; lb ft)
Pounds-force feet (lbf ft; lb ft)	X 1.356	= Newton metres (Nm)	X 0.738	= Pounds-force feet (lbf ft; lb ft)
Newton metres (Nm)	X 0.102	= Kilograms-force metres (kgf m; kg m)	X 9.804	= Newton metres (Nm)

Vacuum
Inches mercury (in. Hg)	X 3.377	= Kilopascals (kPa)	X 0.2961	= Inches mercury
Inches mercury (in. Hg)	X 25.4	= Millimeters mercury (mm Hg)	X 0.0394	= Inches mercury

Power
Horsepower (hp)	X 745.7	= Watts (W)	X 0.0013	= Horsepower (hp)

Velocity (speed)
Miles per hour (miles/hr; mph)	X 1.609	= Kilometres per hour (km/hr; kph)	X 0.621	= Miles per hour (miles/hr; mph)

*Fuel consumption**
Miles per gallon, Imperial (mpg)	X 0.354	= Kilometres per litre (km/l)	X 2.825	= Miles per gallon, Imperial (mpg)
Miles per gallon, US (mpg)	X 0.425	= Kilometres per litre (km/l)	X 2.352	= Miles per gallon, US (mpg)

Temperature
Degrees Fahrenheit = (°C x 1.8) + 32   Degrees Celsius (Degrees Centigrade; °C) = (°F - 32) x 0.56

*It is common practice to convert from miles per gallon (mpg) to litres/100 kilometres (l/100km), where mpg (Imperial) x l/100 km = 282 and mpg (US) x l/100 km = 235

Safety first!

Regardless of how enthusiastic you may be about getting on with the job at hand, take the time to ensure that your safety is not jeopardized. A moment's lack of attention can result in an accident, as can failure to observe certain simple safety precautions. The possibility of an accident will always exist, and the following points should not be considered a comprehensive list of all dangers. Rather, they are intended to make you aware of the risks and to encourage a safety conscious approach to all work you carry out on your vehicle.

Essential DOs and DON'Ts

DON'T rely on a jack when working under the vehicle. Always use approved jackstands to support the weight of the vehicle and place them under the recommended lift or support points.

DON'T attempt to loosen extremely tight fasteners (i.e. wheel lug nuts) while the vehicle is on a jack - it may fall.

DON'T start the engine without first making sure that the transmission is in Neutral (or Park where applicable) and the parking brake is set.

DON'T remove the radiator cap from a hot cooling system - let it cool or cover it with a cloth and release the pressure gradually.

DON'T attempt to drain the engine oil until you are sure it has cooled to the point that it will not burn you.

DON'T touch any part of the engine or exhaust system until it has cooled sufficiently to avoid burns.

DON'T siphon toxic liquids such as gasoline, antifreeze and brake fluid by mouth, or allow them to remain on your skin.

DON'T inhale brake lining dust - it is potentially hazardous (see *Asbestos* below).

DON'T allow spilled oil or grease to remain on the floor - wipe it up before someone slips on it.

DON'T use loose fitting wrenches or other tools which may slip and cause injury.

DON'T push on wrenches when loosening or tightening nuts or bolts. Always try to pull the wrench toward you. If the situation calls for pushing the wrench away, push with an open hand to avoid scraped knuckles if the wrench should slip.

DON'T attempt to lift a heavy component alone - get someone to help you.

DON'T rush or take unsafe shortcuts to finish a job.

DON'T allow children or animals in or around the vehicle while you are working on it.

DO wear eye protection when using power tools such as a drill, sander, bench grinder, etc. and when working under a vehicle.

DO keep loose clothing and long hair well out of the way of moving parts.

DO make sure that any hoist used has a safe working load rating adequate for the job.

DO get someone to check on you periodically when working alone on a vehicle.

DO carry out work in a logical sequence and make sure that everything is correctly assembled and tightened.

DO keep chemicals and fluids tightly capped and out of the reach of children and pets.

DO remember that your vehicle's safety affects that of yourself and others. If in doubt on any point, get professional advice.

Asbestos

Certain friction, insulating, sealing, and other products - such as brake linings, brake bands, clutch linings, torque converters, gaskets, etc. - may contain asbestos. Extreme care must be taken to avoid inhalation of dust from such products, since it is hazardous to health. If in doubt, assume that they do contain asbestos.

Fire

Remember at all times that gasoline is highly flammable. Never smoke or have any kind of open flame around when working on a vehicle. But the risk does not end there. A spark caused by an electrical short circuit, by two metal surfaces contacting each other, or even by static electricity built up in your body under certain conditions, can ignite gasoline vapors, which in a confined space are highly explosive. Do not, under any circumstances, use gasoline for cleaning parts. Use an approved safety solvent.

Always disconnect the battery ground (-) cable at the battery before working on any part of the fuel system or electrical system. Never risk spilling fuel on a hot engine or exhaust component. It is strongly recommended that a fire extinguisher suitable for use on fuel and electrical fires be kept handy in the garage or workshop at all times. Never try to extinguish a fuel or electrical fire with water.

Fumes

Certain fumes are highly toxic and can quickly cause unconsciousness and even death if inhaled to any extent. Gasoline vapor falls into this category, as do the vapors from some cleaning solvents. Any draining or pouring of such volatile fluids should be done in a well ventilated area.

When using cleaning fluids and solvents, read the instructions on the container carefully. Never use materials from unmarked containers.

Never run the engine in an enclosed space, such as a garage. Exhaust fumes contain carbon monoxide, which is extremely poisonous. If you need to run the engine, always do so in the open air, or at least have the rear of the vehicle outside the work area.

If you are fortunate enough to have the use of an inspection pit, never drain or pour gasoline and never run the engine while the vehicle is over the pit. The fumes, being heavier than air, will concentrate in the pit with possibly lethal results.

The battery

Never create a spark or allow a bare light bulb near a battery. They normally give off a certain amount of hydrogen gas, which is highly explosive.

Always disconnect the battery ground (-) cable at the battery before working on the fuel or electrical systems.

If possible, loosen the filler caps or cover when charging the battery from an external source (this does not apply to sealed or maintenance-free batteries). Do not charge at an excessive rate or the battery may burst.

Take care when adding water to a non maintenance-free battery and when carrying a battery. The electrolyte, even when diluted, is very corrosive and should not be allowed to contact clothing or skin.

Always wear eye protection when cleaning the battery to prevent the caustic deposits from entering your eyes.

Household current

When using an electric power tool, inspection light, etc., which operates on household current, always make sure that the tool is correctly connected to its plug and that, where necessary, it is properly grounded. Do not use such items in damp conditions and, again, do not create a spark or apply excessive heat in the vicinity of fuel or fuel vapor.

Secondary ignition system voltage

A severe electric shock can result from touching certain parts of the ignition system (such as the spark plug wires) when the engine is running or being cranked, particularly if components are damp or the insulation is defective. In the case of an electronic ignition system, the secondary system voltage is much higher and could prove fatal.

Troubleshooting

Contents

Symptom	Section

Engine
Engine backfires	15
Engine diesels (continues to run) after switching off	18
Engine hard to start when cold	3
Engine hard to start when hot	4
Engine lacks power	14
Engine lopes while idling or idles erratically	8
Engine misses at idle speed	9
Engine misses throughout driving speed range	10
Engine rotates but will not start	2
Engine runs with oil pressure light on	17
Engine stalls	13
Engine starts but stops immediately	6
Engine stumbles on acceleration	11
Engine surges while holding accelerator steady	12
Engine will not rotate when attempting to start	1
Oil puddle under engine	7
Pinging or knocking engine sounds during acceleration or uphill	16
Starter motor noisy or excessively rough in engagement	5

Engine electrical system
Alternator light fails to go out	20
Battery will not hold a charge	19
Alternator light fails to come on when key is turned on	21

Fuel system
Excessive fuel consumption	22
Fuel leakage and/or fuel odor	23

Cooling system
Coolant loss	28
External coolant leakage	26
Internal coolant leakage	27
Overcooling	25
Overheating	24
Poor coolant circulation	29

Clutch
Clutch pedal stays on floor	36
Clutch slips (engine speed increases with no increase in vehicle speed)	32
Grabbing (chattering) as clutch is engaged	33
High pedal effort	37
Noise in clutch area	35
Pedal travels to floor - no pressure or very little resistance	30
Transaxle rattling (clicking)	34
Unable to select gears	31

Manual transaxle
Clicking noise in turns	41
Clunk on acceleration or deceleration	40
Knocking noise at low speeds	38
Leaks lubricant	47
Hard to shift	48
Noise most pronounced when turning	39
Noisy in all gears	45
Noisy in neutral with engine running	43
Noisy in one particular gear	44
Slips out of gear	46
Vibration	42

Automatic transaxle
Engine will start in gears other than Park or Neutral	53
Fluid leakage	49
General shift mechanism problems	51
Transaxle fluid brown or has burned smell	50
Transaxle slips, shifts roughly, is noisy or has no drive in forward or reverse gears	54
Transaxle will not downshift with accelerator pedal pressed to the floor	52

Driveaxles
Clicking noise in turns	55
Shudder or vibration during acceleration	56
Vibration at highway speeds	57

Brakes
Brake pedal feels spongy when depressed	65
Brake pedal travels to the floor with little resistance	66
Brake roughness or chatter (pedal pulsates)	60
Dragging brakes	63
Excessive brake pedal travel	62
Excessive pedal effort required to stop vehicle	61
Grabbing or uneven braking action	64
Noise (high-pitched squeal when the brakes are applied)	59
Parking brake does not hold	67
Vehicle pulls to one side during braking	58

Suspension and steering systems
Abnormal or excessive tire wear	69
Abnormal noise at the front end	74
Cupped tires	79
Erratic steering when braking	76
Excessive pitching and/or rolling around corners or during braking	77
Excessive play or looseness in steering system	83
Excessive tire wear on inside edge	81
Excessive tire wear on outside edge	80
Hard steering	72
Poor returnability of steering to center	73
Rattling or clicking noise in steering gear	84
Shimmy, shake or vibration	71
Suspension bottoms	78
Tire tread worn in one place	82
Vehicle pulls to one side	68
Wander or poor steering stability	75
Wheel makes a thumping noise	70

This section provides an easy reference guide to the more common problems which may occur during the operation of your vehicle. These problems and their possible causes are grouped under headings denoting various components or systems, such as Engine, Cooling system, etc. They also refer you to the chapter and/or section which deals with the problem.

Remember that successful troubleshooting is not a mysterious black art practiced only by professional mechanics. It is simply the result of the right knowledge combined with an intelligent, systematic approach to the problem. Always work by a process of elimination, starting with the simplest solution and working through to the most complex - and never overlook the obvious. Anyone can run the gas tank dry or leave the lights on overnight, so don't assume that you are exempt from such oversights.

Finally, always establish a clear idea of why a problem has occurred and take steps to ensure that it doesn't happen again. If the electrical system fails because of a poor connection, check the other connections in the system to make sure that they don't fail as well. If a particular fuse continues to blow, find out why - don't just replace one fuse after another. Remember, failure of a small component can often be indicative of potential failure or incorrect functioning of a more important component or system.

Engine

1 Engine will not rotate when attempting to start

1   Battery terminal connections loose or corroded (Chapter 1).
2   Battery discharged or faulty (Chapter 1).
3   Automatic transaxle not completely engaged in Park (Chapter 7) or clutch pedal not completely depressed (Chapter 8).
4   Broken, loose or disconnected wiring in the starting circuit (Chapters 5 and 12).
5   Starter motor pinion jammed in flywheel ring gear (Chapter 5).
6   Starter solenoid faulty (Chapter 5).
7   Starter motor faulty (Chapter 5).
8   Ignition switch faulty (Chapter 12).
9   Starter pinion or flywheel teeth worn or broken (Chapter 5).
10   Defective BATTERY (140-amp) fusible link (see Chapter 12)

2 Engine rotates but will not start

1   Fuel tank empty.
2   Battery discharged (engine rotates slowly) (Chapter 5).
3   Battery terminal connections loose or corroded (Chapter 1).
4   Leaking fuel injector(s), faulty fuel pump, pressure regulator, etc. (Chapter 4).
5   Broken or stripped timing chain (Chapter 2).
6   Ignition components damp or damaged (Chapter 5).
7   Worn, faulty or incorrectly gapped spark plugs (Chapter 1).
8   Broken, loose or disconnected wiring in the starting circuit (Chapter 5).
9   Loose distributor is changing ignition timing (Chapter 5).
10   Broken, loose or disconnected wires at the ignition coil or faulty coil (Chapter 5).
11   Defective MAF sensor (see Chapter 6).

3 Engine hard to start when cold

1   Battery discharged or low (Chapter 1).
2   Malfunctioning fuel system (Chapter 4).
3   Faulty coolant temperature sensor or intake air temperature sensor (Chapter 6).
4   Injector(s) leaking (Chapter 4B).
5   Faulty ignition system (Chapter 5).
6   Defective MAF sensor (see Chapter 6).

4 Engine hard to start when hot

1   Air filter clogged (Chapter 1).
2   Fuel not reaching the fuel injection system (Chapter 4).
3   Corroded battery connections, especially ground (Chapter 1).
4   Faulty coolant temperature sensor or intake air temperature sensor (Chapter 6).

5 Starter motor noisy or excessively rough in engagement

1   Pinion or flywheel gear teeth worn or broken (Chapter 5).
2   Starter motor mounting bolts loose or missing (Chapter 5).

6 Engine starts but stops immediately

1   Loose or faulty electrical connections at distributor, coil or alternator (Chapter 5).
2   Insufficient fuel reaching the fuel injector(s) (Chapters 1 and 4).
3   Vacuum leak at the gasket between the intake manifold/plenum and throttle body (Chapters 1 and 4).
4   Idle speed incorrect (Chapter 4).
5   Intake air leaks, broken vacuum lines (see Chapter 4).

7 Oil puddle under engine

1   Oil pan gasket and/or oil pan drain bolt washer leaking (Chapter 2).
2   Oil pressure sending unit leaking (Chapter 2).
3   Valve covers leaking (Chapter 2).
4   Engine oil seals leaking (Chapter 2).
5   Oil pump housing leaking (Chapter 2).

8 Engine lopes while idling or idles erratically

1   Vacuum leakage (Chapters 2 and 4).
2   Leaking EGR valve (Chapter 6).
3   Air filter clogged (Chapter 1).
4   Fuel pump not delivering sufficient fuel to the fuel injection system (Chapter 4).
5   Leaking head gasket (Chapter 2).
6   Timing chain and/or sprockets worn (Chapter 2).
7   Camshaft lobes worn (Chapter 2).

9 Engine misses at idle speed

1   Spark plugs worn or not gapped properly (Chapter 1).
2   Faulty spark plug wires, 1993 and 1994 models (Chapter 1).
3   Vacuum leaks (Chapters 2 and 4).
4   Incorrect ignition timing (Chapter 5).
5   Uneven or low compression (Chapter 2).
6   Problem with the fuel injection system (Chapter 4).
7   Faulty individual ignition coils, 1995 and later models (Chapter 5).

10 Engine misses throughout driving speed range

1   Fuel filter clogged and/or impurities in the fuel system (Chapter 1).
2   Low fuel output at the fuel injector(s) (Chapter 4).
3   Faulty or incorrectly gapped spark plugs (Chapter 1).
4   Incorrect ignition timing (Chapter 5).
5   Cracked distributor cap, disconnected distributor wires or damaged distributor components, 1993 and 1994 models (Chapters 1 and 5).
6   Leaking spark plug wires, 1993 and 1994 models (Chapters 1 or 5).
7   Faulty emission system components (Chapter 6).
8   Low or uneven cylinder compression pressures (Chapter 2).
9   Weak or faulty ignition system (Chapter 5).
10   Vacuum leak in fuel injection system, throttle body, intake manifold, IAC/AAC valve or vacuum hoses (Chapter 4).

11 Engine stumbles on acceleration

1   Spark plugs fouled (Chapter 1).
2   Problem with fuel injection system (Chapter 4).
3   Fuel filter clogged (Chapters 1 and 4).

Troubleshooting

4 Incorrect ignition timing (Chapter 5).
5 Intake manifold air leak (Chapters 2 and 4).
6 EGR system malfunction (Chapter 6).

12 Engine surges while holding accelerator steady

1 Intake air leak (Chapter 4).
2 Fuel pump or fuel pressure regulator faulty (Chapter 4).
3 Problem with fuel injection system (Chapter 4).
4 Problem with the emissions control system (Chapter 6).

13 Engine stalls

1 Idle speed incorrect (Chapter 4).
2 Fuel filter clogged and/or water and impurities in the fuel system (Chapters 1 and 4).
3 Distributor components damp or damaged (Chapter 5).
4 Faulty emissions system components (Chapter 6).
5 Faulty or incorrectly gapped spark plugs (Chapter 1).
6 Faulty spark plug wires, 1993 and 1994 models (Chapter 1).
7 Vacuum leak in the fuel injection system, intake manifold or vacuum hoses (Chapters 2 and 4).
8 Valve clearances incorrectly set (Chapter 2).

14 Engine lacks power

1 Incorrect ignition timing (Chapter 5).
2 Excessive play in distributor shaft, 1993 and 1994 models (Chapter 5).
3 Worn rotor, distributor cap, spark plug wires or faulty coil (Chapters 1 and 5).
4 Faulty or incorrectly gapped spark plugs (Chapter 1).
5 Problem with the fuel injection system (Chapter 4).
6 Plugged air filter (Chapter 1).
7 Brakes binding (Chapter 9).
8 Automatic transaxle fluid level incorrect (Chapter 1).
9 Clutch slipping (Chapter 8).
10 Fuel filter clogged and/or impurities in the fuel system (Chapters 1 and 4).
11 Emission control system not functioning properly (Chapter 6).
12 Low or uneven cylinder compression pressures (Chapter 2).
13 Obstructed exhaust system (Chapters 2 and 4).

15 Engine backfires

1 Emission control system not functioning properly (Chapter 6).
2 Ignition timing incorrect (Chapter 5).
3 Faulty secondary ignition system, cracked spark plug insulator, faulty plug wires, distributor cap and/or rotor, 1993 and 1994 models (Chapters 1 and 5).
4 Problem with the fuel injection system (Chapter 4).
5 Vacuum leak at fuel injector(s), intake manifold, air control valve or vacuum hoses (Chapters 2 and 4).
6 Valve clearances incorrectly set and/or valves sticking (Chapter 2).

16 Pinging or knocking engine sounds during acceleration or uphill

1 Incorrect grade of fuel.
2 Ignition timing incorrect (Chapter 5).
3 Fuel injection system faulty (Chapter 4).
4 Improper or damaged spark plugs or wires (Chapter 1).
5 Worn or damaged distributor components, 1993 and 1994 models (Chapter 5).
6 EGR valve not functioning (Chapter 6).
7 Vacuum leak (Chapters 2 and 4).

17 Engine runs with oil pressure light on

1 Low oil level (Chapter 1).
2 Idle rpm below specification (Chapter 4).
3 Short in wiring circuit (Chapter 12).
4 Faulty oil pressure sender (Chapter 2).
5 Worn engine bearings and/or oil pump (Chapter 2).

18 Engine diesels (continues to run) after switching off

1 Idle speed too high (Chapter 4).
2 Excessive engine operating temperature (Chapter 3).
3 Ignition timing in need of adjustment (Chapter 5).
4 Excessive carbon deposits on valves and pistons (Chapter 2).

Engine electrical system

19 Battery will not hold a charge

1 Alternator drivebelt defective or not adjusted properly (Chapter 1).
2 Battery electrolyte level low (Chapter 1).
3 Battery terminals loose or corroded (Chapter 1).
4 Alternator not charging properly (Chapter 5).
5 Loose, broken or faulty wiring in the charging circuit (Chapter 5).
6 Short in vehicle wiring (Chapter 12).
7 Internally defective battery (Chapters 1 and 5).

20 Alternator light fails to go out

1 Faulty alternator or charging circuit (Chapter 5).
2 Alternator drivebelt defective or out of adjustment (Chapter 1).
3 Alternator voltage regulator inoperative (Chapter 5).

21 Alternator light fails to come on when key is turned on

1 Warning light bulb defective (Chapter 12).
2 Fault in the printed circuit, dash wiring or bulb holder (Chapter 12).

Fuel system

22 Excessive fuel consumption

1 Dirty or clogged air filter element (Chapter 1).
2 Incorrectly set ignition timing (Chapter 5).
3 Emissions system not functioning properly (Chapter 6).
4 Fuel injection system not functioning properly (Chapter 4).
5 Low tire pressure or incorrect tire size (Chapter 1).

23 Fuel leakage and/or fuel odor

1 Leaking fuel feed or return line (Chapters 1 and 4).
2 Tank overfilled.
3 Charcoal canister filter clogged (Chapters 6).
4 Problem with fuel injection system (Chapter 4).

Cooling system

24 Overheating

1 Insufficient coolant in system (Chapter 1).
2 Water pump drivebelt defective or out of adjustment (Chapter 1).
3 Radiator core blocked or grille restricted (Chapter 3).
4 Thermostat faulty (Chapter 3).
5 Electric coolant fan inoperative or blades broken (Chapter 3).

Troubleshooting

6 Radiator cap not maintaining proper pressure (Chapter 3).
7 Ignition timing incorrect (Chapter 5).

25 Overcooling

1 Faulty thermostat (Chapter 3).
2 Inaccurate temperature gauge sending unit (Chapter 3)

26 External coolant leakage

1 Deteriorated/damaged hoses; loose clamps (Chapters 1 and 3).
2 Water pump defective (Chapter 3).
3 Leakage from radiator core or coolant reservoir tank (Chapter 3).
4 Engine drain or water jacket core plugs leaking (Chapter 2).

27 Internal coolant leakage

1 Leaking cylinder head gasket (Chapter 2).
2 Cracked cylinder bore or cylinder head (Chapter 2).

28 Coolant loss

1 Too much coolant in system (Chapter 1).
2 Coolant boiling away because of overheating (Chapter 3).
3 Internal or external leakage (Chapter 3).
4 Faulty radiator cap (Chapter 3).

29 Poor coolant circulation

1 Inoperative water pump (Chapter 3).
2 Restriction in cooling system (Chapters 1 and 3).
3 Water pump drivebelt defective/out of adjustment (Chapter 1).
4 Thermostat sticking (Chapter 3).

Clutch

30 Pedal travels to floor - no pressure or very little resistance

1 Hydraulic release system leaking or air in the system (Chapter 8).
2 Broken release bearing or fork (Chapter 8).

31 Unable to select gears

1 Faulty transaxle (Chapter 7).
2 Faulty clutch disc or pressure plate (Chapter 8).
3 Faulty release lever or release bearing (Chapter 8).
4 Faulty shift lever assembly or rods (Chapter 8).

32 Clutch slips (engine speed increases with no increase in vehicle speed)

1 Clutch plate worn (Chapter 8).
2 Clutch plate is oil soaked by leaking rear main seal (Chapter 8).
3 Clutch plate not seated (Chapter 8).
4 Warped pressure plate or flywheel (Chapter 8).
5 Weak diaphragm springs in pressure plate (Chapter 8).
6 Clutch plate overheated. Allow to cool.
7 Piston stuck in bore of clutch release cylinder, preventing clutch from fully engaging (Chapter 8).

33 Grabbing (chattering) as clutch is engaged

1 Oil on clutch plate lining, burned or glazed facings (Chapter 8).
2 Worn or loose engine or transaxle mounts (Chapters 2 and 7).
3 Worn splines on clutch plate hub (Chapter 8).
4 Warped pressure plate or flywheel (Chapter 8).
5 Burned or smeared resin on flywheel or pressure plate (Chapter 8).

34 Transaxle rattling (clicking)

1 Release fork loose (Chapter 8).
2 Low engine idle speed (Chapter 1).

35 Noise in clutch area

Faulty bearing (Chapter 8).

36 Clutch pedal stays on floor

1 Broken release bearing or fork (Chapter 8).
2 Hydraulic release system leaking or air in the system (Chapter 8).

37 High pedal effort

1 Piston binding in bore of release cylinder (Chapter 8).
2 Pressure plate faulty (Chapter 8).
3 Incorrect size master or release cylinder (Chapter 8).

Manual transaxle

38 Knocking noise at low speeds

Worn driveaxle constant velocity (CV) joints (Chapter 8).

39 Noise most pronounced when turning

Differential gear noise (Chapter 7A).*

40 Clunk on acceleration or deceleration

1 Loose engine or transaxle mounts (Chapters 2 and 7A).
2 Worn differential pinion shaft in case.*
3 Worn or damaged driveaxle inboard CV joints (Chapter 8).

41 Clicking noise in turns

Worn or damaged outboard CV joint (Chapter 8).

42 Vibration

1 Rough wheel bearing (Chapter 10).
2 Damaged driveaxle (Chapter 8).
3 Out of round tires (Chapter 1).
4 Tire out of balance (Chapters 1 and 10).
5 Worn CV joint (Chapter 8).

43 Noisy in neutral with engine running

1 Damaged input gear bearing (Chapter 7A).*
2 Damaged clutch release bearing (Chapter 8).

44 Noisy in one particular gear

1 Damaged or worn constant mesh gears (Chapter 7A).*
2 Damaged or worn synchronizers (Chapter 7A).*
3 Bent reverse fork (Chapter 7A).*
4 Damaged fourth speed gear or output gear (Chapter 7A).*
5 Worn or damaged reverse idler gear or idler bushing (Chapter 7A).*

45 Noisy in all gears

1 Insufficient lubricant (Chapter 7A).
2 Damaged or worn bearings (Chapter 7A).*

Troubleshooting 0-25

3   Worn or damaged input gear shaft and/or output gear shaft (Chapter 7A).*

46   Slips out of gear

1   Worn or improperly adjusted linkage (Chapter 7A).
2   Transaxle loose on engine (Chapter 7A).
3   Shift linkage does not work freely, binds (Chapter 7A).
4   Input gear bearing retainer broken or loose (Chapter 7A).*
5   Dirt between clutch cover and engine housing (Chapter 7A).
6   Worn shift fork (Chapter 7A).*

47   Leaks lubricant

1   Side gear shaft seals worn (Chapter 7A).
2   Excessive amount of lubricant in transaxle (Chapters 1 and 7A).
3   Loose or broken input gear shaft bearing retainer (Chapter 7A).*
4   Input gear bearing retainer O-ring and/or lip seal damaged (Chapter 7A).*
5   Striking rod seal leaking (Chapter 7A).
6   Vehicle speed sensor O-ring leaking (Chapter 7A).

48   Hard to shift

Shift linkage loose or worn (Chapter 7A).
* Although the corrective action necessary to remedy the symptoms described is beyond the scope of this manual, the above information should be helpful in isolating the cause of the condition so that the owner can communicate clearly with a professional mechanic.

Automatic transaxle

Note: *Due to the complexity of the automatic transaxle, it is difficult for the home mechanic to properly diagnose and service this component. For problems other than the following, the vehicle should be taken to a dealer or transaxle shop.*

49   Fluid leakage

1   Automatic transaxle fluid is a deep red color. Fluid leaks should not be confused with engine oil, which can easily be blown onto the transaxle by air flow.
2   To pinpoint a leak, first remove all built-up dirt and grime from the transaxle housing with degreasing agents and/or steam cleaning. Then drive the vehicle at low speeds so air flow will not blow the leak far from its source. Raise the vehicle and determine where the leak is coming from. Common areas of leakage are:
  a) *Pan (Chapters 1 and 7B).*
  b) *Dipstick tube (Chapters 1 and 7B).*
  c) *Transaxle oil lines (Chapter 7B).*
  d) *Speed sensor (Chapter 7B).*
  e) *Driveaxle oil seals (Chapter 7B).*

50   Transaxle fluid brown or has a burned smell

Transaxle fluid overheated (Chapter 1).

51   General shift mechanism problems

1   Chapter 7, Part B, deals with checking and adjusting the shift linkage on automatic transaxles. Common problems which may be attributed to poorly adjusted linkage are:
  a) *Engine starting in gears other than Park or Neutral.*
  b) *Indicator on shifter pointing to a gear other than the one actually being used.*
  c) *Vehicle moves when in Park.*
2   Refer to Chapter 7B for the shift linkage adjustment procedure.

52   Transaxle will not downshift with accelerator pedal pressed to the floor

The transaxle is electronically controlled. This type of problem - which is caused by a malfunction in the control unit, a sensor or solenoid, or the circuit itself - is beyond the scope of this book. Take the vehicle to a dealer service department or a competent automatic transmission shop.

53   Engine will start in gears other than Park or Neutral

Neutral start switch out of adjustment or malfunctioning (Chapter 7B).

54   Transaxle slips, shifts roughly, is noisy or has no drive in forward or reverse gears

There are many probable causes for the above problems, but the home mechanic should be concerned with only one possibility - fluid level. Before taking the vehicle to a repair shop, check the level and condition of the fluid as described in Chapter 1. Correct the fluid level as necessary or change the fluid and filter if needed. If the problem persists, have a professional diagnose the cause.

Driveaxles

55   Clicking noise in turns

Worn or damaged outboard CV joint (Chapter 8).

56   Shudder or vibration during acceleration

1   Excessive toe-in (Chapter 10).
2   Incorrect spring heights (Chapter 10).
3   Worn or damaged inboard or outboard CV joints (Chapter 8).
4   Sticking inboard CV joint assembly (Chapter 8).

57   Vibration at highway speeds

1   Out-of-balance front wheels and/or tires (Chapters 1 and 10).
2   Out-of-round front tires (Chapters 1 and 10).
3   Worn CV joint(s) (Chapter 8).

Brakes

Note: *Before assuming that a brake problem exists, make sure that:*
  a) *The tires are in good condition and properly inflated (Chapter 1).*
  b) *The front end alignment is correct (Chapter 10).*
  c) *The vehicle is not loaded with weight in an unequal manner.*

58   Vehicle pulls to one side during braking

1   Incorrect tire pressures (Chapter 1).
2   Front end out of alignment (have the front end aligned).
3   Front, or rear, tire sizes not matched to one another.
4   Restricted brake lines or hoses (Chapter 9).
5   Malfunctioning drum brake or caliper assembly (Chapter 9).
6   Loose suspension parts (Chapter 10).
7   Loose calipers (Chapter 9).
8   Excessive wear of brake shoe or pad material or disc/drum on one side.

59   Noise (high-pitched squeal when the brakes are applied)

Front and/or rear disc brake pads worn out. The noise comes from the wear sensor rubbing against the disc (does not apply to all

Troubleshooting

vehicles). Replace pads with new ones immediately (Chapter 9).

60 Brake roughness or chatter (pedal pulsates)

1. Excessive disc lateral runout (Chapter 9).
2. Uneven pad wear (Chapter 9).
3. Defective disc (Chapter 9).
4. Drum out-of-round (Chapter 9).

61 Excessive brake pedal effort required to stop vehicle

1. Malfunctioning power brake booster (Chapter 9).
2. Partial system failure (Chapter 9).
3. Excessively worn pads or shoes (Chapter 9).
4. Piston in caliper or wheel cylinder stuck or sluggish (Chapter 9).
5. Brake pads or shoes contaminated with oil or grease (Chapter 9).
6. Brake disc grooved and/or glazed (Chapter 1).
7. New pads or shoes installed and not yet seated. It will take a while for the new material to seat against the disc or drum.

62 Excessive brake pedal travel

1. Partial brake system failure (Chapter 9).
2. Insufficient fluid in master cylinder (Chapters 1 and 9).
3. Air trapped in system (Chapters 1 and 9).

63 Dragging brakes

1. Incorrect adjustment of brake light switch (Chapter 9).
2. Master cylinder pistons not returning correctly (Chapter 9).
3. Restricted brakes lines or hoses (Chapters 1 and 9).
4. Incorrect parking brake adjustment (Chapter 9).

64 Grabbing or uneven braking action

1. Malfunction of proportioning valve (Chapter 9).
2. Malfunction of power brake booster unit (Chapter 9).
3. Binding brake pedal mechanism (Chapter 9).

65 Brake pedal feels spongy when depressed

1. Air in hydraulic lines (Chapter 9).

2. Master cylinder mounting bolts loose (Chapter 9).
3. Master cylinder defective (Chapter 9).

66 Brake pedal travels to the floor with little resistance

1. Little or no fluid in the master cylinder reservoir caused by leaking caliper piston(s) (Chapter 9).
2. Loose, damaged or disconnected brake lines (Chapter 9).

67 Parking brake does not hold

Parking brake linkage improperly adjusted (Chapters 1 and 9).

Suspension and steering systems

Note: Before attempting to diagnose the suspension and steering systems, perform the following preliminary checks:

a) Tires for wrong pressure and uneven wear.
b) Steering universal joints from the column to the rack and pinion for loose connectors or wear.
c) Front and rear suspension and the rack-and-pinion assembly for loose or damaged parts.
d) Out-of-round or out-of-balance tires, bent rims and loose and/or rough wheel bearings.

68 Vehicle pulls to one side

1. Mismatched or uneven tires (Chapter 10).
2. Broken or sagging springs (Chapter 10).
3. Wheel alignment out-of-specifications (Chapter 10).
4. Front brake dragging (Chapter 9).

69 Abnormal or excessive tire wear

1. Wheel alignment out-of-specifications (Chapter 10).
2. Sagging or broken springs (Chapter 10).
3. Tire out-of-balance (Chapter 10).
4. Worn strut damper (Chapter 10).
5. Overloaded vehicle.
6. Tires not rotated regularly.

70 Wheel makes a thumping noise

1. Blister or bump on tire (Chapter 10).
2. Improper strut damper action (Chapter 10).

71 Shimmy, shake or vibration

1. Tire or wheel out-of-balance or out-of-round (Chapter 10).
2. Loose or worn wheel bearings (Chapters 1, 8 and 10).
3. Worn tie-rod ends (Chapter 10).
4. Worn lower balljoints (Chapters 1 and 10).
5. Excessive wheel runout (Chapter 10).
6. Blister or bump on tire (Chapter 10).

72 Hard steering

1. Lack of lubrication at balljoints, tie-rod ends and rack and pinion assembly (Chapter 10).
2. Front wheel alignment out-of-specifications (Chapter 10).
3. Low tire pressure(s) (Chapters 1 and 10).

73 Poor returnability of steering to center

1. Lack of lubrication at balljoints and tie-rod ends (Chapter 10).
2. Binding in balljoints (Chapter 10).
3. Binding in steering column (Chapter 10).
4. Lack of lubricant in steering gear assembly (Chapter 10).
5. Front wheel alignment out-of-specifications (Chapter 10).

74 Abnormal noise at the front end

1. Lack of lubrication at balljoints and tie-rod ends (Chapters 1 and 10).
2. Damaged strut mounting (Chapter 10).
3. Worn control arm bushings or tie-rod ends (Chapter 10).
4. Loose stabilizer bar (Chapter 10).
5. Loose wheel nuts (Chapters 1 and 10).
6. Loose suspension bolts (Chapter 10).

75 Wander or poor steering stability

1. Mismatched or uneven tires (Chapter 10).
2. Lack of lubrication at balljoints and tie-rod ends (Chapters 1 and 10).
3. Worn strut assemblies (Chapter 10).
4. Loose stabilizer bar (Chapter 10).
5. Broken or sagging springs (Chapter 10).
6. Wheels out of alignment (Chapter 10).

76 Erratic steering when braking

1. Wheel bearings worn (Chapter 10).
2. Broken or sagging springs (Chapter 10).
3. Leaking wheel cylinder or caliper (Chapter 10).
4. Warped discs or drums (Chapter 10).

Troubleshooting

77 Excessive pitching and/or rolling around corners or during braking

1 Loose stabilizer bar (Chapter 10).
2 Worn strut dampers or mountings (Chapter 10).
3 Broken or sagging springs (Chapter 10).
4 Overloaded vehicle.

78 Suspension bottoms

1 Overloaded vehicle.
2 Worn strut dampers (Chapter 10).
3 Incorrect, broken or sagging springs (Chapter 10).

79 Cupped tires

1 Front wheel or rear wheel alignment out-of-specifications (Chapter 10).
2 Worn struts or shock absorbers (Chapter 10).
3 Wheel bearings worn (Chapter 10).
4 Excessive tire or wheel runout (Chapter 10).
5 Worn balljoints (Chapter 10).

80 Excessive tire wear on outside edge

1 Inflation pressures incorrect (Chapter 1).
2 Excessive speed in turns.
3 Front end alignment incorrect (excessive toe-in). Have professionally aligned.
4 Suspension arm bent or twisted (Chapter 10).

81 Excessive tire wear on inside edge

1 Inflation pressures incorrect (Chapter 1).
2 Front end alignment incorrect (toe-out). Have professionally aligned.
3 Loose or damaged steering components (Chapter 10).

82 Tire tread worn in one place

1 Tires out-of-balance.
2 Damaged or buckled wheel. Inspect and replace if necessary.
3 Defective tire (Chapter 1).

83 Excessive play or looseness in steering system

1 Wheel bearing(s) worn (Chapter 10).
2 Tie-rod end loose (Chapter 10).
3 Steering gear loose (Chapter 10).
4 Worn or loose steering intermediate shaft (Chapter 10).

84 Rattling or clicking noise in steering gear

1 Steering gear loose (Chapter 10).
2 Steering gear defective.

Notes

Chapter 1
Tune-up and routine maintenance

Contents

Section		Section	
Air filter replacement	20	Manual transaxle lubricant change	27
Automatic transaxle fluid change	26	Manual transaxle lubricant level check	19
Automatic transmission fluid level check	7	Nissan Maxima Maintenance schedule	1
Battery check, maintenance and charging	11	Positive Crankcase Ventilation (PCV) valve and hose	
Brake check	17	check and replacement	24
Check Engine light	See Chapter 6	Power steering fluid level check	6
Cooling system check	14	Seat belt check	9
Cooling system servicing (draining, flushing and refilling)	25	Spark plug check and replacement	22
Drivebelt check, adjustment and replacement	12	Spark plug wire, distributor cap and rotor check	
Engine oil and filter change	8	and replacement - 1993 and 1994 models	23
Exhaust Gas Recirculation (EGR) valve check	29	Suspension, steering and driveaxle boot check	28
Exhaust system check	18	Tire and tire pressure checks	5
Fluid level checks	4	Tire rotation	15
Fuel filter replacement	21	Tune-up general information	3
Fuel system check	16	Underhood hose check and replacement	13
Introduction	2	Wiper blade inspection and replacement	10

Specifications

Recommended lubricants and fluids

Note: *Listed here are manufacturer recommendations at the time this manual was written. Manufacturers occasionally upgrade their fluid and lubricant specifications, so check with your local auto parts store for current recommendations.*

Engine oil
  Type .... API grade "certified for gasoline engines"
  Viscosity .... See accompanying chart

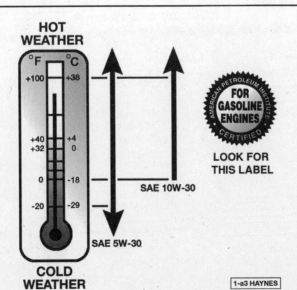

Recommended engine oil viscosity

Recommended lubricants and fluids (continued)

Automatic transaxle fluid	Nissan-Matic "D" automatic transmission fluid or equivalent
Manual transaxle lubricant	API GL-4 80W-90 gear oil
Brake fluid	DOT 3 brake fluid
Power steering system	DEXRON III automatic transmission fluid
Wheel bearings	NLGI no. 2 lithium-base grease

Capacities*

Engine oil (including filter)
- 1993 and 1994 engines .... 4-1/8 qts (3.9 liters)
- 1995 and later engines .... 4-1/4 qts (4.0 liters)

Coolant
- 1993 and 1994 engines .... 8-3/4 qts (8.27 liters)
- 1995 and later engines .... 9 qts (8.5 liters)

Automatic transaxle .... The best way to determine the amount of fluid to add during a routine fluid charge is to measure the amount drained.

Manual transaxle
- 1993 and 1994 engines .... 8 7/8 to 9 1/2 pints (4.2 to 4.5 liters)
- 1995 and later engines .... 9 1/4 to 10 1/8 pints (4.5 to 4.8 liters)

All capacities approximate. Add as necessary to bring up to appropriate level.

Ignition system

Spark plug type and gap
- 1993 and 1994 engines
  - Type .... NGK BKR5ES-11 or equivalent platinum type
  - Gap .... 0.039 to 0.043 inch (1.0 to 1.1 mm)
- 1995 and later engines
  - Type .... NGK PFR5G-11 (platinum) or equivalent
  - Gap .... 0.043 inch (1.1 mm)

Spark plug wire resistance .... less than 9,100 ohms per foot (0.305 meters)

Engine firing order 1-2-3-4-5-6

Cylinder location and distributor (1993 and 1994 models only) rotation - 1995 and later models do not have a distributor

Valve clearances, engine cold (1995 and later)

Intake	0.010 to 0.013 inch (0.254 to 0.330 mm)
Exhaust	0.011 to 0.015 inch (0.279 to 0.381 mm)

Brakes

Disc brake pad lining thickness (minimum)	1/16 inch (1.59 mm)
Parking brake adjustment	10 to 11 clicks

Suspension and steering

Steering wheel freeplay limit	1-3/16 inch or less (30.16 mm)
Balljoint allowable movement	0 inch (0 mm)

Torque specifications

	Nm	Ft-lbs (unless otherwise indicated)
Automatic transaxle drain plug	29.8 to 39.3	22 to 29
Automatic transaxle pan bolts	7.0 to 8.9	62 to 79 in-lbs
Engine oil drain plug	29.8 to 39.3	22 to 29
Manual transaxle drain plug	24.4 to 33.9	18 to 25
Spark plugs	18.9 to 29.8	14 to 22
Wheel lug nuts	97.6 to 118.0	72 to 87

Chapter 1 Tune-up and routine maintenance

Engine compartment components (1995 model shown)

1. PCV valve
2. Spark plugs (under covers)
3. Fuel filter
4. Brake fluid reservoir
5. Air filter housing
6. Battery
7. Fusible link and fuse box
8. Relay box #2
9. Automatic transaxle fluid dipstick
10. Engine oil dipstick
11. Radiator cap
12. Coolant reservoir
13. Windshield washer fluid reservoir
14. Oil filler cap
15. Power steering fluid reservoir
16. Relay box #1

Chapter 1 Tune-up and routine maintenance

Engine compartment underside components

1. Air conditioning compressor
2. Lower radiator hose
3. Radiator drain plug
4. Automatic transaxle drain plug
5. Lower balljoint
6. Driveaxle boots
7. Exhaust system
8. Engine oil pan drain plug
9. Disc brake caliper

Chapter 1 Tune-up and routine maintenance

Typical rear underside components

1. Exhaust system
2. Muffler
3. Fuel tank
4. Rear spring and shock absorber
5. Rear axle
6. Rear axle lateral link
7. Rear disc brake
8. Rear axle trailing arm

1 Nissan Maxima Maintenance schedule

The maintenance intervals in this manual are provided with the assumption that you, not the dealer, will be doing the work. These are the minimum maintenance intervals recommended by the factory for Maximas that are driven daily. If you wish to keep your vehicle in peak condition at all times, you may wish to perform some of these procedures even more often. Because frequent maintenance enhances the efficiency, performance and resale value of your car, we encourage you to do so. If you drive in dusty areas, tow a trailer, idle or drive at low speeds for extended periods or drive for short distances (less than four miles) in below freezing temperatures, shorter intervals are also recommended.

When your vehicle is new, follow the maintenance schedule to the letter, record the maintenance performed in your owners manual and keep all receipts to protect the new vehicle warranty. In many cases, the initial maintenance check is done at no cost to the owner.

Every 250 miles or weekly, whichever comes first

Check the engine oil level (Section 4)
Check the engine coolant level (Section 4)
Check the windshield washer fluid level (Section 4)
Check the brake fluid level (Section 4)
Check the tires and tire pressures (Section 5)

Every 3000 miles or 3 months, whichever comes first

All items listed above plus:
Check the power steering fluid level (Section 6)
Check the automatic transaxle fluid level (Section 7)
Change the engine oil and oil filter (Section 8)

Every 7500 miles or 6 months, whichever comes first

All items listed above plus:
Inspect the seat belts (Section 9)
Inspect and replace if necessary the windshield wiper blades (Section 10)
Check and service the battery (Section 11)
Check and adjust if necessary the engine drivebelts (Section 12)
Inspect and replace if necessary all underhood hoses (Section 13)
Check the cooling system (Section 14)
Rotate the tires (Section 15)

Every 15,000 miles or 12 months, whichever comes first

All items listed above plus:
Inspect the fuel system (Section 16)
Inspect the brake system (Section 17)*
Inspect the exhaust system (Section 18)
Inspect the manual transaxle lubricant level (Section 19)

Every 30,000 miles or 24 months, whichever comes first

All items listed above plus:
Replace the air filter (Section 20)
Replace the fuel filter (Section 21)
Check and replace if necessary the spark plugs (conventional spark plugs) (Section 22)
Inspect and replace if necessary the spark plug wires, distributor cap and rotor - 1993 and 1994 engines (Section 23)
Check and replace if necessary the PCV valve (Section 24)
Service the cooling system (drain, flush and refill) (Section 25)
Change the automatic transaxle fluid (Section 26)**
Change the manual transaxle lubricant (Section 27)**
Suspension, steering and driveaxle boot check (Section 28)

Every 60,000 miles or 48 months, whichever comes first

Check and adjust if necessary, the valve clearances (1995 and later models - only if the valves are making excessive noise) (see Chapter 2)
Exhaust Gas Recirculation (EGR) valve (Section 29)
Replace the spark plugs (platinum spark plugs) (Section 22)
Check or replace the timing belt - 1993 and 1994 engines (see Chapter 2)

* This item is affected by "severe" operating conditions as described below. If your vehicle is operated under "severe" conditions, perform all maintenance indicated with an asterisk (*) at 3000 mile/3 month intervals. Severe conditions are indicated if you mainly operate your vehicle under one or more of the following conditions:
Operating in dusty areas
Towing a trailer
Idling for extended periods and/or low speed operation
Operating when outside temperatures remain below freezing and when most trips are less than 4 miles

** If operated under one or more of the following conditions, change the manual or automatic transaxle fluid lubricant every 15,000 miles:
In heavy city traffic where the outside temperature regularly reaches 90-degrees F (32-degrees C) or higher
In hilly or mountainous terrain
Frequent trailer pulling

Chapter 1 Tune-up and routine maintenance

2 Introduction

This Chapter is designed to help the home mechanic maintain the Nissan Maxima for peak performance, economy, safety and long life.

On the following pages is a master maintenance schedule, followed by sections dealing specifically with each item on the schedule. Visual checks, adjustments, component replacement and other helpful items are included. Refer to the accompanying illustrations of the engine compartment and the underside of the vehicle for the location of various components.

Servicing your Maxima in accordance with the mileage/time maintenance schedule and the following Sections will provide it with a planned maintenance program that should result in a long and reliable service life. This is a comprehensive plan, so maintaining some items but not others at the specified service intervals will not produce the same results.

As you service your Maxima, you will discover that many of the procedures can - and should - be grouped together because of the nature of the particular procedure you're performing or because of the close proximity of two otherwise unrelated components to one another.

For example, if the vehicle is raised for any reason, you should inspect the exhaust, suspension, steering and fuel systems while you're under the vehicle. When you're rotating the tires, it makes good sense to check the brakes and wheel bearings since the wheels are already removed.

Finally, let's suppose you have to borrow or rent a torque wrench. Even if you only need to tighten the spark plugs, you might as well check the torque of as many critical fasteners as time allows.

The first step of this maintenance program is to prepare yourself before the actual work begins. Read through all sections pertinent to the procedures you're planning to do, then make a list of and gather together all the parts and tools you will need to do the job. If it looks as if you might run into problems during a particular segment of some procedure, seek advice from your local parts man or dealer service department.

3 Tune-up general information

The term tune-up is used in this manual to represent a combination of individual operations rather than one specific procedure.

If, from the time the vehicle is new, the routine maintenance schedule is followed closely and frequent checks are made of fluid levels and high wear items, as suggested throughout this manual, the engine will be kept in relatively good running condition and the need for additional work will be minimized.

More likely than not, however, there will be times when the engine is running poorly due to lack of regular maintenance. This is even more likely if a used vehicle, which has not received regular and frequent maintenance checks, is purchased. In such cases, an engine tune-up will be needed outside of the regular routine maintenance intervals.

The first step in any tune-up or engine diagnosis to help correct a poor running engine would be a cylinder compression check. A check of the engine compression (Chapter 2, Part C) will give valuable information regarding the overall performance of many internal components and should be used as a basis for tune-up and repair procedures. If, for instance, a compression check indicates serious internal engine wear, a conventional tune-up will not help the running condition of the engine and would be a waste of time and money.

The following series of operations are those most often needed to bring a generally poor running engine back into a proper state of tune.

Minor tune-up

Check all engine-related fluids (Section 4)
Clean, inspect and test the battery
   (Section 11)
Check and adjust the drivebelts
   (Section 12)
Check all underhood hoses (Section 13)
Check the cooling system (Section 14)
Check the air filter (Section 20)
Replace the spark plugs (Section 22)
Inspect the distributor cap and rotor
   (1993 and 1994 models, Section 23)
Inspect the spark plug and coil wires
   (1993 and 1994 models, Section 23)

Major tune-up

All items listed under Minor tune-up, plus . . .
Check the charging system (Chapter 5)
Check the fuel system (Section 16)
Replace the air filter (Section 20)
Replace the spark plug wires (1993 and
   1994 models, Section 23)
Replace the distributor cap and rotor
   (1993 and 1994 models, Section 23)

Check the idle speed and ignition timing
   (See Chapter 6)

4 Fluid level checks (every 250 miles or weekly)

Note: *The following are fluid level checks to be done on a 250 mile or weekly basis. Additional fluid level checks can be found in specific maintenance procedures that follow. Regardless of intervals, be alert to fluid leaks under the vehicle, which would indicate a fault to be corrected immediately.*

1 Fluids are an essential part of the lubrication, cooling, brake, clutch and windshield washer systems. Because the fluids gradually become depleted and/or contaminated during normal operation of the vehicle, they must be periodically replenished. See *Recommended lubricants and fluids* at the beginning of this Chapter before adding fluid to any of the following components. **Note:** *The vehicle must be on level ground when fluid levels are checked.*

Engine oil

Refer to illustrations 4.2, 4.4 and 4.6

2 The engine oil level is checked with a dipstick that extends through a tube and into the oil pan at the bottom of the engine **(see illustration).**

3 The oil level should be checked before the vehicle has been driven, or about 5 minutes after the engine has been shut off. If the oil is checked immediately after driving the vehicle, some of the oil will remain in the upper engine components, resulting in an inaccurate reading on the dipstick.

4 Pull the dipstick out of the tube and wipe all the oil from the end with a clean rag or paper towel. Insert the clean dipstick all the way back into the tube, then pull it out again. Note the oil at the end of the dipstick. Add oil as necessary to keep the level between the MIN and MAX marks or within the SAFE zone on the dipstick **(see illustration).**

4.2 The engine oil dipstick is clearly marked

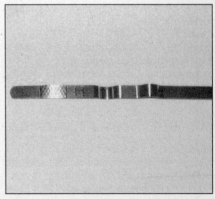

4.4 The oil level must be maintained between the marks at all times - it takes one quart of oil to raise the level from the MIN to MAX mark

1-8   Chapter 1   Tune-up and routine maintenance

4.6 Oil is added to the engine after unscrewing the oil filler cap - always make sure the area around the opening is clean before removing the cap to prevent dirt from contaminating the engine

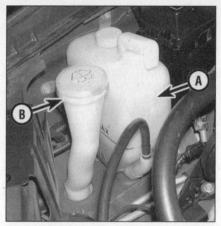

4.8 The coolant reservoir (A) is located at the front of the right fenderwell - keep the level near the MAX mark (arrow) on the side (B is the windshield washer tank)

4.19 The brake fluid level should be kept at the top of the slotted window - never let it drop below the MIN mark; unscrew the reservoir cap to add fluid

5   Do not overfill the engine by adding too much oil since this may result in oil-fouled spark plugs, oil leaks or oil seal failures.
6   Oil is added to the engine after unscrewing a cap from the valve cover **(see illustration)**. A funnel may help to reduce spills.
7   Checking the oil level is an important preventive maintenance step. A consistently low oil level indicates oil leakage through damaged seals, defective gaskets or past worn rings or valve guides. If the oil looks milky or has water droplets in it, the cylinder head gasket(s) may be blown or the head(s) or block may be cracked. The engine should be checked immediately. The condition of the oil should also be checked. Whenever you check the oil level, slide your thumb and index finger up the dipstick before wiping off the oil. If you see small dirt or metal particles clinging to the dipstick, the oil should be changed (see Section 8).

Engine coolant

Refer to illustration 4.8
Warning: *Do not allow antifreeze to come in contact with your skin or painted surfaces of the vehicle. Flush contaminated areas immediately with plenty of water. Don't store new coolant or leave old coolant lying around where it's accessible to children or pets - they're attracted by its sweet smell. Ingestion of even a small amount of coolant can be fatal! Wipe up garage floor and drip pan coolant spills immediately. Keep antifreeze containers covered and repair leaks in the cooling system as soon as they are noted.*

8   All vehicles covered by this manual are equipped with a pressurized coolant recovery system. A white plastic coolant reservoir located in the engine compartment is connected by a hose to the radiator filler neck **(see illustration)**. If the engine overheats, coolant escapes through a valve in the radiator cap and travels through the hose into the reservoir. As the engine cools, the coolant is automatically drawn back into the cooling system to maintain the correct level. **Warning:** *Do not remove the radiator cap to check the coolant level when the engine is warm.*

9   The coolant level in the reservoir should be checked regularly. The level in the reservoir varies with the temperature of the engine. When the engine is cold, the coolant level should be below the FULL mark on the dipstick. Once the engine has warmed up, the level should be at or near the FULL mark. If it isn't, allow the engine to cool, then remove the cap from the reservoir and add a 50/50 mixture of ethylene glycol-based antifreeze and water.
10   Drive the vehicle and recheck the coolant level. If only a small amount of coolant is required to bring the system up to the proper level, water can be used. However, repeated additions of water will dilute the antifreeze and water solution. In order to maintain the proper ratio of antifreeze and water, always top up the coolant level with the correct mixture. Do not use rust inhibitors or additives.
11   If the coolant level drops consistently, there may be a leak in the system. Inspect the radiator, hoses, filler cap, drain plugs and water pump (see Section 14). If no leaks are noted, have the radiator cap pressure-tested.
12   If you have to remove the radiator cap, wait until the engine has cooled, then wrap a thick cloth around the cap and turn it to the first stop. If coolant or steam escapes, let the engine cool down longer, then remove the cap.
13   Check the condition of the coolant as well. It should be relatively clear. If it's brown or rust colored, the system should be drained, flushed and refilled. Even if the coolant appears to be normal, the corrosion inhibitors wear out, so it must be replaced at the specified intervals.

Windshield washer fluid

14   Fluid for the windshield washer system is located in a plastic reservoir in the right side of engine compartment **(see illustration 4.8)**.
15   In milder climates, plain water can be used in the reservoir, but it should be kept no more than 2/3 full to allow for expansion if the water freezes. In colder climates, use windshield washer system antifreeze, available at any auto parts store, to lower the freezing point of the fluid. Mix the antifreeze with water in accordance with the manufacturer's directions on the container. **Caution:** *Don't use cooling system antifreeze - it will damage the vehicle's paint.*
16   To help prevent icing in cold weather, warm the windshield with the defroster before using the washer.

Battery electrolyte

17   These vehicles are equipped with a battery which is permanently sealed (except for vent holes) and has no filler caps. Water doesn't have to be added to these batteries at any time. If a maintenance-type battery is installed, the caps on the top of the battery should be removed periodically to check for a low electrolyte level. This check is most critical during the warm summer months. Add only distilled water to any battery.

Brake and clutch fluid

Refer to illustration 4.19
18   The brake master cylinder is mounted on the upper left of the engine compartment firewall. The clutch cylinder used on manual transmission models is mounted near the brake master cylinder.
19   The translucent plastic reservoir allows the fluid inside to be checked without removing the cap **(see illustration)**. Note that the clutch system is a sealed unit and it shouldn't be necessary to add fluid under most conditions (see Chapter 8 for more information). Be sure to wipe the top of either reservoir cap with a clean rag to prevent contamination of the brake and/or clutch system before removing the cap.
20   When adding fluid, pour it carefully into

Chapter 1 Tune-up and routine maintenance

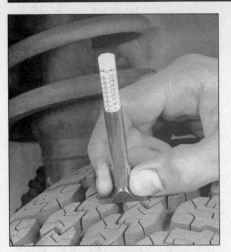

5.2 Use a tire tread depth indicator to monitor tire wear - they are available at auto parts stores and service stations and cost very little

manual. **Warning:** *Brake fluid can harm your eyes and damage painted surfaces, so use extreme caution when handling or pouring it. Do not use brake fluid that has been standing open or is more than one year old. Brake fluid absorbs moisture from the air. Moisture in the system can cause a dangerous loss of brake performance.*

21 At this time, the fluid and master cylinder can be inspected for contamination. The system should be drained and refilled if deposits, dirt particles or water droplets are seen in the fluid.

22 After filling the reservoir to the proper level, make sure the cover or cap is on tight to prevent fluid leakage.

23 The brake fluid level in the master cylinder will drop slightly as the pads at the front wheels wear down during normal operation. If the master cylinder requires repeated additions to keep it at the proper level, it's an indication of leakage in the brake system, which should be corrected immediately. Check all brake lines and connections (see Section 17 for more information).

24 If, upon checking the master cylinder fluid level, you discover one or both reservoirs empty or nearly empty, the brake system should be bled and thoroughly inspected (see Chapter 9).

the reservoir to avoid spilling it on surrounding painted surfaces. Be sure the specified fluid is used, since mixing different types of brake fluid can cause damage to the system. See *Recommended lubricants and fluids* at the front of this Chapter or your owner's

5 Tire and tire pressure checks (every 250 miles or weekly)

Refer to illustrations 5.2, 5.3, 5.4a, 5.4b and 5.8

1 Periodic inspection of the tires may spare you the inconvenience of being stranded with a flat tire. It can also provide you with vital information regarding possible problems in the steering and suspension systems before major damage occurs.

2 The original tires on this vehicle are equipped with 1/2-inch wear bands that will appear when tread depth reaches 1/16-inch, at which point the tires should be considered worn out. Tread wear can be monitored with a simple, inexpensive device known as a tread depth indicator **(see illustration)**.

3 Note any abnormal tread wear **(see illustration)**. Tread pattern irregularities such as cupping, flat spots and more wear on one side than the other are indications of front end alignment and/or balance problems. If any of these conditions are noted, take the vehicle to a tire shop or service station to correct the problem.

4 Look closely for cuts, punctures and embedded nails or tacks. Sometimes a tire

UNDERINFLATION

CUPPING

Cupping may be caused by:
- Underinflation and/or mechanical irregularities such as out-of-balance condition of wheel and/or tire, and bent or damaged wheel.
- Loose or worn steering tie-rod or steering idler arm.
- Loose, damaged or worn front suspension parts.

OVERINFLATION

INCORRECT TOE-IN OR EXTREME CAMBER

FEATHERING DUE TO MISALIGNMENT

5.3 This chart will help you determine the condition of the tires and the probable cause(s) of abnormal wear

5.4a If a tire loses air on a steady basis, check the valve stem core first to make sure it's snug (special inexpensive wrenches are commonly available at auto parts stores)

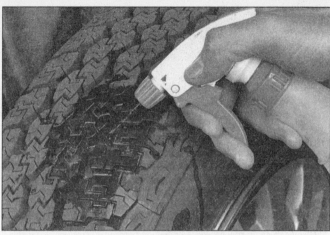

5.4b If the valve stem core is tight, raise the corner of the vehicle with the low tire and spray a soapy water solution onto the tread as the tire is turned slowly - leaks will cause small bubbles to appear

will hold air pressure for a short time or leak down very slowly after a nail has embedded itself in the tread. If a slow leak persists, check the valve stem core to make sure it's tight **(see illustration)**. Examine the tread for an object that may have embedded itself in the tire or for a "plug" that may have begun to leak (radial tire punctures are repaired with a plug that's installed in a puncture). If a puncture is suspected, it can be easily verified by spraying a solution of soapy water onto the puncture area **(see illustration)**. The soapy solution will bubble if there's a leak. Unless the puncture is unusually large, a tire shop or service station can usually repair the tire.

5 Carefully inspect the inner sidewall of each tire for evidence of brake fluid leakage. If you see any, inspect the brakes immediately.

6 Correct air pressure adds miles to the lifespan of the tires, improves mileage and enhances overall ride quality. Tire pressure cannot be accurately estimated by looking at a tire, especially if it's a radial. A tire pressure gauge is essential. Keep an accurate gauge in the vehicle. The pressure gauges attached to the nozzles of air hoses at gas stations are often inaccurate.

7 Always check tire pressure when the tires are cold. Cold, in this case, means the vehicle has not been driven over a mile in the three hours preceding a tire pressure check. A pressure rise of four to eight pounds is not uncommon once the tires are warm.

8 Unscrew the valve cap protruding from the wheel or hubcap and push the gauge firmly onto the valve stem **(see illustration)**. Note the reading on the gauge and compare the figure to the recommended tire pressure shown on the placard on the driver's side door pillar. Be sure to reinstall the valve cap to keep dirt and moisture out of the valve stem mechanism. Check all four tires and, if necessary, add enough air to bring them up to the recommended pressure.

9 Don't forget to keep the spare tire inflated to the specified pressure (refer to your owner's manual or the tire sidewall).

6 Power steering fluid level check (every 3000 miles or 3 months)

Refer to illustration 6.2

1 Unlike manual steering, the power steering system relies on fluid which may, over a period of time, require replenishing.

2 The fluid reservoir for the power steering system is remote from the pump and located at the front of the right shock tower **(see illustration)**.

3 For the check, the front wheels should be pointed straight ahead and the engine should be off.

4 Use a clean rag to wipe off the reservoir cap and the area around the cap. This will help prevent any foreign matter from entering the reservoir during the check.

5 Twist off the cap and check the temperature of the fluid at the end of the dipstick with your finger.

6 Wipe off the fluid with a clean rag, reinsert the dipstick, then withdraw it and read

5.8 To extend the life of the tires, check the air pressure at least once a week with an accurate gauge (don't forget the spare!)

6.2 The power steering fluid reservoir, located ahead of the right shock tower, has marks on it so the fluid level can be checked hot or cold

Chapter 1 Tune-up and routine maintenance

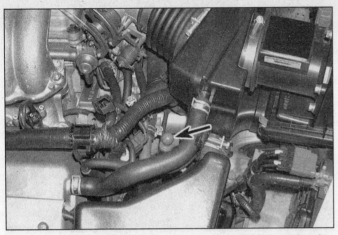

7.3 The automatic transmission dipstick (arrow) is located at the left front of the engine

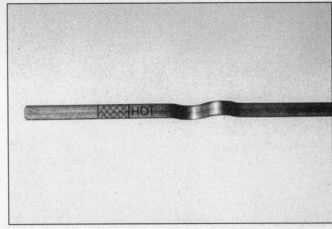

7.6 The automatic transmission fluid must be kept in the crosshatched area on the dipstick at normal operating temperature

the fluid level. The fluid should be at the proper level, depending on whether it was checked hot or cold **(see illustration 6.2)**. Never allow the fluid level to drop below the lower mark on the dipstick.

7   If additional fluid is required, pour the specified type directly into the reservoir, using a funnel to prevent spills.

8   If the reservoir requires frequent fluid additions, all power steering hoses, hose connections, steering gear and the power steering pump should be carefully checked for leaks.

7 Automatic transmission fluid level check (every 3000 miles or 3 months)

Refer to illustrations 7.3 and 7.6

1   The automatic transmission fluid level should be carefully maintained. Low fluid level can lead to slipping or loss of drive, while overfilling can cause foaming and loss of fluid.

2   With the parking brake set, start the engine, then move the shift lever through all the gear ranges, ending in Park. The fluid level must be checked with the vehicle level and the engine running at idle. **Note:** *Incorrect fluid level readings will result if the vehicle has just been driven at high speeds for an extended period, in hot weather in city traffic, or if it has been pulling a trailer. If any of these conditions apply, wait until the fluid has cooled (about 30 minutes).*

3   With the transmission at normal operating temperature, remove the dipstick from the filler tube. The dipstick is located at the front of the engine compartment on the driver's side **(see illustration)**.

4   Wipe the fluid from the dipstick with a clean rag and push it back into the filler tube until the cap seats.

5   Pull the dipstick out again and note the fluid level.

6   If the fluid is warm, the level should be between the two dimples **(see illustration)**. If it's hot, the level should be in the crosshatched area, near the MAX line. If additional fluid is required, add it directly into the tube using a funnel. It takes about one pint to raise the level from the bottom of the crosshatched area to the MAX line with a hot transmission, so add the fluid a little at a time and keep checking the level until it's correct.

7   The condition of the fluid should also be checked along with the level. If the fluid at the end of the dipstick is a dark reddish-brown color, or if it smells burned, it should be changed. If you are in doubt about the condition of the fluid, purchase some new fluid and compare the two for color and smell.

8 Engine oil and filter change (every 3000 miles or 3 months)

Refer to illustrations 8.3, 8.9, 8.14 and 8.18

1   Frequent oil changes are the most important preventive maintenance procedures that can be done by the home mechanic. As engine oil ages, it becomes diluted and contaminated, which leads to premature engine wear.

2   Although some sources recommend oil filter changes every other oil change, we feel that the minimal cost of an oil filter and the relative ease with which it is installed dictate that a new filter be installed every time the oil is changed.

3   Gather together all necessary tools and materials before beginning this procedure **(see illustration)**.

4   You should have plenty of clean rags and newspapers handy to mop up any spills. Access to the under side of the vehicle may be improved if the vehicle can be lifted on a hoist, driven onto ramps or supported by jackstands. **Warning:** *Do not work under a vehicle which is supported only by a bumper, hydraulic or scissors-type jack.*

5   If this is your first oil change, familiarize yourself with the locations of the oil drain plug and the oil filter.

6   Warm the engine to normal operating

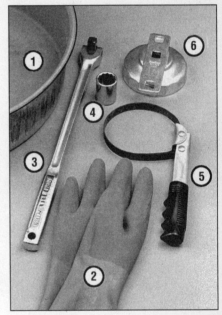

8.3 These tools are required when changing the engine oil and filter

1 *Drain pan* - It should be fairly shallow in depth, but wide to prevent spills
2 *Rubber gloves* - When removing the drain plug and filter, you will get oil on your hands (the gloves will prevent burns)
3 *Breaker bar* - Sometimes the oil drain plug is tight, and a long breaker bar is needed to loosen it
4 *Socket* – To be used with the breaker bar or a ratchet (must be the correct size to fit the drain plug - six-point preferred)
5 *Filter wrench* - This is a metal band-type wrench, which requires clearance around the filter to be effective
6 *Filter wrench* - This type fits on the bottom of the filter and can be turned with a ratchet or breaker bar (different-size wrenches are available for different types of filters)

Chapter 1 Tune-up and routine maintenance

8.9 Use a proper size box-end wrench or socket to remove the oil drain plug and avoid rounding it off

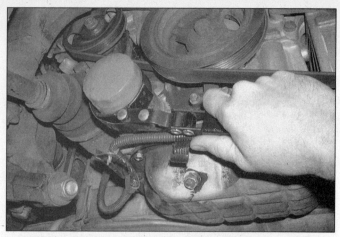

8.14 Since the oil filter is on very tight, you'll need a special wrench for removal - DO NOT use the wrench to tighten the new filter

temperature. If the new oil or any tools are needed, use this warm-up time to gather everything necessary for the job. The correct type of oil for your application can be found in *Recommended lubricants and fluids* at the beginning of this Chapter.

7 With the engine oil warm (warm engine oil will drain better and more built-up sludge will be removed with it), raise and support the vehicle. Make sure it's safely supported!

8 Move all necessary tools, rags and newspapers under the vehicle. Set the drain pan under the drain plug. Keep in mind that the oil will initially flow from the pan with some force; position the pan accordingly.

9 Being careful not to touch any of the hot exhaust components, use a wrench to remove the drain plug near the bottom of the oil pan **(see illustration)**. Depending on how hot the oil is, you may want to wear gloves while unscrewing the plug the final few turns.

10 Allow the oil to drain into the pan. It may be necessary to move the pan as the oil flow slows to a trickle.

11 After all the oil has drained, wipe off the drain plug with a clean rag. Small metal particles may cling to the plug and would immediately contaminate the new oil.

12 Clean the area around the drain plug opening and reinstall the plug. Tighten the plug securely with the wrench. If a torque wrench is available, use it to tighten the plug to the torque listed in this Chapter's Specifications.

13 Move the drain pan into position under the oil filter.

14 Use the oil filter wrench to loosen the oil filter **(see illustration)**.

15 Completely unscrew the old filter. Be careful: it's full of oil. Empty the oil inside the filter into the drain pan, then lower the filter.

16 Compare the old filter with the new one to make sure they're the same type.

17 Use a clean rag to remove all oil, dirt and sludge from the area where the oil filter mounts to the engine. Check the old filter to make sure the rubber gasket isn't stuck to the engine. If the gasket is stuck to the engine (use a flashlight if necessary), remove it.

18 Apply a light coat of clean oil to the rubber gasket on the new oil filter **(see illustration)**.

19 Attach the new filter to the engine, following the tightening directions printed on the filter canister or packing box. Most filter manufacturers recommend against using a filter wrench for installation, due to the possibility of overtightening and damage to the seal.

20 Remove all tools, rags, etc. from under the vehicle, being careful not to spill the oil in the drain pan, then lower the vehicle.

21 Move to the engine compartment and locate the oil filler cap.

22 Pour the fresh oil through the filler opening. A funnel will be helpful.

23 Refer to the engine oil capacity in this Chapter's Specifications and add the proper amount of fresh oil into the engine. Wait a few minutes to allow the oil to drain into the pan, then check the level on the oil dipstick (see Section 4 if necessary). If the oil level is above the hatched area, start the engine and allow the new oil to circulate.

24 Run the engine for only about a minute and then shut it off. Immediately look under the vehicle and check for leaks at the oil pan drain plug and around the oil filter.

25 With the new oil circulated and the filter now completely full, recheck the level on the dipstick and add more oil as necessary.

26 During the first few trips after an oil change, make it a point to check frequently for leaks and proper oil level.

27 The old oil drained from the engine cannot be reused in its present state and should be disposed of. Check with your local auto parts store, disposal facility or environmental agency to see if they will accept the oil for recycling. After the oil has cooled it can be drained into a container (capped plastic jugs, topped bottles, milk cartons, etc.) for transport to one of these disposal sites. Don't dispose of the oil by pouring it on the ground or down a drain!

8.18 Lubricate the oil filter gasket with clean engine oil before installing the filter on the engine

9 Seat belt check (every 6000 miles or 6 months)

1 Check seat belts, buckles, latch plates and guide loops for obvious damage and signs of wear.

2 Where the seat belt receptacle bolts to the floor of the vehicle, check that the bolts are secure.

3 See if the seat belt Warning light comes on when the key is turned to the Run or Start position. A chime should also sound.

10 Wiper blade inspection and replacement (every 6000 miles or 6 months)

Refer to illustrations 10.3 and 10.4

1 The windshield wiper blade elements should be checked periodically for cracks and deterioration.

2 Lift the wiper blade assembly away from the glass.

Chapter 1 Tune-up and r...

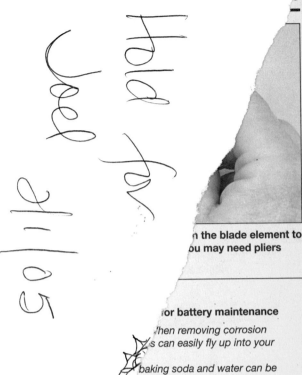

10.3 Depress the release lever (finger is on it here) and slide the wiper assembly down the wiper arm and out of the hook in the end of the arm

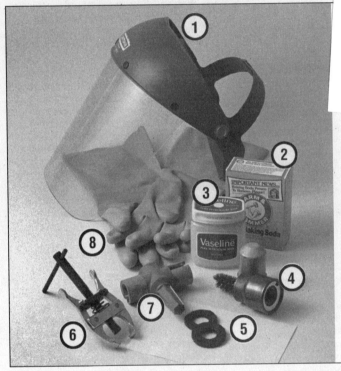

... or battery maintenance

... hen removing corrosion
... s can easily fly up into your
... baking soda and water can be
...
3   *Petroleum j...* ...er of this on the battery posts will help prevent corrosion
4   **Battery post/cable cleaner** - This wire brush cleaning tool will remove all traces of corrosion from the battery posts and cable clamps
5   **Treated felt washers** - Placing one of these on each post, directly under the cable clamps, will help prevent corrosion
6   **Puller** - Sometimes the cable clamps are very difficult to pull off the posts, even after the nut/bolt has been completely loosened. This tool pulls the clamp straight up and off the post without damage
7   **Battery post/cable cleaner** - Here is another cleaning tool that is a slightly different version of Number 4 above, but it does the same thing
8   **Rubber gloves** - Another safety item to consider when servicing the battery; remember that's acid inside the battery!

3   Press the release lever and slide the blade assembly out of the hook in the end of the wiper arm **(see illustration)**.

4   Use needle-nose pliers to squeeze the two metal prongs at the end of the blade element, then slide the element out of the frame **(see illustration)**.

5   Slide the new element into the frame and make sure the prongs snap out behind the last clip of the blade assembly.

6   Installation is the reverse of removal.

11 Battery check, maintenance and charging (every 6000 miles or 6 months)

Refer to illustrations 11.1, 11.5, 11.6a, 11.6b, 11.7a and 11.7b

Warning: *Certain precautions must be fol-* lowed when checking and servicing the battery. Hydrogen gas, which is highly flammable, is always present in the battery cells, so keep lighted tobacco and all other open flames and sparks away from the battery. The electrolyte inside the battery is actually dilute sulfuric acid, which will cause injury if splashed on your skin or in your eyes. It will also ruin clothes and painted surfaces. When removing the battery cables, always detach the negative cable first and hook it up last!

1   A routine preventive maintenance program for the battery in your vehicle is the only way to ensure quick and reliable starts. But before performing any battery maintenance, make sure that you have the proper equipment necessary to work safely around the battery **(see illustration)**.

2   There are also several precautions that should be taken whenever battery mainte- nance is performed. Before servicing the battery, always turn the engine and all accessories off and disconnect the cable from the negative terminal of the battery.

3   The battery produces hydrogen gas, which is both flammable and explosive. Never create a spark, smoke or light a match around the battery. Always charge the battery in a ventilated area.

4   Electrolyte contains poisonous and corrosive sulfuric acid. Do not allow it to get in your eyes, on your skin or on your clothes. Never ingest it. Wear protective safety glasses when working near the battery. Keep children away from the battery.

5   Note the external condition of the battery. If the positive terminal and cable clamp on your vehicle's battery is equipped with a rubber protector, make sure that it's not torn or damaged. It should completely cover the terminal. Look for any corroded or loose con-

Terminal end corrosion or damage.

Insulation cracks.

Chafed insulation or exposed wires.

Burned or melted insulation.

11.5 Typical battery cable problems

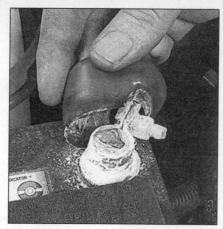

11.6a Battery terminal corrosion usually appears as light, fluffy powder

11.6b Removing the cable from a battery post with a wrench - sometimes special battery pliers are required for this procedure if corrosion has caused deterioration of the nut hex (always remove the ground cable first and hook it up last!)

nections, cracks in the case or cover or loose hold-down clamps. Also check the entire length of each cable for cracks and frayed conductors **(see illustration)**.

6 If corrosion, which looks like white, fluffy deposits is evident, particularly around the terminals, the battery should be removed for cleaning **(see illustration)**. Loosen the cable clamp bolts with a wrench, being careful to remove the ground cable first, and slide them off the terminals **(see illustration)**. Then disconnect the hold-down clamp bolt and nut, remove the clamp and lift the battery from the engine compartment.

7 Clean the cable clamps thoroughly with a battery brush or a terminal cleaner and a solution of warm water and baking soda **(see illustration)**. Wash the terminals and the top of the battery case with the same solution but make sure that the solution doesn't get into the battery. When cleaning the cables, terminals and battery top, wear safety goggles and rubber gloves to prevent any solution from coming in contact with your eyes or hands. Wear old clothes too - even diluted, sulfuric acid splashed onto clothes will burn holes in them. If the posts have been extensively corroded, clean them up with a terminal cleaner **(see illustration)**. Thoroughly wash all cleaned areas with plain water.

8 Make sure that the battery tray is in good condition and the hold-down clamp bolts are tight. If the battery is removed from the tray, make sure no parts remain in the bottom of the tray when the battery is reinstalled. When reinstalling the hold-down clamp bolts, do not overtighten them.

9 Any metal parts of the vehicle damaged by corrosion should be covered with a zinc-based primer, then painted.

10 Information on removing and installing the battery can be found in Chapter 5. Information on jump-starting can be found at the front of this manual. For more detailed battery checking procedures, refer to the *Haynes Automotive Electrical Manual*.

Charging

Warning: *When batteries are being charged, hydrogen gas, which is very explosive and flammable, is produced. Do not smoke or allow open flames near a charging or a recently charged battery. Wear eye protection when near the battery during charging. Also, make sure the charger is unplugged before connecting or disconnecting the battery from the charger.*

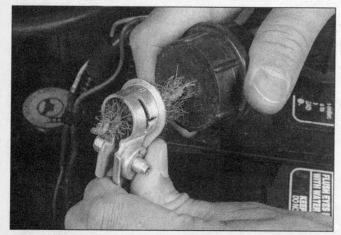

11.7a When cleaning the cable clamps, all corrosion must be removed (the inside of the clamp is tapered to match the taper on the post, so don't remove too much material)

11.7b Regardless of the type of tool used on the battery posts, a clean, shiny surface should be the result

Chapter 1 Tune-up and routine maintenance

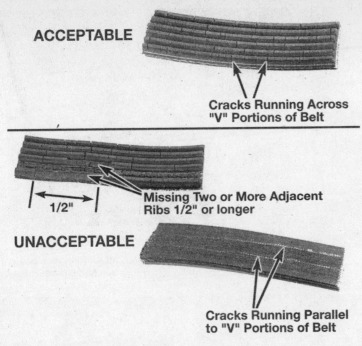

12.2a Check ribbed (serpentine) belts for signs of wear like these - if they look worn, replace them as a set

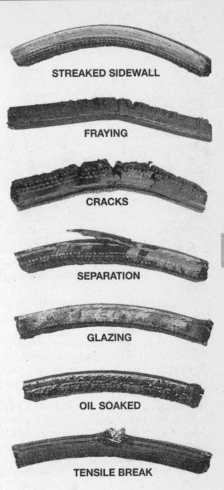

12.2b Look for these signs of wear or damage on V-belt drivebelts

Note: *The manufacturer recommends the battery be removed from the vehicle for charging because the gas that escapes during this procedure can damage the paint. Fast charging with the battery cables connected can result in damage to the electrical system.*

11 Slow-rate charging is the best way to restore a battery that's discharged to the point where it will not start the engine. It's also a good way to maintain the battery charge in a vehicle that's only driven a few miles between starts. Maintaining the battery charge is particularly important in the winter when the battery must work harder to start the engine and electrical accessories that drain the battery are in greater use.

12 It's best to use a one or two-amp battery charger (sometimes called a "trickle" charger). They are the safest and put the least strain on the battery. They are also the least expensive. For a faster charge, you can use a higher amperage charger, but don't use one rated more than 1/10th the amp/hour rating of the battery. Rapid boost charges that claim to restore the power of the battery in one to two hours are hardest on the battery and can damage batteries not in good condition. This type of charging should only be used in emergency situations.

13 The average time necessary to charge a battery should be listed in the instructions that come with the charger. As a general rule, a trickle charger will charge a battery in 12 to 16 hours.

14 Remove all the cell caps (if equipped) and cover the holes with a clean cloth to prevent spattering electrolyte. Disconnect the negative battery cable and hook the battery charger cable clamps up to the battery posts (positive to positive, negative to negative), then plug in the charger. Make sure it is set at 12-volts if it has a selector switch.

15 If you're using a charger with a rate higher than two amps, check the battery regularly during charging to make sure it doesn't overheat. If you're using a trickle charger, you can safely let the battery charge overnight after you've checked it regularly for the first couple of hours.

16 If the battery has removable cell caps, measure the specific gravity with a hydrometer every hour during the last few hours of the charging cycle. Hydrometers are available inexpensively from auto parts stores - follow the instructions that come with the hydrometer. Consider the battery charged when there's no change in the specific gravity reading for two hours and the electrolyte in the cells is gassing (bubbling) freely. The specific gravity reading from each cell should be very close to the others. If not, the battery probably has a bad cell(s).

17 Some batteries with sealed tops have built-in hydrometers on the top that indicate the state of charge by the color displayed in the hydrometer window. Normally, a bright-colored hydrometer indicates a full charge and a dark hydrometer indicates the battery still needs charging.

18 If the battery has a sealed top and no built-in hydrometer, you can hook up a digital voltmeter across the battery terminals to check the charge. A fully charged battery should read 12.6 volts or higher.

19 Further information on the battery and jump-starting can be found in Chapter 5 and at the front of this manual.

12 Drivebelt check, adjustment and replacement (every 6000 miles or 6 months)

Refer to illustrations 12.2a, 12.2b, 12.4, 12.5a, 12.5b and 12.5c

1 Drivebelts are located at the front of the engine and play an important role in the overall operation of the engine and its components. Due to their function and material make up, the belts are prone to wear and should be periodically inspected. Early models have three belts, including serpentine and V-belt types, while 1995 and later models have two belts, both of the serpentine type. The drivebelts drive the alternator, power steering pump, water pump and air conditioning compressor (if equipped).

2 With the engine off, open the hood and use your fingers (and a flashlight, if necessary), to move along the belt checking for cracks and separation of the belt plies. Also check for fraying and glazing, which gives the belt a shiny appearance **(see illustrations)**. Both sides of the belt should be inspected, which means you will have to twist the belt to check the underside.

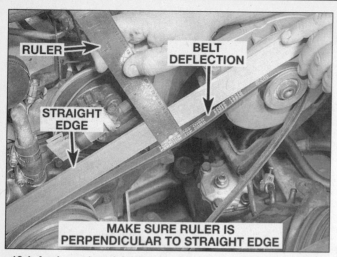

12.4 A ruler and straightedge can be used to determine the belt deflection (tension) between two pulleys

12.5a Loosen/tighten the idler pulley adjuster bolt (arrow) to adjust the main serpentine belt (1995 model shown)

3   Check the ribs on the underside of the belt. They should all be the same depth, with none of the surface uneven.

4   Belt tension must be checked manually, by pushing on the belt at a distance halfway between two pulleys. Push firmly with your thumb and see how much the belt moves (deflects) **(see illustration)**. As rule of thumb, if the distance from pulley center-to-pulley center is between 7 and 11 inches, the belt should deflect 1/4-inch. If the belt travels between pulleys spaced 12 to 16 inches apart, the belt should deflect 1/4 to 1/2-inch.

5   On three-belt models, one belt drives the water pump and power steering pump, and is adjusted by a idler accessed from below the engine. Another short belt drives the alternator, and is adjusted by moving the alternator adjuster. A longer belt drives the air-conditioning compressor and is adjusted by a top-mounted idler. Later models have one serpentine belt adjusted at the power steering pump, and a larger serpentine belt adjusted at the upper idler **(see illustrations)**.

6   To replace the belts, loosen the adjuster or component until the belt can be removed from the various pulleys. In multiple-belt applications, outer belts will have to be removed to access inner belts, but as a general rule, all belts should be replaced at the same time anyway.

7   Route the new belt over the various pulleys, then adjust the tension. **Note:** *Refer to your owner's manual for the belt routing diagram for your specific model.*

13 Underhood hose check and replacement (every 6000 miles or 6 months)

General

Caution: *Replacement of air conditioning hoses must be left to a dealer service department or air conditioning shop that has the*

12.5b From below, loosen the bolt (arrow) at the back of the power steering pump . . .

equipment to depressurize the system safely and recover the refrigerant. Never remove air conditioning components or hoses until the system has been depressurized.

1   High temperatures in the engine compartment can cause the deterioration of the rubber and plastic hoses used for engine, accessory and emission systems operation. Periodic inspection should be made for cracks, loose clamps, material hardening and leaks. Information specific to the cooling system hoses can be found in Section 14.

2   Some, but not all, hoses are secured to their fittings with clamps. Where clamps are used, check to be sure they haven't lost their tension, allowing the hose to leak. If clamps aren't used, make sure the hose has not expanded and/or hardened where it slips over the fitting, allowing it to leak.

Vacuum hoses

3   It's quite common for vacuum hoses, especially those in the emissions system, to be color-coded or identified by colored stripes molded into them. Various systems require hoses with different wall thickness,

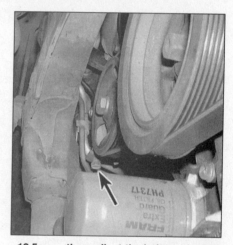

12.5c . . . then adjust the belt tension at the adjuster bolt (arrow) - when proper tension is achieved, retighten the bolt on the back of the pump

collapse resistance and temperature resistance. When replacing hoses, be sure the new ones are made of the same material.

4   Often the only effective way to check a hose is to remove it completely from the vehicle. If more than one hose is removed, be sure to label the hoses and fittings to ensure correct installation.

5   When checking vacuum hoses, be sure to include any plastic T-fittings in the check. Inspect the fittings for cracks and the hose where it fits over the fitting for distortion, which could cause leakage.

6   A small piece of vacuum hose (1/4-inch inside diameter) can be used as a stethoscope to detect vacuum leaks. Hold one end of the hose to your ear and probe around vacuum hoses and fittings, listening for the "hissing" sound characteristic of a vacuum leak. **Warning:** *When probing with the vacuum hose stethoscope, be very careful not to come into contact with moving engine components such as the drivebelt, cooling fan, etc.*

Chapter 1 Tune-up and routine maintenance

Check for a chafed area that could fail prematurely.

Check for a soft area indicating the hose has deteriorated inside.

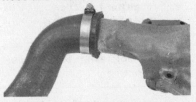

Overtightening the clamp on a hardened hose will damage the hose and cause a leak.

Check each hose for swelling and oil-soaked ends. Cracks and breaks can be located by squeezing the hose.

14.4 Hoses, like drivebelts, have a habit of failing at the worst possible time - to prevent the inconvenience of a blown radiator or heater hose, inspect them carefully as shown here

Fuel hose

Warning: *There are certain precautions that must be taken when inspecting or servicing fuel system components. Work in a well-ventilated area and do not allow open flames (cigarettes, appliance pilot lights, etc.) or bare light bulbs near the work area. Mop up any spills immediately and do not store fuel soaked rags where they could ignite. The fuel system is under high pressure, so if any fuel lines are to be disconnected, the pressure in the system must be relieved first (see Chapter 4 for more information).*

7 Check all rubber fuel lines for deterioration and chafing. Check especially for cracks in areas where the hose bends and just before fittings, such as where a hose attaches to the fuel filter.
8 High quality fuel line, made specifically for high-pressure fuel-injection systems, must be used for fuel line replacement. Never, under any circumstances, use unreinforced vacuum line, clear plastic tubing or water hose for fuel lines.
9 Spring-type clamps are commonly used on fuel lines. These clamps often lose their tension over a period of time, and can be "sprung" during removal. Replace all spring-type clamps with screw clamps whenever a hose is replaced.

Metal lines

10 Sections of metal line are routed along the frame, between the fuel tank and the engine. Check carefully to be sure the line has not been bent or crimped and that cracks have not started in the line.
11 If a section of metal fuel line must be replaced, only seamless steel tubing should be used, since copper and aluminum tubing don't have the strength necessary to withstand normal engine vibration.
12 Check the metal brake lines where they enter the master cylinder and brake-proportioning unit for cracks in the lines or loose fittings. Any sign of brake fluid leakage calls for an immediate and thorough inspection of the brake system.

14 Cooling system check (every 6000 miles or 6 months)

Refer to illustration 14.4

1 Many major engine failures can be attributed to a faulty cooling system. If the vehicle is equipped with an automatic transmission, the cooling system also cools the transmission fluid and thus plays an important role in prolonging transmission life.
2 The cooling system should be checked with the engine cold. Do this before the vehicle is driven for the day or after it has been shut off for at least three hours.
3 Remove the radiator cap by turning it to the left until it reaches a stop. If you hear a hissing sound (indicating there is still pressure in the system), wait until this stops. Now press down on the cap with the palm of your hand and continue turning to the left until the cap can be removed. Thoroughly clean the cap, inside and out, with clean water. Also clean the filler neck on the radiator. All traces of corrosion should be removed. The coolant inside the radiator should be relatively transparent. If it is rust-colored, the system should be drained, flushed and refilled (see Section 25). If the coolant level is not up to the top, add additional anti-freeze/coolant mixture (see Section 4).
4 Carefully check the large upper and lower radiator hoses along with the smaller diameter heater hoses that run from the engine to the firewall. Inspect each hose along its entire length, replacing any hose that is cracked, swollen or shows signs of deterioration. Cracks may become more apparent if the hose is squeezed **(see illustration)**. Regardless of condition, it's a good idea to replace hoses with new ones every two years.
5 Make sure all hose connections are tight. A leak in the cooling system will usually show up as white or rust-colored deposits on the areas adjoining the leak. If wire-type clamps are used at the ends of the hoses, it may be a good idea to replace them with more secure screw-type clamps.
6 Use compressed air or a soft brush to remove bugs, leaves, etc. from the front of the radiator or air conditioning condenser. Be careful not to damage the delicate cooling fins or cut yourself on them.
7 Every other inspection, or at the first indication of cooling system problems, have the cap and system pressure tested. If you don't have a pressure tester, most repair shops will do this for a minimal charge.

15 Tire rotation (every 6000 miles or 6 months)

Refer to illustrations 15.2a and 15.2b

1 The tires should be rotated at the specified intervals and whenever uneven wear is noticed.
2 Tires must be rotated in the recommended pattern **(see illustrations)**. **Note:** *Most vehicles are sold with non-directional*

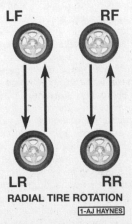

15.2a The recommended four-tire rotation pattern for directional radial tires

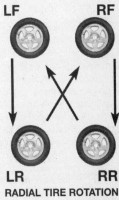

15.2b The recommended four-tire rotation pattern for non-directional radial tires

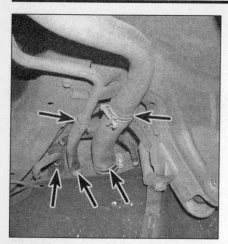

16.5 Inspect fuel filler hoses for cracks and make sure the clamps (arrows) are tight on all hoses in the system

17.7a You will find an inspection hole like this in each caliper - placing a ruler across the hole should enable you to determine the thickness of remaining pad material for both inner and outer pads

17.7b If the outer pad's thickness is hard to see through the inspection hole, look up at the pad from below the caliper

radial tires, but some replacement performance tires are available that are directional, and have a different rotation pattern. Directional tires have an arrow on the sidewall indicating the direction they must turn when mounted on the vehicle.

3 Refer to the information in *Jacking and towing* at the front of this manual for the proper procedures to follow when raising the vehicle and changing a tire. If the brakes are to be checked, don't apply the parking brake as stated. Make sure the tires are blocked to prevent the vehicle from rolling as it's raised.

4 Preferably, the entire vehicle should be raised at the same time. This can be done on a hoist or by jacking up each corner and then lowering the vehicle onto jackstands placed under the frame rails. Always use four jackstands and make sure the vehicle is safely supported.

5 After rotation, check and adjust the tire pressures as necessary. Tighten the lug nuts to the torque listed in this Chapter's Specifications.

16 Fuel system check (every 15,000 miles or 12 months)

Refer to illustration 16.5

Warning: *Gasoline is extremely flammable, so take extra precautions when you work on any part of the fuel system. Don't smoke or allow open flames or bare light bulbs near the work area, and don't work in a garage where a natural gas-type appliance (such as a water heater or clothes dryer) with a pilot light is present. Since gasoline is carcinogenic, wear latex gloves when there's a possibility of being exposed to fuel, and, if you spill any fuel on your skin, rinse it off immediately with soap and water. Mop up any spills immediately and do not store fuel-soaked rags where they could ignite. The fuel system is under constant pressure, so, if any fuel lines are to be disconnected, the fuel pressure in the system must be relieved first (see Chapter 4 for more information). When you perform any kind of work on the fuel system, wear safety glasses and have a Class B type fire extinguisher on hand.*

1 If you smell gasoline while driving or after the vehicle has been sitting in the sun, inspect the fuel system immediately.

2 Remove the gas filler cap and inspect it for damage and corrosion. The gasket should have an unbroken sealing imprint. If the gasket is damaged or corroded, remove it and install a new one.

3 Inspect the fuel feed and return lines for cracks. Make sure the connections between the fuel lines and the fuel injection system are tight.

4 Since some components of the fuel system - the fuel tank and part of the fuel feed and return lines, for example - are underneath the vehicle, they can be inspected more easily with the vehicle raised on a hoist. If that's not possible, raise the vehicle and support it securely on jackstands.

5 With the vehicle raised and safely supported, inspect the gas tank and filler neck for punctures, cracks and other damage. The connection between the filler neck and the tank is particularly critical. Sometimes a rubber filler neck will leak because of loose clamps or deteriorated rubber **(see illustration)**. These are problems a home mechanic can usually rectify. **Warning:** *Do not, under any circumstances, try to repair a fuel tank (except rubber components). A welding torch or any open flame can easily cause fuel vapors inside the tank to explode.*

6 Carefully check all rubber hoses and metal lines leading away from the fuel tank. Check for loose connections, deteriorated hoses, crimped lines and other damage. Carefully inspect the lines from the tank to the fuel injection system. Repair or replace damaged sections as necessary (see Chapter 4).

17 Brake check (every 15,000 miles or 12 months)

Warning: *The dust created by the brake system may contain asbestos, which is harmful to your health. Never blow it out with compressed air and don't inhale any of it. An approved filtering mask should be worn when working on the brakes. Do not, under any circumstances, use petroleum-based solvents to clean brake parts. Use brake system cleaner only! Try to use non-asbestos replacement parts whenever possible.*

Note: This Section applies to both front and rear disc brakes on the covered models. *For detailed photographs of the brake system, refer to Chapter 9.*

Refer to illustrations 17.7a, 17.7b, 17.9 and 17.11

1 In addition to the specified intervals, the brakes should be inspected every time the wheels are removed or whenever a defect is suspected.

2 Any of the following symptoms could indicate a potential brake system defect: The vehicle pulls to one side when the brake pedal is depressed; the brakes make squealing or dragging noises when applied; brake pedal travel is excessive; the pedal pulsates; or brake fluid leaks, usually onto the inside of the tire or wheel.

3 Loosen the wheel lug nuts.

4 Raise the vehicle and place it securely on jackstands.

5 Remove the wheels (see *Jacking and towing* at the front of this book, or your owner's manual, if necessary).

6 There are two pads (an outer and an inner) in each caliper. The pads are visible with the wheels removed.

7 Check the pad thickness by looking at each end of the caliper and through the inspection window in the caliper body **(see illustrations)**. If the lining material is less than the thickness listed in this Chapter's Specifications, replace the pads. **Note:** *Keep in*

Chapter 1 Tune-up and routine maintenance 1-19

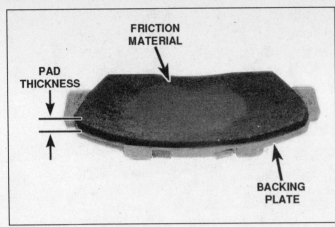

17.9 An exact measurement of the pad material remaining can be measured when the pads are removed

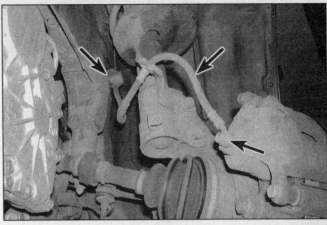

17.11 Inspect the flexible brake hoses (A) at each caliper, and the connections at the caliper (B) and where the flexible line joins the steel line (C)

mind that the lining material is riveted or bonded to a metal backing plate and the metal portion is not included in this measurement.

8 If it is difficult to determine the exact thickness of the remaining pad material by the above method, or if you are at all concerned about the condition of the pads, remove the caliper(s), then remove the pads from the calipers for further inspection (refer to Chapter 9).
9 Once the pads are removed from the calipers, clean them with brake cleaner and re-measure them with a ruler or a vernier caliper **(see illustration)**.
10 Measure the disc thickness with a micrometer to make sure that it still has service life remaining. If any disc is thinner than the specified minimum thickness, replace it (refer to Chapter 9). Even if the disc has service life remaining, check its condition. Look for scoring, gouging and burned spots. If these conditions exist, remove the disc and have it resurfaced (see Chapter 9).
11 Before installing the wheels, check all brake lines and hoses for damage, wear, deformation, cracks, corrosion, leakage, bends and twists, particularly in the vicinity of the rubber hoses at the calipers **(see illustration)**. Check the clamps for tightness and the connections for leakage. Make sure that all hoses and lines are clear of sharp edges, moving parts and the exhaust system. If any of the above conditions are noted, repair, reroute or replace the lines and/or fittings as necessary (see Chapter 9).

Brake booster check

12 Sit in the driver's seat and perform the following sequence of tests.
13 With the brake fully depressed, start the engine - the pedal should move down a little when the engine starts.
14 With the engine running, depress the brake pedal several times - the travel distance should not change.
15 Depress the brake, stop the engine and hold the pedal in for about 30 seconds - the pedal should neither sink nor rise.
16 Restart the engine, run it for about a minute and turn it off. Then firmly depress the brake several times - the pedal travel should decrease with each application.
17 If your brakes do not operate as described, the brake booster has failed. Refer to Chapter 9 for the replacement procedure.

Parking brake

18 One method of checking the parking brake is to park the vehicle on a steep hill with the parking brake set and the transmission in Neutral. If the parking brake cannot prevent the vehicle from rolling, it's in need of adjustment (see Chapter 9).

18 Exhaust system check (every 30,000 miles or 24 months)

Refer to illustrations 18.2a and 18.2b

1 With the engine cold (at least three hours after the vehicle has been driven), check the complete exhaust system from its starting point at the engine to the end of the tailpipe. This should be done on a hoist where unrestricted access is available.
2 Check the pipes and connections for evidence of leaks, severe corrosion or damage. Make sure that all brackets and hangers are in good condition and tight **(see illustrations)**.

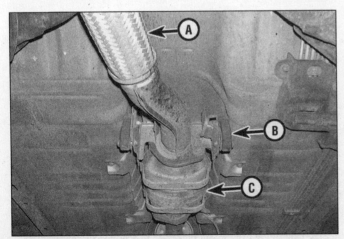

18.2a Inspect for rustout or signs of damage or leakage at the front pipe (A), exhaust hangers (B), and the catalytic converter (C)

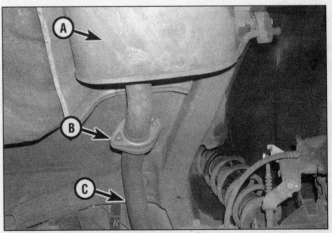

18.2b At the rear of the vehicle, inspect the muffler (A), the rear pipe flange (B), and the tailpipe (C)

1-20 Chapter 1 Tune-up and routine maintenance

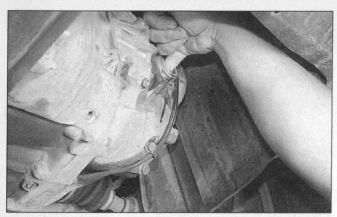

19.2 Remove the filler plug from the side of the manual transaxle - place your finger in the hole to check the lubricant level

20.1a Release the four spring clips (arrows) securing the air filter cover

3   At the same time, inspect the underside of the body for holes, corrosion, open seams, etc. which may allow exhaust gases to enter the passenger compartment. Seal all body openings with silicone or body putty.
4   Rattles and other noises can often be traced to the exhaust system, especially the mounts and hangers. Try to move the pipes, muffler and catalytic converter. If the components can come in contact with the body or suspension parts, secure the exhaust system with new mounts.
5   Check the running condition of the engine by inspecting inside the end of the tailpipe. The exhaust deposits here are an indication of engine state-of-tune. If the pipe is black and sooty or coated with white deposits, the engine is in need of a tune-up, including a thorough fuel system inspection.

19 Manual transaxle lubricant level check (every 15,000 miles or 12 months)

Refer to illustration 19.2
1   The manual transaxle does not have a dipstick. To check the fluid level, raise the vehicle and support it securely on jackstands.

2   Remove the oil filler plug from the side of the transaxle. Measure with your finger to feel that the oil level is at the plug opening **(see illustration)**.
3   If the transaxle needs more lubricant (if the level is not up to the opening, use a gear oil pump to add more and bring it up to the proper level. Stop filling the transaxle when the lubricant begins to run out the hole.
4   Clean and install the plug and tighten it securely. Drive the vehicle a short distance, then check for leaks.

20 Air filter replacement (every 30,000 miles or 24 months)

Refer to illustrations 20.1a and 20.1b
1   The air filter is located inside a housing at the left side of the engine compartment. To remove the air filter, release the four spring clips that keep the two halves of the air cleaner housing together, then lift the cover up and remove the air filter element **(see illustrations)**.
2   Inspect the outer surface of the filter element. If it is dirty, replace it. If it is only moderately dusty, it can be reused by blowing it clean from the back to the front surface with com-

pressed air. Because it is a pleated paper type filter, it cannot be washed or oiled. If it cannot be cleaned satisfactorily with compressed air, discard and replace it. While the cover is off, be careful not to drop anything down into the housing. **Caution:** *Never drive the vehicle with the air cleaner removed. Excessive engine wear could result and backfiring could even cause a fire under the hood.*
3   Wipe out the inside of the air cleaner housing.
4   Place the new filter into the air cleaner housing, making sure it seats properly.
5   Installation of the cover is the reverse of removal.

21 Fuel filter replacement (every 30,000 miles or 24 months)

Refer to illustration 21.3
Note: *The 2000 and later models are not equipped with a separate fuel filter in the engine compartment. On these models, the fuel filter is an integral part of the fuel pump assembly located in the fuel tank (see Chapter 4).*
1   The canister filter is mounted in a clip on the firewall near the brake master cylinder.

20.1b Lift the cover enough to pull the air filter out - when reinstalling, make sure the housing is clean and the lip of the filter fits the housing all the way around

21.3 Disconnect the inlet and outlet fuel hoses (A indicates the outlet hose clamp) from the fuel filter (B), then pull it out of the spring-clip bracket (C)

Chapter 1 Tune-up and routine maintenance

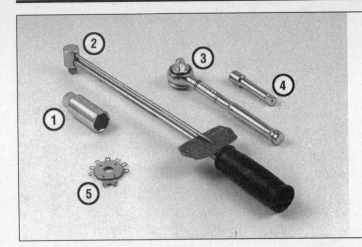

22.1 Tools required for changing spark plugs

1. **Spark plug socket** - This will have special padding inside to protect the spark plug porcelain insulator
2. **Torque wrench** - Although not mandatory, use of this tool is the best way to ensure that the plugs are tightened properly
3. **Ratchet** - Standard hand tool to fit the plug socket
4. **Extension** - Depending on model and accessories, you may need special extensions and universal joints to reach one or more of the plugs
5. **Spark plug gap gauge** - This gauge for checking the gap comes in a variety of styles. Make sure the gap for your engine is included

2   Depressurize the fuel system (see Chapter 4), then disconnect the cable from the negative terminal of the battery.

3   Detach the filter from the bracket, loosen the screw clamps, then detach the hoses from the top and bottom of the fuel filter and remove it **(see illustration)**.

4   Note that the inlet and outlet pipes are clearly labeled on their respective ends of the filter. Make sure the new filter is installed so that it's facing the proper direction as noted above. When correctly installed, the filter should be installed so the outlet pipe faces up and the inlet pipe faces down.

5   Install the inlet and outlet hoses and tighten the screw clamps securely. Reconnect the battery cable, start the engine and check for leaks.

22 Spark plug check and replacement (see maintenance schedule for service intervals)

Refer to illustrations 22.1, 22.4a, 22.4b, 22.6, 22.8a, 22.8b, 22.8c, 22.10a and 22.10b

Note: *On 2000 and later models, the manufacturer suggests that checking and adjusting the spark plug gap is no longer necessary.*

1   Spark plug replacement requires a spark plug socket that fits onto a ratchet handle. This socket is lined with a rubber grommet to protect the porcelain insulator of the spark plug and to hold the plug while you insert it into the spark plug hole. You will also need a wire-type feeler gauge to check and adjust the spark plug gap and a torque wrench to tighten the new plugs to the specified torque **(see illustration)**.

2   If you are replacing the plugs, purchase the new plugs, adjust them to the proper gap and then replace each plug one at a time. **Note:** *When buying new spark plugs, it's essential that you obtain the correct plugs for your specific vehicle. This information can be found in the Specifications Section at the beginning of this Chapter, on the Vehicle Emissions Control Information (VECI) label located on the underside of the hood or in the owner's manual. If these sources specify different plugs, purchase the spark plug type specified on the VECI label because that information is provided specifically for your engine.*

3   Inspect each of the new plugs for defects. If there are any signs of cracks in the porcelain insulator of a plug, don't use it.

4   Check the electrode gaps of the new plugs. Check the gap by inserting the wire gauge of the proper thickness between the electrodes at the tip of the plug **(see illustration)**. The gap between the electrodes should be identical to that listed in this Chapter's Specifications or on the VECI label. If the gap is incorrect, use the notched adjuster on the feeler gauge body to bend the curved side electrode slightly **(see illustration)**.

5   If the side electrode is not exactly over the center electrode, use the notched adjuster to align them. **Caution:** *If the gap of a new plug must be adjusted, bend only the base of the ground electrode - do not touch the tip.*

Removal

6   To prevent the possibility of mixing up spark plug wires on 1993 and 1994 models, work on one spark plug at a time. Remove the wire and boot from one spark plug. Grasp

22.4a Spark plug manufacturers recommend using a wire-type gauge when checking the gap - if the wire does not slide between the electrodes with a slight drag, adjustment is required

22.4b To change the gap, bend the side electrode only, as indicated by the arrows, and be very careful not to crack or chip the porcelain insulator surrounding the center electrode

the boot as shown - not the cable, give it a half twisting motion and pull straight up **(see illustration)**.

7   If compressed air is available, blow any dirt or foreign material away from the spark plug area before proceeding (a common bicycle pump will also work).

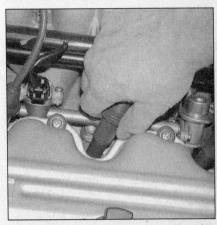

22.6 When removing the spark plug wires on 1993 and 1994 engines, pull only on the boot and use a twisting/pulling motion

Chapter 1 Tune-up and routine maintenance

22.8a On 1995 and later engines, remove the screws (arrows) and the coil cover from each valve cover ...

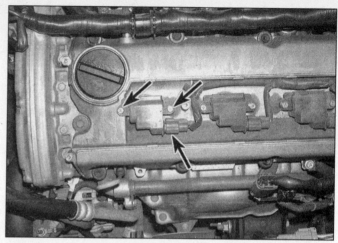

22.8b ... then disconnect the coil electrical connector (lower arrow), remove the two bolts (upper arrows) and pull the coil out

22.8c Use a socket wrench with a long extension to unscrew the spark plugs (later DOHC engine, shown; earlier engine similar, but plugs are to the intake side of the valve cover)

8 Remove the spark plug (see illustrations). Work on one plug at a time, so as not to mix up wires (1993 and 1994) or individual coils (later engines).
9 Whether you are replacing the plugs at this time or intend to reuse the old plugs, compare each old spark plug with those shown in the photos on the inside-back cover of this book to determine the overall running condition of the engine.

Installation

10 Prior to installation, apply a thin coat of anti-seize compound to the plug threads. It's often difficult to insert spark plugs into their holes without cross-threading them. To avoid this possibility, fit a short piece of snug-fitting rubber hose over the end of the spark plug (see illustrations). The flexible hose acts as a universal joint to help align the plug with the plug hole. Should the plug begin to cross-thread, the hose will slip on the spark plug, preventing thread damage. Tighten the plug to the torque listed in this Chapter's Specifications.
11 Attach the plug wire to the new spark plug, again using a twisting motion on the boot until it is firmly seated on the end of the spark plug (1993 and 1994 models). Individual coils on 1995 and later models have their own short plug wires, but when reinstalling them on new plugs, just push the assembly straight down onto the plug, without twisting.
12 Follow the above procedure for the remaining spark plugs, replacing them one at a time to prevent mixing up the spark plug wires.

23 Spark plug wire, distributor cap and rotor check and replacement - 1993 and 1994 models (every 30,000 miles or 24 months)

Refer to illustrations 23.8, 23.11 and 23.12
1 The spark plug wires should be checked whenever new spark plugs are installed.
2 Begin this procedure by making a visual check of the spark plug wires while the engine is running. In a darkened garage (make sure there is ventilation) start the engine and observe each plug wire. Be careful not to come into contact with any moving engine parts. If there is a break in the wire, you will see arcing or a small spark at the damaged area. If arcing is noticed, make a note to obtain new wires, then allow the engine to cool and check the distributor cap and rotor.
3 The spark plug wires should be inspected one at a time to prevent mixing up the order, which is essential for proper engine operation. Each original plug wire should be numbered to help identify its location. If the number is illegible, a piece of tape can be marked with the correct number and wrapped around the plug wire.
4 Disconnect the plug wire from the spark plug. A removal tool can be used for this purpose or you can grasp the rubber boot, twist the boot half a turn and pull the boot free. Do not pull on the wire itself.
5 Check inside the boot for corrosion, which will look like a white crusty powder.

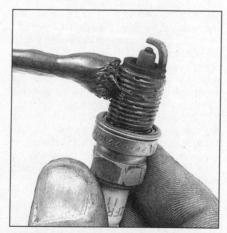

22.10a Apply a coat of anti-seize compound to the spark plug threads

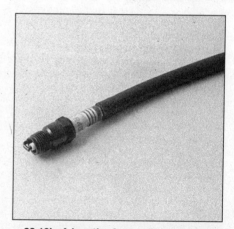

22.10b A length of snug-fitting rubber hose will save time and prevent damaged threads when installing the spark plugs

Chapter 1 Tune-up and routine maintenance

23.8 Remove the wires from the distributor cap one at a time, by pulling only on the boots

6 Push the wire and boot back onto the end of the spark plug. It should fit tightly onto the end of the plug. If it doesn't, remove the wire and use pliers to carefully crimp the metal connector inside the wire boot until the fit is snug.

7 Using a clean rag, wipe the entire length of the wire to remove built-up dirt and grease. Once the wire is clean, check for burns, cracks and other damage. Do not bend the wire sharply, because the conductor inside might break.

8 Disconnect the wire from the distributor. Again, pull only on the rubber boot. Check for corrosion and a tight fit **(see illustration)**. Replace the wire in the distributor.

9 Inspect the remaining spark plug wires, making sure that each one is securely fastened at the distributor and spark plug when the check is complete.

10 If new spark plug wires are required, purchase a set for your specific engine model. Pre-cut wire sets with the boots already installed are available. Remove and replace the wires one at a time to avoid mix-ups in the firing order.

11 Detach the distributor cap by loosening the cap retaining screws. Remove it and look inside for cracks, carbon tracks and worn, burned or loose contacts **(see illustration)**.

12 Pull the rotor off the distributor shaft and examine it for cracks and carbon tracks **(see illustration)**. Replace the cap and rotor if any damage or defects are noted.

13 It is common practice to install a new cap and rotor whenever new spark plug wires are installed, but if you wish to continue using the old cap, check the resistance between the spark plug wires and the cap first. If the indicated resistance is more than the maximum value listed in this Chapter's Specifications, replace the cap and/or wires.

14 When installing a new cap, remove the wires from the old cap one at a time and attach them to the new cap in the exact same location - do not simultaneously remove all the wires from the old cap or firing order mix-ups may occur.

24 Positive Crankcase Ventilation (PCV) valve and hose check and replacement (every 30,000 miles or 24 months)

Refer to illustration 24.2

1 The PCV valve is located in the rear valve cover, just behind the throttle body.

2 Grasp the PCV valve and pull it out of the valve cover **(see illustration)**.

3 With the engine idling at normal operat-

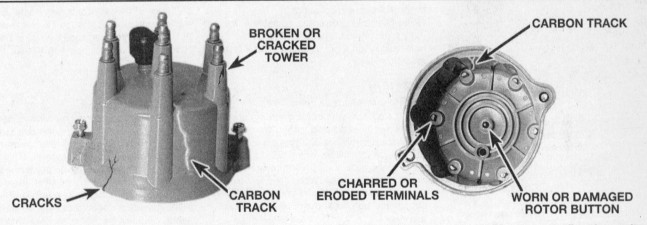

23.11 Check the distributor cap terminals and center electrode (typical examples shown) for wear and damage as well as the cap itself for carbon tracks (if in doubt about its condition, install a new one)

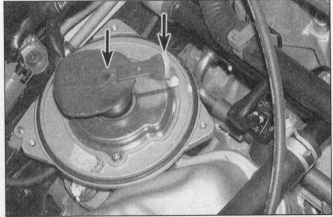

23.12 The ignition rotor should be checked for wear and corrosion of the center contact and tip (arrows) and any other cracks or damage (if in doubt about its condition, buy a new one)

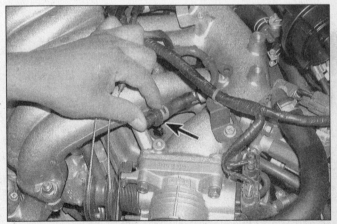

24.2 Pull the PVC valve (arrow) out of the rear valve cover and verify that vacuum is felt at the valve with the engine running

Chapter 1 Tune-up and routine maintenance

25.3 Unscrew the radiator drain fitting (arrow) located at the bottom of the radiator

25.4 The block coolant drain (arrow) is accessible from below (front side shown, later model)

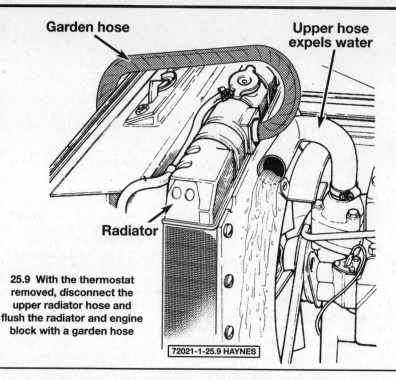

25.9 With the thermostat removed, disconnect the upper radiator hose and flush the radiator and engine block with a garden hose

ing temperature, place your finger over the end of the valve. If there's no vacuum at the valve, check for a plugged hose or valve. Replace any plugged or deteriorated hoses.
4   When purchasing a replacement PCV valve, make sure it's for your particular vehicle and engine size. Compare the old valve with the new one to make sure they're the same.

25 Cooling system servicing (draining, flushing and refilling) (every 30,000 miles or 24 months)

Warning: *Do not allow antifreeze to come in contact with your skin or painted surfaces of the vehicle. Flush contacted areas immediately with plenty of water. Do not store new coolant or leave old coolant lying around where it is easily accessible to children and pets, because they are attracted by its sweet taste. Ingestion of even a small amount can be fatal. Wipe up the garage floor and drip pan coolant spills immediately. Keep antifreeze containers covered and repair leaks in your cooling system immediately.*

Antifreeze is flammable - be sure to read the precautions on the container.
Note: *Non-toxic coolant is available at local auto parts stores. Although the coolant is non-toxic when fresh, proper disposal is still required.*

Draining

Refer to illustrations 25.3 and 25.4

1   Periodically, the cooling system should be drained, flushed and refilled to replenish the antifreeze mixture and prevent formation of rust and corrosion, which can impair the performance of the cooling system and cause engine damage. When the cooling system is serviced, all hoses and the radiator cap should be checked and replaced if necessary.
2   Apply the parking brake and block the wheels. **Warning:** *If the vehicle has just been driven, wait several hours to allow the engine to cool down before beginning this procedure.*
3   Move a large container under the radiator drain to catch the coolant. The drain plug is located on the lower left side of the radiator **(see illustration).** Attach a 3/8-inch diameter hose to the drain fitting (if possible) to direct the coolant into the container, then open the drain fitting (a pair of pliers may be required to turn it). Remove the radiator cap.
4   After coolant stops flowing out of the radiator, move the container under the engine block drain plugs - there's one on each side of the block **(see illustration).** Remove the plugs and allow the coolant in the block to drain. **Note:** *Frequently, the coolant will not drain from the block after the plug is removed. This is due to a rust layer that has built up behind the plug. Insert a Phillips screwdriver into the hole to break the rust barrier.*

5   While the coolant is draining, check the condition of the radiator hoses, heater hoses and clamps (refer to Section 13 if necessary).
6   Replace any damaged clamps or hoses. Reinstall the drain plugs and tighten them securely.

Flushing

Refer to illustration 25.9

7   Once the system is completely drained, remove the thermostat from the engine (see Chapter 3). Then reinstall the thermostat housing without the thermostat. This will allow the system to be thoroughly flushed.
8   Reinstall the lower radiator hose and tighten the radiator drain plug. Turn your heating system controls to Hot, so that the heater core will be flushed at the same time as the rest of the cooling system.
9   Disconnect the upper radiator hose, then place a garden hose in the upper radiator inlet and flush the system until the water runs clear at the upper radiator hose **(see illustration).**
10   In severe cases of contamination or clogging of the radiator, remove the radiator (see Chapter 3) and have a radiator repair facility clean and repair it if necessary.
11   Many deposits can be removed by the chemical action of a cleaner available at auto parts stores. Follow the procedure outlined in the manufacturer's instructions. **Note:** *When the coolant is regularly drained and the system refilled with the correct antifreeze/water mixture, there should be no need to use chemical cleaners or descalers.*

Refilling

Refer to illustration 25.14

12   To refill the system, install the thermostat, reconnect any radiator hoses and install

Chapter 1 Tune-up and routine maintenance 1-25

25.14 Location of the engine air relief plug (arrow) on 1993 and 1994 engines

26.7 The automatic transaxle drain plug (arrow) is located on the bottom of the fluid pan - after draining the fluid, remove the pan bolts and the pan

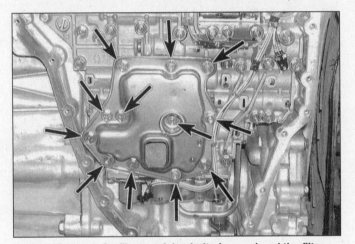

26.8 Remove the filter retaining bolts (arrows) and the filter

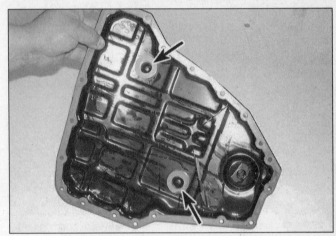

26.10 After cleaning the pan, place the magnets in position and install the new gasket

the reservoir and the overflow hose.
13   Place the heater temperature control in the maximum heat position.
14   Make sure to use the proper coolant listed in this Chapter's Specifications. Slowly fill the radiator with the recommended mixture of antifreeze and water to the base of the filler neck. Then add coolant to the reservoir until it reaches the FULL COLD mark. Wait five minutes and recheck the coolant level in the radiator, adding if necessary. **Note:** *On 1993, 1994 and 2000 and later models, loosen the air relief plug on the engine (see illustration). When coolant comes out of the plug, tighten it and continue until the radiator is full.*
15   Leave the radiator cap off and run the engine in a well-ventilated area until the thermostat opens (coolant will begin flowing through the radiator and the upper radiator hose will become hot).
16   Turn the engine off and let it cool. Add more coolant mixture to bring the level back up to the base of the filler neck.
17   Squeeze the upper radiator hose to expel air, then add more coolant mixture if necessary. Replace the radiator cap.

18   Place the heater temperature control and the blower motor speed control to their maximum setting.
19   Start the engine, allow it to reach normal operating temperature and check for leaks.

26 Automatic transaxle fluid change (every 30,000 miles or 24 months)

Refer to illustrations 26.7, 26.8, 26.10 and 26.12

1   At the specified time intervals, the automatic transaxle fluid should be drained and replaced.
2   Before beginning work, purchase the specified transmission fluid (see *Recommended fluids and lubricants* at the front of this Chapter).
3   Other tools necessary for this job include jackstands to support the vehicle in a raised position, a wrench, a drain pan capable of holding at least eight quarts, newspapers and clean rags.
4   The fluid should be drained immediately after the vehicle has been driven. Hot fluid is

more effective than cold fluid at removing built up sediment. **Warning:** *Fluid temperature can exceed 350-degrees F in a hot transaxle. Wear protective gloves.*
5   After the vehicle has been driven to warm up the fluid, raise it and place it on jackstands for access to the transaxle drain plugs.
6   Move the necessary equipment under the vehicle, being careful not to touch any of the hot exhaust components.
7   Place the drain pan under the drain plug in the transaxle housing or fluid pan and remove the drain plug **(see illustration)**. Be sure the drain pan is in position, as fluid will come out with some force. Once the fluid is drained, reinstall the drain plug securely.
8   The transaxle filter should be replaced at the time of the fluid change. Remove the filter bolts and pull the filter down **(see illustration).**
9   Install the new filter.
10   Clean the oil pan mounting surface on the transaxle with lacquer thinner to remove all traces of gasket and sealer. Clean the inside and outside of the transaxle fluid pan. Most models will have one or two magnets in place in the bottom of the pan to collect

1-26 Chapter 1 Tune-up and routine maintenance

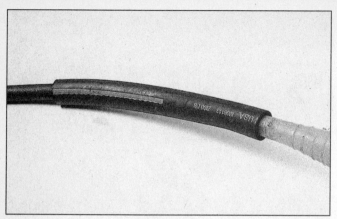

26.12 To adapt your funnel to the transaxle dipstick tube, attach lengths of two different size rubber hoses

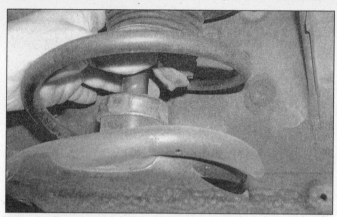

28.6 Check the front struts for leakage at the point where the rod meets the strut body; it may be necessary to lift the rubber boot, as shown

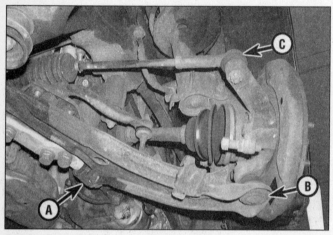

28.9a On the front suspensions, check the A-arm bushings (A), the lower balljoint (B) and the tie-rod ends (C)

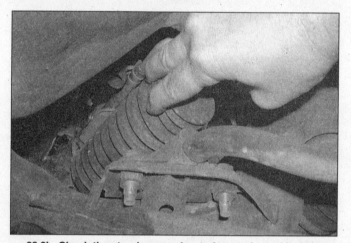

28.9b Check the steering gear boots for cracks and leaking steering fluid

metal particles. Remove the magnets, clean them and reattach them to the pan in the original positions **(see illustration)**.

11 Install the pan with a new gasket and the bolts finger-tight. Tighten each bolt a little at a time until the final torque figure listed in this Chapter's Specifications is reached. Don't overtighten the bolts!

12 With the engine off, add new fluid to the transaxle through the dipstick tube (see *Recommended fluids and lubricants* for the recommended fluid type and capacity). Use a funnel to prevent spills **(see illustration)**. It is best to add a little fluid at a time, continually checking the level with the dipstick (see Section 7). Allow the fluid time to drain into the pan. **Note:** *The fluid capacity figures given in the Specifications are for a total replacement, including the fluid in the converter. When you use the fluid replacement procedure above, not all of this fluid will come out. When adding new fluid to the transaxle, never add more fluid than was drained out, until you have checked the fluid level as in Step 14.*

13 Start the engine and shift the selector into all positions from P through L, then shift into P and apply the parking brake.

14 With the engine idling, check the fluid level. Add fluid up to the Cool level on the dipstick.

27 Manual transaxle lubricant change (every 30,000 miles or 24 months)

1 Remove the drain plug at the bottom of the transaxle and drain the fluid.
2 Reinstall the drain plug securely.
3 Remove the filler plug **(see illustration 19.2)**. See *Recommended lubricants and fluids* for the specified lubricant type. Use a siphon pump or squeeze bottle to fill the transaxle until the level reaches the fill-plug hole.
4 Install the plug and tighten it securely.

28 Suspension, steering and driveaxle boot check (every 30,000 miles or 24 months)

Note: *The steering linkage and suspension components should be checked periodically. Worn or damaged suspension and steering linkage components can result in excessive and abnormal tire wear, poor ride quality and vehicle handling and reduced fuel economy. For detailed illustrations of the steering and suspension components, refer to Chapter 10.*

Shock absorber check

Refer to illustration 28.6

1 Park the vehicle on level ground, turn the engine off and set the parking brake. Check the tire pressures.
2 Push down at one corner of the vehicle, then release it while noting the movement of the body. It should stop moving and come to rest in a level position within one or two bounces.
3 If the vehicle continues to move up-and-down or if it fails to return to its original position, a worn or weak shock absorber is probably the reason.
4 Repeat the above check at each of the three remaining corners of the vehicle.
5 Raise the vehicle and support it securely on jackstands.
6 Check the shock absorbers for evidence of fluid leakage **(see illustration)**. A light film of fluid is no cause for concern. Make sure

Chapter 1 Tune-up and routine maintenance

28.11 With the steering wheel in the locked position and the vehicle raised, grasp the front tire and try to move it back-and-forth to check for play in the steering

28.14 Flex the inner and outer driveaxle boots by hand to check for cracks and/or leaking grease

that any fluid noted is from the shocks and not from some other source. If leakage is noted, replace the shocks as a set.

7   Check the shocks to be sure that they are securely mounted and undamaged. Check the upper mounts for damage and wear. If damage or wear is noted, replace the shocks as a set (front or rear).

8   If the shocks must be replaced, refer to Chapter 10 for the procedure.

Steering and suspension check

Refer to illustrations 28.9a, 28.9b and 28.11

9   Visually inspect the steering and suspension components (front and rear) for damage and distortion. Look for damaged seals, boots and bushings and leaks of any kind Examine the bushings where the lower control arms meet the chassis **(see illustrations)**.

10   Clean the lower end of the steering knuckle. Have an assistant grasp the lower edge of the tire and move the wheel in-and-out while you look for movement at the steering knuckle-to-control arm balljoint. If there is any movement the suspension balljoint(s) must be replaced.

11   Grasp each front tire at the front and rear edges, push in at the front, pull out at the rear and feel for play in the steering system components If any freeplay is noted, check the idler arm and the tie-rod ends for looseness **(see illustration)**.

12   Additional steering and suspension system information and illustrations can be found in Chapter 10.

Driveaxle boot check

Refer to illustration 28.14

13   The driveaxle boots are very important because they prevent dirt, water and foreign material from entering and damaging the constant velocity (CV) joints. Oil and grease on the exterior can cause the boot material to deteriorate prematurely, so it's a good idea to clean the boots with soap and water. Because it constantly pivots back and forth following the steering action of the front hub, the outer CV boot wears out sooner and should be inspected regularly.

14   Inspect the boots for tears and cracks as well as loose clamps **(see illustration)**. If there is any evidence of cracks or leaking lubricant, they must be replaced as described in Chapter 8.

29 Exhaust Gas Recirculation (EGR) valve check (every 60,000 miles or 48 months)

Refer to illustration 29.2

1   The EGR valve is bolted to the rear cylinder head, at the transaxle end. The most common problem with this system is usually a stuck or corroded EGR valve.

2   With the engine cold to prevent burns, reach under the EGR valve and push up on

29.2 Push up on the underside of the EGR valve to check for free movement of the diaphragm

the diaphragm. Using moderate pressure you should be able to press the diaphragm up-and-down within the housing **(see illustration)**.

3   If the diaphragm doesn't move, or moves only with much effort, replace the EGR valve with a new one. If in doubt about the condition of the valve, compare the free movement of your EGR valve with a new valve.

4   Refer to Chapter 6 for more information on the EGR system.

Notes

Chapter 2 Part A
Single Overhead Cam (SOHC) engine

Contents

	Section		Section
Camshafts, lifters and seals - removal and installation	8	Oil pan - removal and installation	14
Crankshaft front oil seal - replacement	13	Oil pump - removal, inspection and installation	15
Crankshaft pulley - removal and installation	12	Powertrain mounts - check and replacement	18
Cylinder compression check	See Chapter 2C	Rear main oil seal - replacement	17
Cylinder heads - removal and installation	11	Repair operations possible with the engine in the vehicle	2
Drivebelt check, adjustment and replacement	See Chapter 1	Rocker arm assembly - removal, inspection and installation	5
Engine - removal and installation	See Chapter 2C	Spark plug replacement	See Chapter 1
Engine oil and filter change	See Chapter 1	Timing belt and sprockets - removal and installation	7
Engine overhaul - general information	See Chapter 2C	Top Dead Center (TDC) for number one piston - locating	3
Exhaust manifolds - removal and installation	10	Valve covers - removal and installation	4
Flywheel/driveplate - removal and installation	16	Valve springs, retainers and seals - replacement	6
General information	1	Valves - servicing	See Chapter 2C
Intake manifold - removal and installation	9		

Specifications

General
Cylinder numbers (timing belt end-to-transaxle end)	
Rear (firewall) side	1-3-5
Front (radiator) side	2-4-6
Firing order	1-2-3-4-5-6
Bore	3.43 inches (87mm)
Stroke	3.27 inches (83 mm)
Displacement	181 cubic inches (3.0 liters)

Cylinder location and distributor rotation diagram

Camshaft and rocker arms
Camshaft endplay	0.0012 to 0.0024 inch (0.03 to 0.06 mm)
Rocker arm shaft diameter	0.7082 to 0.7087 inch (17.988 to 18.000 mm)
Rocker arm bore diameter	0.7089 to 0.7098 inch (18.007 to 18.028 mm)
Rocker arm-to-shaft oil clearance	0.0003 to 0.0019 inch (0.007 to 0.049 mm)

Oil pump

Outer rotor-to-body clearance	0.0043 to 0.0079 inch (0.11 to 0.20 mm)
Inner gear to crescent clearance	0.0047 to 0.0091 inch (0.12 to 0.23 mm)
Outer gear to crescent clearance	0.0083 to 0.0126 inch (0.21 to 0.32 mm)
Housing to inner gear side clearance	0.0020 to 0.0035 inch (0.05 to 0.09 mm)
Housing to outer gear side clearance	0.0020 to 0.0043 inch (0.05 to 0.11 mm)

Torque specifications

	Nm	Ft-lbs (unless otherwise indicated)
Camshaft thrust plate bolt	78 to 88	58 to 65
Camshaft sprocket bolt	78 to 88	58 to 65
Crankshaft pulley bolt	123 to 132	90 to 98
Cylinder head bolts (in sequence - **see illustration 11.21**)		
Step one	29	22
Step two	59	43
Step three	Loosen all bolts (in reverse of tightening sequence)	
Step four	29	22
Step five	54 to 64	40 to 47
Valve cover bolts	1 to 3	9 to 26 in-lbs
Driveplate bolts	83 to 93	61 to 69
Exhaust manifold nuts	18 to 22	13 to 16
Intake manifold		
Upper intake manifold (plenum) bolts	18 to 22	13 to 16
Lower intake manifold bolts/nuts		
Step one (all)	3 to 5	26 to 43 in-lbs
Step two		
Bolts	16 to 20	144 to 168 in-lbs
Step three		
Nuts	24 to 27	17 to 20
Oil pan bolts/nuts	7 to 8	62 to 70 in-lbs
Oil pan drain plug	29 to 39	22 to 29
Oil pick-up screen mounting bolts	16 to 21	12 to 15
Oil pick-up screen support bracket bolt	6.3 to 8.3	55 to 73 in-lbs
Oil pump mounting bolts		
Long	12 to 16	108 to 144 in-lbs
Short	6 to 7	53 to 62 in-lbs
Oil pump cover screws	4 to 5	35 to 44 in-lbs
Rocker arm shaft bolts	18 to 22	13 to 16
Timing belt tensioner nut	43 to 58	32 to 43
Timing belt cover bolts	3 to 5	26 to 44 in-lbs
Transaxle-to-engine brace bolts	43 to 55	32 to 41
Rear main oil seal retainer bolts	3 to 5	26 to 44 in-lbs
Right front engine mount through-bolts	77 to 98	57 to 72
Right rear engine mount through-bolts	77 to 98	57 to 72
Left front engine mount through-bolt	64 to 78	47 to 58
Left front engine mount bracket bolts	41 to 52	30 to 38
Left rear engine mount through-bolt	43 to 55	32 to 41
Left rear engine mount lower nuts	43 to 55	32 to 41

1 General information

This Part of Chapter 2 is devoted to in-vehicle repair procedures for the VG30E 3.0L Single Overhead Camshaft (SOHC) V6 engine equipped in 1993 and 1994 vehicles. All information concerning engine removal and installation and engine block and cylinder head overhaul can be found in Chapter 2, Part C.

The following repair procedures are based on the assumption that the engine is installed in the vehicle. If the engine has been removed from the vehicle and mounted on a stand, many of the steps outlined in this Part of Chapter 2 will not apply.

The Specifications included in this Part of Chapter 2 apply only to the procedures contained in this Part. Part C of Chapter 2 contains the Specifications necessary for cylinder head and engine block rebuilding.

2 Repair operations possible with the engine in the vehicle

Many major repair operations can be accomplished without removing the engine from the vehicle.

Clean the engine compartment and the exterior of the engine with some type of degreaser before any work is done. It will make the job easier and help keep dirt out of the internal areas of the engine.

Depending on the components involved, it may be helpful to remove the hood to improve access to the engine as repairs are performed (refer to Chapter 11 if necessary). Cover the fenders to prevent damage to the paint. Special pads are available, but an old bedspread or blanket will also work.

If vacuum, exhaust, oil or coolant leaks develop, indicating a need for gasket or seal replacement, the repairs can generally be

Chapter 2 Part A Single Overhead Cam (SOHC) engine

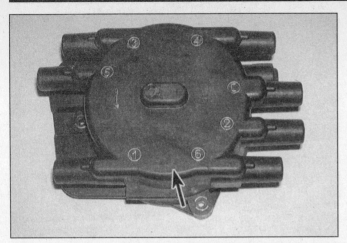

3.5 The inner post (arrow) for the number one spark plug is offset from the number one terminal on the outside of the cap - turn the cap over to verify the exact location of the inner post

3.8 Align the zero notch on the crankshaft pulley with the pointer on the timing belt cover (arrow) - the zero notch is the one farthest to the left when facing the front of the engine and is typically yellow in color

made with the engine in the vehicle. The intake and exhaust manifold gaskets, oil pan gasket, crankshaft oil seals and cylinder head gaskets are all accessible with the engine in place.

Exterior engine components, such as the intake and exhaust manifolds, the oil pan, the oil pump, the water pump (see Chapter 3), the starter motor, the alternator, the distributor (see Chapter 5) and the fuel system components (see Chapter 4) can be removed for repair with the engine in place.

Since the cylinder heads can be removed without pulling the engine, valve component servicing can also be accomplished with the engine in the vehicle. Replacement of the camshafts, timing belt and sprockets is also possible with the engine in the vehicle, although the cylinder heads must be removed from the engine to replace the camshafts.

In extreme cases caused by a lack of necessary equipment, repair or replacement of piston rings, pistons, connecting rods and rod bearings is possible with the engine in the vehicle. However, this practice is not recommended because of the cleaning and preparation work that must be done to the components involved.

3 Top Dead Center (TDC) for number one piston - locating

Refer to illustrations 3.5, 3.8 and 3.9

Note: The following procedure is based on the assumption that the distributor is correctly installed. If you are trying to locate TDC to install the distributor correctly, piston position must be determined by feeling for compression at the number one spark plug hole, then aligning the ignition timing marks as described in Step 8.

1 Top Dead Center (TDC) is the highest point in the cylinder that each piston reaches as it travels up-and-down when the crankshaft turns. Each piston reaches TDC on the compression stroke and again on the exhaust stroke, but TDC generally refers to piston position on the compression stroke.

2 Positioning the piston(s) at TDC is an essential part of several procedures such as camshaft and timing belt/sprocket removal and distributor removal.

3 Before beginning this procedure, be sure to place the transaxle in Neutral and apply the parking brake or block the rear wheels. Also, disable the ignition system by detaching the coil wire from its terminal on the distributor cap and grounding it on the engine block with a jumper wire. Remove the spark plugs (see Chapter 1).

4 In order to bring any piston to TDC, the crankshaft must be turned using one of the methods outlined below. When looking at the front (drivebelt end) of the engine, normal crankshaft rotation is clockwise.

   a) *The preferred method is to remove the lower splash shield on the passenger side and turn the crankshaft with a socket and ratchet attached to the bolt threaded into the front of the crankshaft.*
   b) *A remote starter switch, which may save some time, can also be used. Follow the instructions included with the switch. Once the piston is close to TDC, use a socket and ratchet as described in the previous paragraph.*
   c) *If an assistant is available to turn the ignition switch to the Start position in short bursts, you can get the piston close to TDC without a remote starter switch. Make sure your assistant is out of the vehicle, away from the ignition switch, then use a socket and ratchet as described in Paragraph a) to complete the procedure.*

5 Note the position of the terminal for the number one spark plug wire on the distributor cap **(see illustration)**. If the plug wire isn't marked, follow the plug wire from the number one cylinder spark plug to the cap.

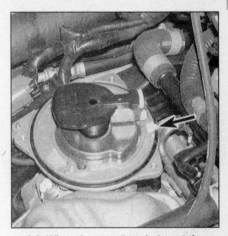

3.9 When the rotor is pointing at the number one spark plug wire terminal on the inside of the distributor cap (which is indicated by the mark on the distributor body) and the timing marks are aligned, the number one piston is at TDC on the compression stroke

6 Use a felt-tip pen or chalk to make a mark on the distributor body directly under the inner post of the number 1 terminal.

7 Detach the cap from the distributor and set it aside (see Chapter 1 if necessary).

8 Turn the crankshaft (see Step 4) until the TDC mark in the crankshaft pulley is aligned with the pointer on the timing belt cover **(see illustration)**. *Note: There are several marks on the pulley starting from zero or TDC to 30 degrees in 5 degree increments. The zero notch is the one farthest to the left when facing the front of the engine and is typically yellow in color.*

9 Look at the distributor rotor - it should be pointing directly at the mark you made on the distributor body **(see illustration)**.

10 If the rotor is 180-degrees off, the number one piston is at TDC on the exhaust stroke.

11 To get the piston to TDC on the com-

2A-4   Chapter 2 Part A   Single Overhead Cam (SOHC) engine

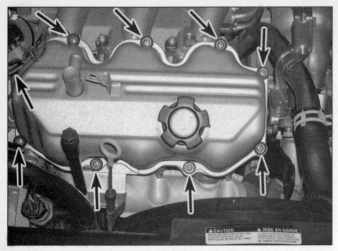

4.8 Valve cover retaining bolts (arrows)

5.2 Loosen the rocker arm shaft bolts (arrows) a little at a time to avoid distorting the shaft

pression stroke if the rotor is 180-degrees off, turn the crankshaft one complete turn (360-degrees) clockwise. The rotor should now be pointing at the mark on the distributor. When the rotor is pointing at the number one spark plug wire terminal in the distributor cap and the ignition timing marks are aligned, the number one piston is at TDC on the compression stroke.

12 After the number one piston has been positioned at TDC on the compression stroke, TDC for any of the remaining pistons can be located by turning the crankshaft and following the firing order. Mark the remaining spark plug wire terminal locations on the distributor body just like you did for the number one terminal, then number the marks to correspond with the cylinder numbers. As you turn the crankshaft, the rotor will also turn. The crankshaft must be turned 120-degrees to move from one cylinder to the next one in the firing order. When it's pointing directly at one of the marks on the distributor, the piston for that particular cylinder is at TDC on the compression stroke.

4   Valve covers - removal and installation

Removal

Refer to illustration 4.8

1 Relieve the fuel system pressure (see Chapter 4).
2 Disconnect the cable from the negative terminal of the battery.
3 Label and detach the spark plug wires. Mark them clearly with pieces of masking tape to prevent confusion during installation.
4 Remove the distributor (see Chapter 5).
5 Remove the upper intake plenum (see Section 9).
6 Remove the breather hose by sliding the hose clamp back and pulling the hose off the fitting on the valve cover.
7 Detach any remaining hoses and wiring which would interfere with valve cover removal.
8 Remove the valve cover bolts and washers **(see illustration)**.
9 Detach the valve cover. **Note:** *If the cover is stuck to the cylinder head, bump one end with a block of wood and a hammer to jar it loose. If that doesn't work, try to slip a flexible putty knife between the cylinder head and cover to break the gasket seal. Don't pry at the cover-to-cylinder head joint or damage to the sealing surfaces may occur (leading to oil leaks in the future).*

Installation

10 The mating surfaces of each cylinder head and valve cover must be perfectly clean when the covers are installed. Use a gasket scraper to remove all traces of sealant and old gasket material, then clean the mating surfaces with lacquer thinner or acetone. If there's sealant or oil on the mating surfaces when the cover is installed, oil leaks may develop.
11 If necessary, clean the mounting bolt threads with a die to remove any corrosion and restore damaged threads. Make sure the threaded holes in the cylinder head are clean - run a tap into them to remove corrosion and restore damaged threads.
12 The gaskets should be mated to the covers before the covers are installed. Apply a thin coat of RTV sealant to the cover groove, then position the gasket inside the cover and allow the sealant to set up so the gasket adheres to the cover. If the sealant isn't allowed to set, the gasket may fall out of the cover as it's installed on the engine.
13 Carefully position the cover on the cylinder head and install the bolts.
14 Tighten the bolts in three or four steps to the torque listed in this Chapter's Specifications.
15 The remaining installation steps are the reverse of removal.
16 Start the engine and check carefully for oil leaks.

5   Rocker arm assembly - removal, inspection and installation

Removal

Refer to illustrations 5.2, 5.3 and 5.4

1 Remove the valve covers (see Section 4).
2 Loosen the rocker arm shaft retaining bolts **(see illustration)** in two or three stages, working from the ends toward the middle of the shafts. **Caution:** *Some of the valves will be open when you loosen the rocker arm shaft bolts and the rocker arm shafts will be under a certain amount of valve spring pressure. Therefore, the bolts must be loosened gradually. Loosening a bolt all at once near a rocker arm under spring pressure could distort the rocker arm shaft.*
3 Prior to removal, scribe or paint identifying marks on the rockers to ensure they will be installed in their original locations **(see illustration)**.
4 Remove the bolts and lift off the rocker arm shaft assemblies one at a time. Lay them

5.3 Mark the rockers to identify their locations - these are marked FI for Front Intake and FE for Front Exhaust

Chapter 2 Part A Single Overhead Cam (SOHC) engine

2A-5

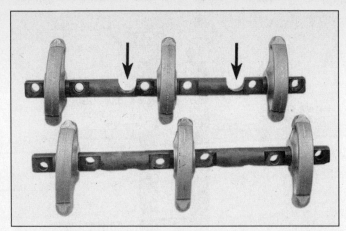

5.4 The rocker arm shaft assemblies are installed with the large notches (arrows) on the intake manifold side - the small notches on the other shaft must face the exhaust manifold

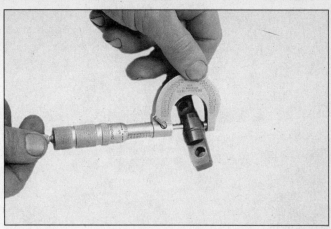

5.6 Measure the rocker arm shaft diameter at each journal where a rocker arm rides on the shaft

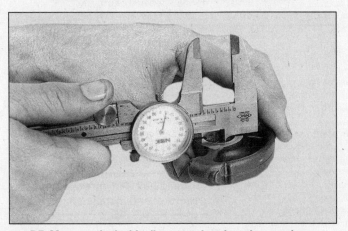

5.7 Measure the inside diameter of each rocker arm bore, subtract the corresponding rocker arm shaft diameter to obtain the clearance and compare the results to Specifications

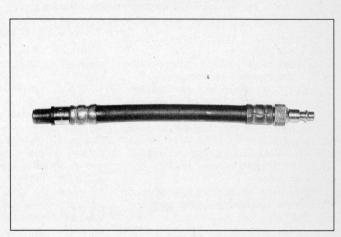

6.5 The air hose adapter threads into the spark plug hole - they're commonly available from auto parts stores

down on a nearby workbench in the same relationship to each other that they're in when installed. They must be reinstalled on the same cylinder head. Note that the shafts with the larger notches go on the intake manifold side **(see illustration)**.

Inspection

Refer to illustrations 5.6 and 5.7

5   Check the rocker arms and shafts for abnormal wear, pits, galling, score marks and rough spots. Don't attempt to restore rocker arms by grinding the pad surfaces.

6   Measure the outside diameter of the rocker arm shaft at each rocker arm journal **(see illustration)**. Compare the measurements to the rocker arm shaft outside diameter specified in this Chapter.

7   Measure the inside diameter of each rocker arm with either an inside micrometer or a dial caliper **(see illustration)**. Compare the measurements to the rocker arm bore diameter specified in this Chapter.

8   Subtract the outside diameter of each rocker arm shaft journal from the corresponding rocker arm bore diameter to compute the clearance between the rocker arm shaft and the rocker arm. Compare the measurements to the clearance specified in this Chapter. If any of them fall outside the specified limits, replace either the rocker arms or the shaft, or both.

Installation

9   Installation is the reverse of the removal procedure. Tighten the rocker arm shaft retaining bolts, in several steps, to the torque listed in this Chapter's Specifications. Work from the ends of the shafts toward the middle.

6   Valve springs, retainers and seals - replacement

Refer to illustrations 6.5, 6.7, 6.13 and 6.15

Note: *Broken valve springs and defective valve stem seals can be replaced without removing the cylinder heads. Two special tools and a compressed air source are normally required to perform this operation, so read through this Section carefully. The universal shaft-type valve spring compressor required for the tight valve spring pockets of this vehicle may not be available at all tool rental yards, so check on the availability before beginning the job.*

1   Remove the valve cover(s) (see Section 4).

2   Refer to Section 5 and remove the rocker arm assembly, then refer to Section 8 and remove the lifter guide assembly.

3   Remove the spark plug from the cylinder that has the defective component. If all of the valve stem seals are being replaced, all of the spark plugs should be removed.

4   Turn the crankshaft until the piston in the affected cylinder is at Top Dead Center on the compression stroke (refer to Section 3). If you're replacing all of the valve stem seals, begin with cylinder number one and work on the valves for one cylinder at a time. Move from cylinder-to-cylinder following the firing order sequence (see this Chapter's Specifications).

5   Thread a long adapter into the spark plug hole and connect an air hose from a compressed air source to it **(see illustration)**. Most auto parts stores can supply the air hose adapter. **Note:** *Because of the length of the spark plug tubes, it will be necessary to use a long spark plug adapter with a length of hose attached (as used on many cylinder*

6.7 Compress the valve spring enough to release the valve stem locks and lift them out with a magnet or needle-nose pliers

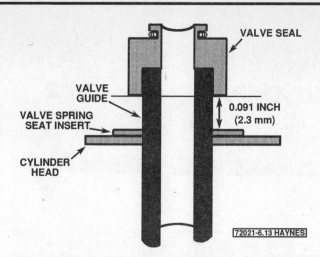

6.13 Exhaust seals should be installed by hand, not with a tool, and only to the specified depth

compression gauges) utilizing a quick-disconnect fitting to hook to your air source.

6   Apply compressed air to the cylinder. **Warning:** *The piston may be forced down by the compressed air, causing the crankshaft to turn suddenly. If the wrench used when positioning the number one piston at TDC is still attached to the bolt in the crankshaft nose, it could cause damage or injury when the crankshaft moves.*

7   Stuff shop rags into the cylinder head holes around the valves to prevent parts and tools from falling into the engine, then use a valve spring compressor to compress the spring **(see illustration)**. Remove the valve stem locks with small needle-nose pliers or a magnet. **Note:** *The valves should be held in place by the air pressure. If the valve faces or seats are in poor condition, leaks may prevent air pressure from retaining the valves. If the valves cannot hold air, the cylinder head should be removed for a valve job at a machine shop.*

8   Remove the spring retainer, shield and valve spring, then remove the valve stem seal.

9   Wrap a rubber band or tape around the top of the valve stem so the valve won't fall into the combustion chamber, then release the air pressure.

10   Inspect the valve stem for damage. Rotate the valve in the guide and check the end for eccentric movement, which would indicate that the valve is bent.

11   Move the valve up-and-down in the guide and make sure it doesn't bind. If the valve stem binds, either the valve is bent or the guide is damaged. In either case, the cylinder head will have to be removed for repair.

12   Reapply air pressure to the cylinder to retain the valve in the closed position, then remove the tape or rubber band from the valve stem.

13   Lubricate the valve stems with engine oil and install a new valve stem seals. Intake valve seals can be installed with a special tool, or a deep socket and hammer - tap the seal only until seated. Exhaust seals should be installed by hand, not with a tool, and only to the specified depth **(see illustration)**. **Caution:** *Intake and exhaust seals are different, do not mix them up.*

14   Install the inner and outer springs in position over the valve, with the more closely-wound spring coils toward the cylinder head.

15   Install the valve spring retainer. Compress the valve springs and carefully position the valve stem locks in the groove. Apply a small dab of grease to the inside of each valve stem lock to hold it in place **(see illustration)**.

16   Remove the pressure from the spring tool and make sure the valve stem locks are seated.

17   Disconnect the air hose and remove the adapter from the spark plug hole.

18   Refer to Section 8 and install the lifter assembly, then refer to Section 5 and install the rocker arm assembly.

19   Refer to Section 4 and install the valve cover.

20   Install the spark plug(s) and hook up the wire(s).

21   Start and run the engine, then check for oil leaks and unusual sounds coming from the valve cover area.

7   Timing belt and sprockets - removal and installation

Removal

Refer to illustrations 7.9, 7.11, 7.12, 7.13a, 7.13b, 7.14, 7.15, 7.17 and 7.18

1   Disconnect the cable from the negative terminal of the battery and drain the cooling system (see Chapter 1).

2   Loosen the lug nuts on the right front wheel.

3   Raise the front of the vehicle and support it securely on jackstands. Apply the parking brake.

4   Remove the right front wheel and detach the splash shield from the inner fenderwell (see Chapter 11).

5   Remove all of the drivebelts (see Chapter 1).

6   Remove the splash pan and drain the cooling system (see Chapter 1).

7   Remove the spark plugs (see Chapter 1). Position the number one piston at TDC on the compression stroke (see Section 3).

8   Disconnect the radiator hose and bypass hose from the thermostat housing. Remove the water pump pulley.

9   Remove the air conditioning compressor idler pulley and bracket **(see illustration)**.

10   Remove the crankshaft pulley (see Section 12). **Note:** *Don't allow the crankshaft to rotate during removal of the pulley. If the crankshaft moves, the number one piston will no longer be at TDC.*

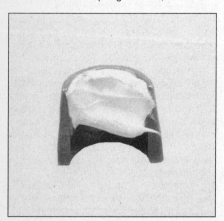

6.15 Apply a small dab of grease to each valve stem lock as shown here before installation - it will hold them in place on the valve stem as the spring is released

Chapter 2 Part A Single Overhead Cam (SOHC) engine

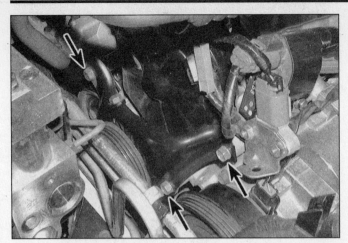

7.9 Remove the three bolts (arrows) and remove the air-conditioning belt idler pulley and bracket

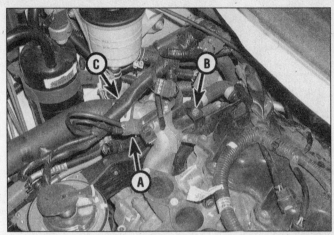

7.11 Disconnect the electrical connectors at the front of the intake manifold (A), remove the bolt (B) holding the wiring harness and move it aside, then disconnect the upper radiator hose (C)

7.12 Remove the outer timing belt guide (arrow) from the crankshaft – be sure to install it with the curved lip facing out

7.13a Make sure the marks on the camshaft sprockets align with the marks on the rear timing belt cover (arrows)

11 Disconnect the hoses and wiring harness at the top of the upper timing belt cover **(see illustration)**.

12 Remove outer timing belt guide from the crankshaft **(see illustration)**, then remove the bolts securing the timing belt upper and lower covers. Note that various types and sizes of bolts are used. They must be reinstalled in their original locations. Mark each bolt or make a sketch to help remember where they go.

13 Confirm that the number one piston is still at TDC on the compression stroke by verifying that the timing marks on the camshaft and crankshaft sprockets are aligned with their respective stationary alignment marks **(see illustrations)**.

14 Relieve tension on the timing belt by loosening the retaining nut one half turn and rotating the timing belt tensioner 70 to 80 degrees clockwise **(see illustration)**.

7.13b The mark on the crankshaft sprocket aligns with the mark on the oil pump housing (arrows)

7.14 Loosen the lock nut and rotate the timing belt tensioner 70 to 80 degrees clockwise to relieve belt tension

7.15 The timing belt should be marked (arrow) to indicate which side faces out - if not, use chalk to make an arrow on the belt before removal

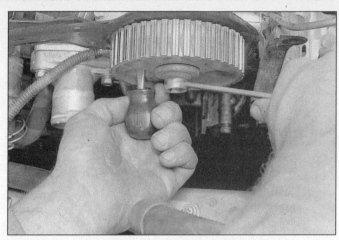

7.17 Insert a screwdriver through the camshaft sprocket to hold it while loosening the bolt

7.18 When installing the camshaft timing belt sprockets, note the R and L marks (arrows) which designate the right (rear cylinder bank) and left (front cylinder bank) camshaft sprockets - don't mix them up!

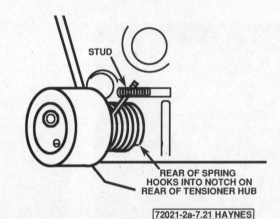

7.21 Belt tensioner spring mounting details (if the stud is removed, use a thread locking compound on the threads during installation)

15   Check to see if the timing belt is marked with an arrow indicating which side faces out **(see illustration)**. If there isn't a mark, paint one on (only if the same belt will be reinstalled). Slide the timing belt off the sprockets. If the belt is cracked, worn or contaminated with oil or coolant, replace it with a new one. Check the belt tensioner and spring for wear and damage, the pulley should rotate smoothly.

16   Make sure the camshaft and crankshaft sprockets are in good condition - if they're worn or damaged, replace them as described in Steps 17 through 19. **Note:** *It is not necessary to remove the camshaft and crankshaft sprockets during the timing belt replacement procedure unless they're damaged or to allow access to and removal of other components such as the cylinder head, oil pump, front crankshaft oil seal or rear timing cover.*

17   Insert a screwdriver through a hole in the camshaft sprocket to lock it in place while loosening the mounting bolt **(see illustration)**.

18   Once the bolt is out, the sprocket can be removed by hand. **Note:** *Each sprocket is marked with either an R or L* **(see illustration)**. *If you're removing both camshaft sprockets, don't mix them up. They must be installed on the same cam they were removed from.*

19   To replace the crankshaft sprocket, refer to Section 13.

Installation

Refer to illustrations 7.21, 7.23, 7.25 and 7.27

20   Verify that you have the correct belt for your vehicle. A new factory belt will have three white marks that ease installation by aligning exactly with the two camshaft timing marks and the crankshaft timing marks. Aftermarket belts may or may not have these marks. **Note:** *Also check the tooth design on the camshaft or crankshaft sprockets. Some earlier models may have teeth with a SQUARE edge at the bottom of the sprocket groove, while other models are ROUNDED. The replacement belts are available as either straight-tooth or rounded, and they are NOT interchangeable. Use only a belt that matches the tooth design of your sprockets. Use of the wrong belt will cause whining noise and premature failure.*

21   If the tensioner was removed, reinstall it and make sure the spring is positioned properly **(see illustration)**. Prepare to install the timing belt by turning the tensioner clockwise with an Allen wrench away from the belt and temporarily tightening the locking nut as described in Step 14.

22   Install the timing belt while aligning the belt marks with the punch marks on the sprockets in a counterclockwise direction. Making sure there is no slack, place the belt around the crankshaft sprocket first, then around left (front) camshaft sprocket, over the right (rear) camshaft sprocket and finally around the tensioner.

23   Remove any slack from the tensioner side of the belt by hand, loosen the tensioner retaining nut and rotate the tensioner 70 to 80 degrees counterclockwise, this will allow the tensioner to operate and remove all slack from the belt, then retighten the nut. Make sure all three sets of timing marks are properly aligned with the directional arrow pointing away from the engine **(see illustration)**.

24   Slowly turn the crankshaft clockwise

Chapter 2 Part A Single Overhead Cam (SOHC) engine

7.23 When correctly installed, the punch marks on the crankshaft and camshaft sprockets must align with the marks on the rear timing belt covers and the white marks on the belt (arrows) must align with the punch marks on the camshaft sprockets - don't continue with belt installation until you're sure it's correct!

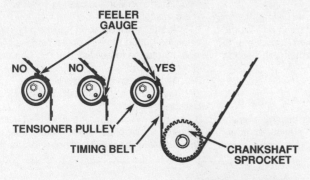

7.27 Position the feeler gauge between the tensioner pulley and the belt, then turn the crankshaft to move the feeler gauge to the point shown here (it must be exact, so work carefully)

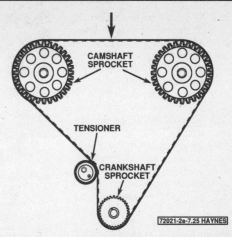

7.25 The deflection of the timing belt is checked exactly half-way between the front and rear camshaft sprockets

8.2 Wrap each lifter with a rubber band so it can't fall out of the lifter guide

two full revolutions, returning the number one piston to TDC on the compression stroke. **Caution:** *If excessive resistance is felt while turning the crankshaft, it's an indication that the pistons are coming into contact with the valves. Go back over the procedure to correct the situation before proceeding.*

25 Midway between the front and rear camshaft sprockets, push downward on the belt with 22 lbs. (97.8 N) of force and measure the belt deflection, which should be 0.51 to 0.59 inch (13 to 15 mm) **(see illustration)**.

26 If the deflection is as specified, the belt tension is adjusted properly. If not, loosen the tensioner locking nut while keeping the tensioner steady with the Allen wrench. **Note:** *Another person to help with the following procedure will be helpful.*

27 Place a 0.0138-inch (0.35 mm) thick and 1/2-inch (12.7 mm) wide feeler gauge (or a combination of gauges to obtain this thickness) adjacent to the tensioner pulley and slowly turn the crankshaft clockwise until the feeler gauge is between the belt and the tensioner pulley **(see illustration)**.

28 Keeping the tensioner steady against the belt and feeler gauges with the Allen wrench, tighten the tensioner locking nut.

29 Turn the crankshaft to remove the feeler gauge and continue turning for two revolutions and return the number one piston to TDC.

30 Recheck the belt tension as in Step 25, and readjust the belt if necessary. **Note:** *If proper tension with and old belt can't be achieved, installation of a new belt will be necessary.*

31 Install the various components removed during disassembly, referring to the appropriate Sections as necessary.

8 Camshafts, lifters and seals - removal and installation

Lifters

Refer to illustrations 8.2 and 8.3

1 Remove the valve cover (see Section 4) and the rocker arm shaft assemblies (see Section 5).

2 Secure the lifters by raising them slightly and wrapping a rubber band around each one to prevent them from falling out of the guides **(see illustration)**. **Note:** *If a lifter*

8.3 With the lifters retained by rubber bands, the lifter guide assembly can be removed from the cylinder head

should fall out of the guide, immediately put it back in its original location.

3 Remove the lifter guide assembly **(see illustration)**.

4 Remove the lifters from the bores one at a time. Keep them in order. Each lifter must

be reinstalled in its original bore. **Caution:** *Do not lay the lifters on their side or upside down, or air can become trapped inside and the lifter will have to be bled as follows. The lifters can be laid on their side only if they are submerged in a pan of clean engine oil until reassembly.*

5 With the lifter in its bore, push down on the plunger at the top of the lifter. If it moves more than 0.040-inch (1 mm), air may be trapped inside the lifter.

6 If you think air is trapped inside a valve lifter, reinstall the rocker arm shaft assemblies and valve cover.

7 Bleed air from the lifters by running the engine at 1,000 rpm under no load for about 10 minutes.

8 Remove the valve cover and rocker arm shaft assemblies again. Repeat the procedure in Step 5 once more. If there's still air in the lifter, replace it with a new one.

9 While the lifters are out of the engine, inspect them for wear. Refer to Chapter 2C for inspection procedures.

10 Installation is the reverse of removal. Be sure to lubricate each lifter with liberal amounts of clean engine oil prior to installation.

Camshafts

Removal

Refer to illustrations 8.12, 8.13, 8.14 and 8.15

11 Remove the cylinder heads from the engine (see Section 11).

12 Remove the bolts **(see illustration)** and gently pry off the camshaft cover plate at the transaxle end of the cylinder head.

13 Use the holding lugs to secure the camshaft while loosening the retaining bolt **(see illustration)**. Remove the bolt and the thrust plate.

14 Carefully pry the camshaft oil seal out of the cylinder head with a small screwdriver **(see illustration)**. Don't scratch or nick the camshaft in the process! **Note:** *Pushing on the rear of the camshaft through the thrust plate opening may help facilitate removal of the oil seal.*

15 Carefully pull the camshaft out the front

8.12 Remove the cover plate bolts (arrows) and gently pry off the cover (front cylinder head shown, rear cylinder head similar)

8.13 Hold the camshaft lug with pliers or a wrench to prevent the camshaft from moving while loosening the bolt

of the cylinder head using a twisting motion **(see illustration)**. **Caution:** *Don't scratch the bearing surfaces with the cam lobes.* Refer to Chapter 2, Part C for camshaft and bearing inspection procedures.

Installation

Refer to illustration 8.19

16 Lubricate the camshaft bearing journals and lobes with moly-based engine assembly lube, then install it carefully in the cylinder head. Don't scratch the bearing surfaces with the camshaft lobes!

17 Install the camshaft thrust plate and retaining bolt at the rear of the camshaft and tighten it to the torque listed in this Chapter's Specifications.

18 With the camshaft installed in the cylinder head, refer to Chapter 2C and check the camshaft end play.

19 Endplay outside the specified range requires thrust plate replacement. Measure the old plate **(see illustration)** and obtain a

8.14 Carefully pry the camshaft oil seal out with a small screwdriver

8.15 Withdraw the camshaft from the cylinder head, using both hands to support it to avoid damage to the bearing surfaces in the cylinder head

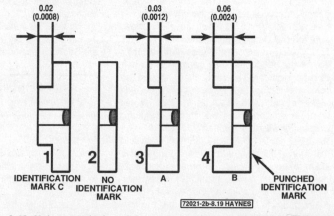

8.19 If the camshaft endplay exceeds the specified limit, select a different thrust plate to bring the endplay within specification

Chapter 2 Part A Single Overhead Cam (SOHC) engine

8.22 Use a seal driver to press the new camshaft seal squarely into place

new one that will produce endplay as close to the specification as possible.

20 Install the camshaft oil seal as described in Step 22. The remainder of the cylinder head assembly is the reverse of the disassembly procedure. Refer to Section 11 for cylinder head installation.

Camshaft oil seals

Refer to illustration 8.22

21 In the course of replacing a camshaft, the old seal is removed. If you are replacing the seals only, the seals may be replaced in vehicle without removal of the cylinder head and camshafts, by removing the timing belt and camshaft sprockets and prying the seal outward. When performing this procedure be extremely careful not to nick or gouge the camshaft journal in the process (see illustration 8.14).

22 After the camshaft has been installed, use a seal installation tool (see illustration 13.7a), deep socket or piece of pipe of the appropriate diameter to press the new seal squarely into the cylinder head (see illustration). Press the seal in only until the seal bottoms.

9 Intake manifold - removal and installation

Warning: *The engine must be completely cool before beginning this procedure.*

Upper intake manifold (plenum)

Refer to illustrations 9.4 and 9.7

1 Disconnect the cable from the negative terminal of the battery.
2 Refer to Chapter 4 and remove the air intake duct.
3 Disconnect the accelerator cable, cruise control cable (if equipped), hoses and electrical connectors from the throttle body. Be prepared for some coolant spillage and plug the hoses immediately to prevent excessive coolant loss.
4 Detach the spark plug wires from the spark plugs and remove the spark plug wires from the retainers on the plenum. Remove the distributor cap from the distributor and position the distributor cap and spark plug wires aside. Label and disconnect the hoses and electrical connectors attached to the plenum (see illustration).
5 Remove the ground strap attached to the upper plenum and loosen the upper intake manifold bolts in the reverse order of the tightening sequence (see illustration 9.7). Remove the upper plenum with the throttle body attached.
6 To install the upper manifold, clean the mounting surfaces of the intake manifold, and the upper plenum with lacquer thinner and remove all traces of the old gasket material or sealant.
7 Install the new gasket over the intake manifold studs, then install the upper plenum onto the intake manifold and tighten the bolts in the proper sequence (see illustration) to the torque listed in this Chapter's Specifications. The remainder of installation is the reverse of removal. Check the coolant level and add some, if necessary (see Chapter 1).

Lower intake manifold

Refer to illustrations 9.14, 9.18 and 9.19

8 Relieve the fuel pressure (see Chapter 4), then disconnect the cable from the negative terminal of the battery.
9 Remove the upper intake manifold (see Steps 1 through 5).
10 Drain the cooling system (see Chapter 1).
11 Disconnect the electrical connectors from the coolant temperature switch and the coolant temperature sensor. Also remove the radiator hose from the thermostat housing.
12 Label and detach any vacuum hoses which would interfere with the removal of the lower intake manifold.
13 Refer to Chapter 4 and remove the fuel

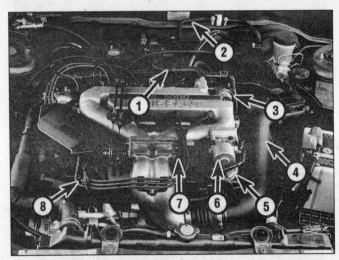

9.4 Label and disconnect the electrical connectors and hoses from the following components before removing the upper intake manifold

1 PCV valve
2 Power brake booster
3 Power valve actuator
4 Air intake duct
5 EGR valve and flare tube
6 IAC/air cut valve assembly
7 Throttle position sensor/closed throttle position switch
8 Distributor cap and spark plug wires

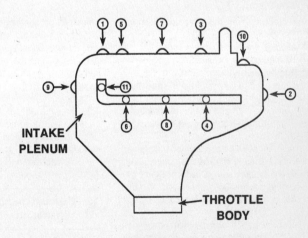

9.7 Upper plenum TIGHTENING sequence

9.14 Remove the bolts and disconnect the coolant pipes (arrow) from the rear of the lower intake manifold

9.18 Check the condition of the coolant pipes and the gasket (arrow) between the pipes and the thermostat housing - they are only accessible when the lower intake manifold is off

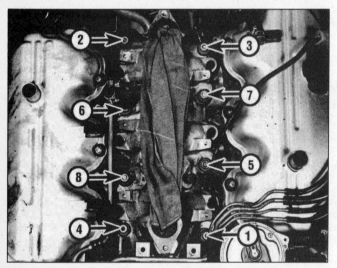

9.19 Lower intake manifold TIGHTENING sequence

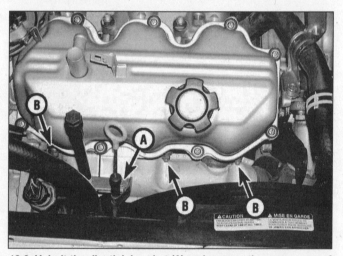

10.6 Unbolt the dipstick bracket (A) and remove the upper row of exhaust manifold nuts (B)

rail and injectors from the lower intake manifold.

14 Remove the bolts and disconnect the coolant pipes at the transmission end of the intake manifold **(see illustration)**.

15 Loosen the manifold mounting bolts/nuts in 1/4-turn increments until they can be removed by hand in the reverse order of the tightening sequence **(see illustration 9.19)**.

16 The manifold will probably be stuck to the cylinder heads and force may be required to break the gasket seal. **Caution:** *Don't pry between the manifold and the heads or damage to the gasket sealing surfaces may occur, leading to vacuum leaks.*

17 Carefully use a scraper to remove all traces of old gasket material and sealant from the manifold and cylinder heads, then clean the mating surfaces with lacquer thinner or acetone.

18 Inspect the coolant pipes below the intake before installing the new gaskets and lower intake manifold **(see illustration)**. Install new gaskets, then position the lower manifold on the engine. Make sure the gaskets and manifold are aligned over the dowels in the cylinder heads and install the nuts.

19 Following the recommended tightening sequence, tighten the nuts/bolts, in three equal steps, to the torque listed in this Chapter's Specifications **(see illustration)**.

20 The remainder of the installation is the reverse of the removal procedure. Refill the cooling system and change the engine oil (see Chapter 1). Run the engine and check for fuel, vacuum and coolant leaks.

10 Exhaust manifolds - removal and installation

Caution: *The engine must be completely cool before beginning this procedure.*

Removal

Refer to illustrations 10.6 and 10.7

1 Disconnect the cable from the negative terminal of the battery and support the vehicle on jackstands.

2 Spray penetrating oil on the exhaust manifold fasteners and allow it to soak in.

3 Working under the vehicle, disconnect the exhaust Y-pipe from the exhaust manifolds (see Chapter 4).

Front (radiator side) manifold

4 Remove the air intake duct and the top half of air cleaner housing as an assembly from the throttle body (see Chapter 4).

5 Disconnect the EGR flare tube from the exhaust manifold.

6 Remove the bracket holding the engine oil dipstick to provide access to the manifold fasteners, and remove the upper row of manifold nuts **(see illustration)**. Work from the ends toward the middle when removing the manifold fasteners.

7 Working from below, remove the lower row of manifold fasteners and remove the manifold from the engine **(see illustration)**.

Rear (firewall side) manifold

8 Working underneath the vehicle, remove the rear manifold nuts and remove the manifold.

Chapter 2 Part A Single Overhead Cam (SOHC) engine

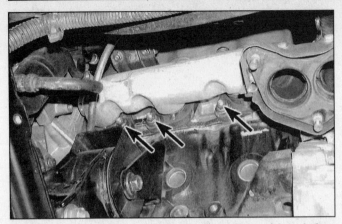

10.7 Remove the lower row of manifold nuts (arrows) from below

11.14 Remove the old gaskets and clean the engine block and cylinder head mating surfaces thoroughly

Installation

9 Carefully inspect the manifolds and fasteners for cracks and damage.
10 Use a scraper to remove all traces of old gasket material and carbon deposits from the manifold and cylinder head mating surfaces. If the gasket was leaking, have the manifold checked for warpage at an automotive machine shop and resurfaced if necessary.
11 Position new gaskets over the cylinder head studs.
12 Install the manifold and thread the mounting nuts into place.
13 Working from the center out, tighten the nuts to the torque listed in this Chapter's Specifications in three or four equal steps.
14 Reinstall the remaining parts in the reverse order of removal. Use new gaskets when connecting the exhaust pipe to the front manifold.
15 Run the engine and check for exhaust leaks.

11 Cylinder heads - removal and installation

Warning: *Allow the engine to cool completely before beginning this procedure.*

Removal

1 Relieve the fuel system pressure (see Chapter 4), then disconnect the cable from the negative terminal of the battery. Drain the cooling system (see Chapter 1). Remove the timing belt, camshaft sprocket(s) and rear timing belt cover (see Section 7). Also remove the bolts holding the coolant pipes to the rear (transaxle end) of each cylinder head at the camshaft end plates.
2 Remove the intake manifold (see Section 9), and set aside the fuel injectors and their harnesses.
3 Remove the rocker arm components (see Section 5) and lifters (see Section 8).
4 Remove the exhaust manifold(s) (see Section 10).
5 Label and disconnect the hoses and electrical harness connectors at the timing belt end of the engine.

11.18 Position the gasket over the dowel pins (arrows) so that all the holes line up

Front (radiator side) cylinder head

6 Remove the spark plug wires, distributor and coil (see Chapter 5).
7 Remove the air conditioning compressor from the bracket without disconnecting the refrigerant hoses (see Chapter 3) and set it aside. Secure the compressor to the vehicle with rope or wire to make sure it doesn't hang by its hoses. **Note:** *There isn't room to fully remove the upper compressor bolts until the compressor is moved aside.*
8 Remove the compressor and alternator bracket.
9 Remove the bracket holding the engine oil dipstick tube to the cylinder head **(see illustration 10.6)**.

Rear (firewall side) cylinder head

10 Detach the heater hoses and brackets from the transaxle end of the cylinder head.

Both cylinder heads

11 Loosen the cylinder head bolts with a hex drive tool in 1/4-turn increments until they can be removed by hand. Be sure to loosen the bolts in the reverse order of the tightening sequence, removing the small bolt, located outside of the cylinder head first **(see illustration 11.21)**.
12 Remove the washers from the cylinder head bolts and <u>discard</u> the bolts (except the small bolt on the outside of the cylinder head). NEW cylinder head bolts must be used on reassembly **Note:** *The cylinder head bolt washers can be reused*.

13 Lift the cylinder head off the engine block. If resistance is felt, dislodge the cylinder head by striking it with a wood block and hammer. If prying is required, be very careful not to damage the cylinder head or engine block!

Installation

Refer to illustrations 11.14, 11.18 and 11.21

14 Remove the old cylinder head gaskets **(see illustration)**. The mating surfaces of the cylinder heads and engine block must be perfectly clean when the heads are installed.
15 Use a gasket scraper to remove all traces of carbon and old gasket material, then clean the mating surfaces with lacquer thinner or acetone. If there's oil on the mating surfaces when the heads are installed, the gaskets may not seal correctly and leaks may develop. Use a vacuum cleaner to remove any debris that falls into the cylinders.
16 Check the engine block and cylinder head mating surfaces for nicks, deep scratches and other damage. If damage is slight, it can be removed with a file - if it's excessive, machining may be the only alternative.
17 Use a tap of the correct size to chase the threads in the cylinder head bolt holes. Dirt, corrosion, sealant and damaged threads will affect torque readings. Ensure that the threaded holes in the engine block are clean and dry.
18 Position the new gaskets over the dowel pins in the engine block **(see illustration)**.

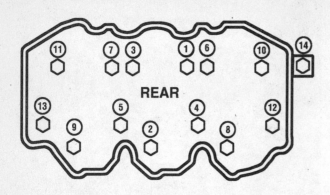

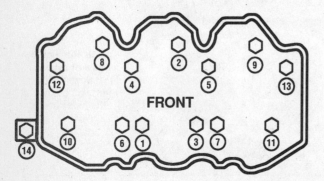

11.21 Cylinder head bolt TIGHTENING sequence - note that bolts 4, 5, 12 and 13 in the sequence are 5.0 inches long, the rest (except no. 14) are 4.17 inches in length; make sure the bolts are installed in the correct locations

12.4 Use a strap wrench to hold the crankshaft pulley while removing the center bolt (a chain-type wrench may be used if you wrap a section of old drivebelt around the crankshaft pulley first)

12.5 Use two screwdrivers or a pry bar to carefully pry the crankshaft pulley off

19 Carefully position the heads on the engine block without disturbing the gaskets.
20 Lightly oil the threads of the NEW cylinder head bolts and install in the proper locations **(see illustration 11.21)**. Tighten them finger tight. **Caution:** *Make sure the washers are in place on the bolts - the chamfered side of the washer must be against the bolt head, which means the flat side must be against the cylinder head surface.*
21 Follow the recommended sequence and tighten the bolts (except the small bolt on the outside of the cylinder head) in five steps to the torque specified in this Chapter **(see illustration)**. **Caution:** *Bolts 4, 5, 12 and 13 in the sequence are longer than the others - be sure all bolts are in their proper locations!*
22 Tighten the small bolt on the outside of the cylinder head to 78 to 104 in-lbs (9 to 12 Nm).
23 The remaining installation steps are the reverse of removal.
24 Add coolant and change the engine oil and filter (see Chapter 1), then start the engine and check carefully for oil and coolant leaks.

12 Crankshaft pulley - removal and installation

Refer to illustrations 12.4 and 12.5
1 Disconnect the negative cable from the battery. Raise the front of the vehicle, remove the right front wheel, and secure the vehicle on jackstands.
2 Remove the drivebelts (see Chapter 1).
3 Remove right front wheel and the inner splash shield from the right fenderwell.
4 Use a strap wrench around the crankshaft pulley to hold it while using a breaker bar and socket to remove the crankshaft pulley center bolt **(see illustration)**.
5 Wedge a prybar or two screwdrivers behind the crankshaft pulley and carefully pry it off the crankshaft **(see illustration)**. If the pulley is difficult to remove, use a bolt-type puller and pull it off.
6 Installation is the reverse of removal.

13 Crankshaft front oil seal - replacement

Refer to illustrations 13.5, 13.6, 13.7a and 13.7b
1 Remove the crankshaft pulley (see Section 12), timing belt covers and timing belt (see Section 7).
2 Carefully remove the crankshaft sprocket with a prybar or two screwdrivers, be very careful not to damage the oil pump body.
3 If the sprocket cannot be pried off, drill and tap two holes into the face of the sprocket and use a bolt-type puller to pull it off the crankshaft. **Caution:** *Do not reuse a drilled sprocket - replace it.*
4 Remove the inner timing belt guide, noting the side facing out (mark it if necessary).
5 Carefully pry the oil seal out with a screwdriver **(see illustration)**. Don't scratch or nick the crankshaft in the process!
6 Before installation, apply a thin coat of

Chapter 2 Part A Single Overhead Cam (SOHC) engine

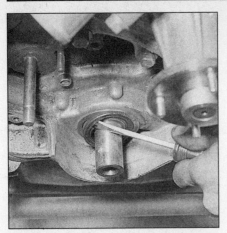

13.5 Pry the seal out very carefully with a seal removal tool or screwdriver - if the crankshaft is nicked or otherwise damaged, the new seal will leak!

13.6 Apply multi-purpose grease or clean engine oil to the lips of the new seal before installing it (if you apply a small amount of grease to the outer edge, it will be easier to press into the bore)

13.7a Fabricate a seal installation tool from a piece of pipe, a long bolt and a large washer - the outside diameter of the pipe must be the same size or slightly smaller than the outer diameter of the seal

13.7b Install the seal installation tool and press the seal into the bore by tightening the bolt

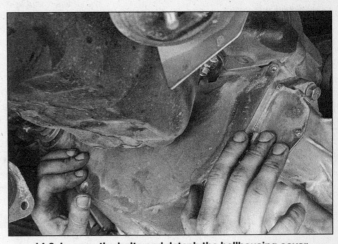

14.6 Loosen the bolts and detach the bellhousing cover

multi-purpose grease to the inside of the seal **(see illustration)**.

7   Fabricate a seal installation tool with a short length of pipe of equal or slightly smaller outside diameter than the seal itself. File the end of the pipe that will bear down on the seal until it's free of sharp edges. You'll also need a long bolt of the same thread pitch as the crankshaft pulley bolt and a large washer, slightly larger in diameter than the pipe, on which the bolt head can seat **(see illustration)**. Install the oil seal by pressing it into position with the seal installation tool **(see illustration)**. When the seal is bottomed in the housing, don't turn the bolt any more or you'll damage the seal.

8   Install the inner timing belt guide onto the nose of the crankshaft.

9   Make sure the Woodruff key is in place in the crankshaft.

10   Apply a thin coat of assembly lube to the inside of the timing belt sprocket and slide it onto the crankshaft.

11   Installation of the remaining components is the reverse of removal. Refer to Section 7 for the timing belt installation and adjustment procedure. Tighten all bolts to the torque listed in this Chapter's Specifications.

14 Oil pan - removal and installation

Removal

Refer to illustration 14.6

1   Disconnect the cable from the negative terminal of the battery.
2   Raise the vehicle and support it securely on jackstands.
3   Remove the under-vehicle splash pan (if equipped).
4   Drain the engine oil and remove the oil filter (see Chapter 1).
5   Unbolt and remove exhaust Y-pipe from the exhaust system (see Chapter 4).
6   Detach the bellhousing cover **(see illustration)**.

7   Support the engine/transaxle securely from above with a hoist or a three-bar engine support fixture. **Warning:** *Be absolutely certain the engine/transaxle is securely supported! DO NOT place any part of your body under the engine/transaxle - it could crush you if the support or hoist fails!*

8   Unbolt the front and rear engine mounts from the crossmember (see Section 18).
9   Remove the engine-mount crossmember from beneath the oil pan.
10   Remove the oil pan bolts, following the reverse of the tightening sequence **(see illustration 14.16)**.
11   Detach the oil pan. Don't pry between the pan and engine block or damage to the sealing surfaces may result and oil leaks could develop. If the pan is stuck, dislodge it with a soft-face hammer.
12   Use a gasket scraper to remove all traces of old gasket material and sealant from the engine block and pan. Clean the mating surfaces with lacquer thinner or acetone.

Chapter 2 Part A Single Overhead Cam (SOHC) engine

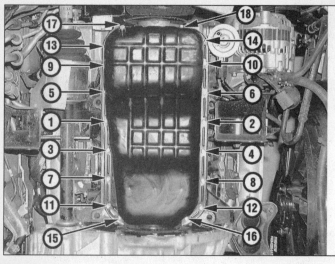

14.16 Oil pan bolt TIGHTENING sequence

Installation

Refer to illustration 14.16

13 Ensure that the threaded holes in the engine block are clean (use a tap to remove any sealant or corrosion from the threads).
14 Apply RTV sealant to the ends of the seals and position them on the oil pump and rear seal housings.
15 Apply a continuous 5/32-inch (3.5 mm) bead of RTV sealant to the inner sealing surface of the oil pan. **Note:** *Install the oil pan within five minutes of sealant application.*
16 Install the oil pan and tighten the bolts in three or four steps following the sequence shown **(see illustration)** to the torque listed in this Chapter's Specifications.
17 The remaining installation steps are the reverse of removal.
18 Allow at least 30 minutes for the sealant to dry, add oil and a new oil filter, start the engine and check for oil pressure and leaks.

15 Oil pump - removal, inspection and installation

Removal

Refer to illustrations 15.3 and 15.4

1 Remove the timing belt and the crankshaft sprocket (see Section 7). Remove the oil pan (see Section 14).
2 Refer to Chapter 5 and remove the alternator adjusting bar and the bar-to-oil pump bolt.
3 Unbolt the power steering pump (see Chapter 10) and without disconnecting the hoses, position it aside. Remove the power steering pump bracket **(see illustration)**.
4 Remove the oil pump-to-engine block bolts from the front of the engine **(see illustration)**. **Note:** *The pickup tube and oil filter adapter can remain attached to the pump at this time if desired. One of the two long oil pump mounting bolts is the bolt removed in Step 2 that held the alternator adjusting bar.*
5 Use a block of wood and a hammer to break the oil pump gasket seal.
6 Pull out on the oil pump to remove it from the engine block.
7 Use a scraper to remove old gasket material and sealant from the oil pump and engine block mating surfaces. Clean the mating surfaces with lacquer thinner or acetone.

Inspection

Refer to illustrations 15.8, 15.10, 15.11a, 15.11b, 15.11c, 15.11d and 15.13

8 Use a large Phillips screwdriver to remove the screws holding the rear cover on the oil pump **(see illustration)**.
9 Clean all components with solvent, then inspect them for wear and damage.
10 Remove the oil pressure regulator cap, washer, spring and valve **(see illustration)**. Check the oil pressure regulator valve sliding surface and valve spring. If either the spring or the valve is damaged, they must be replaced as a set.
11 Check the clearance of the following oil pump components with a feeler gauge **(see illustrations)** and compare the measurements to the clearance listed in this Chapter's Specifications:

 a) *Cover-to-inner rotor clearance*
 b) *Cover-to-outer rotor clearance*
 c) *Outer rotor-to-body clearance*
 d) *Outer gear-to-crescent clearance*
 e) *Inner gear-to-crescent clearance*

If any clearance is excessive, replace the entire oil pump assembly.

12 **Note:** *Pack the pump with petroleum jelly to prime it.* Assemble the oil pump and tighten the screws securely. Install the oil pressure regulator valve, spring and washer, then tighten the oil pressure regulator valve cap.

15.3 Remove the power steering pump bracket bolts (arrows)

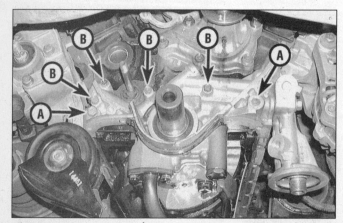

15.4 Remove the oil pump mounting bolts (A indicates the two long bolts, B the shorter bolts) and detach the pump from the engine

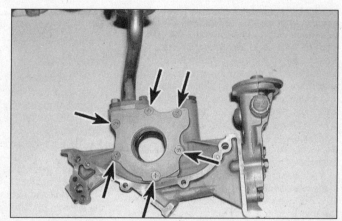

15.8 Remove the screws and lift the cover off

Chapter 2 Part A Single Overhead Cam (SOHC) engine 2A-17

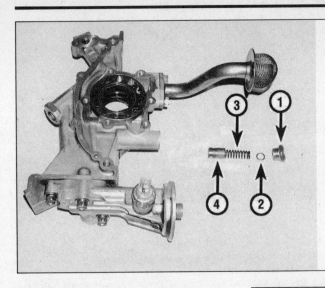

15.10 Oil pressure relief valve components

1. Plug
2. Washer
3. Spring
4. Relief valve

15.11a Measure the cover-to-rotor end clearance with a straightedge and feeler gauge - measure above the outer rotor and above the inner rotor

13 Replace the O-ring on the flange of the oil pick-up tube **(see illustration)** and reinstall the tube. Tighten the pick-up tube bolts to the torque listed in this Chapter's Specifications.

Installation

Refer to illustration 15.15

14 Apply RTV sealant to the oil pump mounting surface.

15 Use new gaskets on all disassembled parts and reverse the removal procedure for installation. Align the flats on the crankshaft **(see illustration)** with the flats on the oil pump gear. Tighten all fasteners to the torque listed in this Chapter's Specifications.

16 Flywheel/driveplate - removal and installation

Refer to illustration 16.4

1 Raise the vehicle and support it securely on jackstands, then refer to Chapter 7 and remove the transaxle. **Warning:** *The engine must be supported from above with an engine hoist or three-bar support fixture*

15.11b Measuring the outer gear-to-body clearance

before working underneath the vehicle with the transaxle removed.

2 If the vehicle is equipped with a manual transaxle, remove the pressure plate and clutch disc (see Chapter 8). Now is a good time to check/replace the clutch components and pilot bearing if necessary.

If the vehicle is equipped with automatic

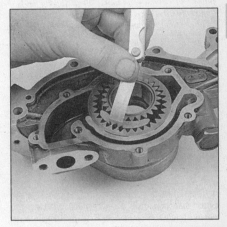

15.11c Measuring the outer gear-to-crescent clearance

transaxle, now would be a good time to check and replace the front pump seal/O-ring.

3 Use paint or a center-punch to make alignment marks on the flywheel/driveplate and crankshaft to ensure correct alignment during reinstallation.

4 Remove the bolts that secure the fly-

15.11d Measuring the inner gear-to-crescent clearance

15.13 Remove the oil pick-up tube bolts and replace the rubber O-ring

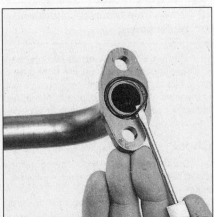

15.15 There is a flat surface (arrow) on each side of the crankshaft - align them with the flats on the gear

Chapter 2 Part A Single Overhead Cam (SOHC) engine

16.4 Wedge a screwdriver or pry bar between the flywheel/driveplate and the engine block while the mounting bolts are removed - note the painted marks made at the crank and flywheel/driveplate for alignment

17.2 Pry the seal out very carefully with a seal removal tool or screwdriver - if the crankshaft is damaged the new seal will leak!

wheel/driveplate to the crankshaft **(see illustration)**. If the crankshaft turns, hold the flywheel with a pry bar or wedge a screwdriver into the ring gear teeth to jam the flywheel.

5 Remove the flywheel/driveplate from the crankshaft. Since the flywheel is fairly heavy, be sure to support it while removing the last bolt.

6 Clean the flywheel to remove grease and oil. Inspect the surface for cracks, rivet grooves, burned areas and score marks. Light scoring can be removed with emery cloth. Check for cracked and broken ring gear teeth or a loose ring gear. Lay the flywheel on a flat surface and use a straightedge to check for warpage.

7 Clean and inspect the mating surfaces of the flywheel/driveplate and the crankshaft. If the crankshaft rear seal is leaking, replace it before reinstalling the flywheel/driveplate.

8 Position the flywheel/driveplate against the crankshaft. Be sure to align the marks made during removal. Note that some engines have an alignment dowel or staggered bolt holes to ensure correct installation. Before installing the bolts, apply thread locking compound to the threads.

9 Wedge a screwdriver into the ring gear teeth to keep the flywheel/driveplate from turning as you tighten the bolts to the torque listed in this Chapter's Specifications. **Caution:** *The flywheel mounting bolts should only be used once. Replace the bolts if they were used before.*

10 The remainder of installation is the reverse of the removal.

17 Rear main oil seal - replacement

Refer to illustrations 17.2 and 17.3

1 The transaxle must be removed from the vehicle for this procedure (see Chapter 7). **Warning:** *The engine must be supported* *from above with an engine hoist or three-bar support fixture before working underneath the vehicle with the transaxle removed.* Remove the flywheel/driveplate (see Section 16).

2 Carefully pry out the old seal out of the retainer with a seal removal tool or screwdriver **(see illustration)**.

3 Apply multi-purpose grease to the crankshaft seal journal and the lip of the new seal. Preferably, a seal installation tool should be used to press the new seal into place. If the proper seal installation tool is unavailable, use a large socket, section of pipe or a blunt tool and carefully drive the new seal into place **(see illustration)**. The lip is stiff so carefully work it onto the seal journal of the crankshaft. Don't rush it or you may damage the seal. **Note:** *Install the seal squarely and only until flush with the back of the seal plate, no further.*

4 The remaining steps are the reverse of removal.

18 Powertrain mounts - check and replacement

1 There are four powertrain mounts; front and rear engine mounts on the right side of the vehicle attached to the engine block and to the lower crossmember, and front and rear transaxle mounts on the left side attached to the transaxle and the frame.

Check

2 During the check, the engine must be raised slightly to remove the weight from the mounts.

3 Raise the vehicle and support it securely on jackstands. Support the engine/transaxle from above using a hoist or an engine support fixture.

4 Check the mounts to see if the rubber is

17.3 If you don't have a seal installation tool, use a blunt tool (such as a brass punch) to carefully work the edge of the seal evenly into the bore and around the crankshaft

cracked, hardened or separated from the bushing in the center of the mount.

5 Check for relative movement between the mounts and the engine or frame (use a large screwdriver or prybar to attempt to move the mounts). If movement is noted, lower the engine and tighten the mount fasteners.

6 Rubber preservative should be applied to the mounts to slow deterioration.

Replacement

Refer to illustrations 18.9, 18.14, 18.16 and 18.17

7 Disconnect the cable from the negative terminal of the battery, then set the parking brake, block the rear wheels, raise the front of the vehicle and support it securely on jackstands. Remove the splash shields from under the vehicle.

Chapter 2 Part A Single Overhead Cam (SOHC) engine

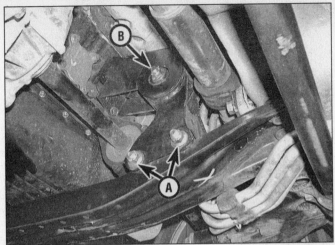

18.9 Engine mount-to-crossmember through-bolts (A) and engine mount-to-engine bracket through-bolt (B) - rear engine mount shown, front engine mount similar

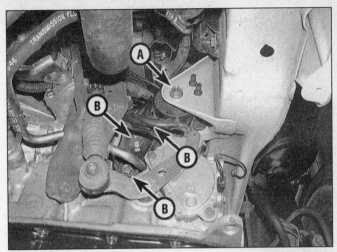

18.14 To remove the left-front transaxle mount, remove the through-bolt (A), then the three bolts (B) retaining the mount to the transaxle

18.16 Remove the three nuts (arrows) at the rear transaxle bracket from below (driveaxle removed for clarity)

18.17 With the air cleaner and air inlet removed, there is access at the top to remove the left-rear transaxle mount through-bolt (arrow).

Front and rear engine mounts

8 Support the engine from above using a hoist or an engine support fixture.
9 Remove the through-bolts from the right-front and right rear mounts where they attach to the crossmember (see illustration).
10 Remove the crossmember-to-chassis bolts and detach the crossmember from the vehicle.
11 Remove the mount-to-engine bracket through-bolts and remove the mounts (see illustration 18.9).
12 Installation is the reverse of removal.
Note: *Tighten the bolts to Specifications only after the engine weight is back onto the mounts and the jack is removed. If more than one mount has been replaced, see Final tightening (Step 20).*

Front and rear transaxle mounts

13 There are two transaxle mounts, one at the radiator side of the transaxle (left-front mount) and one at the top of the transaxle (left-rear mount).

Left-front mount

14 Support the transaxle with a jack placed under the transaxle bellhousing. Remove the through-bolt at the chassis, then the bolts holding the mount to the transaxle and remove the mount (see illustration).
15 Installation is the reverse of removal.
Note: *Tighten the bolts to Specifications only after the powertrain weight is back onto the mounts and the jack is removed. If more than one mount has been replaced, see Final tightening (Step 20).*

Left-rear mount

16 From below, remove the nuts at the mount bracket on the transaxle (see illustration).
17 From above, remove the through-bolt at the chassis bracket (see illustration).
18 Lower the transaxle enough for the studs to clear the transaxle bracket and remove the mount.
19 Installation is the reverse of removal.
Note: *Tighten the bolts to Specifications only after the powertrain weight is back onto the mounts and the jack is removed. If more than one mount has been replaced, see Final tightening below.*

Final tightening

20 To ensure maximum bushing life and prevent excessive noise and vibration, the vehicle should be level and the engine should be on the mounts during the final tightening stage. **Note:** *Use thread-locking compound on the nuts/bolts. Ensure that the bushings are not twisted or offset.* If you have replaced more than one mount, or when you are installing the engine, tighten the mounts in the following order: crossmember bolts, right-rear engine mount, left-front transaxle mount, right-front engine mount, and left-rear transaxle mount.

Notes

Chapter 2 Part B
Dual Overhead Cam (DOHC) engine

Contents

Section		Section	
Camshafts and lifters - removal and installation	8	Oil pump - removal, inspection and installation	15
Crankshaft front oil seal - replacement	13	Powertrain mounts - check and replacement	18
Crankshaft pulley – removal and installation	12	Rear main oil seal - replacement	17
Cylinder compression check	See Chapter 2C	Repair operations possible with the engine in the vehicle	2
Cylinder head - removal and installation	11	Spark plug replacement	See Chapter 1
Drivebelt check, adjustment and replacement	See Chapter 1	Timing chain and sprockets - removal, inspection and installation	7
Engine - removal and installation	See Chapter 2C	Top Dead Center (TDC) for number one piston – locating	3
Engine oil and filter change	See Chapter 1	Valve clearance check and adjustment	5
Engine overhaul - general information	See Chapter 2C	Valve cover - removal and installation	4
Exhaust manifold - removal and installation	10	Valve springs, retainers and seals - replacement	6
Flywheel/driveplate - removal and installation	16	Valves - servicing	See Chapter 2C
General information	1	Water pump - removal and installation	See Chapter 3
Intake manifold - removal and installation	9		
Oil pan - removal and installation	14		

Specifications

General

Cylinder numbers (timing belt end-to-transaxle end)
- Rear (firewall) side 1-3-5
- Front (radiator) side 2-4-6
Firing order 1-2-3-4-5-6
Bore .... 3.66 inches (93 mm)
Stroke .... 2.886 inches (73.3 mm)
Displacement .... 182.33 cubic inches (3.0 liters)

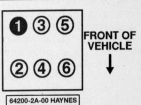

Cylinder location diagram

Camshaft

Camshaft endplay .... 0.0045 to 0.0074 inch (0.115 to 0.188 mm)
Valve clearance (cold)
- Intake .... 0.010 to 0.013 inch (0.26 to 0.34 mm)
- Exhaust .... 0.011 to 0.015 inch (0.29 to 0.37 mm)

Oil pump

Outer gear-to-body clearance	
1995 through 1997	0.0045 to 0.0102 inch (0.114 to 0.260 mm)
1998 and later	0.0045 to 0.0079 inch (0.114 to 0.200 mm)
Inner gear-to-outer gear tip clearance	0.0071 inch max (0.18 mm)
Inner gear-to-housing side clearance	
1995 through 1997	0.0020 to 0.0035 inch (0.05 to 0.09 mm)
1998 and later	0.0012 to 0.0028 inch (0.03 to 0.07 mm)
Outer gear-to-housing side clearance	
1995 through 1997	0.0012 to 0.0075 inch (0.03 to 0.19 mm)
1998 and later	0.0020 to 0.0043 inch (0.05 to 0.11 mm)
Inner rotor hub-to-housing clearance	0.0018 to 0.0036 inch (0.045 to 0.091 mm)

Torque specifications

	Nm	Ft-lbs (unless otherwise indicated)
Camshaft sprocket bolts	119 to 128	88 to 95
Camshaft bearing cap bolts (in sequence; **see illustration 8.13**)		
Step one (tighten the No.1 bearing cap bolts first, [bolts 13 through 16] then follow the tightening sequence)	1.96	17 in-lbs
Step two	6	52 in-lbs
Step three	9 to 11.8	80 to 104 in-lbs
Crankshaft pulley bolt		
Step 1	39 to 49	29 to 36
Step 2	Tighten an additional 60 degrees	
Cylinder head bolts (in sequence; **see illustration 11.20**)		
Step one	98	72
Step two	Loosen all bolts (in reverse of tightening sequence)	
Step three	34 to 44	25 to 33
Step four	Tighten all bolts an additional 90 to 95 degrees	
Step five	Tighten all bolts an additional 90 to 95 degrees	
Valve cover bolts		
Step one	1 to 3	9 to 26 in-lbs
Step two	5.4 to 7.4	48 to 65 in-lbs
Driveplate bolts	83 to 93	1 to 69
Exhaust manifold nuts	30 to 32	22 to 24
Exhaust manifold heat shield bolts	5 to 6.5	45 to 57 in-lbs
Intake manifold		
Upper intake manifold	18 to 22	13 to 16
Lower intake manifold bolts/nuts		
Step one	5 to 10	43 to 86 in-lbs
Step two	26 to 31	20 to 23
Step three	Repeat Step 2 several more times	
Oil pan bolts		
Aluminum oil pan	16 to 19	144 to 168 in-lbs
Steel oil pan	6.4 to 7.5	57 to 66 in-lbs
Oil pan to transaxle	35 to 47	26 to 35
Oil pan to rear cover	6.4 to 7.5	57 to 66 in-lbs
Oil pan drain plug	29 to 39	22 to 29
Oil pick-up tube mounting bolts	16 to 19	144 to 168 in-lbs
Oil pump mounting bolts		
1995 and 1996	6.4 to 6.6	57 to 58 in-lbs
1997 and later	8.4 to 10.8	75 to 95 in-lbs
Oil pump cover screws	4 to 5	35 to 44 in-lbs
Front timing chain cover bolts		
6 mm	11.8 to 13.7	105 to 120 in-lbs
8 mm	26 to 31	19 to 23
Rear timing chain cover bolts	11.8 to 13.7	105 to 120 in-lbs
Upper timing chain guide(s) bolts	13 to 19	108 to 168 in-lbs
Main timing chain tensioner bolts	8.4 to 10.8	75 to 95 in-lbs
Main timing chain guide pivot bolt	13 to 19	108 to 168 in-lbs
Secondary timing chain tensioner bolts	8.4 to 10.8	5 to 95 in-lbs
Rear main oil seal retainer bolts	8.4 to 10.8	75 to 95 in-lbs
Front and rear engine mount bolts	77 to 98	57 to 72
Front and rear engine mount brackets to engine block bolts	43 to 55	32 to 41
Lower crossmember to frame bolts	77 to 98	57 to 72
Right side engine mount upper bracket bolts	43 to 55	32 to 41
Right side engine mount bolts/nuts	43 to 55	32 to 41
Right side engine mount lower bracket bolt/nut	78 to 90	58 to 67
Left side transaxle mount bolts/nuts	43 to 55	32 to 41

Chapter 2 Part B Dual Overhead Cam (DOHC) engine

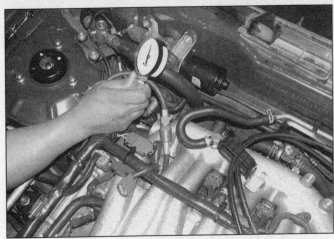

3.5 A compression gauge can be used in the number one plug hole to assist in finding TDC on DOHC engines

3.6 Align the TDC notch on the crankshaft pulley with the pointer on the timing chain cover (arrow) - the TDC notch is the one farthest to the left when facing the front of the engine and is typically yellow in color

1 General information

This Part of Chapter 2 is devoted to in-vehicle repair procedures for the VQ30DE 3.0L Dual Overhead Camshaft (DOHC) V6 engine equipped on 1995 and later vehicles. All information concerning engine removal and installation and engine block and cylinder head overhaul can be found in Part C of this Chapter.

The following repair procedures are based on the assumption that the engine is installed in the vehicle. If the engine has been removed from the vehicle and mounted on a stand, many of the steps outlined in this Part of Chapter 2 will not apply.

The Specifications included in this Part of Chapter 2 apply only to the procedures contained in this Part. Part C of Chapter 2 contains the Specifications necessary for cylinder head and engine block rebuilding.

2 Repair operations possible with the engine in the vehicle

Many major repair operations can be accomplished without removing the engine from the vehicle.

Clean the engine compartment and the exterior of the engine with some type of degreaser before any work is done. It will make the job easier and help keep dirt out of the internal areas of the engine.

Depending on the components involved, it may be helpful to remove the hood to improve access to the engine as repairs are performed (refer to Chapter 11 if necessary). Cover the fenders to prevent damage to the paint. Special pads are available, but an old bedspread or blanket will also work.

If vacuum, exhaust, oil or coolant leaks develop, indicating a need for gasket or seal replacement, the repairs can generally be made with the engine in the vehicle. The intake and exhaust manifold gaskets, oil pan gasket, crankshaft oil seals and cylinder head gaskets are all accessible with the engine in place.

Exterior engine components, such as the intake and exhaust manifolds, the oil pan, the oil pump, the water pump (see Chapter 3), the starter motor, the alternator and the fuel system components (see Chapter 4) can be removed for repair with the engine in place.

Since the cylinder heads can be removed without pulling the engine, valve component servicing can also be accomplished with the engine in the vehicle. Replacement of the camshafts, timing chains and sprockets are also possible with the engine in the vehicle.

In extreme cases caused by a lack of necessary equipment, repair or replacement of piston rings, pistons, connecting rods and rod bearings is possible with the engine in the vehicle. However, this practice is not recommended because of the cleaning and preparation work that must be done to the components involved.

3 Top Dead Center (TDC) for number one piston – locating

Refer to illustrations 3.5 and 3.6

1 Top Dead Center (TDC) is the highest point in the cylinder that each piston reaches as it travels up the cylinder bore. Each piston reaches TDC on the compression stroke and again on the exhaust stroke, but TDC generally refers to piston position on the compression stroke.
2 Positioning the piston(s) at TDC is an essential part of many procedures such as valve timing, camshaft and timing chain/sprocket removal.
3 Before beginning this procedure, be sure to place the transaxle in Park (automatic) or Neutral (manual) and apply the parking brake or block the rear wheels. Also, disable the ignition system by disconnecting the primary electrical connectors at the ignition coil packs, then remove the coil packs and spark plugs (see Chapter 1).
4 In order to bring any piston to TDC, the crankshaft must be turned using one of the methods outlined below. When looking at the front of the engine, normal crankshaft rotation is clockwise.

a) The preferred method is to turn the crankshaft with a socket and ratchet attached to the bolt threaded into the front of the crankshaft. Apply pressure on the bolt in a clockwise direction only. Never turn the bolt counterclockwise.
b) A remote starter switch, which may save some time, can also be used. Follow the instructions included with the switch. Once the piston is close to TDC, use a socket and ratchet as described in the previous paragraph.
c) If an assistant is available to turn the ignition switch to the Start position in short bursts, you can get the piston close to TDC without a remote starter switch. Make sure your assistant is out of the vehicle, away from the ignition switch, then use a socket and ratchet as described in Paragraph (a) to complete the procedure.

5 Install a compression pressure gauge in the number one spark plug hole (refer to Chapter 2C). It should be a gauge with a screw-in fitting and a hose at least six inches long **(see illustration)**.
6 Rotate the crankshaft using one of the methods described above while observing the compression gauge. When TDC for the compression stroke of number one cylinder is reached, compression pressure will show on the gauge as the marks are beginning to line up on the crankshaft pulley **(see illustration)**. If you go past the marks, release the gauge pressure and rotate the crankshaft around two more revolutions.

2B-4   Chapter 2 Part B Dual Overhead Cam (DOHC) engine

4.2 Remove the ignition coil cover (A) and the breather hoses (B) from the front cover

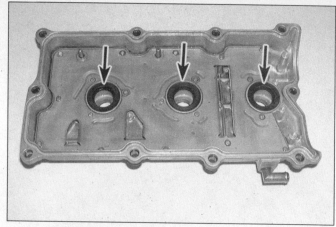

4.14 Install new spark plug tube seals (arrows) into the valve cover

4   Valve cover - removal and installation

Removal

Refer to illustration 4.2

1   Disconnect the cable from the negative terminal of the battery.

Front cover

2   Remove the breather hoses and the ignition coil cover from the valve cover **(see illustration)**.
3   Remove the ignition coils from the front valve cover (see Chapter 5).
4   Remove the valve cover bolts and washers in the reverse order of the tightening sequence **(see illustration 4.17)**.
5   Detach the valve cover. **Note:** *If the cover is stuck to the cylinder head, bump one end with a block of wood and a hammer to jar it loose. If that doesn't work, try to slip a flexible putty knife between the cylinder head and cover to break the gasket seal. Don't pry at the cover-to-cylinder head joint or damage to* the sealing surfaces may occur (leading to oil leaks in the future).

Rear cover

6   Remove the upper intake manifold (plenum) (see Section 9).
7   Remove the ignition coils from the rear valve cover (see Chapter 5).
8   Remove the breather hose by sliding the hose clamp back and pulling the hose off the fitting on the valve cover.
9   Detach PCV hose and any wiring which would interfere with valve cover removal.

Both covers

10   Remove the valve cover bolts and washers in the reverse order of the tightening sequence **(see illustration 4.17)**.
11   Detach the valve cover. **Note:** *If the cover is stuck to the cylinder head, bump one end with a block of wood and a hammer to jar it loose. If that doesn't work, try to slip a flexible putty knife between the cylinder head and cover to break the gasket seal. Don't pry at the cover-to-cylinder head joint or damage to the sealing surfaces may occur (leading to oil leaks in the future).*

Installation

Refer to illustrations 4.14 and 4.17

12   The mating surfaces of each cylinder head and valve cover must be perfectly clean when the covers are installed. Use a gasket scraper to remove all traces of sealant and old gasket material, then clean the mating surfaces with lacquer thinner or acetone. If there's sealant or oil on the mating surfaces when the cover is installed, oil leaks may develop.
13   If necessary, clean the mounting bolt threads with a die to remove any corrosion and restore damaged threads. Make sure the threaded holes in the cylinder head are clean - run a tap into them to remove corrosion and restore damaged threads.
14   Inspect and replace if necessary the spark plug tube sealing washers **(see illustration)**.
15   The valve cover gaskets should be mated to the covers before the covers are installed. Apply a thin coat of RTV sealant to the cover groove and to the corners on the front camshaft journal cap, then position the gasket inside the cover and allow the sealant

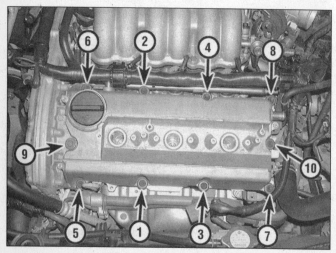

4.17 Valve cover tightening sequence

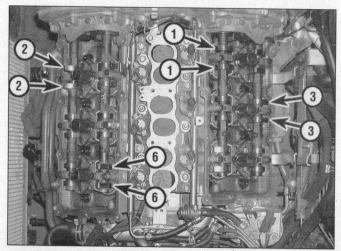

5.6a When the no. 1 piston is at TDC on the compression stroke, the valve clearance for the no. 1 and no. 6 cylinder intake valves and the no. 2 and no. 3 cylinder exhaust valves can be measured

Chapter 2 Part B Dual Overhead Cam (DOHC) engine

5.6b Measure the clearance for each valve with a feeler gauge of the specified thickness - if the clearance is correct, you should feel a slight drag on the gauge as you pull it out

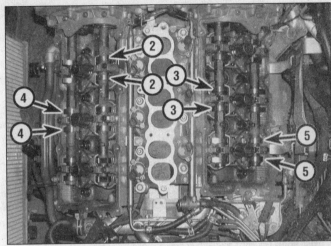

5.7 When the no. 3 piston is at TDC on the compression stroke, the valve clearance for the no. 2 and no. 3 cylinder intake valves and the no. 4 and no. 5 cylinder exhaust valves can be measured

to set up so the gasket adheres to the cover. If the sealant isn't allowed to set, the gasket may fall out of the cover as it's installed on the engine.
16 Carefully position the cover on the cylinder head and install the bolts.
17 Following the recommended tightening sequence, tighten the bolts, in two equal steps, to the torque listed in this Chapter's Specifications **(see illustration)**.
18 The remaining installation steps are the reverse of removal.
19 Start the engine and check carefully for oil leaks.

5 Valve clearance check and adjustment

Refer to illustrations 5.6a, 5.6b, 5.7, 5.8, 5.10a, 5.10b, 5.10c and 5.11
Note: *The following procedure requires the use of special valve lifter tools. The tools are available from specialty tool manufacturers and some auto parts stores. It is impossible to perform this task without them.*
1 Disconnect the cable from the negative terminal of the battery.
2 Remove the valve cover (see Section 4).
3 On manual transaxle vehicles set the parking brake and place the transaxle in the neutral position.
4 Remove the spark plugs (see Chapter 1).
5 Position the number 1 piston at TDC on the compression stroke and align the timing marks (see Section 3).
6 Measure the clearance of the indicated valves with a feeler gauge **(see illustrations)**. Record each measurement and compare your measurements with the desired valve clearance found in this Chapter's Specifications. Note which are out of specification, as this data will be used later to determine the required replacement shims.
7 Turn the crankshaft 240 degrees clockwise and position the number 3 cylinder at TDC on the compression stroke. Measure and record the clearances of the indicated valves **(see illustration)**.
8 Turn the crankshaft an additional 240 degrees clockwise and position the number 5 cylinder at TDC on the compression stroke. Measure and record the clearances of the indicated valves **(see illustration)**.
9 After the clearance of all the valves have been measured, rotate the crankshaft pulley until the camshaft lobe above the first valve which you intend to adjust is pointing up, away from the lifter.
10 Rotate the hole in the valve shim toward the center of cylinder head casting. Place the special valve lifter tool in position as shown, with the upper jaw over the camshaft, next to the lobe and the lower jaw on top of the shim **(see illustration)**. Depress the valve lifter by squeezing the handles of the valve lifter tool together and rotating the tool away from the camshaft. Insert the small tool between the edge of the lifter and the camshaft and release the lifter. Remove the adjusting shim with a small screwdriver and a magnet or a

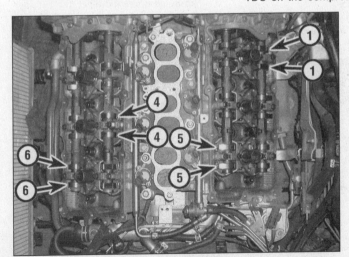

5.8 When the no. 5 piston is at TDC on the compression stroke, the valve clearance for the no. 4 and no. 5 cylinder intake valves and the no. 1 and no. 6 cylinder exhaust valves can be measured

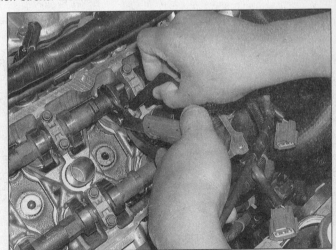

5.10a Install the valve lifter tool as shown - squeeze the handles together and rotate the tool away from the camshaft to depress the valve lifter

5.10b With the small tool wedged between the lifter and the camshaft pry the shim up with a small screwdriver at the hole in the shim . . .

5.10c . . . and remove the shim with a pair of tweezers or a magnet as shown

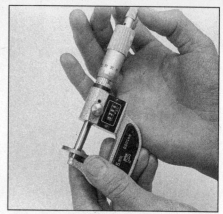

5.11 Measure the shim thickness with a micrometer or a dial caliper

pair of tweezers **(see illustrations)**. **Note:** *Blowing compressed air into the valve shim hole may help facilitate removal of the shim.*

11   Measure the thickness of the shim with a micrometer **(see illustration)**. To calculate the correct thickness of a replacement shim that will place the valve clearance within the specified value, use the following formula:

Intake side: N = R + (M − 0.0118-inch [0.30 mm])
Exhaust side: N = R + (M − 0.0130-inch [0.33 mm])
R = thickness of the old shim
M = valve clearance measured
N = thickness of the new shim

12   Select a shim with a thickness as close as possible to the valve clearance calculated. Shims are available in 64 sizes, in increments of 0.0004-inch (0.010 mm). Available shims range in sizes from 0.0913 inch (2.32 mm) to 0.1161 inch (2.95 mm). **Note:** *Through careful analysis of the shim sizes needed to bring all the out-of-specification valve clearances within specification, it is often possible to simply move a shim that has to come out anyway to another valve lifter requiring a shim of that particular size, thereby reducing the number of new shims that must be purchased.*

13   Place the special valve lifter tools in position as shown in **illustration 5.10a**, depress the valve lifter and install the new adjusting shim. Measure the clearance with a feeler gauge to make sure that your calculations are correct.

14   Repeat this procedure until all the valves which are out of specification have been corrected.

15   The remainder of installation is the reverse of removal.

6 Valve springs, retainers and seals - replacement

Refer to illustrations 6.5, 6.7a, 6.7b, 6.13 and 6.15

Note: *Broken valve springs and defective valve stem seals can be replaced without removing the cylinder heads. Two special tools and a compressed air source are normally required to perform this operation, so read through this Section carefully. The universal shaft-type valve spring compressor required for the tight valve spring pockets of this vehicle may not be available at all tool rental yards, so check on the availability before beginning the job.*

1   Remove the upper intake manifold (see Section 9) and the valve cover(s) (see Section 4).

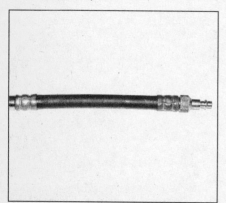

6.5 The air hose adapter threads into the spark plug hole - they're commonly available from auto parts stores

6.7a Compress the valve spring enough to release the valve stem locks . . .

2   Refer to Section 7 and remove the timing chain, then refer to Section 8 and remove the camshafts and lifters from the affected cylinder head.

3   Remove the spark plug from the cylinder that has the defective component. If all of the valve stem seals are being replaced, all of the spark plugs should be removed.

4   Turn the crankshaft until the piston in the affected cylinder is at Top Dead Center on the compression stroke (refer to Section 3). If you're replacing all of the valve stem seals, begin with cylinder number one and work on the valves for one cylinder at a time. Move from cylinder-to-cylinder following the firing order sequence (see this Chapter's Specifications).

5   Thread a long adapter into the spark plug hole and connect an air hose from a compressed air source to it **(see illustration)**. Most auto parts stores can supply the air hose adapter. **Note:** *Because of the length of the spark plug tubes, it will be necessary to use a long spark plug adapter with a length of hose attached (as used on many cylinder compression gauges) utilizing a quick-disconnect fitting to hook to your air source.*

6   Apply compressed air to the cylinder. **Warning:** *The piston may be forced down by the compressed air, causing the crankshaft to turn suddenly. If the wrench used when positioning the number one piston at TDC is still attached to the bolt in the crankshaft nose, it could cause damage or injury when the crankshaft moves.*

7   Stuff shop rags into the cylinder head holes around the valves to prevent parts and tools from falling into the engine, then use a valve spring compressor to compress the spring **(see illustrations)**. Remove the valve stem locks with small needle-nose pliers or a magnet. **Note:** *The valves should be held in place by the air pressure. If the valve faces or seats are in poor condition, leaks may prevent air pressure from retaining the valves. If the valves cannot hold air, the cylinder head should be removed for a valve job at a machine shop.*

8   Remove the spring retainer and valve spring, then remove the valve stem seal.

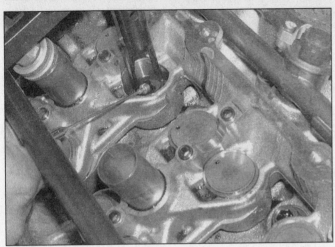

6.7b ... and lift them out with a magnet or needle-nose pliers

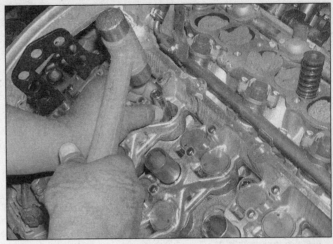

6.13 Using a deep socket and hammer, gently tap the new seals onto the valve guide only until seated

9   Wrap a rubber band or tape around the top of the valve stem so the valve won't fall into the combustion chamber, then release the air pressure.

10   Inspect the valve stem for damage. Rotate the valve in the guide and check the end for eccentric movement, which would indicate that the valve is bent.

11   Move the valve up-and-down in the guide and make sure it doesn't bind. If the valve stem binds, either the valve is bent or the guide is damaged. In either case, the cylinder head will have to be removed for repair.

12   Reapply air pressure to the cylinder to retain the valve in the closed position, then remove the tape or rubber band from the valve stem.

13   Lubricate the valve stems with engine oil and install a new valve stem seals. Valve stem seals can be installed with a special tool, or a deep socket and hammer - tap the seal only until seated (see illustration).

14   Install the valve spring in position over the valve, with the more closely-wound spring coils toward the cylinder head.

15   Install the valve spring retainer. Compress the valve springs and carefully position the valve stem locks in the groove. Apply a small dab of grease to the inside of each valve stem lock to hold it in place (see illustration).

16   Remove the pressure from the spring tool and make sure the valve stem locks are seated.

17   Disconnect the air hose and remove the adapter from the spark plug hole.

18   Refer to Section 8 and install the camshaft and lifters, then refer to Section 7 and install the timing chain.

19   Refer to Section 4 and install the valve covers.

20   Install the spark plug(s), ignition coils and the upper intake manifold referring to the appropriate sections as necessary.

21   Start and run the engine, then check for oil leaks and unusual sounds coming from the valve cover area.

7   Timing chain and sprockets - removal, inspection and installation

Caution: *The engine must be completely cool before beginning this procedure.*

Removal

Refer to illustrations 7.19a, 7.19b, 7.19c, 7.20, 7.21, 7.23, 7.24a, 7.24b, 7.24c and 7.25

1   Disconnect the cable from the negative terminal of the battery.

2   Block the rear wheels and set the parking brake.

3   Loosen the lug nuts on the right front wheel and raise the vehicle. Support the front of the vehicle securely on jackstands. Remove the right-front wheel.

4   Remove the lower splash cover and right side inner fenderwell splash shield.

5   Relieve the system fuel pressure (see Chapter 4) and drain the cooling system (see Chapter 1).

6   Remove the upper intake plenum (see Section 9) and the valve covers (see Section 4).

7   Remove the spark plugs (see Chapter 1). Position the number one piston at TDC on the compression stroke (see Section 3).

8   Remove the drivebelts (see Chapter 1) and the idler pulley bracket.

9   Remove the power steering pump and bracket (see Chapter 10).

10   Remove the air conditioning compressor (see Chapter 3).

11   Remove the crankshaft pulley (see Section 12). **Note:** *Don't allow the crankshaft to rotate during removal of the pulley. If the crankshaft moves, the number one piston will no longer be at TDC.*

12   Disconnect the oxygen sensor connectors and remove the front exhaust pipe from the exhaust manifolds and the catalytic converter (see Chapter 4).

13   Support the engine/transaxle securely from above with a hoist or a three-bar engine support brace. **Warning:** *Be absolutely certain the engine/transaxle is securely supported! DO NOT place any part of your body under the engine/transaxle - it could crush you if the support or hoist fails!*

14   Detach the front and rear engine mount through-bolts and remove the crossmember and engine mounts from the vehicle as an assembly (see Section 18).

15   Remove the air conditioning compressor bracket and the upper and lower oil pan (see Section 14).

16   Working from above remove the right side engine mount (see Section 18).

17   Remove the camshaft position sensor from the timing chain cover (see Chapter 6).

18   Detach the wiring harness from the brackets at the top of the timing chain cover and remove the bolts securing the front timing chain cover in the reverse order of the tightening sequence (see illustration 7.36b). Note that various types and sizes of bolts are used. They must be reinstalled in their original locations. Mark each bolt or make a sketch to help remember where they go.

6.15 Apply a small dab of grease to each valve stem lock as shown here before installation - it will hold them in place on the valve stem as the spring is released

7.19a Make sure the dark colored links (arrow) on the main timing chain align with the dowel pin slot on the left (front) camshaft sprocket . . .

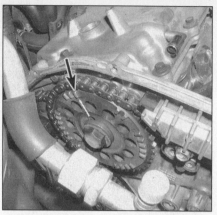

7.19b . . . and with the dowel pin slot on the right (rear) camshaft sprocket – if there are no marks, apply dabs of paint to the timing chain and the sprockets to mark them as an aid to installation

7.19c The light colored timing chain link (arrow) aligns with the mark on the crankshaft sprocket

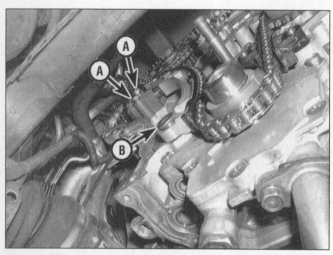

7.20 Main timing chain tensioner mounting bolts (A) and the tensioner arm/chain guide pivot bolt (B)

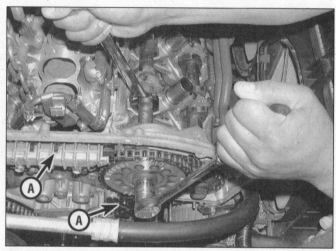

7.21 Detach the upper chain guides (A) and the camshaft sprocket bolts from the main timing chain

19 Remove the front timing chain cover and confirm that the number one piston is still at TDC on the compression stroke by verifying that the timing marks on the camshaft and crankshaft sprockets are aligned with the colored links on the chain (see illustrations).

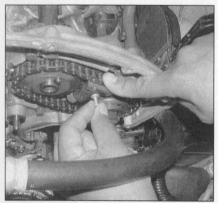

7.23 An ordinary thumb tack (push pin) can be used to lock the secondary timing chain tensioners in place

20 Relieve tension on the main timing chain by removing the timing chain tensioner and tensioner arm/chain guide (see illustration).
21 Remove the upper timing chain guides and the camshaft sprocket bolts from the main timing chain (see illustration).
22 Remove the main intake camshaft sprockets and the main timing chain from the engine. Note: *Mark the main camshaft sprockets with either an R or L. Don't mix them up. They must be installed on the same cam they were removed from.*
23 Depress the secondary timing chain tensioners and lock the tensioners in place by inserting a suitable stopper pin into the hole

Note: *The 3.0L DOHC engine utilizes three timing chains to produce proper valve timing. The main timing chain runs around the crankshaft sprocket, the water pump and around two intake camshaft sprockets. This chain synchronizes the valve timing with the crankshaft and pistons, while two secondary timing chains run around two separate intake and exhaust camshaft sprockets to synchronize the intake and exhaust camshaft events.*

on the front of each tensioner (see illustration).
24 Remove the retaining bolts from the secondary exhaust camshaft sprockets (see illustration). Confirm that the timing marks on the secondary camshaft sprockets are aligned with their respective stationary alignment marks (see illustration) and remove the secondary timing chain and sprocket assemblies from the front and rear cylinder head. Note: *The secondary intake and exhaust sprockets are different thicknesses* (see illustration), *don't mix them up. They must be installed on the same cam they were removed from.*
25 Remove the crankshaft sprocket and the lower timing chain guide (see illustration).

Inspection

Refer to illustrations 7.26a and 7.26b
26 Inspect the camshaft, water pump and crankshaft sprockets for wear on the teeth and keyways. Inspect the chains for cracks or excessive wear of the rollers. Inspect the facing of the chain guides and secondary timing

Chapter 2 Part B Dual Overhead Cam (DOHC) engine

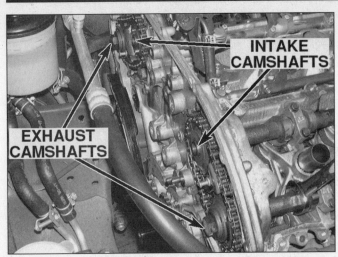

7.24a Remove the retaining bolts from the secondary exhaust camshaft sprockets

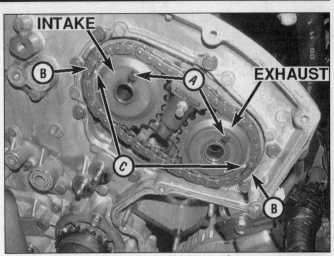

7.24b With the dowel pins (A) facing up (180 degrees from the cylinder head mating surface) and inline with the cylinder bank, the dark colored links (B) should align with the secondary camshaft sprocket marks (C) – the front (left) secondary timing chain is shown, intake and exhaust camshafts are opposite on the right (rear) secondary timing chain

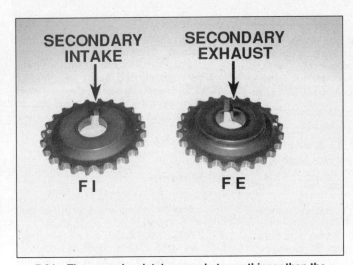

7.24c The secondary intake sprockets are thinner than the secondary exhaust sprockets – make sure to mark the sprockets FI (front intake), FE (front exhaust), RI (rear intake) and RE (rear exhaust) so they can be installed on the same camshaft from which they were removed

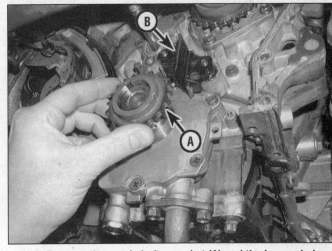

7.25 Remove the crankshaft sprocket (A) and the lower chain guide; make a note that the mark (B) on the chain guide must face up when reinstalling

chain tensioners for excessive wear (see illustration). Note: *If the secondary timing chain tensioners need to be replaced, the front camshaft journal will have to be removed from the affected cylinder head to allow access to the secondary tensioner bolts (see illustration).*

Installation

Refer to illustrations 7.36a and 7.36b

27 Install the crankshaft sprocket and the lower timing chain guide with mark facing up (see illustration 7.25).

28 Verify that you have the correct timing chains for your vehicle by counting the number of links each chain has and comparing the new chains with the old chains. Also compare the position of the colored links in the new chains with the position of the colored links in the old chains.

7.26a Examine the chain guides for deep grooves and excessive wear – replace them if necessary

7.26b Secondary timing chain tensioner(s) mounting bolts (arrows)

Chapter 2 Part B Dual Overhead Cam (DOHC) engine

7.36a Apply RTV sealant to the front timing chain cover at the areas shown – be sure to wipe off any excess sealant

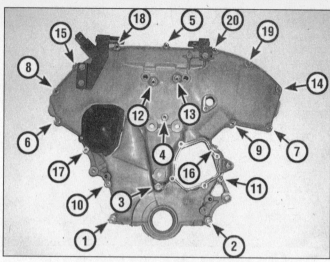

7.36b Front timing chain cover TIGHTENING sequence

29 If the secondary tensioners were removed, reinstall them and make sure the tensioner spring is locked in place. Install the secondary timing chains and sprocket assemblies on the front and rear cylinder head with the timing marks aligned as **shown in illustration 7.24b** and tighten the exhaust camshaft bolts to the torque listed in this Chapter's specifications.

30 Remove the stopper pins from the secondary chain tensioners **(see illustration 7.23)**.

31 Prepare to install the main timing chain by aligning the main intake camshaft sprocket marks with dark colored links on the chain.

32 Install the main intake camshaft sprockets and timing chain onto the engine by looping the chain around the crankshaft sprocket and aligning the light colored chain link with the mark on the crankshaft sprocket. Then place the chain around the water pump sprocket and install the camshaft sprockets onto the intake camshaft dowel pins making sure the colored links align with their respective marks on the sprockets **(see illustrations 7.19a, 7.19b and 7.19c)**.

33 Install the upper timing chain guides **(see illustration 7.21)**.

34 Install the main tensioner arm/chain guide and the timing chain tensioner assembly **(see illustration 7.20)**. Reconfirm that the number one piston is still at TDC on the compression stroke and that the timing marks on the camshaft and crankshaft sprockets are aligned with the colored links on the chain.

35 Remove all traces of old sealant from the front timing chain cover and the cover bolts.

36 Apply a bead of RTV sealant to the timing cover sealing surfaces **(see illustration)**. Place the front timing cover in position on the engine and install the bolts in their original locations. Following the recommended tightening sequence, tighten the bolts to the torque listed in this Chapter's Specifications **(see illustration)**.

37 The remainder of the installation is the reverse of removal.

8 Camshafts and lifters - removal and installation

Note: The camshafts and lifters should always be thoroughly inspected before installation and camshaft endplay should always be checked prior to camshaft removal. Refer to Chapter 2C for the camshaft and lifter inspection procedures.

Removal

Refer to illustrations 8.4, 8.5, 8.6a and 8.6b

1 Detach the cable from the negative terminal of the battery.

2 Remove the valve covers (see Section 4).

3 Remove the timing chains and camshaft sprockets (see Section 7).

4 Mark the camshaft bearing caps from 1 to 4, and with an "I" or an "E", to indicate intake or exhaust. Also mark arrows indicating the front of the engine **(see illustration)**. Loosen the camshaft bearing caps in two or three steps, in the reverse order of the tightening sequence **(see illustration 8.13)**. **Caution:** *Keep the caps in order. They must go back in the same location they were removed from.*

5 Remove the bearing caps and camshafts. Make a note of the camshaft markings to ensure correct installation **(see illustration)**.

6 Remove the lifters and shims from the cylinder head, keeping the proper shim with each lifter **(see illustrations)**. **Caution:** *Keep the lifters and shims in order. They must go back in the same location they were removed from.*

7 Inspect the camshaft and lifters as described in Chapter 2C.

Installation

Refer to illustrations 8.10, 8.11 and 8.13

8 Install the lifters and shims into their original locations.

9 Apply moly-based engine assembly lubricant to the camshaft lobes and journals.

10 Install the camshafts in their original positions with the dowel pins facing up (180 degrees from the cylinder head mating surface) and inline with the cylinder bank **(see illustration)**.

11 Apply a bead of RTV sealant to the seal-

8.4 The camshaft bearing caps should be marked with a number and letter stamp or a marker to ensure correct reinstallation

Chapter 2 Part B Dual Overhead Cam (DOHC) engine

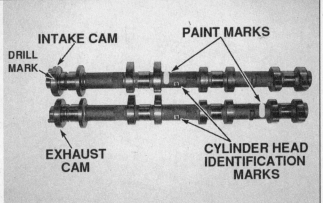

8.5 The ID mark in the center of each camshaft identifies which cylinder head the camshaft belongs to "L" for left (front) and "R" for right (rear) – paint marks between the No. 2 and No. 3 journals indicate that it is an intake camshaft, while paint marks between the No. 3 and No. 4 journals indicate that it is an exhaust camshaft

ing surfaces of the No. 1 bearing cap(s) **(see illustration)**.

12 Install the bearing caps and bolts and tighten them hand tight.

13 Tighten the bearing cap bolts in several steps, to the torque listed in this Chapter's Specifications, using the proper tightening sequence **(see illustration)**.

14 Install the camshaft sprockets and timing chain (see Section 7). Hold the camshafts with a suitable wrench as you tighten the sprocket bolts to the specified torque.

15 The remainder of installation is the reverse of removal. If any part of the valve train was replaced, check and adjust the valve clearance (see Section 5).

9 Intake manifold - removal and installation

Upper intake manifold (plenum)

Refer to illustrations 9.4, 9.5a, 9.5b, 9.8a and 9.8b

1 Relieve the fuel pressure (see Chapter 4).

2 Disconnect the cable from the negative terminal of the battery.

8.6b The lifters and shims can be stored in individually marked plastic bags or a divided box as shown

8.6a Pull straight up to remove each lifter and shim

8.10 Install the camshafts with the dowel pins facing up (180 degrees from the cylinder head mating surface) and inline with the cylinder bank

8.11 Apply RTV sealant to the rear timing chain cover and the cylinder head at the areas shown - be sure to wipe off any excess sealant

8.13 Camshaft bearing cap TIGHTENING sequence

2B-12 Chapter 2 Part B Dual Overhead Cam (DOHC) engine

3 Refer to Chapter 4 and remove the air intake duct.

4 Remove the ignition coils (see Chapter 5). Label and disconnect the hoses and electrical connectors attached to the plenum and throttle body **(see illustration)**.

5 Remove the EGR guide tube and the upper manifold support brackets from the rear of the upper plenum **(see illustrations)**. **Note:** *The EGR guide tube and the manifold support bracket bolts are difficult to access.*

6 Loosen the upper intake manifold bolts in the reverse order of the tightening sequence **(see illustration 9.8b)** and remove the upper plenum with the throttle body attached.

7 To install the upper manifold, clean the mounting surfaces of the intake manifold and the upper plenum with lacquer thinner and remove all traces of the old gasket material or sealant.

8 Install the new gasket over the intake manifold studs with the mark facing forward **(see illustration)**, then install the upper plenum onto the lower intake manifold and tighten the bolts in the recommended tightening sequence **(see illustration)** to the torque listed in this Chapter's Specifications. The remainder the installation is the reverse of removal.

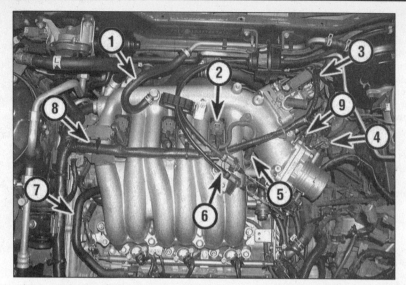

9.4 Label and disconnect the following components required for upper intake manifold removal (1995 through 1999 model shown)

1 Power brake booster hose
2 Ignition coil(s)
3 IAC/fast idle solenoid connectors
4 Throttle position sensor/closed throttle position switch connectors
5 PCV valve hose
6 Accelerator/cruise control cables
7 Crankcase breather hose
8 Wiring harness
9 Coolant hose

9.5a The EGR guide tube (A) is attached to the rear of the upper intake manifold by two bolts (B) (upper manifold removed for clarity)

9.5b Upper intake manifold support brackets (arrows) – access from above is difficult so be patient!

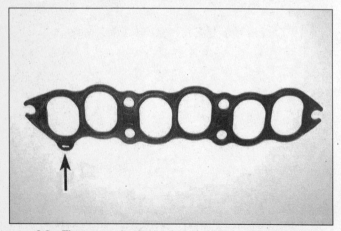

9.8a The upper plenum gasket must be positioned with the mark (arrow) facing the timing chain end of the engine

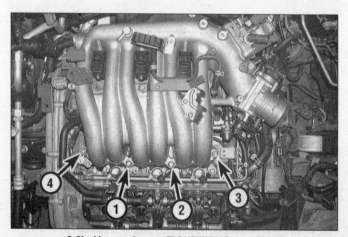

9.8b Upper plenum TIGHTENING sequence

Chapter 2 Part B Dual Overhead Cam (DOHC) engine

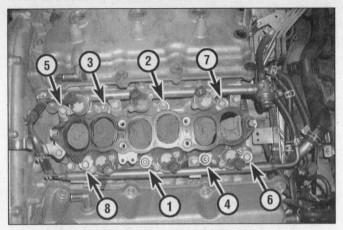

9.16 Lower intake manifold TIGHTENING sequence

10.5 Front exhaust manifold heat shield mounting bolts (arrows) - rear exhaust manifold heat shield similar

Lower intake manifold

Refer to illustration 9.16

9 Remove the upper intake manifold (see Steps 1 through 6).
10 Label and detach any remaining hoses which would interfere with the removal of the lower intake manifold.
11 Refer to Chapter 4 and remove the fuel rail and injectors from the lower intake manifold.
12 Loosen the manifold mounting bolts/nuts in 1/4-turn increments until they can be removed by hand In the reverse order of the tightening sequence **(see illustration 9.16)**.
13 The manifold will probably be stuck to the cylinder heads and force may be required to break the gasket seal. **Caution:** *Don't pry between the manifold and the heads or damage to the gasket sealing surfaces may occur, leading to vacuum leaks.*
14 Carefully use a scraper to remove all traces of old gasket material and sealant from the manifold and cylinder heads, then clean the mating surfaces with lacquer thinner or acetone.
15 Install new gaskets, then position the lower manifold on the engine. Make sure the gaskets and manifolds are aligned over the studs in the cylinder heads and install the nuts.
16 Following the recommended tightening sequence, tighten the nuts/bolts, in several steps, to the torque listed in this Chapter's Specifications **(see illustration)**.
17 The remainder of the installation is the reverse of the removal procedure. Run the engine and check for fuel, vacuum and coolant leaks.

10 Exhaust manifold - removal and installation

Warning: *The engine must be completely cool before beginning this procedure.*

Removal

Refer to illustrations 10.5, 10.9 and 10.11

1 Disconnect the cable from the negative terminal of the battery.
2 Block the rear wheels, set the parking brake.
3 Raise the front of the vehicle and support it securely on jackstands and remove the lower splash cover from below the engine.
4 If the rear exhaust manifold is to be removed refer to Section 9 and remove the upper intake manifold.
5 Remove the heat shield from the manifold(s) **(see illustration)**.
6 Support the engine/transaxle securely from above with a hoist or a three bar engine support brace. **Warning:** *Be absolutely certain the engine/transaxle is securely supported! DO NOT place any part of your body under the engine/transaxle - it could crush you if the support or hoist fails!*
7 Working underneath the vehicle disconnect the oxygen sensor connectors and remove the front exhaust pipe from the vehicle (see Chapter 4).
8 Detach the front and rear engine mount through bolts and remove the crossmember and engine mounts from the vehicle as an assembly (see Section 18).
9 Remove the engine mount support brackets from each side of the engine block **(see illustration)**.
10 Disconnect the EGR tube from the rear exhaust manifold.
11 Remove the manifold-to-head nuts/bolts and detach the manifold and gaskets **(see illustration)**.

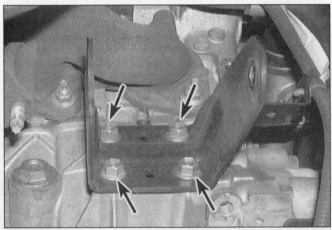

10.9 Remove the bolts (arrows) from the engine mount support brackets on each side of the engine block

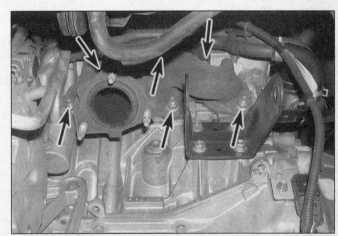

10.11 Exhaust manifold retaining nuts (arrows) (front exhaust manifold shown, rear exhaust manifold similar)

Chapter 2 Part B Dual Overhead Cam (DOHC) engine

11.6 Remove the coolant tube retaining nuts (arrows) from the end of each cylinder head

11.8 Pry on a casting protrusion to break the head loose

Installation

12 Use a scraper to remove all traces of old gasket material and carbon deposits from the manifold and cylinder head mating surfaces.
13 Position the new exhaust manifold gaskets over the cylinder head studs.
14 Install the manifold and thread the mounting nuts/bolts into place.
15 Working from the center out, tighten the nuts/bolts to the torque listed in this Chapter's Specifications in three or four equal steps.
16 Reinstall the remaining parts in the reverse order of removal.
17 Run the engine and check for exhaust leaks.

11 Cylinder head - removal and installation

Caution: *The engine must be completely cool before beginning this procedure.*

Removal

Refer to illustrations 11.6 and 11.8

1 Refer to Section 7, Steps 1 through 24 and remove the timing chain(s) and sprockets. **Caution:** *Be careful not to disturb the crankshaft from TDC on the compression stroke of the No.1 cylinder during the remainder of this procedure.*
2 Remove the rear timing chain cover bolts in the reverse order of the tightening sequence **(see illustration 11.22c)**.
3 Detach the rear timing cover from the engine. **Note:** *If the cover is stuck to the cylinder head or engine block, bump one end with a block of wood and a hammer to jar it loose. If that doesn't work, try to slip a flexible putty knife between the cover and the engine to break the gasket seal. Don't pry at the cover-to-cylinder head joint or damage to the sealing surfaces may occur (leading to oil leaks in the future.*
4 Remove the lower intake manifold (see Section 9).
5 Remove the camshafts and lifters from the cylinder head (see Section 8).
6 Label and remove any remaining items attached to the cylinder head, such as coolant fittings, tubes, cables, hoses, wires or brackets **(see illustration)**.
7 Using a breaker bar and the appropriate sized Allen-head socket, loosen the cylinder head bolts in 1/4-turn increments until they can be removed by hand. Loosen the bolts in the reverse order of the tightening sequence **(see illustration 11.20)** to avoid warping or cracking the head.
8 Lift the cylinder head off the engine block. If it's stuck, very carefully pry up at the transaxle end, beyond the gasket surface, at a casting protrusion **(see illustration)**.

9 Remove all external components from the head to allow for thorough cleaning and inspection. **Note:** *See Chapter 2, Part C, for cylinder head inspection and servicing procedures.*

Installation

Refer to illustrations 11.11, 11.14, 11.20, 11.22a, 11.22b and 11.22c

10 The mating surfaces of the cylinder head and block must be perfectly clean when the head is installed.
11 Use a gasket scraper to remove all traces of carbon and old gasket material from the cylinder head and engine block, then clean the mating surfaces with lacquer thinner or acetone **(see illustration)**. If there's oil on the mating surfaces when the head is installed, the gasket may not seal correctly and leaks could develop. When working on the block, stuff the cylinders with clean shop rags to keep out debris. Use a vacuum cleaner to remove material that falls into the cylinders.
12 Check the block and head mating surfaces for nicks, deep scratches and other damage. If damage is slight, it can be removed with a file; if it's excessive, machining may be the only alternative.
13 Use a tap of the correct size to chase the threads in the head bolt holes, then clean the holes with compressed air - make sure that nothing remains in the holes. **Warning:** *Wear eye protection when using compressed air!*
14 Measure each cylinder head bolt for

11.11 Carefully remove all traces of old gasket material from the sealing surfaces

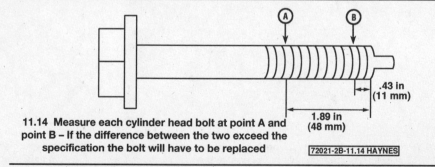

11.14 Measure each cylinder head bolt at point A and point B – If the difference between the two exceed the specification the bolt will have to be replaced

Chapter 2 Part B Dual Overhead Cam (DOHC) engine

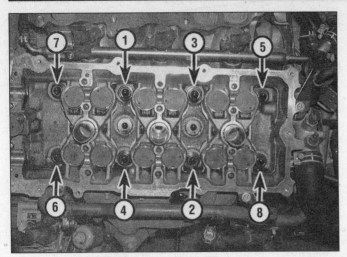

11.20 Cylinder head TIGHTENING sequence

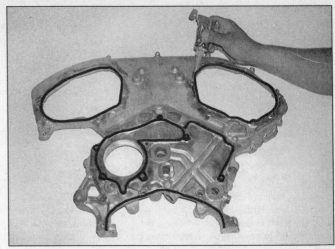

11.22a Apply RTV sealant to the rear timing chain cover at the areas shown – be sure to wipe off any excess sealant

stretching **(see illustration)**. If the diameter of the bolt threads at point A and the diameter of the bolt threads at point B differ more than 0.0043 inch (0.11mm), the bolts have exceeded the maximum amount of stretch and will need to be replaced.
15 Check the cylinder head for warpage (see Chapter 2C): Check the head gasket, intake and exhaust manifold surfaces.
16 Install the components that were removed from the head.
17 Position the new cylinder head gasket over the dowel pins on the block noting which direction on the gasket faces up.
18 Carefully set the head on the block without disturbing the gasket.
19 Before installing the head bolts, apply a small amount of clean engine oil to the threads and hardened washers (if equipped). The chamfered side of the washers must face the bolt heads.
20 Install the bolts in their original locations and tighten them finger tight. Then tighten all the bolts in several steps, following the proper sequence **(see illustration)**, to the torque listed in this Chapter's Specifications.
21 Remove all traces of old sealant from the rear timing chain cover and the cover bolts.
22 Apply a bead of RTV sealant to the rear timing cover sealing surfaces **(see illustration)**. Install new O-rings in the front of the engine block **(see illustration)**. Place the rear timing chain cover in position over the dowels on the engine and install the bolts in their original locations. Following the recommended tightening sequence, tighten the bolts to the torque listed in this Chapter's Specifications **(see illustration)**.
23 Install the camshafts as described in Section 8, then install the timing chains and sprockets as described in Section 7. The remaining installation steps are the reverse of removal.
24 If any part of the valve train was replaced, check and adjust the valve clearance (see Section 5).
25 Refill the cooling system, and check the engine oil adding if necessary (see Chapter 1).
26 Start the engine and check for oil and coolant leaks.

12 Crankshaft pulley – removal and installation

Refer to illustration 12.7

1 Disconnect the cable from the negative terminal of the battery.
2 Block the rear wheels and set the parking brake.
3 Loosen the lug nuts on the right front wheel and raise the vehicle. Support the front of the vehicle securely on jackstands. Remove the right-front wheel.
4 Remove the lower splash cover and the right side inner fenderwell splash shield.
5 Remove the drivebelts (see Chapter 1).
6 Remove the crankshaft position sensor from the front timing cover (see Chapter 6).
7 Use a strap wrench around the crankshaft pulley to hold it while using a breaker bar and socket to remove the crankshaft pulley center bolt **(see illustration)**.

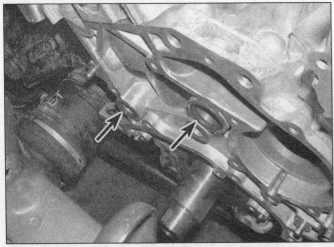

11.22b Install new O-rings (arrows) in the front of the engine block

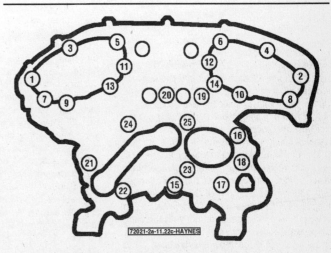

11.22c Rear timing chain cover TIGHTENING sequence

12.7 Use strap wrench to hold the crankshaft pulley while removing the center bolt (a chain-type wrench may be used if you wrap a section of old drivebelt around the crankshaft pulley first)

13.2 Pry the seal out very carefully with a seal removal tool or screwdriver, being careful not to nick or gouge the seal bore or the crankshaft

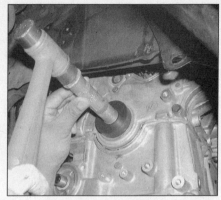

13.4 Use a large socket, seal driver or large-diameter pipe to drive the new seal into the cover

8 Wedge a prybar or two screwdrivers behind the crankshaft pulley and carefully pry it off the crankshaft. If the pulley is difficult to remove, place a three jaw type puller around the center hub and pull it off. **Caution:** *DO NOT place the puller jaws around the outside of the crankshaft pulley or damage to the pulley will occur.*

9 To install the crankshaft pulley align the pulley groove with the key on the crankshaft and slide the pulley onto the crankshaft.

10 Install the crankshaft pulley retaining bolt and tighten it to the torque listed in this Chapter's Specifications.

11 The remainder of installation is the reverse of the removal.

13 Crankshaft front oil seal - replacement

Refer to illustrations 13.2, 13.4 and 13.5

1 Remove the crankshaft pulley from the engine (see Section 12).

2 Carefully pry the seal out of the cover with a seal removal tool or a large screwdriver **(see illustration). Caution:** *Be careful not to scratch, gouge or distort the area that the seal fits into or an oil leak will develop.*

3 Clean the bore to remove any old seal material and corrosion. Position the new seal in the bore with the seal lip (usually the side with the spring) facing IN (toward the engine). A small amount of oil applied to the outer edge of the new seal will make installation easier.

4 Drive the seal into the bore with a large socket and hammer until it's completely seated **(see illustration)**. Select a socket that's the same outside diameter as the seal and make sure the new seal is pressed into place until it bottoms against the cover flange.

5 Check the surface of the damper that the oil seal rides on. If the surface has been grooved from long-time contact with the seal, a press-on sleeve may be available to renew the sealing surface **(see illustration)**. This sleeve is pressed into place with a hammer and a block of wood and is commonly available from most auto parts stores.

6 Lubricate the seal lips with engine oil and reinstall the crankshaft pulley. Install the crankshaft pulley retaining bolt and tighten it to the torque listed in this Chapter's Specifications.

7 The remainder of installation is the reverse of the removal. Run the engine and check for oil leaks.

14 Oil pan - removal and installation

Removal

Refer to illustrations 14.8, 14.10, 14.11 and 14.13

1 Disconnect the cable from the negative terminal of the battery.

2 Set the parking brake and block the rear wheels.

3 Raise the front of the vehicle and support it securely on jackstands.

4 Remove the splash shields under the engine and the right side cover.

5 Disconnect the front exhaust pipe from the vehicle (see Chapter 4).

6 Drain the engine oil (see Chapter 1).

7 The oil pan is a two piece design. A lower steel pan is attached to the upper aluminum section of the oil pan which is bolted to the engine block. Remove the lower steel pan from the upper aluminum section of the oil pan **(see illustration 14.21). Caution:** *Do not pry between the steel pan and the aluminum flange or damage to the sealing surface may result.*

8 Remove the air conditioning compressor (see Chapter 3) and set it aside without disconnecting the refrigerant lines. Remove the air conditioning compressor bracket-to-oil pan brace **(see illustration)**.

9 Support the engine/transaxle securely from above with a hoist or a three-bar engine support fixture. **Warning:** *Be absolutely certain the engine/transaxle is securely supported! DO NOT place any part of your body under the engine/transaxle - it could crush you if the support or hoist fails!*

10 Detach the front and rear engine mount

13.5 If the sealing surface of the damper hub has a wear groove from contact with the seal, repair sleeves are available at most auto parts stores

through bolts and remove the crossmember and engine mounts from the vehicle as an assembly **(see illustration)**. Remove the oil filter (see Chapter 1).

11 Remove the rear cover plate and the transaxle mounting bolts from the upper aluminum section of the oil pan **(see illustration)**.

12 Remove the bolts attaching the upper aluminum section of the oil pan to the block, working from the ends toward the center in the reverse order of the tightening sequence to prevent warpage.

13 To loosen the aluminum section of the oil pan, wedge a flathead screwdriver or pry bar into the notch on the front side of the oil pan being careful not to damage the sealing surfaces of the oil pan and engine block **(see illustration)**.

Installation

Refer to illustrations 14.17, 14.18 and 14.21

14 Use a scraper to remove all traces of old gasket material and sealant from the upper aluminum section of the oil pan, the lower steel pan and the engine block. Clean the mating surfaces with lacquer thinner or acetone. **Caution:** *Be careful not to scratch or gouge the gasket surface of the block or oil pan. A leak could develop after the repairs have been completed.*

15 Make sure the threaded bolt holes in the

Chapter 2 Part B Dual Overhead Cam (DOHC) engine

14.8 Air conditioning compressor mounting bracket retaining bolts (arrows)

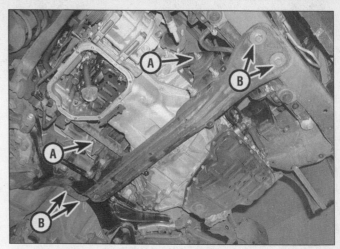

14.10 With the engine supported from above, remove the engine mount through bolts (A) and the crossmember retaining bolts (B)

14.11 Remove the rear cover plate and the transaxle mounting bolts (arrows)

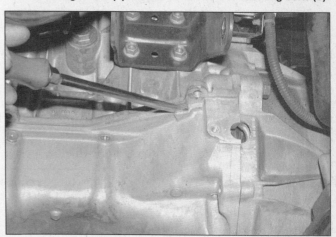

14.13 Insert a flathead screwdriver or small pry bar into the notch on the front side of the oil pan to break it loose - be careful not to damage the sealing surfaces!

block and aluminum section of the oil pan are clean.

16 Apply a bead of RTV sealant to the ends of the front timing cover gasket and the rear oil seal retainer gasket, then place the gaskets in position on the oil pan. Apply a bead of RTV sealant around the upper aluminum oil pan flange. **Note:** *The oil pan must be installed within 15 minutes once the sealant has been applied.*

17 Install new O-rings in the engine block and the oil pump body **(see illustration)**.

18 Carefully position the upper aluminum section of the oil pan on the engine block and install the bolts. Following the recommended sequence, tighten the fasteners in three or four steps to the torque listed in this Chapter's Specifications **(see illustration)**.

14.17 Install new O-rings in the block and the oil pump housing (arrows)

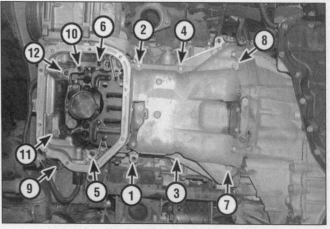

14.18 Aluminum oil pan TIGHTENING sequence

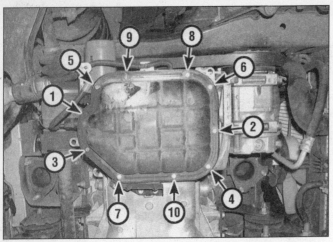

14.21 Steel oil pan TIGHTENING sequence

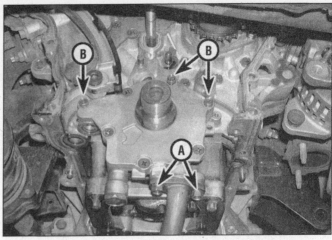

15.3 Oil pump pick up tube mounting bolts (A) and the oil pump housing retaining bolts (B)

15.5 Remove the screws (arrows) and lift the cover off

19 Install the rear cover plate and the transaxle mounting bolts (see illustration 14.11).
20 Check the lower steel oil pan flange for distortion, particularly around the bolt holes. If necessary, place the pan on a wood block and use a hammer to flatten and restore the gasket surface.
21 Apply a bead of RTV sealant around the steel oil pan flange and install the steel oil pan. **Note:** *The oil pan must be installed within 15 minutes once the sealant has been applied.* Following the recommended sequence, tighten the fasteners in several steps to the torque listed in this Chapter's Specifications **(see illustration)**.
22 The remainder of installation is the reverse of removal. Be sure to install a new oil filter (see Chapter 1) and wait at least thirty minutes before adding oil.

15 Oil pump - removal, inspection and installation

Removal

Refer to illustration 15.3

1 Refer to Section 7 and remove the main timing chain and the crankshaft sprocket. **Note:** *It is not necessary to remove the secondary timing chains or the main timing chain tensioner/chain guide assembly during this procedure. Simply remove the tensioner and pivot the chain guide over to the left side to allow removal of the oil pump housing.*
2 Remove the oil pans (see Section 14).

Remove the oil pump pick-up tube.
3 Remove the oil pump-to-engine block bolts from the front of the engine **(see illustration)**.
4 Gently pry the oil pump housing outward enough to clear the dowel pins on the engine block and remove it from the engine.

Inspection

Refer to illustrations 15.5, 15.7, 15.8a, 15.8b, 15.8c, 15.8d and 15.8e

5 Use a large Phillips screwdriver to remove the screws holding the front cover on the oil pump housing **(see illustration)**.
6 Clean all components with solvent, then inspect them for wear and damage.
7 Remove the oil pressure regulator cap, washer, springs and valve **(see illustration)**. Check the oil pressure regulator valve sliding surface and valve spring. If either the spring or the valve is damaged, they must be replaced as a set.
8 Check the clearance of the following oil pump components with a feeler gauge **(see illustrations)** and compare the measurements to the clearance listed in this Chapter's Specifications:

a) *Rotor tooth tip clearance*
b) *Outer rotor-to-body clearance*

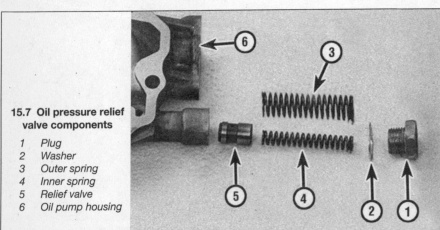

15.7 Oil pressure relief valve components

1 Plug
2 Washer
3 Outer spring
4 Inner spring
5 Relief valve
6 Oil pump housing

15.8a Use feeler gauges to measure the rotor tooth tip clearance . . .

Installation

Refer to illustration 15.10

10 Use new gaskets (where applicable) on all disassembled parts and reverse the removal procedure for installation. Align the flats on the crankshaft (see illustration) with the flats on the oil pump gear. Tighten all fasteners to the torque listed in this Chapter's Specifications. **Note:** *Before installing the oil pan, be sure to replace the O-rings on the oil pump housing and engine block* **(see illustration 14.17).**

16 Flywheel/driveplate - removal and installation

Refer to illustration 16.4

1 Raise the vehicle and support it securely on jackstands, then refer to Chapter 7 and remove the transaxle. **Warning:** *The engine must be supported from above with an engine hoist or three-bar support fixture before working underneath the vehicle with the transaxle removed.*

2 If the vehicle is equipped with a manual transaxle, remove the pressure plate and clutch disc (see Chapter 8). Now is a good time to check/replace the clutch components and pilot bearing if necessary. If the vehicle is equipped with an automatic transaxle, now would be a good time to check and replace the front pump seal/O-ring.

3 Use paint or a center-punch to make alignment marks on the flywheel/driveplate and crankshaft to ensure correct alignment during reinstallation.

4 Remove the bolts that secure the flywheel/driveplate to the crankshaft (see illustration). If the crankshaft turns, hold the flywheel with a pry bar or wedge a screwdriver into the ring gear teeth to jam the flywheel.

5 Remove the flywheel/driveplate from the crankshaft. Since the flywheel is fairly heavy, be sure to support it while removing the last bolt.

15.8b ... and the outer rotor-to-body clearance

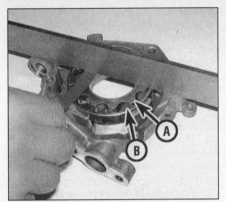

15.8c Measure the cover-to-rotor end clearance with a straightedge and feeler gauge - measure (A) above the inner rotor and (B) above the outer rotor

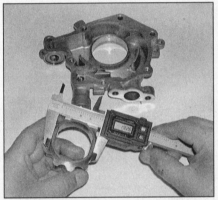

15.8d Use calipers to measure the diameter of the inner rotor ridge (the part of the inner rotor that rides in the pump body) ...

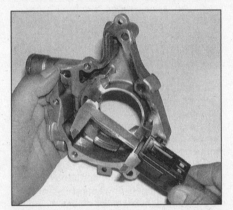

15.8e ... and subtract the inner rotor ridge diameter from the opening in the pump body where the inner rotor rides to obtain the inner rotor ridge-to-body clearance

  c) Cover-to-inner rotor clearance
  d) Cover-to-outer rotor clearance
  e) Inner rotor ridge clearance

If any clearance is excessive, replace the entire oil pump assembly.

9 **Note:** *Pack the pump with petroleum jelly to prime it.* Assemble the oil pump and tighten the screws securely. Install the oil pressure regulator valve, spring and washer, then tighten the oil pressure regulator valve cap.

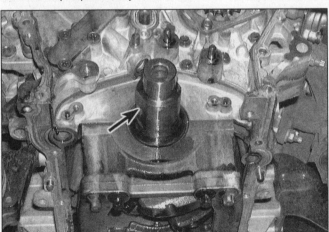

15.10 There is a flat surface (arrow) on each side of the crankshaft - align them with the flats on the inner gear

16.4 Hold a lever against a casting protrusion on the engine block or place a screwdriver through a hole in the driveplate to hold the driveplate while the mounting bolts are removed - note the painted marks made at the crank and driveplate for alignment

6 Clean the flywheel to remove grease and oil. Inspect the surface for cracks, rivet grooves, burned areas and score marks. Light scoring can be removed with emery cloth. Check for cracked and broken ring gear teeth or a loose ring gear. Lay the flywheel on a flat surface and use a straightedge to check for warpage.

7 Clean and inspect the mating surfaces of the flywheel/driveplate and the crankshaft. If the crankshaft rear seal is leaking, replace it before reinstalling the flywheel/driveplate.

8 Position the flywheel/driveplate against the crankshaft. Be sure to align the marks made during removal. Note that some engines have an alignment dowel or staggered bolt holes to ensure correct installation. Before installing the bolts, apply thread locking compound to the threads.

9 Wedge a screwdriver into the ring gear teeth to keep the flywheel/driveplate from turning as you tighten the bolts to the torque listed in this Chapter's Specifications.

10 The remainder of installation is the reverse of the removal.

17 Rear main oil seal - replacement

Refer to illustrations 17.2 and 17.3

1 The transaxle must be removed from the vehicle for this procedure (see Chapter 7). **Warning:** *The engine must be supported from above with an engine hoist or three-bar support fixture before working underneath the vehicle with the transaxle removed.* Remove the flywheel/driveplate (see Section 16).

2 Carefully pry the old seal out of the retainer with a seal removal tool or screwdriver **(see illustration)**.

3 Apply multi-purpose grease to the crankshaft seal journal and the lip of the new seal. Preferably, a seal installation tool should be used to press the new seal into place. If the proper seal installation tool is unavailable, use a large socket, section of pipe or a blunt tool and carefully drive the new seal into place **(see illustration)**. The lip is stiff so carefully work it onto the seal journal of the crankshaft. Don't rush it or you may damage the seal. **Note:** *Install the seal squarely and only until flush with the back of the seal plate, no further.*

4 The remaining steps are the reverse of removal.

18 Powertrain mounts - check and replacement

1 There are four powertrain mounts; front and rear mounts located at the center of the vehicle attached to the engine block and to the lower crossmember, a right side mount attached to the front of the timing cover and the frame and a left side mount attached to the transaxle and the frame.

Check

2 During the check, the engine must be raised slightly to remove the weight from the mounts.

3 Raise the vehicle and support it securely on jackstands. Support the engine/transaxle from above using a hoist or an engine support fixture.

4 Check the mounts to see if the rubber is cracked, hardened or separated from the bushing in the center of the mount.

5 Check for relative movement between the mounts and the engine or frame (use a large screwdriver or prybar to attempt to move the mounts). If movement is noted, lower the engine and tighten the mount fasteners.

6 Rubber preservative should be applied to the mounts to slow deterioration.

17.2 Pry the seal out very carefully with a seal removal tool or screwdriver - if the crankshaft is damaged the new seal will leak!

Replacement

Refer to illustrations 18.9, 18.14 and 18.18

7 Disconnect the cable from the negative terminal of the battery, then set the parking brake, block the rear wheels, raise the front of the vehicle and support it securely on jackstands. Remove the splash shields from under the vehicle.

Front and rear engine mounts

8 Position a floor jack under the transaxle case, close to the crossmember. Place a wood block between the jack head and the transaxle and raise the jack just enough to support the weight of the powertrain.

9 Remove the through-bolts from the front and rear mounts where they attach to the crossmember **(see illustration)**.

10 Remove the crossmember-to-chassis bolts and detach the crossmember from the vehicle.

11 Remove the mount-to-engine bracket

17.3 If you don't have a seal installation tool, use a blunt tool (such as a brass punch) to carefully work the edge of the seal evenly into the bore and around the crankshaft

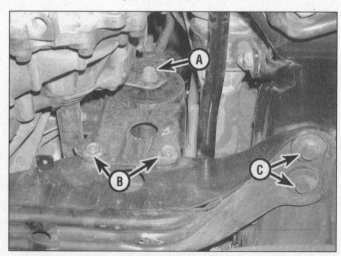

18.9 Engine mount-to-engine bracket through-bolt (A), engine mount-to-crossmember through-bolts (B) and crossmember-to-frame bolts (C) - front engine mount shown, rear engine mount similar

Chapter 2 Part B Dual Overhead Cam (DOHC) engine

18.14 To remove the right side mount, remove the upper mount bracket bolts (A), the engine mount through bolt (B) and the engine mount-to-lower bracket retaining nuts (C) (not visible in this photo)

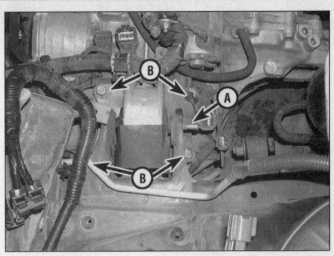

18.18 With the air cleaner housing and air intake duct removed, there is access to the left side mount through-bolt (A) and the engine mount-to-transaxle bolts (B)

through-bolts and remove the mounts **(see illustration 18.9)**.

12 Installation is the reverse of removal.
Note: *Tighten the bolts to Specifications only after the engine weight is back onto the mounts and the jack is removed. If more than one mount has been replaced, see Final tightening (Step 21).*

Right side engine mount

13 Support the engine from above using a hoist or an engine support fixture.
14 Remove the upper engine mount bracket and the engine mount through-bolt **(see illustration)**.
15 Raise the engine slightly higher, remove the nuts securing the mount to the lower engine mount bracket and remove the mount from the vehicle.
16 Installation is the reverse of removal.
Note: *Tighten the bolts to Specifications only after the powertrain weight is back onto the mounts and the jack is removed. If more than one mount has been replaced, see Final tightening (Step 21).*

Left side transaxle mount

17 Position a floor jack under the transaxle housing (not the fluid pan on automatic transaxle models). Place a wood block between the jack head and the transaxle and raise the jack just enough to support the weight of the transaxle.
18 Working from above, remove the air cleaner housing and air intake duct (see Chapter 5). Remove the engine mount through-bolt at the chassis bracket **(see illustration)**.
19 Lower the transaxle slightly, remove the transaxle-to-mount bolts and remove the mount.
20 Installation is the reverse of removal.

Note: *Tighten the bolts to Specifications only after the powertrain weight is back onto the mounts and the jack is removed. If more than one mount has been replaced, see Final tightening below.*

Final tightening

21 To ensure maximum bushing life and prevent excessive noise and vibration, the vehicle should be level and the powertrain weight should be on the mounts during the final tightening stage. **Note:** *Use thread-locking compound on the nuts/bolts.* Ensure that the bushings are not twisted or offset. If you have replaced more than one mount, or when you are installing the engine, tighten the mounts in the following order: crossmember bolts, rear engine mount, front engine mount, left-side transaxle mount and right side engine mount.

Notes

Chapter 2 Part C
General engine overhaul procedures

Contents

	Section		Section
Camshafts, lifters and bearings - inspection	21	Engine removal - methods and precautions	6
Crankshaft - inspection	19	General information - engine overhaul	1
Crankshaft - installation and main bearing oil clearance check	24	Initial start-up and break-in after overhaul	27
Crankshaft - removal	14	Main and connecting rod bearings - inspection and main bearing selection	20
Cylinder compression check	3	Oil pressure check	2
Cylinder head - cleaning and inspection	10	Piston rings - installation	23
Cylinder head - disassembly	9	Pistons/connecting rods - inspection	18
Cylinder head - reassembly	12	Pistons/connecting rods - installation and rod bearing oil clearance check	26
Cylinder honing	17	Pistons/connecting rods - removal	13
Engine - removal and installation	7	Rear main oil seal installation	25
Engine block - cleaning	15	Vacuum gauge diagnostic checks	4
Engine block - inspection	16	Valves - servicing	11
Engine overhaul - disassembly sequence	8		
Engine overhaul - reassembly sequence	22		
Engine rebuilding alternatives	5		

Specifications

General

Bore
  SOHC engine .... 3.43 inches (87 mm)
  DOHC engine .... 3.66 inches (93 mm)
Stroke
  SOHC engine .... 3.27 inches (83 mm)
  DOHC engine .... 2.886 inches (73.3 mm)
Displacement
  SOHC engine .... 180.6 cubic inches (3.0 liters)
  DOHC engine .... 182.3 cubic inches (3.0 liters)
Oil pressure
  SOHC engine
    At idle .... More than 9 psi (59 kPa)
    At 3200 rpm .... 53 to 65 psi (363 to 451 kPa)
  DOHC engine
    1995 through 1998
      At idle .... More than 10 psi (69 kPa)
      At 3000 rpm .... 63 to 79 psi (435 to 551 kPa)
    1999 and later
      At idle .... More than 10 psi (69 kPa)
      At 2000 rpm .... 56.6 psi (390 kPa)

Chapter 2 Part C General engine overhaul procedures

General (continued)

Cylinder compression pressure (at 300 rpm)
  SOHC engine
    Standard .... 173 psi (1,196 kPa)
    Minimum .... 128 psi (883 kPa)
    Maximum difference between cylinders .... 14 psi (98 kPa)
  DOHC engine
    Standard .... 185 psi (1,275 kPa)
    Minimum .... 142 psi (981 kPa)
    Maximum difference between cylinders .... 14 psi (98 kPa)

Cylinder head

Warpage limit .... 0.004 inch (0.1 mm)
Cylinder head height
  SOHC .... 4.205 to 4.220 inches (106.8 to 107.2 mm)
  DOHC .... 4.972 to 4.980 inches (126.3 to 126.5 mm)

Valves and related components

Rocker arm-to-shaft oil clearance (SOHC engine) .... 0.0003 to 0.0019 inch (0.007 to 0.049 mm)
Valve stem diameter
  SOHC engine
    Intake .... 0.2742 to 0.2748 inch (6.965 to 6.980 mm)
    Exhaust .... 0.3136 to 0.3138 inch (7.965 to 7.970 mm)
  DOHC engine
    Intake .... 0.2348 to 0.2354 inch (5.965 to 5.980 mm)
    Exhaust .... 0.2341 to 0.2346 inch (5.945 to 5.960 mm)
Valve margin
  SOHC engine
    Intake .... 0.0453 to 0.0571 inch (1.15 to 1.45 mm)
    Exhaust .... 0.0531 to 0.0650 inch (1.35 to 1.65 mm)
    Service limit (minimum) .... 0.020 inch (0.5 mm)
  DOHC engine
    Intake .... 0.0374 to 0.0492 inch (0.95 to 1.25 mm)
    Exhaust .... 0.0453 to 0.0571 inch (1.15 to 1.45 mm)
    Service limit (minimum) .... 0.020 inch (0.5 mm)
Valve spring free length
  SOHC engine
    Outer .... 2.016 inches (51.2 mm)
    Inner .... 1.736 inches (44.1 mm)
  DOHC engine .... 1.8476 inches (46.93 mm)
Valve spring out-of-square limit
  SOHC engine
    Outer .... 0.087 inch (2.2 mm)
    Inner .... 0.075 inch (1.9 mm)
  DOHC engine .... 0.079 inch (2.0 mm)
Valve stem-to-guide clearance
  SOHC engine
    Intake .... 0.0008 to 0.0021 inch (0.020 to 0.053 mm)
    Exhaust .... 0.0016 to 0.0029 inch (0.040 to 0.073 mm)
    Service limit (maximum) .... 0.0039 inch (0.10 mm)
  DOHC engine
    Intake .... 0.0008 to 0.0021 inch (0.020 to 0.053 mm)
    Exhaust .... 0.0016 to 0.0029 inch (0.040 to 0.073 mm)
    Service limit (maximum)
      Intake .... 0.0031 inch (0.08 mm)
      Exhaust .... 0.0040 inch (0.10 mm)

Valve lifters

SOHC engine
  Lifter outside diameter .... 0.6278 to 0.6282 inch (15.947 to 15.957 mm)
  Lifter guide inside diameter .... 0.6299 to 0.6304 inch (16.000 to 16.013 mm)
  Lifter-to-guide clearance .... 0.0017 to 0.0026 inch (0.043 to 0.066 mm)
DOHC engine
  Lifter outside diameter .... 1.3764 to 1.3770 inch (34.960 to 34.975 mm)
  Lifter guide inside diameter .... 1.3780 to 1.3788 inch (35.000 to 35.021 mm)
  Lifter-to-guide clearance .... 0.0010 to 0.0024 inch (0.025 to 0.061 mm)

Chapter 2 Part C General engine overhaul procedures

Camshaft

Inner diameter of camshaft bearing
  SOHC engine
    Journal A .... 1.8504 to 1.8514 inches (47.000 to 47.025 mm)
    Journal B .... 1.6732 to 1.6742 inches (42.500 to 42.525 mm)
    Journal C .... 1.8898 to 1.8907 inches (48.000 to 48.025 mm)
  DOHC engine
    Journal No.1 .... 1.0236 to 1.0244 inches (26.000 to 26.021 mm)
    Journal No. 2, 3 and 4 .... 0.9252 to 0.9260 inches (23.500 to 23.521 mm)

Outer diameter of camshaft journal
  SOHC engine
    Journal A .... 1.8472 to 1.8480 inches (46.920 to 46.940 mm)
    Journal B .... 1.6701 to 1.6709 inches (42.420 to 42.440 mm)
    Journal C .... 1.8866 to 1.8874 inches (47.920 to 47.940 mm)
  DOHC engine
    1995 through 1998
      Journal No.1 .... 1.0211 to 1.0218 inches (25.935 to 25.955 mm)
      Journal No. 2, 3 and 4 .... 0.9226 to 0.9234 inches (23.435 to 23.455 mm)
    1999 and later
      Journal No.1 .... 1.0211 to 1.0218 inches (25.935 to 25.955 mm)
      Journal No. 2, 3 and 4 .... 0.9230 to 0.9238 inches (23.445 to 23.465 mm)

Camshaft bearing oil clearance
  SOHC engine
    Standard .... 0.0018 to 0.0035 inch (0.045 to 0.090 mm)
  DOHC engine
    1995 through 1998
      Standard .... 0.0018 to 0.0034 inch (0.045 to 0.086 mm)
      Service limit (maximum) .... 0.0059 inch (0.15 mm)
    1999 and later
      Journal No.1 .... 0.0018 to 0.0034 inch (0.045 to 0.086 mm)
      Journal No. 2, 3 and 4 .... 0.0014 to 0.0030 inch (0.035 to 0.076 mm)
      Service limit (maximum) .... 0.0059 inch (0.15 mm)

Camshaft endplay
  SOHC engine .... 0.0012 to 0.0024 inch (0.03 to 0.06 mm)
  DOHC engine .... 0.0045 to 0.0074 inch (0.115 to 0.188 mm)

Camshaft runout (total indicator reading)
  SOHC engine
    Standard .... 0.0016 inch (0.04 mm)
    Service limit .... 0.004 inch (0.10 mm)
  DOHC engine
    Standard .... 0.0008 inch (0.02 mm)
    Service limit .... 0.0020 inch (0.05 mm)

Camshaft lobe height, intake and exhaust
  SOHC engine
    Standard .... 1.5566 to 1.5641 inches (39.537 to 39.727 mm)
    Service limit (lobe lift loss) .... 0.0059 inch (0.15 mm)
  DOHC engine
    Standard .... 1.7299 to 1.7374 inches (43.940 to 44.130 mm)
    Service limit (lobe lift loss) .... 0.0080 inch (0.20 mm)

Engine block

Deck warpage limit
  All engines (maximum) .... 0.0039 inch (0.10 mm)
Cylinder bore diameter
  SOHC engine
    Standard .... 3.4252 to 3.4264 inches (87.000 to 87.030 mm)
    Wear limit .... 0.0079 inch (0.20 mm)
  DOHC engine
    Standard .... 3.6614 to 3.6626 inches (93.000 to 93.030 mm)
    Wear limit .... 0.0079 inch (0.20 mm)
Cylinder out-of-round and taper limit
  All engines (maximum) .... 0.0006 inch (0.015 mm)
Main journal bore diameter
  SOHC engine .... 2.6238 to 2.6249 inches (66.645 to 66.672 mm)
  DOHC engine .... 2.5194 to 2.5203 inches (63.993 to 64.017 mm)

2C-4 Chapter 2 Part C General engine overhaul procedures

Pistons and rings

Piston skirt diameter
  SOHC engine .... 3.4238 to 3.4250 inches (86.965 to 86.995 mm)
  DOHC engine .... 3.6606 to 3.6618 inches (92.980 to 93.010 mm)
Piston-to-cylinder clearance
  SOHC engine .... 0.0006 to 0.0014 inch (0.015 to 0.035 mm)
  DOHC engine .... 0.0004 to 0.0012 inch (0.01 to 0.03 mm)
Piston ring side clearance
  SOHC engine
    Top compression ring
      Standard .... 0.0016 to 0.0029 inch (0.040 to 0.073 mm)
      Service limit (maximum) .... 0.004 inch (0.10 mm)
    Second compression ring
      Standard .... 0.0012 to 0.0025 inch (0.030 to 0.063 mm)
      Service limit (maximum) .... 0.004 inch (0.10 mm)
    Oil ring
      Standard .... 0.0006 to 0.0075 inch (0.015 to 0.19 mm)
      Service limit (maximum) .... 0.004 inch (0.10 mm)
  DOHC engine
    Top compression ring
      Standard .... 0.0016 to 0.0031 inch (0.04 to 0.08 mm)
      Service limit (maximum) .... 0.004 inch (0.10 mm)
    Second compression ring
      Standard .... 0.0012 to 0.0028 inch (0.03 to 0.07 mm)
      Service limit (maximum) .... 0.004 inch (0.10 mm)
    Oil ring
      Standard .... 0.0006 to 0.0075 inch (0.015 to 0.19 mm)
      Service limit (maximum) .... 0.004 inch (0.10 mm)
Piston ring end gap
  SOHC engine
    Top compression ring
      Standard .... 0.0083 to 0.0173 inch (0.21 to 0.44 mm)
      Service limit (maximum) .... 0.039 inch (1.0 mm)
    Second compression ring
      Standard .... 0.0071 to 0.0173 inch (0.18 to 0.44 mm)
      Service limit (maximum) .... 0.039 inch (1.0 mm)
    Oil rail
      Standard .... 0.0079 to 0.0299 inch (0.20 to 0.76 mm)
      Service limit (maximum) .... 0.039 inch (1.0 mm)
  DOHC engine
    1995 and 1996
      Top compression ring
        Standard .... 0.0087 to 0.0161 inch (0.22 to 0.41 mm)
        Service limit (maximum) .... 0.0217 inch (0.55 mm)
      Second compression ring
        Standard .... 0.0197 to 0.0291 inch (0.50 to 0.74 mm)
        Service limit (maximum) .... 0.0335 inch (0.85 mm)
      Oil rail
        Standard .... 0.0079 to 0.0272 inch (0.20 to 0.69 mm)
        Service limit (maximum) .... 0.0374 inch (0.95 mm)
    1997 and later
      Top compression ring
        Standard .... 0.0087 to 0.0126 inch (0.22 to 0.32 mm)
        Service limit (maximum) .... 0.0217 inch (0.55 mm)
      Second compression ring
        Standard .... 0.0126 to 0.0185 inch (0.32 to 0.47 mm)
        Service limit (maximum) .... 0.0335 inch (0.85 mm)
      Oil rail
        Standard .... 0.0079 to 0.0236 inch (0.20 to 0.60 mm)
        Service limit (maximum) .... 0.0374 inch (0.95 mm)

Crankshaft

Main journal diameter
  SOHC engine .... 2.4784 to 2.4793 inches (62.951 to 62.975 mm)
  DOHC engine .... 2.3603 to 2.3612 inches (59.951 to 59.975 mm)
Rod journal diameter
  SOHC engine .... 1.9667 to 1.9675 inches (49.955 to 49.974 mm)
  DOHC engine .... 1.7699 to 1.7706 inches (44.956 to 44.974 mm)

Chapter 2 Part C General engine overhaul procedures

Crankshaft journal out-of-round and taper limit
   SOHC engine (maximum) .... 0.0002 inch (0.005 mm)
   DOHC engine (maximum) .... 0.0001 inch (0.002 mm)
Endplay
   SOHC engine
      Standard .... 0.0020 to 0.0067 inch (0.05 to 0.17 mm)
      Service limit (maximum) .... 0.0118 inch (0.30 mm)
   DOHC engine
      Standard .... 0.0039 to 0.0098 inch (0.10 to 0.25 mm)
      Service limit (maximum) .... 0.0118 inch (0.30 mm)
Main bearing oil clearance
   SOHC engine
      Standard .... 0.0011 to 0.0022 inch (0.028 to 0.055 mm)
      Service limit (maximum) .... 0.0035 inch (0.09 mm)
   DOHC engine
      1995 and 1996
         Standard .... 0.0014 to 0.0021 inch (0.035 to 0.053 mm)
         Service limit (maximum) .... 0.0026 inch (0.065 mm)
      1997 and later
         Standard .... 0.0005 to 0.0012 inch (0.012 to 0.030 mm)
         Service limit (maximum) .... 0.0026 inch (0.065 mm)
Rod bearing oil clearance
   SOHC engine
      Standard .... 0.0006 to 0.0021 inch (0.014 to 0.054 mm)
      Service limit (maximum) .... 0.0035 inch (0.09 mm)
   DOHC engine
      1995 and 1996
         Standard .... 0.0013 to 0.0023 inch (0.034 to 0.059 mm)
         Service limit (maximum) .... 0.0028 inch (0.07 mm)
      1997 and later
         Standard .... 0.0008 to 0.0018 inch (0.020 to 0.045 mm)
         Service limit (maximum) .... 0.0028 inch (0.07 mm)
Connecting rod side clearance (endplay)
   All engines
      Standard .... 0.0079 to 0.0138 inch (0.20 to 0.35 mm)
      Service limit (maximum) .... 0.0157 inch (0.40 mm)

Torque specifications*

	Nm	Ft-lbs unless otherwise indicated)
Rear main oil seal retainer bolts		
SOHC engine	6.3 to 8.3	55 to 73 in-lbs
DOHC engine	8.4 to 10.8	74 to 96 in-lbs
Connecting rod nuts		
SOHC engine		
Step one	14 to 16	120 to 144 in-lbs
Step two	59 to 65	43 to 48
DOHC engine		
Step one	19 to 21	14 to 15
Step two	Turn an additional 90 to 95 degrees	
Main bearing cap bolts		
SOHC engine	90 to 100	64 to 74
DOHC engine		
Step one	32 to 38	24 to 28
Step two	Turn an additional 90 to 95 degrees	

*Refer to Part A or B for additional torque specifications

Chapter 2 Part C General engine overhaul procedures

2.2a Oil pressure sending unit location – SOHC engine

2.2b Oil pressure sending unit location – DOHC engine

1 General information - engine overhaul

Included in this portion of Chapter 2 are the general overhaul procedures for the cylinder head and internal engine components.

The information ranges from advice concerning preparation for an overhaul and the purchase of replacement parts to detailed, step-by-step procedures covering removal and installation of internal engine components and the inspection of parts.

The following Sections have been written based on the assumption that the engine has been removed from the vehicle. For information concerning in-vehicle engine repair, as well as removal and installation of the external components necessary for the overhaul, see Chapter 2A or 2B.

The Specifications included in this Part are only those necessary for the inspection and overhaul procedures which follow. Refer to Chapter 2, Part A or Part B for additional Specifications.

It's not always easy to determine when, or if, an engine should be completely overhauled, as a number of factors must be considered.

High mileage is not necessarily an indication that an overhaul is needed, while low mileage doesn't preclude the need for an overhaul. Frequency of servicing is probably the most important consideration. An engine that's had regular and frequent oil and filter changes, as well as other required maintenance, will most likely give many thousands of miles of reliable service. Conversely, a neglected engine may require an overhaul very early in its life.

Excessive oil consumption is an indication that piston rings, valve seals and/or valve guides are in need of attention. Make sure that oil leaks aren't responsible before deciding that the rings and/or guides are bad. Perform a cylinder compression check to determine the extent of the work required (see Section 3). Also check the vacuum readings under various conditions (see Section 4).

Loss of power, rough running, knocking or metallic engine noises, excessive valve train noise and high fuel consumption rates may also point to the need for an overhaul, especially if they're all present at the same time. If a complete tune-up doesn't remedy the situation, major mechanical work is the only solution.

An engine overhaul involves restoring the internal parts to the specifications of a new engine. During an overhaul, the piston rings are replaced and the cylinder walls are reconditioned (re-bored and/or honed). If a re-bore is done by an automotive machine shop, new oversize pistons will also be installed. The main bearings, connecting rod bearings and camshaft bearings are generally replaced with new ones and, if necessary, the crankshaft may be reground to restore the journals. Generally, the valves are serviced as well, since they're usually in less-than-perfect condition at this point. While the engine is being overhauled, other components, such as the distributor (if equipped), the starter and alternator, can be rebuilt as well. The end result should be a like new engine that will give many trouble free miles. **Note:** *Critical cooling system components such as the hoses, drivebelts, thermostat and water pump should be replaced with new parts when an engine is overhauled. The radiator should be checked carefully to ensure that it isn't clogged or leaking (see Chapter 3). If you purchase a rebuilt engine or short block, some rebuilders will not warranty their engines unless the radiator has been professionally flushed. Also, we don't recommend overhauling the oil pump - always install a new one when an engine is rebuilt.*

Before beginning the engine overhaul, read through the entire procedure to familiarize yourself with the scope and requirements of the job. Overhauling an engine isn't difficult, but it is time-consuming. Plan on the vehicle being tied up for a minimum of two weeks, especially if parts must be taken to an automotive machine shop for repair or reconditioning. Check on availability of parts and make sure that any necessary special tools and equipment are obtained in advance. Most work can be done with typical hand tools, although a number of precision measuring tools are required for inspecting parts to determine if they must be replaced. Often an automotive machine shop will handle the inspection of parts and offer advice concerning reconditioning and replacement. **Note:** *Always wait until the engine has been completely disassembled and all components, especially the engine block, have been inspected before deciding what service and repair operations must be performed by an automotive machine shop.* Since the block's condition will be the major factor to consider when determining whether to overhaul the original engine or buy a rebuilt one, never purchase parts or have machine work done on other components until the block has been thoroughly inspected. As a general rule, time is the primary cost of an overhaul, so it doesn't pay to install worn or substandard parts.

As a final note, to ensure maximum life and minimum trouble from a rebuilt engine, everything must be assembled with care in a spotlessly clean environment.

2 Oil pressure check

Refer to illustrations 2.2a and 2.2b

1 Low engine oil pressure can be a sign of an engine in need of rebuilding. A "low oil pressure" indicator (often called an "idiot light") is not a test of the oiling system. Such indicators only come on when the oil pressure is dangerously low. Even a factory oil pressure gauge in the instrument panel is only a relative indication, although much better for driver information than a warning light. A better test is with a mechanical (not electrical) oil pressure gauge. When used in conjunction with an accurate tachometer, an engine's oil pressure performance can be compared to factory Specifications for that

Chapter 2 Part C General engine overhaul procedures

3.6 A compression gauge with a threaded fitting for the spark plug hole is preferred over the type that requires hand pressure to maintain the seal

4.4 A simple vacuum gauge can be very handy in diagnosing engine condition and performance

year and model.
2 Locate the oil pressure sending unit **(see illustrations)**.
3 Remove the oil pressure sending unit and install a fitting which will allow you to directly connect your hand-held, mechanical oil pressure gauge. Use Teflon tape or sealant on the threads of the adapter and the fitting on the end of your gauge's hose.
4 Connect an accurate tachometer to the engine, according to the tachometer manufacturer's instructions.
5 Check the oil pressure with the engine running (full operating temperature) at the specified engine speed, and compare it to this Chapter's Specifications. If it's extremely low, the bearings and/or oil pump are probably worn out.

3 Cylinder compression check

Refer to illustration 3.6
1 A compression check will tell you what mechanical condition the upper end of your engine (pistons, rings, valves, cylinder head gaskets) is in. Specifically, it can tell you if the compression is down due to leakage caused by worn piston rings, defective valves and seats or a blown cylinder head gasket. **Note:** *The engine must be at normal operating temperature and the battery must be fully charged for this check.*
2 Begin by cleaning the area around the spark plugs before you remove them (compressed air should be used, if available). The idea is to prevent dirt from getting into the cylinders as the compression check is being done.
3 Relieve the fuel system pressure (see Chapter 4). The fuel pump must remain disabled throughout this procedure.
4 Remove all of the spark plugs from the engine (see Chapter 1). Block the throttle wide open.
5 On models with a distributor, detach the coil wire from the center of the distributor cap and ground it on the engine block. Use a jumper wire with alligator clips on each end to ensure a good ground. On models without a distributor, unplug the electrical connector from each ignition coil.
6 Install the compression gauge in the spark plug hole **(see illustration)**.
7 Crank the engine over at least seven compression strokes and watch the gauge. The compression should build up quickly in a healthy engine. Low compression on the first stroke, followed by gradually increasing pressure on successive strokes, indicates worn piston rings. A low compression reading on the first stroke, which doesn't build up during successive strokes, indicates leaking valves or a blown cylinder head gasket (a cracked cylinder head could also be the cause). Deposits on the undersides of the valve heads can also cause low compression. Record the highest gauge reading obtained.
8 Repeat the procedure for the remaining cylinders and compare the results to this Chapter's Specifications.
9 Add some engine oil (about three squirts from a plunger-type oil can) to each cylinder, through the spark plug hole, and repeat the test.
10 If the compression increases after the oil is added, the piston rings are definitely worn. If the compression doesn't increase significantly, the leakage is occurring at the valves or cylinder head gasket. Leakage past the valves may be caused by burned valve seats and/or faces or warped, cracked or bent valves.
11 If two adjacent cylinders have equally low compression, there's a strong possibility that the cylinder head gasket between them is blown. The appearance of coolant in the combustion chambers or the crankcase would verify this condition.
12 If one cylinder is slightly lower than the others, and the engine has a slightly rough idle, a worn lobe on the camshaft could be the cause.
13 If the compression is unusually high, the combustion chambers are probably coated with carbon deposits. If that's the case, the cylinder head(s) should be removed and decarbonized.
14 If compression is way down or varies greatly between cylinders, it would be a good idea to have a leak-down test performed by an automotive repair shop. This test will pinpoint exactly where the leakage is occurring and how severe it is.

4 Vacuum gauge diagnostic checks

Refer to illustrations 4.4 and 4.6
A vacuum gauge provides valuable information about what is going on in the engine at a low-cost. You can check for worn rings or cylinder walls, leaking cylinder head or intake manifold gaskets, incorrect carburetor adjustments, restricted exhaust, stuck or burned valves, weak valve springs, improper ignition or valve timing and ignition problems.
Unfortunately, vacuum gauge readings are easy to misinterpret, so they should be used in conjunction with other tests to confirm the diagnosis.
Both the absolute readings and the rate of needle movement are important for accurate interpretation. Most gauges measure vacuum in inches of mercury (in-Hg). The following references to vacuum assume the diagnosis is being performed at sea level. As elevation increases (or atmospheric pressure decreases), the reading will decrease. For every 1,000 foot increase in elevation above approximately 2000 feet, the gauge readings will decrease about one inch of mercury.
Connect the vacuum gauge directly to intake manifold vacuum, not to ported (throttle body) vacuum **(see illustration)**. Be sure no hoses are left disconnected during the test or false readings will result.
Before you begin the test, allow the engine to warm up completely. Block the wheels and set the parking brake. With the

Chapter 2 Part C General engine overhaul procedures

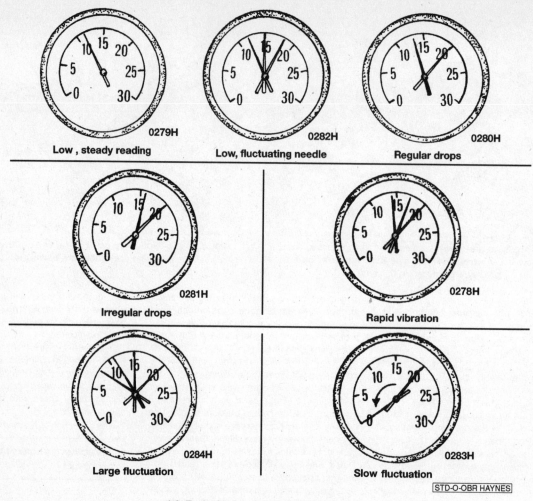

4.6 Typical vacuum gauge readings

transmission in Park, start the engine and allow it to run at normal idle speed. **Warning:** *Keep your hands and the vacuum gauge clear of the fans.*

Read the vacuum gauge; an average, healthy engine should normally produce about 17 to 22 in-Hg of vacuum with a fairly steady needle **(see illustration)**. Refer to the following vacuum gauge readings and what they indicate about the engine's condition:

1   A low steady reading usually indicates a leaking gasket between the intake manifold and cylinder head(s) or throttle body, a leaky vacuum hose, late ignition timing or incorrect camshaft timing. Check ignition timing with a timing light and eliminate all other possible causes, utilizing the tests provided in this Chapter before you remove the timing chain cover to check the timing marks.

2   If the reading is three to eight inches below normal and it fluctuates at that low reading, suspect an intake manifold gasket leak at an intake port or a faulty fuel injector.

3   If the needle has regular drops of about two-to-four in-Hg at a steady rate, the valves are probably leaking. Perform a compression check or leak-down test to confirm this.

4   An irregular drop or down-flick of the needle can be caused by a sticking valve or an ignition misfire. Perform a compression check or leak-down test and read the spark plugs.

5   A rapid vibration of about four in-Hg vibration at idle combined with exhaust smoke indicates worn valve guides. Perform a leak-down test to confirm this. If the rapid vibration occurs with an increase in engine speed, check for a leaking intake manifold gasket or cylinder head gasket, weak valve springs, burned valves or ignition misfire.

6   A slight fluctuation, say one inch up and down, may mean ignition problems. Check all the usual tune-up items and, if necessary, run the engine on an ignition analyzer.

7   If there is a large fluctuation, perform a compression or leak-down test to look for a weak or dead cylinder or a blown cylinder head gasket.

8   If the needle moves slowly through a wide range, check for a clogged PCV system, incorrect idle fuel mixture, carburetor/throttle body or intake manifold gasket leaks.

9   Check for a slow return after revving the engine by quickly snapping the throttle open until the engine reaches about 2,500 rpm and let it shut. Normally the reading should drop to near zero, rise above normal idle reading (about 5 in-Hg over) and then return to the previous idle reading. If the vacuum returns slowly and doesn't peak when the throttle is snapped shut, the rings may be worn. If there is a long delay, look for a restricted exhaust system (often the muffler or catalytic converter). An easy way to check this is to temporarily disconnect the exhaust ahead of the suspected part and redo the test.

5   Engine rebuilding alternatives

The do-it-yourselfer is faced with a number of options when performing an engine overhaul. The decision to replace the engine block, piston/connecting rod assemblies and crankshaft depends on a number of factors, with the number one consideration being the condition of the block. Other considerations are cost, access to machine shop facilities, parts availability, time required to complete the project and the extent of prior mechanical experience on the part of the do-it-yourselfer.

Some of the rebuilding alternatives include:

Individual parts - If the inspection procedures reveal that the engine block and most engine components are in reusable

Chapter 2 Part C General engine overhaul procedures

condition, purchasing individual parts may be the most economical alternative. The block, crankshaft and piston/connecting rod assemblies should all be inspected carefully. Even if the block shows little wear, the cylinder bores should be surface-honed.

Crankshaft kit - This rebuild package consists of a reground crankshaft and a matched set of pistons and connecting rods. The pistons will already be installed on the connecting rods. Piston rings and the necessary bearings will be included in the kit. These kits are commonly available for standard cylinder bores, as well as for engine blocks which have been bored to a regular oversize.

Short block - A short block consists of an engine block with renewed crankshaft and piston/connecting rod assemblies already installed. All new bearings are incorporated and all clearances will be correct. The existing cylinder head(s), camshaft, valve train components and external parts can be bolted to the short block with little or no machine shop work necessary.

Long block - A long block consists of a short block plus an oil pump, oil pan, cylinder heads, valve covers, camshaft and valve train components, timing sprockets, timing chain and timing cover. All components are installed with new bearings, seals and gaskets incorporated throughout. The installation of manifolds and external parts is all that is necessary.

Used engine assembly - While overhaul provides the best assurance of a like-new engine, used engines available from wrecking yards and importers are often a very simple and economical solution. Many used engines come with warranties, but always give any engine a thorough diagnostic check-out before purchase. Check compression, vacuum and also for signs of oil leakage. If possible, have the seller run the engine, ether in the vehicle or on a test stand so you can be sure it runs smoothly with no knocking or other noises.

Give careful thought to which alternative is best for you and discuss the situation with local automotive machine shops, auto parts dealers or parts store countermen before ordering or purchasing replacement parts.

6 Engine removal - methods and precautions

If you've decided that an engine must be removed for overhaul or major repair work, several preliminary steps should be taken.

Locating a suitable place to work is extremely important. Adequate work space, along with storage space for the vehicle, will be needed. If a shop or garage isn't available, at the very least a flat, level, clean work surface made of concrete or asphalt is required.

Cleaning the engine compartment and engine before beginning the removal proce-

7.6 Label the hoses (arrow) and wires to ensure proper assembly

dure will help keep tools clean and organized.

An engine hoist or A-frame will also be necessary. Make sure the equipment is rated in excess of the combined weight of the engine and transaxle. Safety is of primary importance, considering the potential hazards involved in lifting the engine out of the vehicle. The engine and transaxle combinations used in the vehicles covered by this manual are intended to be removed from below, lowered out of the chassis. This is safest and easiest when there is access to a vehicle hoist, since the powertrain can be lowered to the floor with a hoist and the vehicle then raised high enough for powertrain removal.

If the engine is being removed by a novice, a helper should be available. Advice and aid from someone more experienced would also be helpful. There are many instances when one person cannot simultaneously perform all of the operations required when lifting the engine out of the vehicle.

Plan the operation ahead of time. Arrange for or obtain all of the tools and equipment you'll need prior to beginning the job. Some of the equipment necessary to perform engine removal and installation safely and with relative ease are (in addition to an engine hoist) a heavy duty floor jack, complete sets of wrenches and sockets as described in the front of this manual, wooden blocks and plenty of rags and cleaning solvent for mopping up spilled oil, coolant and gasoline. If the hoist must be rented, make sure that you arrange for it in advance and perform all of the operations possible without it beforehand. This will save you money and time.

Plan for the vehicle to be out of use for quite a while. A machine shop will be required to perform some of the work which the do-it-yourselfer can't accomplish without special equipment. These shops often have a busy schedule, so it would be a good idea to consult them before removing the engine in order to accurately estimate the amount of time required to rebuild or repair components that may need work.

Always be extremely careful when

removing and installing the engine. Serious injury can result from careless actions. Plan ahead, take your time and a job of this nature, although major, can be accomplished successfully.

7 Engine - removal and installation

Warning: *The models covered by this manual are equipped with Supplemental Restraint Systems (SRS), more commonly known as airbags. Always disable the airbag system before working in the vicinity of any airbag system components to avoid the possibility of accidental deployment of the airbag(s), which could cause personal injury (see Chapter 12).*

Removal

Refer to illustrations 7.6, 7.13a, 7.13b, 7.13c, 7.17a and 7.17b

Note: *Read through the entire Section before beginning this procedure. The engine and transaxle are removed as a unit from below and then separated outside the vehicle.*

1 Relieve the fuel system pressure (see Chapter 4).
2 Disconnect the cable from the negative terminal of the battery (see Chapter 1).
3 Place protective covers on the fenders and cowl and remove the hood (see Chapter 11).
4 Remove the air cleaner assembly (see Chapter 4).
5 Raise the vehicle and support it securely on jackstands. Drain the cooling system, transaxle and engine oil and remove the drivebelts (see Chapter 1).
6 Clearly label, then disconnect all vacuum lines, coolant and emissions hoses, wiring harness connectors, ground straps and fuel lines. Masking tape and/or a touch up paint applicator work well for marking items **(see illustration)**. Take instant photos or sketch the locations of components and brackets.
7 Remove the cooling fans, and disconnect the radiator hoses and heater hoses (see Chapter 3).
8 Release the residual fuel pressure in the tank by removing the gas cap, then disconnect the fuel lines from the fuel rail (see Chapter 4). Plug or cap all open fittings.
9 Refer to Chapter 3 and unbolt and set aside the air conditioning compressor, without disconnecting the refrigerant lines.
10 Disconnect the throttle linkage (and speed control cable, when equipped) from the engine (see Chapter 4).
11 Unbolt the power steering pump. Tie the pump aside without disconnecting the hoses (see Chapter 10). Remove the alternator (see Chapter 5).
12 On SOHC engines refer to Chapter 5 and remove the ignition coil and distributor cap with spark plug wires.
13 Label and disconnect the main engine electrical harnesses at each end of the engine **(see illustrations)**.

2C-10 Chapter 2 Part C General engine overhaul procedures

7.13a At the transaxle end of the engine, label and disconnect the electrical connectors (arrows indicate some), then pull the main harness away from the powertrain

14 Refer to Chapter 2A or 2B and remove the upper intake manifold to make engine removal easier. Be sure to label and disconnect all hoses, connectors, and any ground straps.

15 Disconnect the front exhaust pipe from the exhaust manifolds and at the converter, then remove the pipe (see Chapter 4).

16 Remove the driveaxles (see Chapter 8). Disconnect the electrical connectors and the shift linkage from the transaxle (see Chapter 7A or 7B).

17 Attach a lifting sling or chain to the lifting eye (if equipped) on the engine **(see illustration)**. If lifting eyes are not equipped attach the lifting sling or chain on a safe place such as the side of each cylinder head. Position a hoist and connect the sling to it. Take up the slack until there is slight tension on the hoist **(see illustration)**.

18 Recheck to be sure nothing except the mounts are still connecting the engine/transaxle to the vehicle. Disconnect anything still remaining.

19 Support the transaxle with a floor jack. Place a block of wood on the jack head to prevent damage to the transaxle. Remove the engine mount bolts and the lower crossmember to frame bolts **(see Chapter 2A or 2B)**. **Warning:** *Do not place any part of your body under the engine/transaxle when it's supported only by a hoist or other lifting device.*

20 Slowly lower the engine/transaxle out of the vehicle.

21 Once the powertrain is on the floor, disconnect the engine lifting hoist and raise the vehicle hoist until the powertrain can be slid out from under the vehicle. **Note:** *A sheet of old hardboard or paneling between the engine and floor makes moving the powertrain easier. A helper will be needed to move the powertrain.*

22 Separate the engine from the transaxle (see Chapter 7).

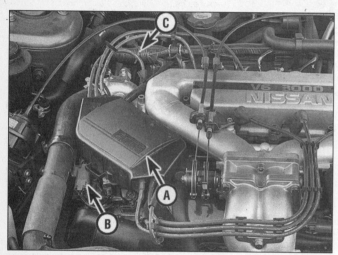

7.13b On SOHC engines, remove the distributor cap cover (A) and the distributor cap with the wires attached, disconnect all the electrical connectors (B indicates some) and pull the main electrical harness (C) away from the engine

7.13c On DOHC engines, disconnect the crankshaft and camshaft position sensors (A) and the connector for the air conditioning compressor, then detach the wiring harness (B) from the brackets at the top of the timing chain cover

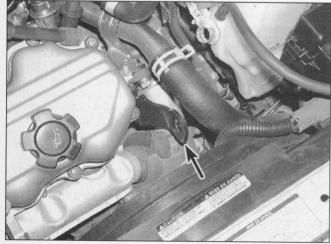

7.17a Attach the chain or sling to a lifting eye (if equipped)

7.17b Attach a lifting sling to the lifting eyes or strong attachment points on the engine - raise the engine enough to remove the mounts and crossmember, then remove the jack and lower the engine/transaxle to the floor (typical)

Chapter 2 Part C General engine overhaul procedures

23 Place the engine on the floor or remove the flywheel/driveplate and mount the engine on an engine stand.

Installation

24 Check the engine/transaxle mounts. If they're worn or damaged, replace them.
25 On automatic transaxle models inspect the converter seal and bushing, and apply a dab of grease to the nose of the converter and to the seal lips. If the vehicle is equipped with a manual transaxle inspect the clutch components and the pilot bearing (see Chapter 8).
26 Carefully guide the transaxle into place, following the procedure outlined in Chapter 7. **Caution:** *Do not use the bolts to force the engine and transaxle into alignment. It may crack or damage major components.*
27 Install the engine-to-transaxle bolts and tighten them securely.
28 Slide the engine/transaxle over a sheet of hardboard or paneling until it is in the approximate position under the vehicle, then lower the vehicle on the vehicle hoist.
29 Roll the engine hoist into position, attach the sling or chain in a position that will allow a good balance, and slowly raise the powertrain until the mounts at the transaxle end can be attached.
30 Support the transaxle with a floor jack for extra security, then reinstall the crossmember and attach the right-side engine mounts. Follow the procedure in Chapter 2, Part A for the final tightening of all engine mount bolts.
31 Reinstall the remaining components and fasteners in the reverse order of removal.
32 Add coolant, oil, power steering and transmission fluids as needed (see Chapter 1).
33 Run the engine and check for proper operation and leaks. Shut off the engine and recheck the fluid levels.

8 Engine overhaul - disassembly sequence

1 It's much easier to disassemble and work on the engine if it's mounted on a portable engine stand. A stand can often be rented quite cheaply from an equipment rental yard. Before the engine is mounted on a stand, the flywheel/driveplate and engine rear plate (if equipped) should be removed from the engine.
2 If a stand isn't available, it's possible to disassemble the engine with it blocked up on the floor. Be extra careful not to tip or drop the engine when working without a stand.
3 If you're going to obtain a rebuilt engine, all external components must come off first, to be transferred to the replacement engine, just as they will if you're doing a complete engine overhaul yourself. These include:

  Alternator and brackets
  Emissions control components
  Distributor, spark plug wires and spark plugs – SOHC engines
  Thermostat and housing cover
  Water pump
  EFI components
  Intake/exhaust manifolds
  Oil filter
  Powertrain mounts
  Flywheel/driveplate
  Engine rear plate – SOHC engines
  Rear main seal retainer

Note: *When removing the external components from the engine, pay close attention to details that may be helpful or important during installation. Note the installed position of gaskets, seals, spacers, pins, brackets, washers, bolts and other small items.*

4 If you're obtaining a short block, which consists of the engine block, crankshaft, pistons and connecting rods all assembled, then the cylinder heads, oil pan and oil pump will have to be removed as well. See *Engine rebuilding alternatives* for additional information regarding the different possibilities to be considered.
5 If you're planning a complete overhaul, the engine must be disassembled and the internal components removed in the following general order:

SOHC engines

  Valve covers
  Rocker arm assemblies
  Valve lifters and guides
  Timing covers
  Timing belt and sprockets
  Cylinder heads
  Camshafts
  Oil pan and pick-up
  Oil pump
  Piston/connecting rod assemblies
  Crankshaft and main bearings

DOHC engines

  Valve covers
  Steel oil pan
  Aluminum oil pan
  Front timing cover
  Timing chains and sprockets
  Rear timing cover
  Camshaft and lifters
  Cylinder heads
  Oil pump and pick-up
  Piston/connecting rod assemblies
  Crankshaft and main bearings

6 Before beginning the disassembly and overhaul procedures, make sure the following items are available. Also, refer to *Engine overhaul - reassembly* sequence for a list of tools and materials needed for engine reassembly.

  Common hand tools
  Small cardboard boxes or plastic bags for storing parts
  Gasket scraper
  Ridge reamer
  Vibration damper puller
  Micrometers
  Telescoping gauges
  Dial indicator set
  Valve spring compressor
  Cylinder surfacing hone
  Piston ring groove cleaning tool
  Electric drill motor
  Tap and die set
  Wire brushes
  Oil gallery brushes
  Cleaning solvent

9 Cylinder head - disassembly

Refer to illustrations 9.2, 9.3 and 9.4

Note: *New and rebuilt cylinder heads are commonly available for most engines at dealerships and auto parts stores. Due to the fact that some specialized tools are necessary for the disassembly and inspection procedures, and replacement parts may not be readily available, it may be more practical and economical for the home mechanic to purchase replacement heads rather than taking the time to disassemble, inspect and recondition the originals.*

1 Cylinder head disassembly involves removal of the intake and exhaust valves and related components. If they're still in place, remove the rocker arms (SOHC engine only), lifters and camshafts (see Chapter 2A or 2B) from the cylinder head. Label the parts or store them separately so they can be reinstalled in their original locations. Refer to Section 21 for camshaft and lifter inspection procedures. **Caution:** *Do not lay the lifters on SOHC engines on their side or upside down, or air can become trapped inside and the lifter will have to be bled (see Chapter 2A). The lifters can be laid on their side only if they are submerged in a pan of clean engine oil until reassembly.*
2 Before the valves are removed, arrange to label and store them, along with their related components, so they can be kept separate and reinstalled in the same valve guides they are removed from **(see illustration).**

9.2 A small plastic bag, with an appropriate label, can be used to store the valve train components so they can be kept together and reinstalled in the original position

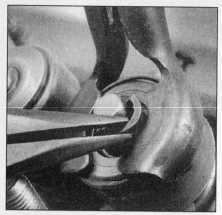

9.3 Use a valve spring compressor to compress the spring, then remove the valve stem locks from the valve stem

10 Cylinder head - cleaning and inspection

1 Thorough cleaning of the cylinder heads and related valve train components, followed by a detailed inspection, will enable you to decide how much valve service work must be done during the engine overhaul. **Note:** *If the engine was severely overheated, the cylinder heads are probably warped (see Step 12).*

Cleaning

2 Scrape all traces of old gasket material and sealing compound off the cylinder head gasket, intake manifold and exhaust manifold sealing surfaces. Be very careful not to gouge the cylinder head. Special gasket removal solvents that soften gaskets and make removal much easier are available at auto parts stores.
3 Remove all built up scale from the coolant passages.
4 Run a stiff wire brush through the various holes to remove deposits that may have formed in them.
5 Run an appropriate-size tap into each of the threaded holes to remove corrosion and thread sealant that may be present. If compressed air is available, use it to clear the holes of debris produced by this operation. **Warning:** *Wear eye protection when using compressed air!*
6 Clean the combustion chambers with a brass wire brush and solvent if carbon has accumulated.
7 Clean the cylinder head with solvent and dry it thoroughly. Compressed air will speed the drying process and ensure that all holes and recessed areas are clean. **Note:** *Decarbonizing chemicals are available and may prove very useful when cleaning cylinder heads and valve train components. They are very caustic and should be used with caution. Be sure to follow the instructions on the container.*

3 Compress the springs on the first valve with a spring compressor and remove the valve stem locks **(see illustration)**. Carefully release the valve spring compressor and remove the retainer, the spring and the spring seat (if used).
4 Pull the valve out of the cylinder head, then remove the oil seal from the guide. If the valve binds in the guide (won't pull through), push it back into the cylinder head and deburr the area around the valve stem lock groove with a fine file or whetstone **(see illustration)**.
5 Repeat the procedure for the remaining valves. Remember to keep all the parts for each valve together so they can be reinstalled in the same locations.
6 Once the valves and related components have been removed and stored in an organized manner, the cylinder head should be thoroughly cleaned and inspected. If a complete engine overhaul is being done, finish the engine disassembly procedures before beginning the cylinder head cleaning and inspection process.

8 On SOHC engines clean the rocker arms and shafts with solvent and dry them thoroughly (don't mix them up during the cleaning process). Compressed air will speed the drying process and can be used to clean out the oil passages.
9 Clean all the valve springs, spring seats, valve stem locks and retainers with solvent and dry them thoroughly. Do the components from one valve at a time to avoid mixing up the parts.
10 Scrape off any heavy deposits that may have formed on the valves, then use a motorized wire brush to remove deposits from the valve heads and stems. Again, make sure the valves don't get mixed up.

Inspection

Note: *Be sure to perform all of the following inspection procedures before concluding that machine shop work is required. Make a list of the items that need attention.*

Cylinder head

Refer to illustrations 10.12 and 10.14

11 Inspect the heads very carefully for cracks, evidence of coolant leakage and other damage. If cracks are found, check with an automotive machine shop concerning repair. If repair isn't possible, a new cylinder head should be obtained.
12 Using a straightedge and feeler gauge, check the cylinder head gasket mating surface for warpage **(see illustration)**. If the warpage exceeds the limit specified in this Chapter, it can be resurfaced at an automotive machine shop. **Note:** *The cylinder heads have a specific MINIMUM height, measured from the cylinder head gasket surface to the valve cover surface. If the cylinder head will fall below the minimum height (see Specifications) after it is machined, a new cylinder head will have to be purchased.*
13 Examine the valve seats in each of the combustion chambers. If they're pitted, cracked or burned, the cylinder head will

9.4 If the valve won't pull through the guide, deburr the edge of the stem end and the area around the top of the valve stem lock groove with a fine file or whetstone

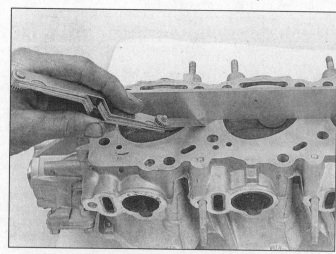

10.12 Check the cylinder head gasket surface for warpage by trying to slip a feeler gauge under the straightedge (see the Specifications for the maximum warpage allowed and use a feeler gauge of that thickness)

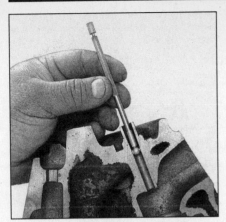

10.14 Use a small hole gauge to determine the inside diameter of the valve guides (the gauge is then measured with a micrometer)

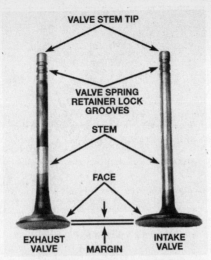

10.15 Check for valve wear at the points shown here

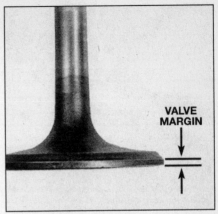

10.16 The margin width on each valve must be as specified (if no margin exists, the valve cannot be reused)

require valve service that's beyond the scope of the home mechanic.

14 Check the valve stem-to-guide clearance with a small hole gauge and micrometer **(see illustration)**, then measure the valve stem diameter with a micrometer and subtract it from the valve guide inside diameter to obtain the stem to guide clearance. When using a small hole gauge or telescoping snap gauge, insert the gauge to the middle portion of the valve guide (where wear should be minimal) and tighten the gauge. Move the gauge up and down in the guide. If the guide isn't worn the clearance should be equal from top to bottom. Loose areas indicate that the guide is tapered. If the measurement exceeds the stem-to-guide clearance limit found in this Chapter's Specifications, the valve guides should be replaced. After this is done, if there's still some doubt regarding the condition of the valve guides they should be checked by an automotive machine shop (the cost should be minimal).

Valves

Refer to illustrations 10.15 and 10.16

15 Carefully inspect each valve face for uneven wear, deformation, cracks, pits and burned areas **(see illustration)**. Check the valve stem for scuffing and galling and the neck for cracks. Rotate the valve and check for any obvious indication that it's bent. Look for pits and excessive wear on the end of the stem. The presence of any of these conditions indicates the need for valve service by an automotive machine shop.

16 Measure the margin width on each valve **(see illustration)**. Any valve with a margin narrower than specified in this Chapter will have to be replaced with a new one.

Valve components

Refer to illustrations 10.17 and 10.18

17 Check each valve spring for wear (on the ends) and pits. Measure the free length and compare it to the Specifications in this Chapter **(see illustration)**. Any springs that are shorter than specified have sagged and should not be reused. The tension of all springs should be checked with a special fixture before deciding that they're suitable for use in a rebuilt engine (take the springs to an automotive machine shop for this check).

18 Stand each spring on a flat surface and check it for squareness **(see illustration)**. If any of the springs are distorted or sagged, replace all of them with new parts.

19 Check the spring retainers and valve stem locks for obvious wear and cracks. Any questionable parts should be replaced with new ones, as extensive damage will occur if they fail during engine operation.

Camshaft, lifters, rocker arms and shafts

20 Refer to Section 21 of this Chapter for the camshaft, lifters and bearing inspection procedures. Be sure to inspect the camshaft bearing journals before the cylinder head is sent to the machine shop to have the valves serviced. If the journals are gouged or scored the cylinder head will have to be replaced regardless of the condition of the valves and related components. If you're working on an SOHC engine refer to Chapter 2A and also inspect the rocker arms and shafts.

21 Any damaged or excessively worn parts must be replaced with new ones.

22 If the inspection process indicates that the valve components are in generally poor condition and worn beyond the limits specified, which is usually the case in an engine that's being overhauled, reassemble the valves in the cylinder head and refer to Section 11 for valve servicing recommendations.

11 Valves - servicing

1 Because of the complex nature of the job and the special tools and equipment needed, servicing of the valves, the valve seats and the valve guides, commonly known as a valve job, should be done by a professional.

2 The home mechanic can remove and disassemble the heads, do the initial cleaning and inspection, then reassemble and deliver them to an automotive machine shop for the actual service work. Doing the inspection will enable you to see what condition the cylinder head and valvetrain components are in and will ensure that you know what work and new parts are required when dealing with an automotive machine shop.

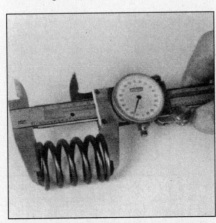

10.17 Measure the free length of each valve spring with a dial or vernier caliper

10.18 Check each valve spring for squareness

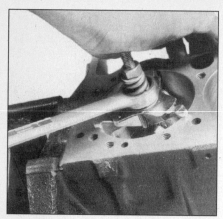

13.1 A ridge reamer is required to remove the ridge from the top of each cylinder - do this before removing the pistons!

13.3 Check the connecting rod side clearance with a feeler gauge as shown

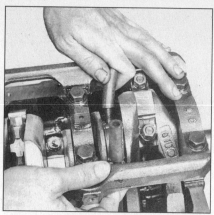

13.6 To prevent damage to the crankshaft journals and cylinder walls, slip sections of rubber or plastic hose over the connecting rod bolts before removing the pistons

3 The machine shop will remove the valves and springs, recondition or replace the valves and valve seats, recondition or replace the valve guides, check and replace the valve springs, spring retainers and valve stem locks (as necessary), replace the valve seals with new ones, reassemble the valve components and make sure the installed spring height is correct. The cylinder head gasket surface will also be resurfaced if it's warped.

4 After the valve job has been performed by a professional, the cylinder head will be in like-new condition. When the cylinder head is returned, be sure to clean it again before installation on the engine to remove any metal particles and abrasive grit that may still be present from the valve service or cylinder head resurfacing operations. Use compressed air, if available, to blow out all the oil holes and passages.

12 Cylinder head - reassembly

1 Regardless of whether or not the cylinder head was sent to an automotive repair shop for valve servicing, make sure it is clean before beginning reassembly.

2 If the cylinder head was sent out for valve servicing, the valves and related components will already be in place. Begin the reassembly procedure with Step 5.

3 Install the valves, with light oiling on the stems into valve guides. Install the valve spring seat(s) in place on the cylinder head. **Note:** *SOHC engines are equipped with an inner and outer valve spring seat while DOHC engines are equipped with a single valve spring seat. Be sure to install the outer valve spring seat first, then install the inner valve spring seat second on SOHC engines.* After the valve spring seat(s) have been positioned on the cylinder head correctly, install the valve seals over the top of the valves tips by hand. Using the stem of the valves as a guide, slide the seals down to the top of each valve guide. To install the oil seals firmly over the guides, you will need a seal installation or an appropriate-size deep socket. Gently tap each seal into place until it is properly seated onto the guide (see Chapter 2A or 2B). **Caution:** *Do not hammer on the guide seal once it is seated or you may damage the seal. Do not twist or cock the seals during installation or they will not seal properly on the valve stems.* **Note:** *On SOHC engines it will only be necessary to use a seal installation tool on the intake valve guide. The exhaust valve seals should simply slide into place on the valve guide by hand.*

4 Slip the valve spring(s) in place on the cylinder head, then use a spring compressor to install the springs, retainers and valve stem locks (see Chapter 2A or 2B). **Note:** *SOHC engines are equipped with an inner and outer valve spring while DOHC engines are equipped with a single valve spring. Be sure to install the inner valve spring into the outer valve spring and install the springs as a set on SOHC engines. On either type of engine always install the end of the outer valve spring with the more closely wound coils or paint marks towards the cylinder head.*

5 On DOHC engines refer to Chapter 2B and install the lifters and the camshafts.

6 On SOHC engines refer to Chapter 2A and install the camshafts, oil seals lifter guide assembly and rocker arms.

7 Always install old cylinder head components in their original locations.

13 Pistons/connecting rods - removal

Refer to illustrations 13.1, 13.3 and 13.6

Note: *Prior to removing the piston/connecting rod assemblies, remove the cylinder heads, the oil pan and the oil pump pick-up by referring to the appropriate Sections in Chapter 2A or 2B, if not already removed.*

1 Use your fingernail to feel if a ridge has formed at the upper limit of ring travel (about 1/4-inch down from the top of each cylinder). If carbon deposits or cylinder wear have produced ridges, they must be completely removed with a special tool **(see illustration)**. Follow the manufacturer's instructions provided with the tool. Failure to remove the ridges before attempting to remove the piston/connecting rod assemblies may result in piston breakage.

2 After the cylinder ridges have been removed, turn the engine upside-down so the crankshaft is facing up.

3 Before the connecting rods are removed, check the side clearance (endplay) with feeler gauges. Slide them between the first connecting rod and the crankshaft throw until the play is removed **(see illustration)**. The side clearance is equal to the thickness of the feeler gauge(s). If the side clearance exceeds the service limit, new connecting rods will be required. If new rods (or a new crankshaft) are installed, the side clearance may fall under the specified minimum (if it does, the rods will have to be machined to restore it - consult an automotive machine shop for advice if necessary). Repeat the procedure for the remaining connecting rods.

4 Check the connecting rods and caps for identification marks. If they aren't plainly marked, use a small center punch to make the appropriate number of indentations on each rod and cap (1, 2, 3, etc.).

5 Loosen each of the connecting rod cap nuts 1/2-turn at a time until they can be removed by hand. Remove the number one connecting rod cap and bearing insert. Don't drop the bearing insert out of the cap.

6 Slip a short length of plastic or rubber hose over each connecting rod cap bolt to protect the crankshaft journal and cylinder wall as the piston is removed **(see illustration)**.

7 Remove the bearing insert and push the connecting rod/piston assembly out through the top of the engine. Use a wooden hammer handle to push on the upper bearing surface in the connecting rod. If resistance is felt, double-check to make sure that all of the

Chapter 2 Part C General engine overhaul procedures

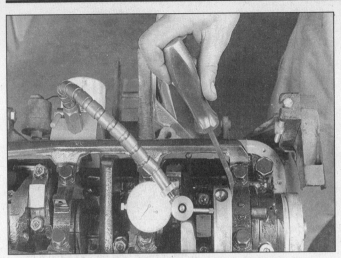

14.1 Checking crankshaft endplay with a dial indicator

14.3 Crankshaft endplay can also be measured with a feeler gauge at the number four main bearing

ridge was removed from the cylinder.
8 Repeat the procedure for the remaining cylinders.
9 After removal, reassemble the connecting rod caps and bearing inserts in their respective connecting rods and install the cap nuts finger tight. Leaving the old bearing inserts in place until reassembly will help prevent the connecting rod bearing surfaces from being accidentally nicked or gouged.
10 Don't separate the pistons from the connecting rods (see Section 18 for additional information).

14 Crankshaft - removal

Refer to illustrations 14.1, 14.3, 14.4 and 14.7
Note: *The crankshaft can be removed only after the engine has been removed from the vehicle. It's assumed that the flywheel/driveplate, crankshaft pulley, timing belt (SOHC engine), timing chain (DOHC engine), sprockets, oil pan, oil pump and piston/connecting rod assemblies have already been removed. The rear main oil seal retainer must be unbolted and separated from the block before proceeding with crankshaft removal.*
1 Before the crankshaft is removed, check the endplay. Mount a dial indicator with the stem in line with the crankshaft and just touching one of the crank throws **(see illustration)**.
2 Push the crankshaft all the way to the rear and zero the dial indicator. Next, pry the crankshaft to the front as far as possible and check the reading on the dial indicator. The distance that it moves is the endplay. If it's greater than specified in this Chapter, check the crankshaft thrust surfaces for wear. If no wear is evident, new main bearings should correct the endplay.
3 If a dial indicator isn't available, feeler gauges can be used. Gently pry or push the crankshaft all the way to the front of the engine. Slip feeler gauges between the crankshaft and the front face of the rear (thrust) main bearing to determine the clearance **(see illustration)**.
4 Loosen the main bearing cap bolts 1/4-turn at a time each, until the bearing cap and brace assembly can be removed by hand.
Note: *Loosen the bearing cap assembly bolts in the reverse of the tightening sequence (see Section 24).* The main bearing cap assembly typically have cast-in arrows, which points to the drivebelt end of the engine **(see illustration)**. If the caps do not have markings, use a number stamp or punch to indicate the direction and location of the main bearing caps.
Note: *On SOHC engines it will only be necessary to mark the direction of the front and rear since the bearing cap and brace assembly are one piece.*
5 Gently tap the cap assembly with a soft-face hammer, then separate it from the engine block. If necessary, use the bolts as levers to remove the cap assembly. Try not to drop the bearing inserts if they come out with the cap assembly.
6 Carefully lift the crankshaft out of the engine. It may be a good idea to have an assistant available, since the crankshaft is quite heavy. With the bearing inserts in place in the engine block and main bearing caps, return the cap assembly to it's location on the engine block and tighten the bolts finger tight.
7 On DOHC engines inspect the main bearing cap bolts for excessive stretching **(see illustration)**. If there is a difference of 0.0043 inch (0.11 mm) or larger in the diameter of the bolts between the indicated areas the bolts must be replaced. Many times it may be cheaper to purchase a complete set of bolts rather than a number of bolts individually.

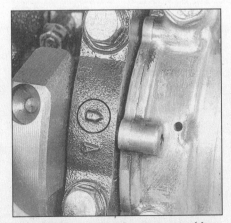

14.4 The main bearing cap assembly typically has cast-in arrows which point toward the drivebelt (front) end of the engine - If no arrows exist use a number stamp or punch to mark the location and direction of the cap assembly

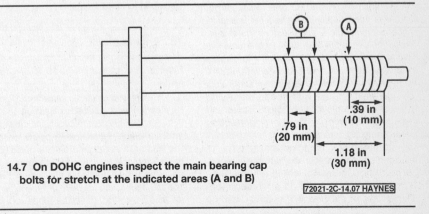

14.7 On DOHC engines inspect the main bearing cap bolts for stretch at the indicated areas (A and B)

Chapter 2 Part C General engine overhaul procedures

15.1a The core plugs can be removed by tapping in one edge until the plug turns sideways . . .

15.1b . . . then remove the core plug with pliers

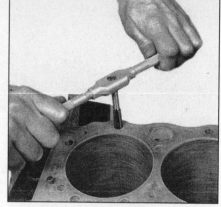

15.8 All bolt holes in the block - particularly the main bearing cap and cylinder head bolt holes - should be cleaned and restored with a tap (be sure to remove debris from the holes after this is done)

15 Engine block - cleaning

Refer to illustrations 15.1a, 15.1b, 15.8 and 15.10

Caution: *The core plugs (also known as freeze or soft plugs) may be difficult or impossible to retrieve if they're driven completely into the block coolant passages.*

1 Using the blunt end of a punch, tap in on the outer edge of the core plug to turn the plug sideways in the bore. Then use pliers to pull the core plug from the engine block **(see illustrations)**.

2 Using a gasket scraper, remove all traces of gasket material from the engine block. Be very careful not to nick or gouge the gasket sealing surfaces.

3 Remove the main bearing cap assembly and separate the bearing inserts from the caps and the engine block. Label the bearings, indicating which cylinder they were removed from and whether they were in the cap or the block, then set them aside.

4 Remove all of the threaded oil gallery plugs from the block. The plugs are usually very tight - they may have to be drilled out and the holes retapped. Use new plugs when the engine is reassembled.

5 If the engine is extremely dirty, it should be taken to an automotive machine shop to be steam cleaned or hot tanked.

6 After the block is returned, clean all oil holes and oil galleries one more time. Brushes specifically designed for this purpose are available at most auto parts stores. Flush the passages with warm water until the water runs clear, dry the block thoroughly and wipe all machined surfaces with a light, rust-preventive oil. If you have access to compressed air, use it to speed the drying process and to blow out all the oil holes and galleries. **Warning:** *Wear eye protection when using compressed air!*

7 If the block isn't extremely dirty or sludged up, you can do an adequate cleaning job with hot soapy water and a stiff brush. Take plenty of time and do a thorough job. Regardless of the cleaning method used, be sure to clean all oil holes and galleries very thoroughly, dry the block completely and coat all machined surfaces with light oil.

8 The threaded holes in the block must be clean to ensure accurate torque readings during reassembly. Run the proper size tap into each of the holes to remove rust, corrosion, thread sealant or sludge and restore damaged threads **(see illustration)**. If possible, use compressed air to clear the holes of debris produced by this operation. Now is a good time to clean the threads on the cylinder head bolts and the main bearing cap bolts as well.

9 Reinstall the main bearing caps and tighten the bolts finger tight.

10 After coating the sealing surfaces of the new core plugs with core plug sealant, install them in the engine block **(see illustration)**. Make sure they're driven in straight and seated properly or leakage could result. Special tools are available for this purpose, but a large socket, with an outside diameter that will just slip into the core plug, a 1/2-inch drive extension and a hammer will work just as well. **Note:** *Make sure the socket only contacts the inside of the core plug, not the rim.*

11 Apply non-hardening thread sealant to the new oil gallery plugs and thread them into the holes in the block. Make sure they're tightened securely.

12 If the engine isn't going to be reassembled right away, cover it with a large plastic trash bag to keep it clean.

16 Engine block - inspection

Refer to illustrations 16.4a, 16.4b, 16.4c and 16.11

1 Before the block is inspected, it should be cleaned as described in Section 15.

2 Visually check the block for cracks, rust and corrosion. Look for stripped threads in the threaded holes. It's also a good idea to have the block checked for hidden cracks by an automotive machine shop that has the special equipment to do this type of work. If defects are found, have the block repaired, if possible, or replaced.

3 Check the cylinder bores for scuffing and scoring.

4 Check the cylinders for taper and out-of-round conditions as follows **(see illustrations)**:

5 Measure the diameter of each cylinder at the top (just under the ridge area), center and bottom of the cylinder bore, parallel to the crankshaft axis.

6 Next measure each cylinder's diameter at the same three locations perpendicular to the crankshaft axis.

7 The taper of the cylinder is the difference between the bore diameter at the top of the cylinder and the diameter at the bottom. The out-of-round specification of the cylinder bore is the difference between the parallel and perpendicular readings. Compare your results to those listed in this Chapter's Specifications.

8 Repeat the procedure for the remaining pistons and cylinders.

9 If the cylinder walls are badly scuffed or scored, or if they're out-of-round or tapered beyond the limits given in this Chapter's

15.10 A large socket on an extension can be used to drive the new core plugs into the bores

Chapter 2 Part C General engine overhaul procedures

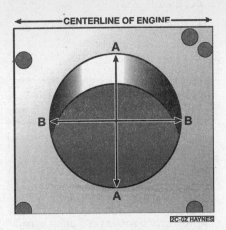

16.4a Measure the diameter of each cylinder at a right angle to the engine centerline (A), and parallel to engine centerline (B) - out-of-round is the distance between A and B; taper is the difference between A and B at the top of the cylinder and A and B at the bottom of the cylinder

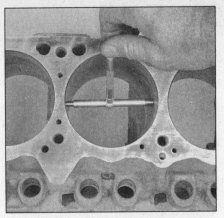

16.4b Use a telescoping gauge to measure the bore - the ability to "feel" when it is at the correct point will be developed over time, so work slowly and repeat the check until you're satisfied the bore measurement is accurate

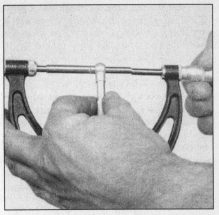

16.4c The gauge is then measured with a micrometer to determine the bore size

Specifications, have the engine block rebored and honed at an automotive machine shop. If a rebore is done, oversize pistons and rings will be required.

10 If the cylinders are in reasonably good condition and not worn to the outside of the limits, and if the piston-to-cylinder clearances can be maintained properly, then they don't have to be rebored. Honing is all that's necessary (see Section 17).

11 Using a precision straightedge and feeler gauge, check the block deck (the surface that mates with the cylinder head) for distortion **(see illustration)**.

17 Cylinder honing

Refer to illustrations 17.3a and 17.3b

1 Prior to engine reassembly, the cylinder bores must be honed so the new piston rings will seat correctly and provide the best possible combustion chamber seal. **Note:** *If you don't have the tools or don't want to tackle the honing operation, most automotive machine shops will do it for a reasonable fee.*

2 Before honing the cylinders, install the main bearing caps and tighten the bolts to the torque specified in this Chapter.

3 Two types of cylinder hones are commonly available - the flex hone or "bottle brush" type and the more traditional surfacing hone with spring-loaded stones. Both will do the job, but for the less experienced mechanic the "bottle brush" hone will probably be easier to use. You'll also need some kerosene or honing oil, rags and an electric drill motor. Proceed as follows:

a) *Mount the hone in the drill motor, compress the stones and slip it into the first cylinder* **(see illustration)**. *Be sure to wear safety goggles or a face shield!*

b) *Lubricate the cylinder with plenty of honing oil, turn on the drill and move the hone up-and-down in the cylinder at a pace that will produce a fine crosshatch pattern on the cylinder walls. Ideally, the crosshatch lines should intersect at approximately a 60-degree angle* **(see illustration)**. *Be sure to use plenty of lubricant and don't take off any more material than is absolutely necessary to produce the desired finish.* **Note:** *Piston ring manufacturers may specify a smaller crosshatch angle than the traditional 60-degrees - read and follow any instructions included with the new rings.*

c) *Don't withdraw the hone from the cylinder while it's running. Instead, shut off the drill and continue moving the hone up-and-down in the cylinder until it comes to a complete stop, then compress the stones and withdraw the hone. If you're using a "bottle brush" type hone, stop the drill motor, then turn the chuck in the normal direction of rotation while withdrawing the hone from the cylinder.*

d) *Wipe the oil out of the cylinder and repeat the procedure for the remaining cylinders.*

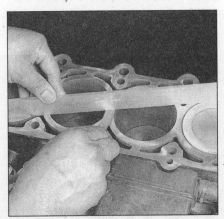

16.11 Check the flatness of the top of the block with a straightedge and feeler gauge - if distortion exceeds Specifications, the block deck will have to be machined

17.3a A "bottle brush" hone is the easiest type of hone to use

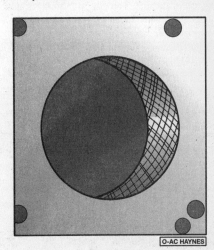

17.3b The cylinder hone should leave a smooth, crosshatch pattern with the lines intersecting at approximately a 60-degree angle

2C-18   Chapter 2 Part C General engine overhaul procedures

4   After the honing job is complete, chamfer the top edges of the cylinder bores with a small file so the rings won't catch when the pistons are installed. Be very careful not to nick the cylinder walls with the end of the file.
5   The entire engine block must be washed again very thoroughly with warm, soapy water to remove all traces of the abrasive grit produced during the honing operation. **Note:** *The bores can be considered clean when a lint-free white cloth - dampened with clean engine oil - used to wipe them out doesn't pick up any more honing residue, which will show up as gray areas on the cloth.* Be sure to run a brush through all oil holes and galleries and flush them with running water.
6   After rinsing, dry the block and apply a coat of light rust-preventive oil to all machined surfaces. Wrap the block in a plastic trash bag to keep it clean and set it aside until reassembly.

18 Pistons/connecting rods - inspection

Refer to illustrations 18.4a, 18.4b, 18.10, 18.11a and 18.11b

1   Before the inspection process can be carried out, the piston/connecting rod assemblies must be cleaned and the original piston rings removed from the pistons. **Note:** *Always use new piston rings when the engine is reassembled.*
2   Using a piston ring installation tool, carefully remove the rings from the pistons. Be careful not to nick or gouge the pistons in the process.
3   Scrape all traces of carbon from the top of the piston. A hand-held brass wire brush or a piece of fine emery cloth can be used (with solvent) once the majority of the deposits have been scraped away. Do not, under any circumstances, use a wire brush mounted in a drill motor to remove deposits from the pistons. The piston material is soft and may be eroded away by the wire brush.

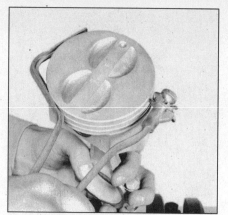

18.4a The piston ring grooves can be cleaned with a special tool, as shown here . . .

18.4b . . . or a section of a broken ring

4   Use a piston ring groove-cleaning tool to remove carbon deposits from the ring grooves. If a tool isn't available, a piece broken off the old ring will do the job. Be very careful to remove only the carbon deposits - don't remove any metal and do not nick or scratch the sides of the ring grooves **(see illustrations)**.
5   Once the deposits have been removed, clean the piston/rod assemblies with solvent and dry them with compressed air (if available). Make sure the oil return holes in the back sides of the ring grooves are clear.
6   If the pistons and cylinder walls aren't damaged or worn excessively, and if the engine block is not rebored, new pistons won't be necessary. Normal piston wear appears as even vertical wear on the piston thrust surfaces and slight looseness of the top ring in its groove. New piston rings, however, should always be used when an engine is rebuilt.
7   Carefully inspect each piston for cracks around the skirt, at the pin bosses and at the ring lands.
8   Look for scoring and scuffing on the thrust faces of the skirt, holes in the piston crown and burned areas at the edge of the crown. If the skirt is scored or scuffed, the engine may have been suffering from overheating and/or abnormal combustion, which caused excessively high operating temperatures. The cooling and lubrication systems should be checked thoroughly. A hole in the piston crown is an indication that abnormal combustion (preignition) was occurring. Burned areas at the edge of the piston crown are usually evidence of spark knock (detonation). If any of the above problems exist, the causes must be corrected or the damage will occur again. The causes may include intake air leaks, incorrect fuel/air mixture, incorrect ignition timing and EGR system malfunctions.
9   Corrosion of the piston, in the form of small pits, indicates that coolant is leaking into the combustion chamber and/or the crankcase. Again, the cause must be corrected or the problem may persist in the rebuilt engine.
10 Measure the piston ring side clearance by laying a new piston ring in each ring groove and slipping a feeler gauge in beside it **(see illustration)**. Check the clearance at three or four locations around each groove. Be sure to use the correct ring for each groove - they are different. If the side clear-

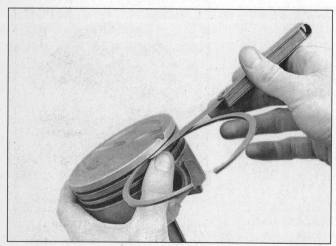

18.10 Check the ring side clearance with a feeler gauge at several points around the groove

18.11a Measure the piston diameter at a 90-degree angle to the piston pin . . .

Chapter 2 Part C General engine overhaul procedures

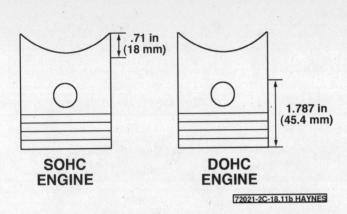

18.11b . . . at the indicated area for your type of engine

19.1 The oil holes should be chamfered so sharp edges don't gouge or scratch the new bearings

ance is greater than specified in this Chapter, new pistons will have to be used.

11   Check the piston-to-bore clearance by measuring the bore (see Section 16) and the piston diameter. Make sure the pistons and bores are correctly matched. Measure the piston across the skirt, at a 90-degree angle to the piston pin at the specified distance from the bottom or top of the piston **(see illustrations)**. Subtract the piston diameter from the bore diameter to obtain the clearance. If it's greater than specified in this Chapter, the block will have to be rebored and new pistons and rings installed.

12   Check the piston-to-rod clearance by twisting the piston and rod in opposite directions. Any noticeable play indicates excessive wear, which must be corrected. The piston/connecting rod assemblies should be taken to an automotive machine shop to have the pistons and rods resized and new pins installed.

13   If the pistons must be removed from the connecting rods for any reason, they should be taken to an automotive machine shop. While they are there have the connecting rods checked for bend and twist, since automotive machine shops have special equipment for this purpose. **Note:** *Unless new pistons and/or connecting rods must be installed, do not disassemble the pistons and connecting rods.*

14   Check the connecting rods for cracks and other damage. Temporarily remove the rod caps, lift out the old bearing inserts, wipe the rod and cap bearing surfaces clean and inspect them for nicks, gouges and scratches. After checking the rods, replace the old bearings, slip the caps into place and tighten the nuts finger tight. **Note:** *If the engine is being rebuilt because of a connecting rod knock, be sure to install new or rebuilt rods.*

19 Crankshaft - inspection

Refer to illustrations 19.1, 19.2, 19.5 and 19.7

1   Remove all burrs from the crankshaft oil holes with a stone, file or scraper **(see illustration)**.

2   Clean the crankshaft with solvent and dry it with compressed air (if available). Be sure to clean the oil holes with a stiff brush **(see illustration)** and flush them with solvent.

3   Check the main and connecting rod bearing journals for uneven wear, scoring, pits and cracks.

4   Check the rest of the crankshaft for cracks and other damage. It should be magnafluxed to reveal hidden cracks - an automotive machine shop will handle the procedure.

5   Using a micrometer, measure the diameter of the main and connecting rod journals **(see illustration)** and compare the results to the Specifications in this Chapter. By measuring the diameter at a number of points around each journal's circumference, you'll be able to determine whether or not the journal is out-of-round. Take the measurement at each end of the journal, near the crank throws, to determine if the journal is tapered.

6   If the crankshaft journals are damaged, tapered, out-of-round or worn beyond the limits given in the Specifications in this Chapter, have the crankshaft reground by an automotive machine shop. Be sure to use the correct size bearing inserts if the crankshaft is reconditioned.

7   Check the oil seal journals at each end of the crankshaft for wear and damage. If the seal has worn a groove in the journal, or if it's

19.2 Use a wire or stiff plastic bristle brush to clean the oil passages in the crankshaft

19.5 Measure the diameter of each crankshaft journal at several points to detect taper and out-of-round conditions

2C-20 Chapter 2 Part C General engine overhaul procedures

19.7 If the seals have worn grooves in the crankshaft journals, or if the seal contact surfaces are nicked or scratched, the new seals will leak

nicked or scratched **(see illustration)**, the new seal may leak when the engine is reassembled. In some cases, an automotive machine shop may be able to repair the journal by pressing on a thin sleeve. If repair isn't feasible, a new or different crankshaft should be installed.

8   Refer to Section 20 and examine the main and rod bearing inserts.

20 Main and connecting rod bearings - inspection and main bearing selection

Inspection

Refer to illustration 20.1

1   Even though the main and connecting rod bearings should be replaced with new ones during the engine overhaul, the old bearings should be retained for close examination, as they may reveal valuable information about the condition of the engine **(see illustration)**.

2   Bearing failure occurs because of lack of lubrication, the presence of dirt or other foreign particles, overloading the engine and corrosion. Regardless of the cause of bearing failure, it must be corrected before the engine is reassembled to prevent it from happening again.

3   When examining the bearings, remove them from the engine block, the main bearing caps, the connecting rods and the rod caps and lay them out on a clean surface in the same general position as their location in the engine. This will enable you to match any bearing problems with the corresponding crankshaft journal.

4   Dirt and other foreign particles get into the engine in a variety of ways. It may be left in the engine during assembly, or it may pass through filters or the PCV system. It may get into the oil, and from there into the bearings. Metal chips from machining operations and normal engine wear are often present. Abrasives are sometimes left in engine components after reconditioning, especially when

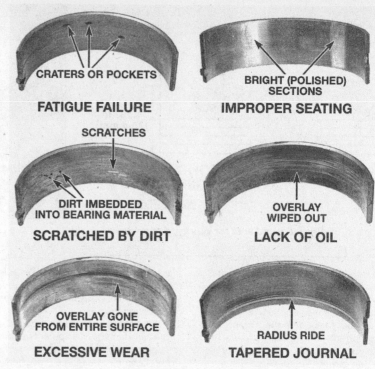

20.1 Typical bearing failures

parts are not thoroughly cleaned using the proper cleaning methods. Whatever the source, these foreign objects often end up embedded in the soft bearing material and are easily recognized. Large particles will not embed in the bearing and will score or gouge the bearing and journal. The best prevention for this cause of bearing failure is to clean all parts thoroughly and keep everything spotlessly clean during engine assembly. Frequent and regular engine oil and filter changes are also recommended.

5   Lack of lubrication (or lubrication breakdown) has a number of interrelated causes. Excessive heat (which thins the oil), overloading (which squeezes the oil from the bearing face) and oil leakage or throw off (from excessive bearing clearances, worn oil pump or high engine speeds) all contribute to lubrication breakdown. Blocked oil passages, which usually are the result of misaligned oil holes in a bearing shell, will also oil starve a bearing and destroy it. When lack of lubrication is the cause of bearing failure, the bearing material is wiped or extruded from the steel backing of the bearing. Temperatures may increase to the point where the steel backing turns blue from overheating.

6   Driving habits can have a definite effect on bearing life. Low-speed operation in too-high a gear (lugging the engine) puts very high loads on bearings, which tends to squeeze out the oil film. These loads cause the bearings to flex, which produces fine cracks in the bearing face (fatigue failure). Eventually the bearing material will loosen in pieces and tear away from the steel backing. Short-trip driving leads to corrosion of bearings because insuf-

ficient engine heat is produced to drive off the condensed water and corrosive gases. These products collect in the engine oil, forming acid and sludge. As the oil is carried to the engine bearings, the acid attacks and corrodes the bearing material.

7   Incorrect bearing installation during engine assembly will lead to bearing failure as well. Tight-fitting bearings leave insufficient bearing oil clearance and will result in oil starvation. Dirt or foreign particles trapped behind a bearing insert result in high spots on the bearing which lead to failure.

Selection

8   If the original bearings are worn or damaged, or if the oil clearances are incorrect (see Sections 24 or 26), new bearings will have to be purchased. It is rare during a thorough rebuild of an engine with many miles on it that new replacement bearings would not be employed. However, if the crankshaft has been reground, new **undersize** bearings must be installed.

9   The automotive machine shop that reconditions the crankshaft will provide or help you select the correct size bearings. Depending on how much material has to be ground from the crankshaft to restore it, different undersize bearings are required. Crankshafts are normally ground in increments of 0.010-inch. Sometimes the amount of material machined on a crankshaft will differ between the mains and rod journals, especially if a rod journal was damaged. Markings on most reground crankshafts indicate how much was machined, such as "10-10", meaning that 0.010-inch was removed from both

Chapter 2 Part C General engine overhaul procedures   2C-21

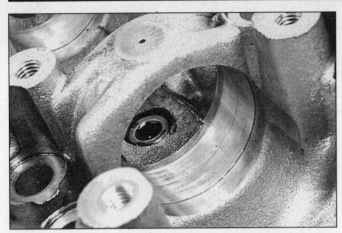

21.1 Inspect the cam bearing surfaces in each cylinder head for pits, score marks and abnormal wear - if wear or damage is noted, the cylinder head must be replaced

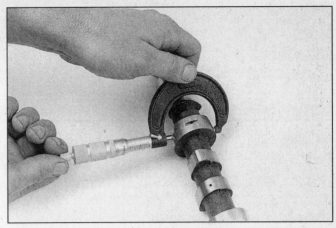

21.2a Measure the outside diameter of each camshaft journal and the inside diameter of each bearing to determine the oil clearance measurement

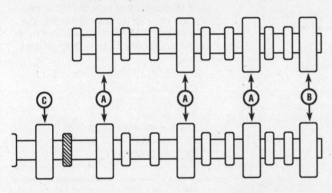

21.2b SOHC engine camshaft journal designations - B are the rearmost journals, while C is the journal ahead of the distributor drive gear

21.4 Measuring cam lobe height with a micrometer; make sure you move the micrometer to get the highest reading (top of cam lobe)

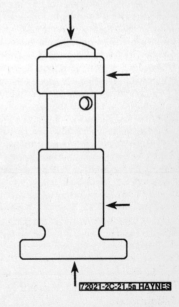

21.5a On SOHC engines check the contact and sliding surfaces of each lifter for wear and damage at the areas shown

the rod and main journals. Such a crankshaft would require 0.010-inch undersize bearings, a common replacement bearing size.

10 Regardless of how the bearing sizes are determined, use the oil clearance, measured with Plastigage, as the final guide to ensure the bearings are the right size. If you have any questions or are unsure which bearings to use, get help from your machine shop or a dealer parts or service department.

21 Camshafts, lifters and bearings - inspection

Refer to illustrations 21.1, 21.2a, 21.2b, 21.4, 21.5a, 21.5b, 21.6 and 21.7

1 Visually check the camshaft bearing surfaces for pitting, score marks, galling and abnormal wear. If the bearing surfaces are damaged, the cylinder head will have to be replaced **(see illustration)**.

2 Measure the outside diameter of each camshaft bearing journal and record your measurements **(see illustrations)**. Compare them to the journal outside diameter specified in this Chapter, then measure the inside diameter of each corresponding camshaft bearing and record the measurements. Subtract each cam journal outside diameter from its respective cam bearing bore inside diameter to determine the oil clearance for each bearing. Compare the results to the specified journal-to-bearing clearance. If any of the measurements fall outside the standard specified wear limits in this Chapter, either the camshaft or the cylinder head, or both, must be replaced.

3 Check camshaft runout by placing the camshaft between two V-blocks or back into the cylinder head and set up a dial indicator on the center journal. Zero the dial indicator. Turn the camshaft slowly and note the dial indicator readings. Record your readings and compare them with the specified runout in this Chapter. If the measured runout exceeds the runout specified in this Chapter, replace the camshaft.

4 Check the camshaft lobe height by measuring each lobe with a micrometer **(see illustration)**. Compare the measurement to the cam lobe height specified in this Chapter. Then subtract the measured cam lobe height from the specified height to compute wear on the cam lobes. Compare it to the specified wear limit. If it's greater than the specified wear limit, replace the camshaft.

5 Inspect the contact and sliding surfaces of each lifter for wear and scratches **(see illustrations)**. **Note:** *If the lifter pad is worn, it's a good idea to check the corresponding camshaft lobe or adjusting shim on DOHC engines, because it will probably be worn too.*

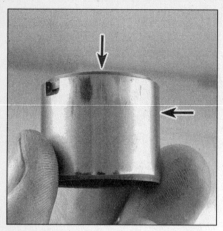

21.5b On DOHC engines inspect the valve lifter at the areas shown – don't forget to also inspect the valve adjusting shims for wear

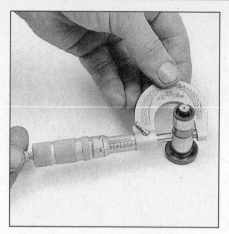

21.6 Measure the outside diameter of each lifter with a micrometer . . .

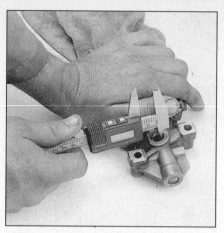

21.7 . . . and the inside diameter of each lifter bore - subtract the lifter diameter from the lifter bore diameter to obtain the lifter-to-guide clearance (SOHC engine shown, on DOHC the procedure is the same except that the lifter bore diameter is an integral part of the cylinder head

Caution: *On SOHC engines do not lay the lifters on their side or upside down, or air can become trapped inside and the lifter will have to be bled (see Chapter 2A). The lifters can be laid on their side only if they are submerged in a pan of clean engine oil until reassembly.*

6   Measure the outside diameter of each lifter with a micrometer **(see illustration)** and compare it to the Specifications in this Chapter. If any lifter is worn beyond the specified limit, replace it.

7   Check each lifter bore diameter **(see illustration)** and compare the results to the Specifications in this Chapter. If any lifter bore is worn beyond the specified limit, the lifter guide assembly (SOHC engines) or the cylinder head (DOHC engines) must be replaced.

8   Subtract the outside diameter of each lifter from the inside diameter of the lifter bore and compare the difference to the clearance specified in this Chapter. If both the lifter and the bore are within acceptable limits, this measurement should fall within tolerance as well. However, if you purchase new parts such as the lifters, the lifter guide assembly on SOHC engines or the cylinder head has been changed on DOHC engines, you may find that this clearance no longer falls within the specified limit.

8   On SOHC engines, inspect the rocker arms and shafts as described in Chapter 2A. If the pads are worn the rocker arms must be replaced. Don't under any circumstances attempt to restore rocker arms by grinding the pad surfaces.

9   In any case make sure all the parts new or old have been thoroughly inspected before reassembly.

22 Engine overhaul - reassembly sequence

1   Before beginning engine reassembly, make sure you have all the necessary new parts (including new cylinder head bolts), gaskets and seals as well as the following items on hand:

Common hand tools
A 1/2-inch drive torque wrench
Piston ring installation tool
Piston ring compressor
Short lengths of rubber or plastic hose to fit over connecting rod bolts
Plastigage
Feeler gauges
A fine-tooth file
New engine oil
Engine assembly lube or moly-base grease
Gasket sealant
Thread locking compound

2   In order to save time and avoid problems, engine reassembly must be done in the following general order:

SOHC engines

Piston rings
Crankshaft and main bearings
Rear main oil seal and retainer
Piston/connecting rod assemblies
Oil pump and pick up tube
Oil pan
Cylinder heads, camshafts, lifters and rocker arms
Timing belt and sprockets
Timing belt covers
Crankshaft pulley
Intake and exhaust manifolds
Valve covers
Engine rear plate
Flywheel/driveplate

DOHC engines

Piston rings
Crankshaft and main bearings
Rear main oil seal and retainer
Piston/connecting rod assemblies
Oil pump and pick up tube
Cylinder heads
Camshaft and lifters
Rear timing cover
Timing chains and sprockets
Front timing cover
Crankshaft pulley
Aluminum oil pan
Steel Oil pan
Intake and exhaust manifolds
Valve covers
Flywheel/driveplate

23 Piston rings - installation

Refer to illustrations 23.3, 23.4, 23.9a, 23.9b and 23.12

1   Before installing the new piston rings, the ring end gaps must be checked. It's assumed that the piston ring side clearance has been checked and verified correct (Section 18).

2   Lay out the piston/connecting rod assemblies and the new ring sets so the ring sets will be matched with the same piston and cylinder during the end gap measurement and engine assembly.

3   Insert the top (number one) ring into the first cylinder and square it up with the cylinder walls by pushing it in with the top of the piston **(see illustration)**. The ring should be near the bottom of the cylinder, at the lower limit of ring travel.

4   To measure the end gap, slip feeler gauges between the ends of the ring until a gauge equal to the gap width is found **(see illustration)**. The feeler gauge should slide between the ring ends with a slight amount of drag. Compare the measurement to the Specifications in this Chapter. If the gap is larger or smaller than specified, double-check to make sure you have the correct rings before proceeding.

5   If the gap is too small, you may have to file the rings to fit or exchange the set. The type of ring set you buy, and the material the rings are faced with, determine whether they

Chapter 2 Part C General engine overhaul procedures

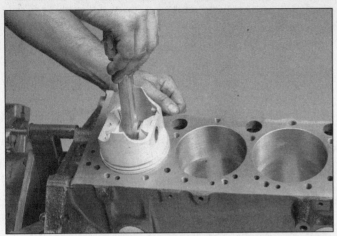

23.3 When checking piston ring end gap, the ring must be square in the cylinder bore (this is done by pushing the ring down with the top of a piston as shown)

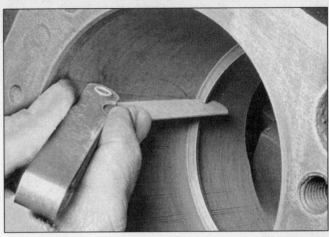

23.4 With the ring square in the cylinder, measure the end gap with a feeler gauge

23.9a Installing the spacer/expander in the oil control ring groove

23.9b DO NOT use a piston ring installation tool when installing the oil ring side rails

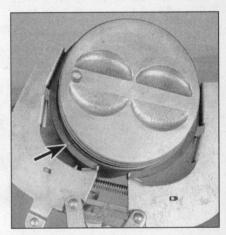

23.12 Installing the compression rings with a ring expander - the mark (arrow) must face up

can be filed. Carefully read the instructions with the ring set.

6 Excess end gap isn't as critical as too little gap, unless the gap is greater than 0.040-inch. Compare your measurements to this Chapter's Specifications for maximum end gap. Again, double-check to make sure you have the correct rings for your engine. If you do file the ring gaps, mount a file in a vise, lubricate the tops of the jaws, and slide the ring back and forth across the file, resting the ring against the top of the jaws and with even pressure on both sides of the ring gap. File a little, then recheck that ring's end gap in the bore before filing any more. When the correct gap is achieved, use a whetstone or fine file to deburr the edges that have been filed.

7 Repeat the procedure for each ring that will be installed in the first cylinder and for each ring in the remaining cylinders. Remember to keep rings, pistons and cylinders matched up.

8 Once the ring end gaps have been checked/corrected, the rings can be installed on the pistons.

9 The oil control ring (lowest one on the piston) is usually installed first. It's composed of three separate components. Slip the spacer/expander into the groove (see illustration). If an anti-rotation tang is used, make sure it's inserted into the drilled hole in the ring groove. Next, install the lower side rail. Don't use a piston ring installation tool on the oil ring side rails, as they may be damaged. Instead, place one end of the side rail into the groove between the spacer/expander and the ring land, hold it firmly in place and slide a finger around the piston while pushing the rail into the groove (see illustration). Next, install the upper side rail in the same manner.

10 After the three oil ring components have been installed, check to make sure that both the upper and lower side rails can be turned smoothly in the ring groove.

11 The number two (middle) ring is installed next. It's usually stamped with a mark which must face up, toward the top of the piston. **Note:** *Always follow the instructions printed on the ring package or box - different manufacturers may require different approaches. Do not mix up the top and middle rings, as they have different cross sections.*

12 Use a piston ring installation tool and make sure the identification mark is facing the top of the piston, then slip the ring into the middle groove on the piston (see illustration). Don't expand the ring any more than necessary to slide it over the piston.

13 Install the number one (top) ring in the same manner. Make sure the mark is facing up. Be careful not to confuse the number one and number two rings.

14 Repeat the procedure for the remaining pistons and rings.

24 Crankshaft - installation and main bearing oil clearance check

Refer to illustrations 24.5, 24.11, 24.13a, 24.13b and 24.15

1 Crankshaft installation is the first step in engine reassembly. It's assumed at this point that the engine block and crankshaft have been cleaned, inspected and repaired or reconditioned.

2 Position the engine on the stand with the crankcase facing up.

Chapter 2 Part C General engine overhaul procedures

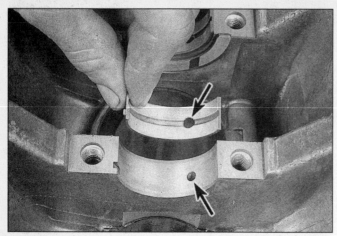

24.5 Make sure the oil holes in the bearings are aligned with the oil holes in the block (arrows)

24.11 Lay the Plastigage strips (arrow) on the main bearing journals, parallel to the crankshaft centerline

24.13a Main bearing cap TIGHTENING sequence (SOHC engines)

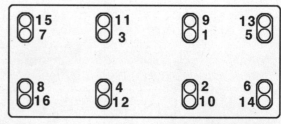

24.13b Main bearing cap TIGHTENING sequence (DOHC engines)

3 Remove the main bearing cap bolts and lift out the bearing cap and brace assembly.
4 If they're still in place, remove the original bearing inserts from the block and the main bearing caps. Wipe the bearing surfaces of the block and caps with a clean, lint-free cloth. They must be kept spotlessly clean.

Main bearing oil clearance check

5 Clean the back sides of the new main bearing inserts and lay one in each main bearing saddle in the block. If any of the bearing inserts have a large groove in it, make sure the grooved inserts are installed in the block. Lay the other bearings from the set in the corresponding main bearing caps. Make sure the tab on the bearing insert fits into the recess in the block or cap. **Caution:** *The oil holes in the block must line up with the oil holes in the bearing inserts* **(see illustration)**. *Do not hammer the bearing into place and don't nick or gouge the bearing faces. No lubrication should be used at this time.*
6 The flanged thrust bearing must be installed in the fourth (rear) cap and saddle on SOHC engines. On DOHC engines the thrust bearing must be installed in the third cap and saddle.
7 Clean the faces of the bearings in the block and the crankshaft main bearing journals with a clean, lint-free cloth.
8 Check or clean the oil holes in the crankshaft, as any dirt here can go only one way - straight through the new bearings.
9 Once you're certain the crankshaft is clean, carefully lay it in position in the main bearings. Do not lubricate the crankshaft with oil at this time.
10 Before the crankshaft can be permanently installed, the main bearing oil clearance must be checked.
11 Cut several pieces of the appropriate size Plastigage (they must be slightly shorter than the width of the main bearings) and place one piece on each crankshaft main bearing journal, parallel with the journal axis **(see illustration)**.
12 Clean the faces of the bearings in the cap and install the bearing caps and brace assembly with the arrows pointing toward the drivebelt end of the engine. Don't disturb the Plastigage.
13 Starting with the center main and working out toward the ends **(see illustrations)**, tighten the main bearing cap assembly bolts, in three steps, to the torque specified in this Chapter. Don't rotate the crankshaft at any time during this operation. **Note:** *Make sure the main bearing cap bolts have been thoroughly inspected as described in Section 14 before reusing any main bearing cap bolts or the torque readings may not be correct.*
14 Remove the bolts and carefully lift off the main bearing caps and brace assembly. Don't disturb the Plastigage or rotate the crankshaft.
15 Compare the width of the crushed Plastigage on each journal to the scale printed on the Plastigage envelope to obtain the main bearing oil clearance **(see illustration)**. Check the Specifications in this Chapter to make sure it's correct.
16 If the clearance is not as specified, the bearing inserts may be the wrong size (which means different ones will be required). Before deciding that different inserts are needed, make sure that no dirt or oil was between the bearing inserts and the caps or block when the clearance was measured. If the Plastigage was wider at one end than the other,

Chapter 2 Part C General engine overhaul procedures

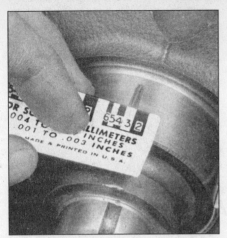

24.15 Compare the width of the crushed Plastigage to the scale on the envelope to determine the main bearing oil clearance (always take the measurement at the widest point of the Plastigage); be sure o use the correct scale - standard and metric ones are included

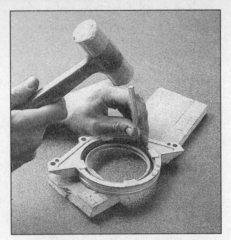

25.3 Place the retainer between two blocks of wood and drive the seal out of the retainer from the rear

25.4 Drive the new seal into the retainer with a block of wood or a section of pipe - make sure that you don't cock the seal in the bore

the journal may be tapered (refer to Section 19).

17 Carefully scrape all traces of the Plastigage material off the main bearing journals and/or the bearing faces. Use your fingernail or the edge of a credit card - don't nick or scratch the bearing faces.

Final crankshaft installation

18 Carefully lift the crankshaft out of the engine.
19 Clean the bearing faces in the block, then apply a thin, uniform layer of moly-base grease or engine assembly lube to each of the bearing surfaces. Be sure to coat the thrust faces as well as the journal face of the thrust bearing.
20 Make sure the crankshaft journals are clean, then lay the crankshaft back in place in the block.
21 Clean the faces of the bearings in the caps, then apply lubricant to them.
22 Install the bearing caps and brace assembly with the arrows pointing toward the drivebelt end of the engine.
23 Tighten the bearing cap bolts to 10-to-12 ft-lbs.
24 Gently tap the ends of the crankshaft forward and backward with a lead or brass hammer to line up the main bearing and crankshaft thrust surfaces.
25 Retighten all main bearing cap bolts to the specified torque, starting with the center main and working out toward the ends **(see illustration 24.13a and 24.13b)**.
26 Rotate the crankshaft a number of times by hand to check for any obvious binding.
27 The final step is to check the crankshaft endplay with a feeler gauge or a dial indicator as described in Section 14. The endplay should be correct if the crankshaft thrust faces aren't worn or damaged and new bearings have been installed.

28 Refer to Section 25 and install the new seal, then bolt the retainer to the block.

25 Rear main oil seal installation

Refer to illustrations 25.3 and 25.4

1 All models are equipped with a one-piece seal that fits into a housing (retainer) attached to the transaxle end of the block. The crankshaft must be installed first and the main bearing caps bolted in place, then the new seal should be installed in the retainer and the retainer bolted to the block.
2 Check the seal contact surface very carefully for scratches and nicks that could damage the new seal lip and cause oil leaks. If the crankshaft is damaged, the only alternative is a new or different crankshaft.
3 The old seal can be removed from the retainer with a hammer and punch by driving it out from the back side **(see illustration)**. Be sure to note how far it's recessed into the retainer bore before removing it; the new seal will have to be recessed an equal amount. Be very careful not to scratch or otherwise damage the bore in the retainer or oil leaks could develop.
4 Make sure the retainer is clean, then apply a thin coat of engine oil to the outer edge of the new seal. The seal must be pressed squarely into the retainer bore, so hammering it into place is not recommended. If you don't have access to a press, sandwich the retainer and seal between two smooth pieces of wood and press the seal into place with the jaws of a large vise. The pieces of wood must be thick enough to distribute the force evenly around the entire circumference of the seal. Work slowly and make sure the seal enters the bore squarely **(see illustration)**.
5 The seal lips must be lubricated with clean engine oil or multi-purpose grease before the seal/retainer is slipped over the crankshaft and bolted to the block. Apply RTV sealant to retainer housing to block surface - and make sure the dowel pins are in place before installing the retainer.
6 Tighten the screws a little at a time until the torque specified in this Chapter is reached.

26 Pistons/connecting rods - installation and rod bearing oil clearance check

Refer to illustrations 26.5, 26.11, 26.13 and 26.17

1 Before installing the piston/connecting rod assemblies, the cylinder walls must be perfectly clean, the top edge of each cylinder must be chamfered, and the crankshaft must be in place.
2 Remove the cap from the end of the number one connecting rod (refer to the marks made during removal). Remove the original bearing inserts and wipe the bearing surfaces of the connecting rod and cap with a clean, lint-free cloth. They must be kept spotlessly clean.

Connecting rod bearing oil clearance check

3 Clean the back side of the new upper bearing insert, then lay it in place in the connecting rod. Make sure the tab on the bearing fits into the recess in the rod. Don't hammer the bearing insert into place and be very careful not to nick or gouge the bearing face. Don't lubricate the bearing at this time.
4 Clean the back side of the other bearing insert and install it in the rod cap. Again, make sure the tab on the bearing fits into the recess in the cap, and don't apply any lubricant. It's critically important that the mating surfaces of the bearing and connecting rod are perfectly clean and oil free when they're assembled for clearance checking.

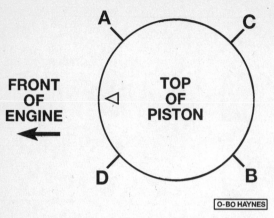

26.5 Stagger the ring end gaps as shown

A Oil ring expander
B Oil ring lower rail
C Top ring gap and upper oil ring rail
D Second ring gap

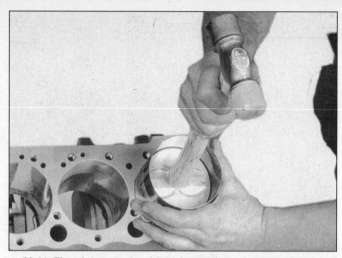

26.11 The piston can be driven (gently) into the cylinder bore with the end of a wooden hammer handle

5   Position the piston ring gaps at the specified intervals around the piston **(see illustration)**.
6   Slip a section of plastic or rubber hose over each connecting rod cap bolt.
7   Lubricate the piston and rings with clean engine oil and attach a piston ring compressor to the piston. Leave the skirt protruding about 1/4-inch to guide the piston into the cylinder. The rings must be compressed until they're flush with the piston.
8   Rotate the crankshaft until the number one connecting rod journal is at BDC (bottom dead center) and apply a coat of engine oil to the cylinder walls.
9   With the notch on top of the piston facing the drivebelt end of the engine, gently insert the piston/connecting rod assembly into the number one cylinder bore and rest the bottom edge of the ring compressor on the engine block.
10 Tap the top edge of the ring compressor to make sure it's contacting the block around its entire circumference.
11 Gently tap on the top of the piston with the end of a wooden hammer handle **(see illustration)** while guiding the end of the connecting rod into place on the crankshaft journal. The piston rings may try to pop out of the ring compressor just before entering the cylinder bore, so keep some downward pressure on the ring compressor. Work slowly, and if any resistance is felt as the piston enters the cylinder, stop immediately. Find out what's hanging up and fix it before proceeding. Do not, for any reason, force the piston into the cylinder - you might break a ring and/or the piston.
12 Once the piston/connecting rod assembly is installed, the connecting rod bearing oil clearance must be checked before the rod cap is permanently bolted in place.
13 Cut a piece of the appropriate size Plastigage slightly shorter than the width of the connecting rod bearing and lay it in place on the number one connecting rod journal, parallel with the journal axis **(see illustration)**.
14 Clean the connecting rod cap bearing face, remove the protective hoses from the connecting rod bolts and install the rod cap. Make sure the mating mark on the cap is on the same side as the mark on the connecting rod.
15 Install the nuts and tighten them to the torque specified in this Chapter (work up to it in three steps). **Note:** *Use a thin-wall socket to avoid erroneous torque readings that can result if the socket is wedged between the rod cap and nut. If the socket tends to wedge itself between the nut and the cap, lift up on it slightly until it no longer contacts the cap. Do not rotate the crankshaft at any time during this operation.*
16 Remove the nuts and detach the rod cap, being very careful not to disturb the Plastigage.
17 Compare the width of the crushed Plastigage to the scale printed on the Plastigage envelope to obtain the oil clearance **(see illustration)**. Compare it to the Specifications in this Chapter to make sure the clearance is correct.
18 If the clearance is not as specified, the bearing inserts may be the wrong size (which means different ones will be required). Before

26.13 Lay the Plastigage strips on each rod bearing journal, parallel to the crankshaft centerline

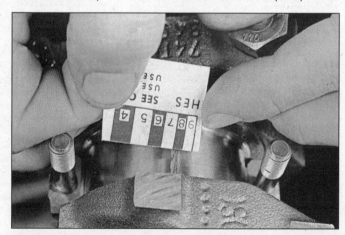

26.17 Measure the width of the crushed Plastigage to determine the rod bearing oil clearance (be sure to use the correct scale - standard and metric ones are included)

deciding that different inserts are needed, make sure that no dirt or oil was between the bearing inserts and the connecting rod or cap when the clearance was measured. Also, recheck the journal diameter. If the Plastigage was wider at one end than the other, the journal may be tapered (refer to Section 19).

Final connecting rod installation

19 Carefully scrape all traces of the Plastigage material off the rod journal and/or bearing face. Be very careful not to scratch the bearing - use your fingernail or the edge of a credit card.
20 Make sure the bearing faces are perfectly clean, then apply a uniform layer of clean moly-base grease or engine assembly lube to both of them. You'll have to push the piston into the cylinder to expose the face of the bearing insert in the connecting rod - be sure to slip the protective hoses over the rod bolts first.
21 Slide the connecting rod back into place on the journal, remove the protective hoses from the rod cap bolts, install the rod cap and tighten the nuts to the specified torque. Again, work up to the torque in three steps.
22 Repeat the entire procedure for the remaining pistons/connecting rods.
23 The important points to remember are . . .
 a) Keep the back sides of the bearing inserts and the insides of the connecting rods and caps perfectly clean when assembling them.
 b) Make sure you have the correct piston/rod assembly for each cylinder.
 c) The "W" mark on the piston and rod must face the drivebelt end of the engine.
 d) Lubricate the cylinder walls with clean oil.
 e) Lubricate the bearing faces when installing the rod caps after the oil clearance has been checked.

24 After all the piston/connecting rod assemblies have been properly installed, rotate the crankshaft a number of times by hand to check for any obvious binding.
25 As a final step, the connecting rod endplay must be checked. Refer to Section 13 for this procedure.
26 Compare the measured endplay to the Specifications to make sure it's correct. If it was correct before disassembly and the original crankshaft and rods were reinstalled, it should still be right. If new rods or a new crankshaft were installed, the endplay may be inadequate. If so, the rods will have to be removed and taken to an automotive machine shop for resizing. If the endplay is too great, new rods may be required.

27 Initial start-up and break-in after overhaul

Warning: *Have a fire extinguisher ready when starting the engine for the first time.*
1 Once the engine has been installed in the vehicle, double-check the engine oil and coolant levels.
2 With the spark plugs out of the engine and the ignition and fuel systems disabled (see Section 3), crank the engine until oil pressure registers on the gauge.
3 Install the spark plugs, and restore the ignition and fuel system functions.
4 Start the engine. It may take a few moments for the fuel system to build up pressure, but the engine should start without a great deal of effort. **Note:** *If backfiring occurs through the throttle body, recheck the valve timing and ignition timing.*
5 After the engine starts, it should be allowed to warm up to normal operating temperature. While the engine is warming up, make a thorough check for fuel, oil and coolant leaks. Also check the automatic transaxle fluid level (if equipped).
6 Shut the engine off and recheck the engine oil and coolant levels.
7 Drive the vehicle to an area with no traffic, accelerate from 30 to 50 mph, then allow the vehicle to slow rapidly to 30 mph with the throttle closed. Repeat the procedure 10 or 12 times. This will load the piston rings and cause them to seat properly against the cylinder walls. Check again for oil and coolant leaks.
8 Drive the vehicle gently for the first 500 miles (no sustained high speeds) and keep a constant check on the oil level. It is not unusual for an engine to use oil during the break-in period.
9 At approximately 500 to 600 miles, change the oil and filter.
10 For the next few hundred miles, drive the vehicle normally. Do not pamper it or abuse it.
11 After 2000 miles, change the oil and filter again and consider the engine broken in.

Notes

Chapter 3
Cooling, heating and air conditioning systems

Contents

	Section
Air conditioning and heating system - check and maintenance	12
Air conditioning compressor - removal and installation	14
Air conditioning condenser - removal and installation	15
Air conditioning receiver/drier - removal and installation	13
Antifreeze - general information	2
Blower motor - removal and installation	9
Blower motor circuit - check	8
CHECK ENGINE light	See Chapter 6
Coolant level check	See Chapter 1
Coolant temperature sending unit - check and replacement	7
Cooling system servicing (draining, flushing and refilling)	See Chapter 1
Drivebelt check, adjustment and replacement	See Chapter 1
Engine cooling fan and circuit - check and component replacement	4
General information	1
Heater and air conditioning control assembly - removal and installation	10
Heater core - replacement	11
Radiator and coolant reservoir - removal and installation	5
Thermostat - check and replacement	3
Underhood hose check and replacement	See Chapter 1
Water pump - check and replacement	6

Specifications

General

Coolant capacity	See Chapter 1
Drivebelt tension	See Chapter 1
Radiator pressure cap rating	11 to 14 psi
Thermostat rating	
Valve opens	180-degrees F
Fully open	194-degrees F
Water control valve rating	
Valve opens	203-degrees F
Fully open	226-degrees F
Valve lift	More than 0.315 in. at 226-degrees F
Refrigerant type (all models)	R-134a

Torque specifications

	Nm	In-lbs
Thermostat housing cover bolts		
SOHC engine	16 to 21	144 to 180
DOHC engine	8.4 to 11.2	75 to 99
Water control valve cover bolts	18 to 23	159 to 203
Water pump retaining bolts		
SOHC engine	16 to 21	144 to 180
DOHC engine	7 to 10	61 to 86
Steering column retaining bolts	15 to 18	132 to 156

Chapter 3 Cooling, heating and air conditioning systems

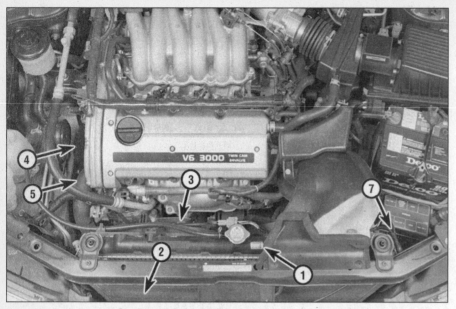

1.1 Cooling and air conditioning system components

1	Radiator	4	Water pump	6	Coolant reservoir
2	Condenser	5	Thermostat	7	Accumulator/drier
3	Engine cooling fan				

1 General information

Refer to illustrations 1.1 and 1.2

Engine cooling system

All vehicles covered by this manual employ a pressurized engine cooling system with thermostatically controlled coolant circulation (see illustration). An impeller-type water pump mounted on the engine block pumps coolant through the engine. The coolant flows around each cylinder and toward the rear of the engine. Cast-in coolant passages direct coolant around the intake and exhaust ports, near the spark plug areas and in close proximity to the exhaust valve guides.

A wax-pellet type thermostat controls engine coolant temperature. During warm up, the closed thermostat prevents coolant from circulating through the radiator. As the engine nears normal operating temperature, the thermostat opens and allows hot coolant to travel through the radiator, where it's cooled before returning to the engine (see illustration).

The cooling system is sealed by a pressure-type radiator cap, which raises the boiling point of the coolant and increases the cooling efficiency of the radiator. If the system pressure exceeds the cap pressure relief value, the excess pressure in the system forces the spring-loaded valve inside the cap off its seat and allows the coolant to escape through the overflow tube into a coolant reservoir. When the system cools the excess coolant is automatically drawn from the reservoir back into the radiator.

The coolant reservoir serves as both the point at which fresh coolant is added to the cooling system to maintain the proper fluid level and as a retaining tank for overheated coolant. This type of cooling system is known as a closed design because coolant that escapes past the pressure cap is saved and reused.

Heating system

The heating system consists of a blower fan and heater core located in the heater unit, the hoses connecting the heater core to the engine cooling system and the heater/air conditioning control panel on the dashboard. Hot engine coolant is circulated through the heater core. When the heater mode is activated, a flap door opens to expose the heater unit to the passenger compartment. A fan switch on the control head activates the blower motor, which forces air through the core, heating the air.

Air conditioning system

The air conditioning system consists of a condenser mounted in front of the radiator, an evaporator mounted adjacent to the heater core, a compressor mounted on the engine, a receiver-drier which contains a high pressure relief valve and the plumbing connecting all of the above components.

A blower fan forces the warmer air of the passenger compartment through the evaporator core (sort of a radiator-in-reverse), transferring the heat from the air to the refrigerant. The liquid refrigerant boils off into low pressure vapor, taking the heat with it when it leaves the evaporator. **Warning:** *The models covered by this manual are equipped with Supplemental Restraint Systems (SRS), more commonly known as airbags. Always disable the airbag system before working in the vicinity of any airbag system components to avoid the possibility of accidental deployment of the airbag(s), which could cause personal injury (see Chapter 12).*

2 Antifreeze - general information

Refer to illustration 2.4

Warning: *Do not allow antifreeze to come in contact with your skin or painted surfaces of the vehicle. Rinse off spills immediately with plenty of water. Antifreeze is highly toxic if ingested. Never leave antifreeze lying around in an open container or in puddles on the floor; children and pets are attracted by its sweet smell and may drink it. Check with local authorities about disposing of used antifreeze. Many communities have collection centers, which will see that antifreeze is disposed of safely. Never dump used anti-freeze on the ground or into drains.*

The cooling system should be filled with a water/ethylene glycol based antifreeze solution, which will prevent freezing down to at least -20-degrees F, or lower if local climate requires it. It also provides protection against corrosion and increases the coolant boiling point.

The cooling system should be drained, flushed and refilled at the specified intervals (see Chapter 1). Old or contaminated antifreeze solutions are likely to cause damage and encourage the formation of rust and scale in the system. Use distilled water with the antifreeze.

Before adding antifreeze, check all hose connections, because antifreeze tends to leak through very minute openings. Engines don't normally consume coolant, so if the level goes down, find the cause and correct it.

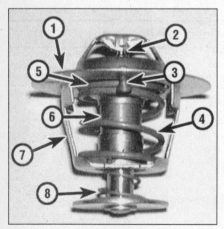

1.2 Typical thermostat

1. Flange
2. Piston
3. Jiggle valve
4. Main coil spring
5. Valve seat
6. Valve
7. Frame
8. Secondary coil spring

Chapter 3 Cooling, heating and air conditioning systems

2.4 The condition of your coolant can easily be checked with this type of hydrometer, available at auto parts stores

3.7 A thermostat can be accurately checked by heating it in a container of water with a thermometer and observing the opening and fully open temperature

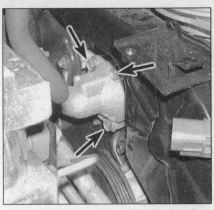

3.14a Thermostat cover retaining bolts (SOHC engine) (arrows indicate three of the four bolts)

The exact mixture of antifreeze-to-water that you should use depends on the relative weather conditions. The mixture should contain at least 50-percent antifreeze, but should never contain more than 70-percent antifreeze. Consult the mixture ratio chart on the antifreeze container before adding coolant. Hydrometers are available at most auto parts stores to test the ratio of antifreeze to water **(see illustration)** or antifreeze test strips are available instead of the hydrometer gauge. Use antifreeze that meets the vehicle manufacturer's specifications.

3 Thermostat and Water Control Valve - check and replacement

Warning: *Do not attempt to remove the radiator cap, coolant, thermostat or water control valve until the engine has cooled completely.* **Note:** *The 2000 and later models are also equipped with a water control valve in addition to the thermostat. The water control valve is mounted on the water outlet pipe assembly at the rear of the cylinder heads and connects to the cylinder block.*

Thermostat Check

Refer to illustration 3.7

1 Before assuming the thermostat is responsible for a cooling system problem, check the coolant level (see Chapter 1), drivebelt tension (see Chapter 1) and temperature gauge (or light) operation.
2 If the engine takes a long time to warm up (as indicated by the temperature gauge or heater operation), the thermostat is probably stuck open. Replace the thermostat with a new one.
3 If the engine runs hot, use your hand to check the temperature of the upper radiator hose. If the hose is not hot, but the engine is, the thermostat is probably stuck in the closed position, preventing the coolant inside the engine from traveling through the radiator. Replace the thermostat. **Caution:** *Do not drive the vehicle without a thermostat. The computer may stay in open loop and emissions and fuel economy will suffer.*

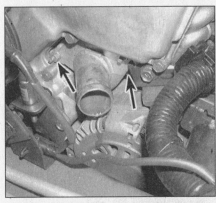

3.14b Thermostat cover retaining bolts (DOHC engine) (arrows indicate two of the three bolts)

4 If the lower radiator hose is hot, it means that the coolant is flowing and the thermostat is open. Consult the *Troubleshooting* Section at the front of this manual for further diagnosis.
5 A more thorough test of the thermostat can only be made when it is removed from the vehicle (see *Replacement*). If the thermostat remains in the open position at room temperature, it is faulty and must be replaced.
6 To test it fully, suspend the (closed) thermostat on a length of string or wire in a container of cold water, with a thermometer (cooking type that reads beyond 212 degrees F).
7 Heat the water on a stove while observing the temperature and the thermostat. Neither should contact the sides of the container **(see illustration)**.
8 Note the temperature when the thermostat begins to open and when it is fully open. Compare the temperatures to the Specifications in this Chapter. The number stamped into the thermostat is generally the fully open temperature. Some manufacturers provide Specifications for the beginning-to-open temperature, the fully open temperature, and sometimes the amount the valve should open.

3.15 On DOHC engines the thermostat is installed in the thermostat cover and is retained by two screws (arrows) - on SOHC engines the thermostat is installed in the thermostat housing on the engine block

9 If the thermostat doesn't open and close as specified, or sticks in any position, replace it.

Thermostat Replacement

Refer to illustrations 3.14a, 3.14b, 3.15 and 3.17

10 Disconnect the negative cable from the battery.
11 Drain the coolant from the radiator and the engine block (see Chapter 1).
12 On SOHC engines detach the upper radiator hose from the coolant outlet at the intake manifold and the radiator, then unbolt the hose clamp at the timing belt cover and remove the radiator hose. Detach the bypass hose from the thermostat cover.
13 On DOHC engines remove the drivebelts and the idler pulley bracket. Also remove the lower radiator hose and the water pump drain plug from the side of the block which faces the front of the vehicle.
14 Remove the thermostat cover from the engine **(see illustrations)**. Be prepared for some coolant to spill as the gasket seal is broken.
15 Remove the thermostat, noting the direction in which it was installed **(see illustration)**.

3-4 Chapter 3 Cooling, heating and air conditioning systems

3.17 On either engine, install the thermostat with the jiggle valve (arrow) UP (SOHC engine shown)

4.2 Typical engine cooling fan relay locations

A Low Speed
B High speed no. 1
C High speed no. 2

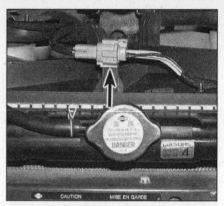

4.5 To test either fan motor, disconnect the electrical connector (arrow) and use jumper wires to connect the black wire terminal directly to ground; apply battery voltage to each of the other terminals in turn - if the fan still doesn't work, or it works at one speed but not both, replace the motor

16 Scrape off any old gasket or sealant on the thermostat housing and the thermostat cover, then clean them with lacquer thinner.

17 Apply a bead of RTV sealant around the perimeter of the cover, install the new thermostat with the jiggle valve UP (see illustration) and bolt the cover in place within 5 minutes of applying the sealant.

18 Installation is the reverse of removal. Tighten the thermostat cover fasteners to the torque listed in this Chapter's Specifications, then reinstall the hoses.

19 Wait at least a half-hour for the sealant to cure. Refill and bleed the cooling system (see Chapter 1). Run the engine and check for leaks and proper operation.

Water control valve check

20 Perform Steps 1 through 6 of *Thermostat check* in this section. Note that the water control valve is fully open at a temperature above boiling (212-degrees F), so you won't be able to see full opening of the valve in boiling water.

Water control valve replacement

21 Disconnect the negative cable from the battery.

22 Drain the coolant from the radiator and engine block (see Chapter 1).

23 Detach the cylinder block hose from the water control valve cover.

24 Remove the water control valve cover and gasket from the water outlet pipe.

25 Remove the water control valve from the water outlet pipe.

26 Scrape off any old gasket or sealant on the water outlet pipe and cover, then clean off with lacquer thinner.

27 Install the new water control valve and new gasket.

28 Installation is the reverse of removal. Tighten the water control valve cover bolts to the torque listed in this Chapter's Specifications, then reinstall the hose.

29 Refill and bleed the cooling system (see Chapter 1). Run the engine and check for leaks and proper operation.

4 Engine cooling fan and circuit - check and component replacement

Warning: *Do not work with your hands near the fans at any time that the engine is running or the key is ON. With the key ON, (even with the engine not running) the fan can start at any time, since it is controlled by coolant temperature.*

Check

Refer to illustrations 4.2 and 4.5

1 All models have a two-speed electric fan mounted in a plastic shroud attached to the back of the radiator.

2 Fan operation is controlled both by the PCM and the high and low-speed fan relays (see illustration). The coolant temperature sensor signals the PCM of engine temperature, and the PCM turns on the appropriate relay(s). At warm idle, the low-speed relay turns the fan on at low speed. When coolant temperature reaches 221 degrees F, the PCM turns the high-speed relays 1 and 2 on, causing the fan to run at high speed. There are two high-speed relays to ensure proper cooling even if one relay fails, and they will also operate if the coolant temperature sensor fails.

3 If the fan operates continuously, the fault could be the coolant temperature sensor or the relays. Refer to Chapter 6 for diagnosis of the sensor, and Chapter 12 for diagnosis of the relays.

4 Warm the engine up until the gauge on the instrument panel indicates the high side of NORMAL. The fan should come on. If not, check the cooling system fan control fuses in the engine compartment fuse panel and in the interior fuse panel (see Chapter 12 for fuse locations).

5 If the fuses checked OK, disconnect the electrical connector from the electric fan motor (see illustration). Attach a fused jumper wire with battery voltage to either of the two power terminals on the fan, and a chassis-ground jumper to the black wire terminal on the fan (see the wiring diagrams in Chapter 12). If the fan doesn't operate, it should be replaced.

6 If it does operate with jumper wires, but doesn't under normal driving conditions, connect a voltmeter to a chassis ground and probe the power terminals of the fan connector on the harness side. If the engine is hot and the temperature gauge shows above

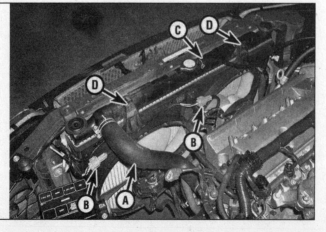

4.12 Remove the upper radiator hose (A), the fan electrical connectors (B), the overflow hose (C) and the fan shroud mounting bolts (D)

Chapter 3 Cooling, heating and air conditioning systems

4.14 Remove the nuts (A) retaining the fan to the motor and the screws (B) retaining the motor to the shroud - four out of six motor-to-shroud screws are not visible in this photo

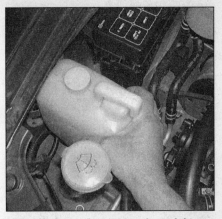

5.2 Lift the coolant reservoir straight up out of its bracket

NORMAL, there should be battery voltage at one of these terminals.

7   Check the ground of the circuit by switching your meter to the ohms scale. Ground one side of the meter and probe the other side at the black wire terminal of the fan connector. Resistance should be no more than 5 ohms. If resistance is high, trace the ground wire circuit to the chassis.

8   If there was no power at the terminals in Step 6, check that power is being supplied to the low-speed fan relay. One of its sockets on the relay panel should exhibit battery voltage at all times, and one only when the key is in the On or Start position.

9   If these sockets check OK, refer to Chapter 12 for checking continuity within the relays themselves.

Replacement

Refer to illustrations 4.12 and 4.14

10   Disconnect the electrical connector from the fan motor.

11   Disconnect the upper radiator hose and overflow hose at the radiator.

12   Remove the two fan shroud bolts and remove the fan/shroud assembly **(see illustration)**. *Note: The bottom of the fan shroud fits into tabs on the radiator.*

13   Remove the nut retaining the fan to the motor shaft.

14   Remove the screws retaining the motor to the shroud **(see illustration)**.

15   With an assistant retaining the fan, hit the shaft of the motor with a hammer and blunt punch to separate the fan from the motor.

16   Installation is the reverse of removal.

5   Radiator and coolant reservoir - removal and installation

Warning: *Wait until the engine is completely cool before beginning this procedure.*

Coolant reservoir

Refer to illustration 5.2

1   The coolant reservoir is mounted adjacent to the battery in the left corner of the engine compartment 1993 and 1994 models. On 1995 and later models the coolant reservoir is mounted on the opposite side of the engine compartment in the front right corner.

2   Detach the hose from the top of the radiator and pinch it off. On 1993 and 1994 models remove the reservoir retaining bolt and remove the reservoir from the vehicle. On 1995 and later models lift the reservoir straight up out of the bracket **(see illustration)**.

3   Pour the coolant into a container.

4   After washing the reservoir inside and out (use a household "bottle" brush to clean inside), inspect the reservoir for cracks and chafing. If it's damaged or so obscured by age as to make reading the water level difficult, replace it. **Warning:** *If you use a brush to clean the coolant reservoir, never again use it for cleaning drinking glasses or bottles.*

5   Installation is the reverse of removal.

Radiator

Refer to illustrations 5.9 and 5.12

6   Disconnect the cable from the negative terminal of the battery.

7   Set the parking brake and block the rear wheels. Raise the front of the vehicle and support it securely on jackstands.

8   Drain the cooling system (see Chapter 1). If the coolant is relatively new or in good condition, save it and reuse it. Read the **Warning** in Section 2.

9   Disconnect the automatic transmission cooler lines from the radiator if equipped **(see illustration)**. Use a drip pan to catch spilled fluid and plug the lines and fittings.

10   Loosen the hose clamps, then detach the radiator hoses from the radiator. If they're stuck, grasp each hose near the end with a pair of slip joint pliers and twist it to break the seal, then pull it off - be careful not to damage the radiator fittings! If the hoses are old or deteriorated, cut them off and install new ones. Also disconnect the small hose to the coolant reservoir.

11   Refer to Section 4 and remove the engine cooling fan assembly.

12   Unbolt the small brackets that attach the top of the radiator to the radiator support **(see illustration)**.

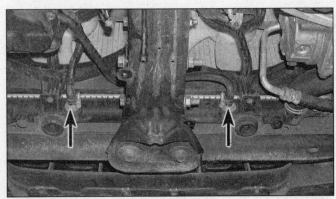

5.9 If equipped with an automatic transmission remove the cooler lines (arrows)

5.12 Remove the bolts (arrow) that attach the upper radiator mounts to the radiator support

6.3 SOHC engines are equipped with a water pump weep hole (arrow) on the underside of the water pump

6.9 On SOHC engines remove the water pump pulley bolts (arrows) while retaining the pulley with a strap wrench

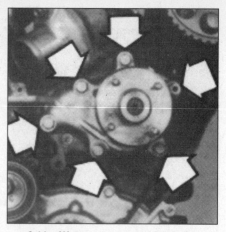

6.11a Water pump mounting bolts (SOHC engine)

13 Carefully lift out the radiator. Don't spill coolant on the vehicle or scratch the paint.

14 Inspect the radiator for leaks and damage. If it needs repair, have a radiator shop or dealer service department perform the work as special techniques are required.

15 Bugs and dirt can be removed from the radiator by spraying with a garden hose nozzle from the back side. The radiator should be flushed out with a garden hose before reinstallation.

16 Check the radiator mounts for deterioration and replace if necessary.

17 Installation is the reverse of the removal procedure. Guide the radiator into the mounts until they seat properly.

18 Refill and bleed the cooling system (see Chapter 1).

19 Start the engine and check for leaks. Allow the engine to reach normal operating temperature, indicated by the upper radiator hose becoming hot. Allow the engine to cool completely, then recheck the coolant level and add more if required.

20 Check and add automatic transmission fluid if needed (see Chapter 1).

6 Water pump - check and replacement

Warning: *Wait until the engine is completely cool before beginning this procedure.*

Check

Refer to illustration 6.3

1 A failure in the water pump can cause serious engine damage due to overheating.

2 There are two ways to check the operation of the water pump while it's installed on the engine. If the pump is found to be defective, it should be replaced with a new or rebuilt unit.

3 Water pumps are equipped with weep (or vent) holes **(see illustration)**. If a failure occurs in the pump seal, coolant will leak from the hole. On SOHC engines you can remove the timing belt cover and use a flashlight and small mirror to find the hole on the water pump from underneath to check for leaks. On DOHC engines, the weep hole directs coolant out from between the timing chain cover and the engine block.

4 If the water pump shaft bearings fail, there may be a howling sound at the pump while it's running. On SOHC engines shaft wear can be felt with the drivebelt removed if the water pump pulley is rocked up and down (with the engine off). On DOHC engines shaft wear can be felt by relieving the main timing chain tension (see Chapter 2B) and removing the water pump access cover from the front timing cover, then rock the water pump sprocket up and down to detect shaft wear. In either case don't mistake drivebelt slippage, which causes a squealing sound, for water pump bearing failure.

5 Even a pump that exhibits no outward signs of a problem, such as noise or leakage, can still be due for replacement. Removal for close examination is the only sure way to tell. Sometimes the fins on the back of the impeller can corrode to the point that cooling efficiency is hampered.

Replacement

Refer to illustrations 6.9, 6.11a and 6.11b

6 Disconnect the cable from the negative terminal of the battery.

7 Drain the cooling system (see Chapter 1). If the coolant is relatively new or in good condition, save it and reuse it.

8 Remove the drivebelts (see Chapter 1).

9 On SOHC engines, immobilize the water pump pulley with a strap wrench, remove the pulley bolts and pulley **(see illustration)**. Remove the crankshaft pulley and timing belt covers (see Chapter 2A).

10 On DOHC engines remove the drivebelt idler pulley bracket and the right side engine mount and brackets (see Chapter 2B). Also remove the main timing chain tensioner cover and the water pump cover from the front timing chain cover. Press the timing chain tensioner piston inward and insert an appropriate size pin into the tensioner hole to lock the tensioner in place. Rotate the crankshaft 20 degrees counterclockwise to loosen the chain around the water pump sprocket.

11 Remove the bolts and detach the water pump from the engine **(see illustrations)**. Check the impeller on the backside for evidence of corrosion or missing fins.

12 Clean the bolt threads and the threaded holes in the engine to remove corrosion and sealant.

13 Compare the new pump to the old one to make sure they're identical.

6.11b Water pump mounting bolts (DOHC engine) - after the mounting bolts have been removed, install two M8 bolts three to four inches long into the holes designated by the letter (A) and tighten them evenly until the water pump is forced out the engine block

Chapter 3 Cooling, heating and air conditioning systems

7.1a The temperature gauge sending unit (arrow) on SOHC engines is located adjacent to the distributor

7.1b The temperature gauge sending unit (A) on DOHC engines is located in the water outlet tube - (B) is the coolant temperature sensor for the computer

14 Remove all traces of old gasket sealant from the engine.
15 Clean the engine and new water pump mating surfaces with lacquer thinner or acetone.
16 On SOHC engines apply a thin layer of RTV sealant to the new pump, then carefully set a new gasket in place. On DOHC engines install new O-rings on the pump, then apply a light coat of grease on the O-rings to ease installation into the engine block.
17 Carefully attach the pump to the engine and thread the bolts into the holes finger tight. Use a small amount of RTV sealant on the bolt threads, and make sure that the dowel pins, if used, are in their original locations.
18 Tighten the bolts to the torque listed in this Chapter's Specifications in 1/4-turn increments. Don't overtighten the bolts or the pump may be distorted. On DOHC engines rotate the crankshaft 20 degrees clockwise to tighten the timing chain around the water pump sprocket. Remove the stopper pin from the timing chain tensioner and install the tensioner cover and water pump access cover after cleaning them and applying RTV sealant to them.
19 Reinstall all parts removed for access to the pump.
20 Refill and bleed the cooling system and check the drivebelt tension (see Chapter 1). Run the engine and check for leaks. **Note:** *Timing chain noise may be apparent after performing this procedure on DOHC engines. This noise is normal and should only last until the air has bled out of the high pressure chamber. Simply start the engine and allow it to run at 3,000 rpm with the transmission in neutral or park until the noise subsides.*

7 Coolant temperature sending unit - check and replacement

Warning 1: *Wait until the engine is completely cool before beginning this procedure.*
Warning 2: *This vehicle is equipped with an electric cooling fan. Stay clear of the fan, which can come on even when the engine is not running, as long as the ignition is ON.*

Check

Refer to illustrations 7.1a and 7.1b
1 The coolant temperature indicator system consists of a temperature gauge mounted in the instrument panel and a coolant temperature sending unit mounted on the engine **(see illustrations)**. There is more than one temperature sensor, but only one is used for the indicator system.
2 If an overheating indication occurs even when the engine is cold, check the wiring between the dash and the sending unit for a short circuit to ground.
3 If the gauge is inoperative, test the circuit by briefly grounding the wire to the sending unit while the ignition is On (engine not running for safety). If the gauge deflects full scale, replace the sending unit.
4 If the gauge doesn't respond in the test outlined in Step 3, check for an open circuit in the gauge wiring.
5 To test the sending unit, disconnect the electrical connector and attach an ohmmeter from the pin on top of the sender to an engine ground. With the engine slightly warm (140 degrees F) resistance should be 70 to 90 ohms for 1998 and earlier engines or 170 to 210 ohms for 1999 engines. When the engine is hot (212 degrees F), the resistance should drop to 21 to 24 ohms for 1998 and earlier engines or 47 to 53 ohms for 1999 engines. If the sender fails the test, replace it.

Replacement

6 If the sending unit must be replaced, unscrew it from the engine and quickly install the replacement. Use a conductive sealant on the threads (not Teflon tape). Make sure the engine is cool before removing the defective sending unit. There will be some coolant loss as the unit is removed, so be prepared to catch it. Check the coolant level after the replacement part has been installed.

8 Blower motor circuit - check

Refer to illustrations 8.4, 8.7, 8.12a and 8.12b
Warning: *The models covered by this manual are equipped with Supplemental Restraint Systems (SRS), more commonly known as airbags. Always disable the airbag system before working in the vicinity of any airbag system components to avoid the possibility of accidental deployment of the airbag(s), which could cause personal injury (see Chapter 12).*
Note: *This procedure applies to vehicles equipped with manual heating and air conditioning systems only. Vehicles equipped with automatic heating and air conditioning systems are very complex and considered beyond the scope of the home mechanic. Vehicles equipped with automatic heating and air conditioning systems should be taken to dealer service department or other qualified repair facility.*
1 If the blower motor speed does not correspond to the setting selected on the blower switch, or the blower motor does not operate at all, the problem could be a bad fuse, relay, switch, blower motor resistor, blower motor or blower motor circuit wiring.
2 Before checking the blower motor or circuit, always check the fuse and relay (if equipped) first (see Chapter 1).
3 Remove the glove compartment and lower dash trim (see Chapter 11) to gain access to the heater case and blower motor.
4 With the ignition key in the ON position, turn the blower switch to the faulty position(s) and, using a test light or voltmeter, check the voltage at the motor electrical connector (see

Chapter 3 Cooling, heating and air conditioning systems

8.4 Remove the blower motor cover screws (A) if equipped and test voltage at the blower motor electrical connector (B)

8.7 The blower resistor (arrow) is located in the blower motor housing

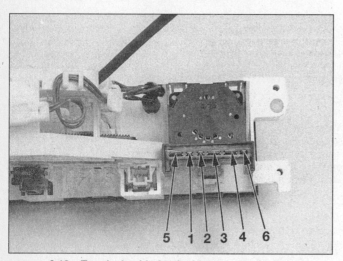

8.12a Terminal guide for the blower motor switch

Terminal	Off	1	2	3	4
1					X
2				X	
3			X		
4		X			
5		X	X	X	X
6		X	X	X	X

8.12b Blower speed control switch continuity chart. There must be continuity between the terminals marked with an X when the switch is in the indicated position

illustration). If the motor is receiving voltage but not operating, either the motor ground is bad (on these models the blower switch and resistor are part of the ground circuit) or the motor itself is faulty or the fan is binding.

5 To check for a bad ground, disconnect the electrical connector from the blower motor, connect a jumper wire between the ground wire terminal on the blower motor and a good ground, then connect a fused jumper wire between the battery positive terminal and the positive terminal on the blower motor. If the motor now operates properly, the ground circuit or resistor is bad. If the motor does not operate, the fan is either binding or faulty.

6 If you suspect the blower motor fan is binding, remove the blower motor (see Section 9) to check for free operation of the fan.

7 If the ground circuit to the motor is bad check the blower resistor for proper operation. The blower motor resistor is located near the blower motor **(see illustration)**.

8 Refer to the wiring diagrams at the end of this manual and probe the wires at the resistor leading from the blower switch. Using an ohmmeter, check for continuity to ground at the resistor electrical connector while the blower switch is rotated through all the switch position(s). Continuity to ground should be present at three of the terminals in all of the switch positions except Off.

9 If a ground signal does not exist at the blower resistor connector repair the blower motor switch and or the circuit leading to the switch.

10 If a ground signal from the switch does exist at the blower resistor connector, check the blower resistor for continuity between the terminals. If any checks indicate infinite resistance replace the blower resistor.

11 To check the blower speed switch refer to Section 10 and pull the heater/air conditioning control unit out from the dash enough to access the rear of the fan switch.

12 Unplug the electrical connector from the blower switch and check the continuity across the indicated switch terminals with an ohmmeter **(see illustrations)**.

13 If continuity isn't as specified, replace the switch.

9 Blower motor - removal and installation

Refer to illustration 9.4

Warning: *The models covered by this manual are equipped with Supplemental Restraint Systems (SRS), more commonly known as airbags. Always disable the airbag system before working in the vicinity of any airbag system components to avoid the possibility of accidental deployment of the airbag(s), which could cause personal injury (see Chapter 12).*

1 Disconnect the cable from the negative terminal of the battery.

2 Remove the glove compartment and lower dash trim (see Chapter 11) to gain access to the heater case and blower motor.

3 Detach the blower motor cover and dis-

Chapter 3 Cooling, heating and air conditioning systems

9.4 Blower motor retaining screws (arrows)

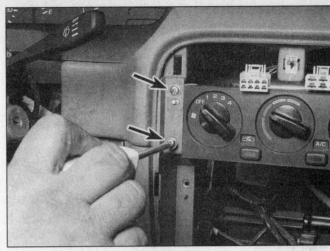

10.2 Remove the screws (arrows) retaining the control panel to the dashboard

connect the electrical connector from the blower motor **(see illustration 8.4)**.
4   Remove the blower motor mounting screws, and pull the blower motor carefully out of the housing **(see illustration)**.
5   If you are replacing the motor, detach the fan and transfer it to the new motor.
6   Installation is the reverse of removal. Run the blower and check for proper operation.

10 Heater and air conditioning control assembly - removal and installation

Refer to illustration 10.2
Warning: *The models covered by this manual are equipped with Supplemental Restraint Systems (SRS), more commonly known as airbags. Always disable the airbag system before working in the vicinity of any airbag system components to avoid the possibility of accidental deployment of the airbag(s), which could cause personal injury (see Chapter 12).*
1   Refer to Chapter 11 for removal of the center dash bezel, around the control assembly and radio.
2   Remove the four screws retaining the control assembly to the instrument panel **(see illustration)**.
3   Pull the control assembly out of the instrument panel and disconnect the electrical connectors.
4   Refer to Section 8 for electrical checks of the blower motor speed switch. The speed switch, function selector, and blend-control switch can all be removed from the control panel (manual air conditioning) by pulling the knob off from the front side, removing the four screws retaining the printed circuit housing, then depressing the plastic tabs on the back of the switch to release it from the control panel.
5   Installation is the reverse of the removal procedure.

11.4 Loosen the two heater hose clamps and disconnect the heater hoses (arrows) from the heater core inlet and outlet pipes at the firewall

11 Heater core - replacement

Refer to illustrations 11.4, 11.6, 11.7, 11.8a, 11.8b, 11.9a, 11.9b, 11.10, 11.11, 11.12, 11.14a and 11.14b
Warning 1: *The models covered by this manual are equipped with Supplemental Restraint Systems (SRS), more commonly known as airbags. Always disable the airbag system before working in the vicinity of any airbag system components to avoid the possibility of accidental deployment of the airbag(s), which could cause personal injury (see Chapter 12).*
Warning 2: *The air conditioning system is under high pressure. DO NOT loosen any fittings or remove any components until after the system has been discharged. Air conditioning refrigerant should be properly discharged into an EPA-approved container at a dealer service department or an automotive air conditioning repair facility. Always wear eye protection when disconnecting air conditioning system fittings.*
Warning 3: *Wait until the engine is completely cool before beginning this procedure.*
Note: *Replacement of the heater core is difficult procedure for the home mechanic, involving removal of the entire dashboard, floor console and many wiring connectors. If you attempt this procedure at home, keep track of the assemblies by taking notes and keeping screws and other hardware in small, marked plastic bags for reassembly.*
1   If the vehicle is equipped with air conditioning, have the air conditioning system discharged at a dealer service department or service station.
2   Turn the heater control setting to HOT. Drain the cooling system (see Chapter 1). If the coolant is relatively new, or tests in good condition (see Section 2), save it and re-use it. **Warning:** *Do not allow antifreeze to come in contact with your skin or painted surfaces of the vehicle. Rinse off spills immediately with plenty of water. Antifreeze is highly toxic if ingested. Never leave antifreeze lying around in an open container or in puddles on the floor; children and pets are attracted by it's sweet smell and may drink it. Check with local authorities about disposing of used antifreeze. Many communities have collection centers which will see that antifreeze is disposed of safely. Never dump used anti-freeze on the ground or into drains.*
3   Disconnect the cable from the negative terminal of the battery.
4   Working in the engine compartment, disconnect the heater hoses where they enter the firewall **(see illustration)**. **Caution:** *If the heater hoses are stuck, it is better to cut off the hoses than to twist them with pliers and risk breaking the heater core tubes.*

Chapter 3 Cooling, heating and air conditioning systems

11.6 Have the air conditioning system discharged (if equipped) then disconnect the refrigerant lines at the firewall

11.7 Release the clip and disconnect the water control valve cable

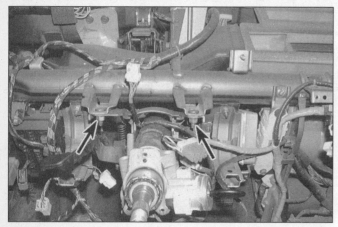

11.8a Remove these two steering column mounting nuts or bolts (arrows) and pivot the steering column down (instrument panel removed for clarity)

11.8b After removing the instrument panel and lowering the steering column, detach the reinforcement bar

5 Remove the rubber grommets where the heater core tubes go through the firewall.
6 Disconnect the air conditioning refrigerant lines and rubber grommet from the evaporator core if the vehicle is so equipped **(see illustration)**. **Warning:** *Always wear eye protection when disconnecting air conditioning system fittings.*
7 Release the clip that retains the water control valve cable and disconnect the cable from the control valve **(see illustration)**. Push the cable through the firewall so it can be removed with the heater core housing.
8 Lower the steering column **(see illustration)** and remove the entire instrument panel (refer to Chapter 11). Remove the instrument panel reinforcement bar **(see illustration)**. **Note:** *Heater core removal is very difficult. Take your time and don't use excessive force - there may be fasteners you haven't found yet.*
9 Remove the floor heater ducting, the steel dash-to-floor braces, any remaining electrical connectors, then refer to chapter 6

11.9a Remove the floor heater ducting . . .

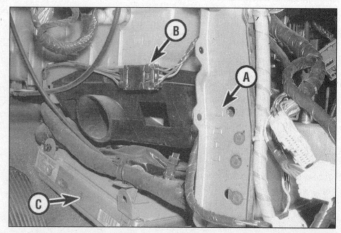

11.9b . . . the dash-to-floor brace (A), any remaining electrical connectors (B) and the PCM (C)

Chapter 3 Cooling, heating and air conditioning systems

11.10 Unsnap the duct from the bottom of the heater core housing

11.11 Evaporator core housing mounting bolts (arrows)

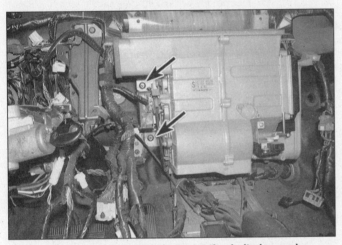

11.12 Heater core housing mounting bolts (arrows)

11.14a Remove the screw (arrow) and plastic brace . . .

to remove the PCM (computer) **(see illustrations)**.

10 Unsnap the duct from the bottom of the heater core housing **(see illustration)**.

11 Remove the evaporator core housing assembly (if equipped with air conditioning) or passenger side heater duct on models without air conditioning **(see illustration)**.

12 Remove bolts/nuts securing the heater unit to the firewall. Disconnect the control cables from the unit **(see illustration)**.

13 Pull down and back on the heater unit, making sure that all fasteners have been removed. Some twisting is required to separate the heater unit from the upper ducts.

14 Remove the heater core from the heater unit **(see illustrations)**.

15 Reassemble the heater unit and check the operation of the control flaps. If any parts bind, correct the problem before installation.

16 Reinstall the remaining parts in the reverse order of removal. When attaching the steering column to the support bracket, tighten the bolts (1993 and 1994 models) or nuts (1995 and later models) to 132 to 156 in-lbs (15 to 18 Nm).

17 Refill the cooling system (see Chapter 1), reconnect the battery and run the engine. Check for leaks and proper operation of the system. Have the air conditioning system recharged if equipped.

12 Air conditioning and heating system - check and maintenance

Air conditioning system

Refer to illustration 12.1

Warning: *The air conditioning system is under high pressure. Do not loosen any hose fittings or remove any components until after the system has been discharged. Air conditioning refrigerant should be properly discharged into an EPA-approved recovery/recycling unit at a dealer service department or an automotive air conditioning repair facility. Always wear eye protection when disconnecting air conditioning system fittings.*

Caution 1: *All models covered by this manual use environmentally friendly R-134a. This refrigerant (and its appropriate refrigerant oils) are not compatible R-12 refrigerant system components and must never be mixed or the components will be damaged.*

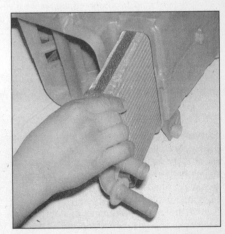

11.14b . . . then slide the heater core out of the case

Caution 2: *When replacing entire components, additional refrigerant oil should be added equal to the amount that is removed with the component being replaced. Be sure to read the can before adding any oil to the system, to make sure it is compatible with the R-134a system.*

Chapter 3 Cooling, heating and air conditioning systems

12.1 Check that the evaporator housing drain tube (arrow) at the firewall is clear of any blockage - the view here is from below the engine

12.9 Insert a thermometer in the center duct while operating the air conditioning system - the output air should be 35-40 degrees F less than the ambient temperature, depending on humidity (but not lower than 40-degrees F)

1 The following maintenance checks should be performed on a regular basis to ensure that the air conditioning continues to operate at peak efficiency.

   a) Inspect the condition of the compressor drivebelt. If it is worn or deteriorated, replace it (see Chapter 1).
   b) Check the drivebelt tension and, if necessary, adjust it (see Chapter 1).
   c) Inspect the system hoses. Look for cracks, bubbles, hardening and deterioration. Inspect the hoses and all fittings for oil bubbles or seepage. If there is any evidence of wear, damage or leakage, replace the hose(s).
   d) Inspect the condenser fins for leaves, bugs and any other foreign material that may have embedded itself in the fins. Use a "fin comb" or compressed air to remove debris from the condenser.
   e) Make sure the system has the correct refrigerant charge.
   f) If you hear water sloshing around in the dash area or have water dripping on the carpet, check the evaporator housing drain tube (see illustration) and insert a piece of wire into the opening to check for blockage.

2 It's a good idea to operate the system for about ten minutes at least once a month. This is particularly important during the winter months because long term non-use can cause hardening, and subsequent failure, of the seals. Note that using the Defrost function operates the compressor.

3 If the air conditioning system is not working properly, proceed to Step 6 and perform the general checks outlined below.

4 Because of the complexity of the air conditioning system and the special equipment necessary to service it, in-depth troubleshooting and repairs beyond checking the refrigerant charge and the compressor clutch operation are not included in this manual. However, simple checks and component replacement procedures are provided in this Chapter. For more complete information on the air conditioning system, refer to the Haynes Automotive Heating and Air Conditioning Manual.

5 The most common cause of poor cooling is simply a low system refrigerant charge. If a noticeable drop in system cooling ability occurs, one of the following quick checks will help you determine whether the refrigerant level is low. Should the system lose its cooling ability, the following procedure will help you pinpoint the cause.

Check

Refer to illustration 12.9

6 Warm the engine up to normal operating temperature.

7 Place the air conditioning temperature selector at the coldest setting and put the blower at the highest setting. Open the doors (to make sure the air conditioning system doesn't cycle off as soon as it cools the passenger compartment).

8 After the system reaches operating temperature, feel the two pipes connected to the evaporator at the firewall.

9 The pipe (thinner tubing) leading from the condenser outlet to the evaporator should be cold, and the evaporator outlet line (the thicker tubing that leads back to the compressor) should be slightly colder (3 to 10 degrees F colder). If the evaporator outlet is considerably warmer than the inlet, the system needs a charge. Insert a thermometer in the center air distribution duct (see illustration) while operating the air conditioning system at its maximum setting - the temperature of the output air should be 35 to 40 degrees F below the ambient air temperature (down to approximately 40 degrees F). If the ambient (outside) air temperature is very high, say 110 degrees F, the duct air temperature may be as high as 60 degrees F, but generally the air conditioning is 35 to 40 degrees F cooler than the ambient air.

10 If the air isn't as cold as it used to be, the system probably needs a charge.

11 If the air is warm and the system doesn't seem to be operating properly check the operation of the compressor clutch.

12 Have an assistant switch the air conditioning On while you observe the front of the compressor. The clutch will make an audible click and the center of the clutch should rotate. If it doesn't, shut the engine off and disconnect the air conditioning system pressure switch (see illustration 12.22). Bridge the terminals of the connector with a jumper wire and turn the air conditioning On again. If it works now, the system pressure is too high or too low. Have your system tested by a dealer service department or air conditioning shop.

13 If the clutch still didn't operate, check the appropriate fuses. Inspect the fuses in the interior fuse panel.

14 Remove the compressor clutch (AC) relay from the engine compartment relay panel and test it (see Chapter 12). With the relay out and the ignition On, check for battery power at two of the relay terminals (refer to the wiring diagrams for wire color designations to determine which terminals to check). There should be battery power with the key On, at the terminals for the relay control and power circuits.

15 Using a jumper wire, connect the terminals in the relay box from the relay power circuit to the terminal that leads to the compressor clutch (refer to the wiring diagrams for wire color designations to determine which terminals to connect). Listen for the clutch to click as you make the connection. If the clutch doesn't respond, disconnect the clutch connector at the compressor and check for battery voltage at the compressor clutch connector. Check for continuity to ground on the black wire terminal of the compressor clutch connector. If power and ground are available and the clutch doesn't operate when connected, the compressor clutch is defective.

16 If the compressor clutch, relay and related circuits are good and the system is fully charged with refrigerant and the compressor does not operate under normal conditions, have the PCM and related circuits checked by a dealer service department or

Chapter 3 Cooling, heating and air conditioning systems

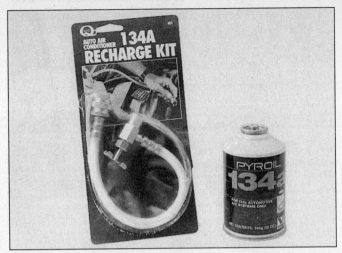

12.18 A basic charging kit for 134a systems is available at most auto parts stores - it must say 134a (not R-12) and so should the can of refrigerant

12.21 Attach the refrigerant kit to the low-side charging port - it's near the right shock tower - the cap should be marked with an "L"

other properly equipped repair facility.

17 Further inspection or testing of the system is beyond the scope of the home mechanic and should be left to a professional.

Adding refrigerant

Refer to illustrations 12.18, 12.21 and 12.22
Caution: *Make sure any refrigerant, refrigerant oil or replacement component your purchase is designated as compatible with environmentally friendly R-134a systems.*

18 Purchase an R-134a automotive charging kit at an auto parts store **(see illustration)**. A charging kit includes a 12-ounce can of refrigerant, a tap valve and a short section of hose that can be attached between the tap valve and the system low side service valve. Because one can of refrigerant may not be sufficient to bring the system charge up to the proper level, it's a good idea to buy an additional can. **Warning:** *Never add more than two cans of refrigerant to the system.*

19 Hook up the charging kit by following the manufacturer's instructions. **Warning:** *DO NOT hook the charging kit hose to the system high side!* The fittings on the charging kit are designed to fit **only** on the low side of the system.

20 Back off the valve handle on the charging kit and screw the kit onto the refrigerant can, making sure first that the O-ring or rubber seal inside the threaded portion of the kit is in place. **Warning:** *Wear protective eyewear when dealing with pressurized refrigerant cans.*

21 Remove the dust cap from the low-side charging and attach the quick-connect fitting on the kit hose **(see illustration)**.

22 Warm up the engine and turn On the air conditioning. Keep the charging kit hose away from the fan and other moving parts. **Note:** *The charging process requires the compressor to be running. If the clutch cycles off, you can put the air conditioning switch on High and leave the car doors open to keep the clutch on and compressor working.* **Note:** *The compressor can be kept on during the charging by removing the connector from the low-pressure switch (combination high-limit and low-limit switch on some models) and bridging it with a paper clip or jumper wire during the procedure* **(see illustration)**.

23 Turn the valve handle on the kit until the stem pierces the can, then back the handle out to release the refrigerant. You should be able to hear the rush of gas. Add refrigerant to the low side of the system, keeping the can upright at all times, but shaking it occasionally. Allow stabilization time between each addition. **Note:** *The charging process will go faster if you wrap the can with a hot-water-soaked shop rag to keep the can from freezing up.*

24 If you have an accurate thermometer, you can place it in the center air conditioning duct inside the vehicle and keep track of the output air temperature **(see illustration 12.9)**. A charged system that is working properly should cool down to approximately 40-degrees F. If the ambient (outside) air temperature is very high, say 110 degrees F, the duct air temperature may be as high as 60 degrees F, but generally the air conditioning is 30-40 degrees F cooler than the ambient air.

25 When the can is empty, turn the valve handle to the closed position and release the connection from the low-side port. Replace the dust cap.

26 Remove the charging kit from the can and store the kit for future use with the piercing valve in the UP position, to prevent inadvertently piercing the can on the next use.

Heating systems

27 If the carpet under the heater core is damp, or if antifreeze vapor or steam is coming through the vents, the heater core is leaking. Remove it (see Section 12) and install a new unit (most radiator shops will not repair a leaking heater core).

28 If the air coming out of the heater vents

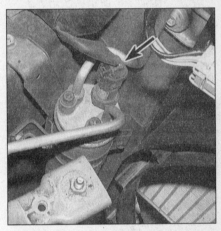

12.22 The air conditioning pressure switch (arrow) is located on top of the receiver/drier - if the compressor will not stay engaged, disconnect the connector and bridge it with a jumper wire during the charging procedure

isn't hot, the problem could stem from any of the following causes:

a) *The thermostat is stuck open, preventing the engine coolant from warming up enough to carry heat to the heater core. Replace the thermostat (see Section 3).*

b) *There is a blockage in the system, preventing the flow of coolant through the heater core. Feel both heater hoses at the firewall. They should be hot. If one of them is cold, there is an obstruction in one of the hoses or in the heater core, or the heater control valve is shut. Detach the hoses and back flush the heater core with a water hose. If the heater core is clear but circulation is impeded, remove the two hoses and flush them out with a water hose.*

c) *If flushing fails to remove the blockage from the heater core, the core must be replaced (see Section 11).*

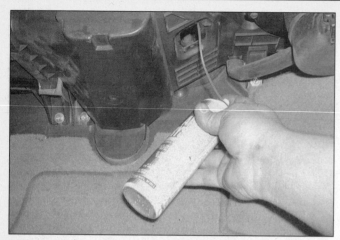

12.32 With the blower motor resistor removed, spray the disinfectant at the evaporator core

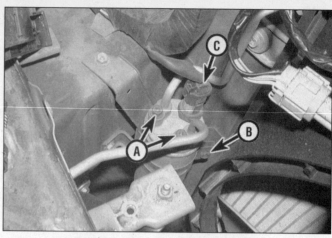

13.4 Receiver/drier mounting details
A Refrigerant lines
B Bracket retaining bolt
C Pressure switch

Eliminating air conditioning odors

Refer to illustration 12.32

29 Unpleasant odors that often develop in air conditioning systems are caused by the growth of a fungus, usually on the surface of the evaporator core. The warm, humid environment there is a perfect breeding ground for mildew to develop.

30 The evaporator core on most vehicles is difficult to access, and factory dealerships have a lengthy, expensive process for eliminating the fungus by opening up the evaporator case and using a powerful disinfectant and rinse on the core until the fungus is gone. You can service your own system at home, but it takes something much stronger than basic household germ-killers or deodorizers.

31 Aerosol disinfectants for automotive air conditioning systems are available in most auto parts stores, but remember when shopping for them that the most effective treatments are also the most expensive. The basic procedure for using these sprays is to start by running the system in the RECIRC mode for ten minutes with the blower on its highest speed. Use the highest heat mode to dry out the system and keep the compressor from engaging by disconnecting the wiring connector at the compressor (see Section 14).

32 The disinfectant can usually comes with a long spray hose. Remove the blower motor resistor (see Section 8), point the nozzle inside the hole and to the left towards the evaporator core, and spray according to the manufacturer's recommendations **(see illustration)**. Try to cover the whole surface of the evaporator core, by aiming the spray up, down and sideways. Follow the manufacturer's recommendations for the length of spray and waiting time between applications.

33 Once the evaporator has been cleaned, the best way to prevent the mildew from coming back again is to make sure your evaporator housing drain tube is clear **(see illustration 12.1)**.

Automatic heating and air conditioning systems

34 Some models are equipped with an optional automatic climate control system. This system has its own computer that receives inputs from various sensors in the heating and air conditioning system. This computer, like the PCM, has self-diagnostic capabilities to help pinpoint problems or faults within the system. Vehicles equipped with automatic heating and air conditioning systems are very complex and considered beyond the scope of the home mechanic. Vehicles equipped with automatic heating and air conditioning systems should be taken to dealer service department or other qualified facility for repair.

13 Air conditioning receiver/drier - removal and installation

Warning: *The air conditioning system is under high pressure. DO NOT loosen any fittings or remove any components until after the system has been discharged. Air conditioning refrigerant should be properly discharged into an EPA-approved container at a dealer service department or an automotive air conditioning repair facility. Always wear eye protection when disconnecting air conditioning system fittings.*

Removal

Refer to illustration 13.4

1 The receiver/drier stores refrigerant and removes moisture from the system. When any major air conditioning component (compressor, condenser, evaporator) is replaced, or the system has been apart and exposed to air for any length of time, the accumulator/drier must be replaced.

2 Take the vehicle to a dealer service department or automotive air conditioning shop and have the air conditioning system discharged and the refrigerant recovered (see the **Warning** at the beginning of this Section). Disconnect the cable from the negative terminal of the battery.

3 Disconnect the electrical connector at the compressor clutch cycling switch on top of the receiver/drier. If the receiver/drier is to be replaced with a new one, remove the cycling switch to transfer to the new drier.

4 Disconnect the refrigerant inlet and outlet lines **(see illustration)**. Cap or plug the open lines immediately.

5 Loosen the clamp-bolt on the mounting bracket and slide the receiver/drier assembly up and out of the mounting bracket **(see illustration 13.4)**.

Installation

6 If you are replacing the receiver/drier, add two ounces of clean refrigerant oil to the new accumulator. This will maintain the correct oil level in the system after the repairs are completed.

7 Place the new a receiver/drier into position, tighten the mounting bracket bolt lightly, still allowing the receiver/drier to be turned to align the line connections.

8 Install the inlet and outlet lines. Lubricate the O-rings using clean refrigerant oil and reconnect the lines. Now tighten the clamp bolt securely and reconnect the electrical connector.

9 Connect the cable to the negative terminal of the battery.

10 Have the system evacuated, recharged and leak tested by a dealer service department or an air conditioning repair facility.

14 Air conditioning compressor - removal and installation

Warning: *The air conditioning system is under high pressure. Do not loosen any hose fittings or remove any components until after the system has been discharged. Air conditioning refrigerant should be properly discharged into an EPA-approved recovery/recycling unit at a*

Chapter 3 Cooling, heating and air conditioning systems

14.5 Remove the bolts (arrows) retaining the refrigerant lines to the rear of the compressor

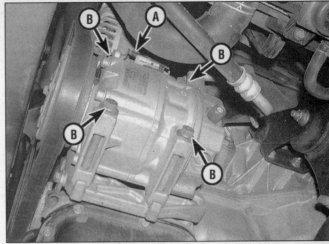

14.6 Disconnect the electrical connector (A) and the compressor mounting bolts (B) (DOHC engine shown, SOHC engine similar)

dealer service department or an automotive air conditioning repair facility. Always wear eye protection when disconnecting air conditioning system fittings.

Note: *The receiver/drier and evaporator orifice tube should be replaced whenever the compressor is replaced (see Sections 13 and 14).*

Removal

Refer to illustrations 14.5 and 14.6

1   Have the air conditioning system refrigerant discharged and recycled by an air conditioning technician (see **Warning** above).
2   Disconnect the cable from the negative terminal of the battery.
3   Set the parking brake, block the rear wheels and raise the front of the vehicle, supporting it securely on jackstands and remove the splash cover from below the engine.
4   Remove the drivebelt (see Chapter 1).
5   Disconnect the refrigerant lines from the compressor. Plug the open fittings to prevent entry of dirt and moisture **(see illustration)**.
6   Disconnect the compressor clutch wiring harness. Unbolt the compressor from the mounting bracket and remove it from the vehicle **(see illustration)**. **Note:** *The upper mounting bolts may not come all the way out of the compressor - leave them in the compressor until it is removed from the vehicle.*

Installation

7   The clutch may have to be transferred from the old compressor to the new unit.
8   Add the proper amount of refrigerant oil to the new compressor using the following calculations:
   a) Drain the refrigerant oil from the old compressor through the suction fitting and measure it in ounces.
   b) Drain any new oil from the new compressor.
   c) If the amount from the old compressor was 3 to 5 ounces, put that amount of clean, new oil in the new compressor.
   d) If the amount from the old compressor was less than 3 ounces, put 3 ounces of clean, new oil in the new compressor.
   e) If the amount from the old compressor was more than 5 ounces, put 5 ounces of clean, new oil in the new compressor.

9   Installation is the reverse of removal, using new O-rings where the lines attach to the compressor. **Note:** *Remember to slip the two upper mounting bolts into the compressor before installing the compressor in the vehicle.*
10 Have the system evacuated, recharged and leak tested by an air conditioning technician.

15 Air conditioning condenser - removal and installation

Warning: *The air conditioning system is under high pressure. Do not loosen any hose fittings or remove any components until after the system has been discharged. Air conditioning refrigerant should be properly discharged into an EPA-approved recovery/recycling unit at a dealer service department or an automotive air conditioning repair facility. Always wear eye protection when disconnecting air conditioning system fittings.*

Removal

Refer to illustration 15.3

1   Have the refrigerant discharged and recycled by an air conditioning technician (see **Warning** above).
2   Disconnect the cable from the negative terminal of the battery and drain the cooling system (see Chapter 1).

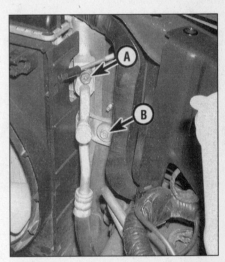

15.3 Disconnect the condenser lines (A) and remove the two condenser mounting bolts (B indicates the right-side bolt, radiator not yet removed in this photo)

3   Disconnect the condenser line and discharge line from the condenser **(see illustration)**. Cap the fittings on the condenser and lines to prevent entry of dirt or moisture.
4   Refer to Section 5 and remove the radiator.
5   Remove the condenser retaining bolts **(see illustration 15.3)**.
6   Lean the condenser back toward the engine and remove it from the vehicle.

Installation

7   Installation is the reverse of removal. If a new condenser was installed, add 1 to 1.7 ounces of fresh refrigerant oil.
8   Have the system evacuated, charged and leak tested by an air conditioning technician.

Notes

Chapter 4
Fuel and exhaust systems

Contents

	Section		Section
Accelerator cable - removal, installation and adjustment	10	Fuel pressure regulator - removal and installation	14
Air filter housing - removal and installation	9	Fuel pressure relief procedure	2
Air filter replacement	See Chapter 1	Fuel pump - removal and installation	5
CHECK ENGINE light	See Chapter 6	Fuel pump/fuel pressure - check	3
Engine idle speed - check and adjustment	16	Fuel rail and injectors - removal and installation	15
Exhaust system check	See Chapter 1	Fuel system check	See Chapter 1
Exhaust system servicing - general information	17	Fuel tank - cleaning and repair	8
Fuel injection system - check	12	Fuel tank - removal and installation	7
Fuel injection system - general information	11	General information	1
Fuel level sending unit - check and replacement	6	Idle control system	See Chapter 6
Fuel lines and fittings - replacement	4	Throttle body - removal and installation	13

Specifications

Engine idle speed
   1993 and 1994 .... 700 rpm
   1995
      Manual transaxle .... 550 to 650 rpm
      Automatic transaxle .... 600 to 700 rpm
   1996 and later
      Manual transaxle .... 525 to 625 rpm
      Automatic transaxle .... 600 to 700 rpm
Fuel system pressure
   Key On, engine Off .... 43 psi
   Engine idling .... 34 to 36 psi
Fuel injector resistance (approximate) .... 10 to 14 ohms
Fuel level sending unit resistance
   1993 and 1994
      Empty .... 929 to 1009 ohms
      1/2 .... 178 to 195 ohms
      Full .... 15 to 18 ohms
   1995 and later
      Empty .... 80 to 83 ohms
      1/2 .... 32 to 35 ohms
      Full .... 5 to 8 ohms

Torque specifications	Nm	Ft-lbs (unless otherwise indicated)
Fuel injector cap screws	2.5 to 3.2	22 to 29 in-lbs
Fuel rail mounting bolts		
1993 and 1994	24 to 27	17 to 20
1995 and later	16 to 20	12 to 14
Throttle body mounting bolts		
Step 1	9 to 11	78 to 96 in-lbs
Step 2	18 to 22	13 to 16

Chapter 4 Fuel and exhaust systems

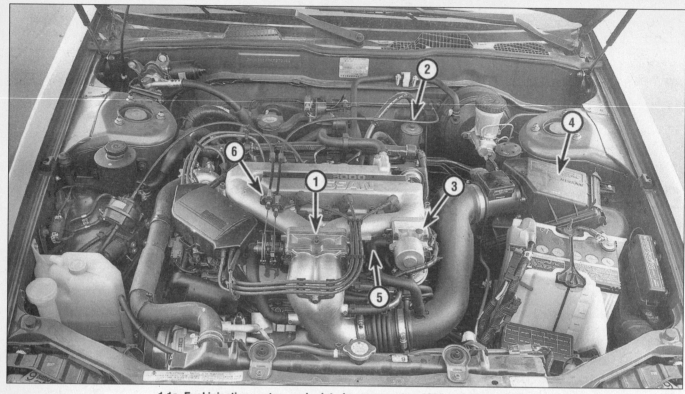

1.1a Fuel injection system and related components - 1993 and 1994 models

1. Throttle body
2. Fuel filter
3. Idle air control valve
4. Air filter housing
5. Fuel rail and fuel injectors (under upper intake manifold)
6. Accelerator cable

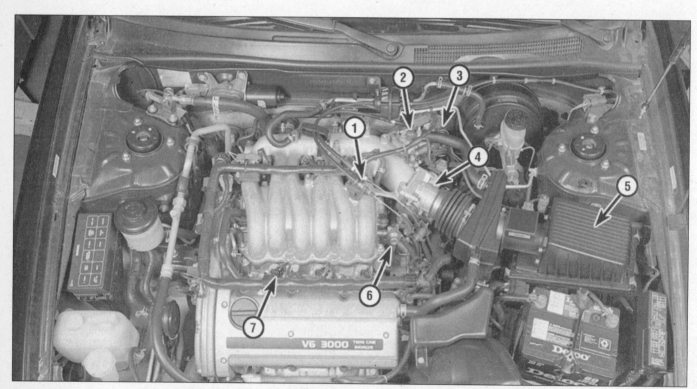

1.1b Fuel injection system and related components - 1995 and later models

1. Accelerator cable
2. Idle air control valve
3. Fuel filter
4. Throttle body
5. Air filter housing
6. Fuel pressure regulator
7. Fuel injector

Chapter 4 Fuel and exhaust systems

1 General information

Refer to illustrations 1.1a and 1.1b

The fuel system consists of a fuel tank, an electric fuel pump (located in the fuel tank), a fuel pressure regulator, a fuel pump relay, the fuel rail and fuel injectors, an air cleaner assembly and a throttle body unit. All models are equipped with a multi-port fuel injection system **(see illustrations)**.

Multi-port fuel injection system

Multi-port fuel injection uses timed impulses to inject the fuel directly into the intake port of each cylinder according to its firing order. The injectors are controlled by the Powertrain Control Module (PCM). The PCM monitors various engine parameters and delivers the exact amount of fuel required into the intake ports. The throttle body serves only to control the amount of air passing into the system. Because each cylinder is equipped with its own injector, much better control of the fuel/air mixture ratio is possible.

Fuel pump and lines

Fuel is circulated from the fuel tank to the fuel injection system, and back to the fuel tank, through a pair of metal lines running along the underside of the vehicle. An electric fuel pump and fuel level sending unit is located inside the fuel tank. A vapor return system routes all vapors back to the fuel tank through a separate return line.

The fuel pump relay is equipped with a primary and secondary voltage circuit. The primary circuit is controlled by the PCM and the secondary circuit is linked directly to battery voltage from the ignition switch. With the ignition switch On (engine not running), the PCM will energize the relay for five seconds. During cranking, the PCM supplies voltage to the fuel pump relay as long as the camshaft position sensor or crankshaft position sensor sends its position signal (see Chapter 6). If there is no signal, the fuel pump will shut off after five seconds.

1995 California emissions models and all 1996 and 1997 models are equipped with a fuel pump control module. The fuel pump control module reduces the voltage at the fuel pump except when the engine is cranking or operating under a heavy load or high speed. The fuel pump control module is located in the trunk behind the left tail light. A dropping resistor is located in the wiring harness near the fuel pump control module.

Exhaust system

The exhaust system includes a pair of exhaust manifolds, a diverter pipe fitted with an oxygen sensor, a catalytic converter, a muffler and a tail pipe. 1995 and later models are equipped with two primary oxygen sensors, two warm-up catalytic converters and a secondary (after catalytic converter) oxygen sensor.

The catalytic converters are an emission control device added to the exhaust system to reduce pollutants. Refer to Chapter 6 for more information regarding the catalytic converters.

2 Fuel pressure relief procedure

Warning: *Gasoline is extremely flammable, so take extra precautions when you work on any part of the fuel system. Don't smoke or allow open flames or bare light bulbs near the work area, and don't work in a garage where a natural gas-type appliance (such as a water heater or a clothes dryer) with a pilot light is present. Since gasoline is carcinogenic, wear latex gloves when there's a possibility of being exposed to fuel, and, if you spill any fuel on your skin, rinse it off immediately with soap and water. Mop up any spills immediately and do not store fuel-soaked rags where they could ignite. The fuel system is under constant pressure, so, if any fuel lines are to be disconnected, the fuel pressure in the system must be relieved first. When you perform any kind of work on the fuel system, wear safety glasses and have a Class B type fire extinguisher on hand.*

1 Remove the fuel pump fuse from the passenger compartment fuse panel (see Chapter 12).
2 Start the engine and allow it to run until it stops. This should take only a few seconds. Crank the engine several more times to ensure the fuel system has been completely relieved. Disconnect the cable from the negative terminal of the battery before working on the fuel system.
3 The fuel system pressure is now relieved. When you're finished working on the fuel system, install the fuel pump fuse back into the fuse panel and connect the negative cable to the battery.
4 After the fuel pressure has been relieved, it's a good idea to lay a shop towel over any fuel connection to be disassembled, to absorb the residual fuel that may leak out when servicing the fuel system.

3 Fuel pump/fuel pressure - check

Warning: *Gasoline is extremely flammable, so take extra precautions when you work on any part of the fuel system. See the **Warning** in Section 2.*
Note: *In order to perform the fuel pressure test, you will need to obtain a fuel pressure gauge capable of measuring high fuel pressure and the necessary fittings to connect the fuel gauge to the fuel line.*

Preliminary check

1 If you suspect insufficient fuel delivery, first inspect all fuel lines to ensure that the problem is not simply a leak in a line. Check that there is adequate fuel in the fuel tank.
2 Set the parking brake and have an assistant turn the ignition switch to the ON

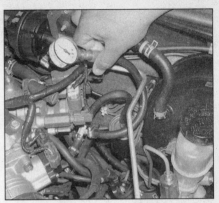

3.6 Using a T-fitting, install the fuel pressure gauge between the fuel filter and the fuel rail

position while you listen to the fuel pump (inside the fuel tank). You should hear a "whirring" sound, lasting for a couple of seconds indicating the fuel pump is operating. If the fuel pump is operating, proceed to the pressure check.
3 If there is no sound, check the fuel pump circuit, referring to Chapter 12 and the wiring diagrams. Check the related fuses, the fuel pump relay and the related wiring to ensure power is reaching the fuel pump connector. Check the ground circuit for continuity.
4 If the power and ground circuits are good and the fuel pump does not operate, replace the fuel pump (see Section 5).

Pressure check

Refer to illustration 3.6

5 Relieve the fuel system pressure (see Section 2).
6 Remove the fuel line from the fuel filter and install a T-fitting between the fuel filter and the fuel rail **(see illustration)**. Connect a fuel pressure gauge to the T-fitting. Make sure the hose clamps are securely tightened.
7 Turn the ignition switch On. The fuel pump should run for about five seconds - pressure should register on the gauge and should hold steady. Compare the pressure reading with the key On, engine Off value listed in this Chapter's Specifications.
8 Start the engine and allow it to idle. Compare the pressure reading with the engine running value listed in this Chapter's Specifications. Disconnect and plug the vacuum hose from the fuel pressure regulator (see Section 14) - the pressure should increase to the value recorded in Step 4. If the pressure readings are correct, the system is operating properly.
9 If the fuel pressure is not within specifications, check the following:

a) *If the pressure is higher than specified, check for vacuum to the fuel pressure regulator. Vacuum must fluctuate with the increase or decrease in the engine rpm. If vacuum is present, check for a pinched or clogged fuel return hose or pipe. If the return line is OK, replace the fuel pressure regulator.*

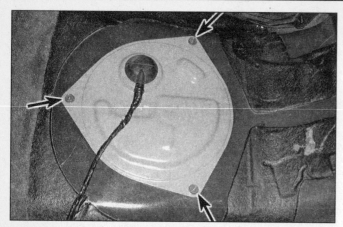

5.3 Remove the screws (arrows) and the fuel pump/fuel level sending unit access cover

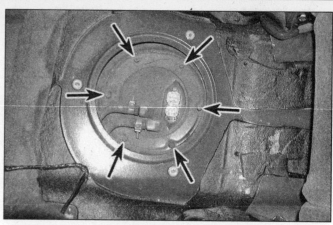

5.4 Disconnect the fuel feed and return lines, then remove the fuel pump/fuel level sending unit mounting screws (arrows) and retaining ring (if equipped)

b) If the pressure is lower than specified, check for a restriction in the fuel filter or fuel line. If the fuel filter and lines are OK, start the engine (if possible) and slowly pinch the return hose shut. If the pressure rises above 43 psi, replace the fuel pressure regulator (see Section 14). **Warning:** *Don't allow the fuel pressure to exceed 60 psi. If the pressure is still low, replace the fuel pump (see Section 5).*

10 Turn the engine Off and place the ignition switch in the On (engine not running) position. Monitor the pressure on the gauge for approximately five minutes - pressure should decrease slowly. If the pressure decreases rapidly, check the following:

a) Turn the ignition key On and pinch the fuel feed line shut between the fuel gauge T-fitting and the fuel tank. Turn the ignition key Off - if the pressure decreases rapidly, an injector (or injectors) may be leaking.

b) Turn the ignition key On and pinch the fuel feed line shut between the fuel gauge T-fitting and the fuel rail. Turn the ignition key Off - if the pressure decreases rapidly, the in-tank fuel pump check valve may be faulty.

11 After completing the testing, relieve the fuel pressure (see Section 2) and remove the fuel pressure gauge.

4 Fuel lines and fittings - replacement

Warning: *Gasoline is extremely flammable, so take extra precautions when you work on any part of the fuel system. See the **Warning** in Section 2.*

1 Always relieve the fuel pressure before servicing fuel lines or fittings (see Section 2).
2 The fuel feed, return and vapor lines extend from the fuel tank to the engine compartment. The lines are secured to the underbody with clip and screw assemblies. These lines must be occasionally inspected for leaks, kinks and dents.
3 If evidence of dirt is found in the system or fuel filter during disassembly, the line should be disconnected and blown out. Check the fuel strainer on the fuel pump pick-up unit (see Section 5) for damage and deterioration.

Steel tubing

4 If replacement of a fuel line or emission line is called for, use welded steel tubing meeting the manufacturer's specifications.
5 Don't use copper or aluminum tubing to replace steel tubing. These materials cannot withstand normal vehicle vibration.
6 Because fuel lines used on fuel-injected vehicles are under high pressure, they require special consideration.
7 If the lines are replaced, always use original equipment parts, or parts that meet original equipment standards.

Flexible hose

Warning: *Use only original equipment replacement hoses or their equivalent. Others may fail from the high pressures generated by this system.*

8 Don't route fuel hose within four inches of any part of the exhaust system or within ten inches of the catalytic converter. Metal lines and rubber hoses must never be allowed to chafe against the frame. A minimum of 1/4-inch clearance must be maintained around a line or hose to prevent contact with the frame.

Replacement

9 In the event of any fuel line damage (metal or flexible lines) it is necessary to replace the damaged lines with factory replacement parts. Others may fail from the high pressures of this system.
10 Relieve the fuel pressure (see Section 2).
11 Remove all fasteners attaching the lines to the vehicle body.
12 Loosen the hose clamp(s), slide the clamp down the hose away from the metal line and pull the hoses off the fitting.
13 Installation is the reverse of removal.

5 Fuel pump - removal and installation

Warning: *Gasoline is extremely flammable, so take extra precautions when you work on any part of the fuel system. See the **Warning** in Section 2.*

Removal

Refer to illustrations 5.3, 5.4, 5.5, 5.6, 5.7, 5.8, 5.9, 5.10, 5.11 and 5.13

1 Relieve the fuel pressure (see Section 2).
2 Remove the rear seat (see Chapter 11).
3 Remove the fuel pump/fuel level sending unit access cover **(see illustration)**. Disconnect the electrical connector from the fuel pump/sending unit assembly and set the cover and wiring harness aside.
4 Detach the fuel feed line and return lines and remove the fuel pump/sending unit mounting screws **(see illustration)**. On 1995 and later models, remove the mounting ring.

1993 and 1994 models

5 Carefully maneuver the fuel pump/sending unit assembly out of the tank **(see illustration)**. **Caution:** *The fuel level float and*

5.5 Carefully angle the fuel pump module out of the fuel tank without damaging the fuel level sending arm and float

Chapter 4 Fuel and exhaust systems

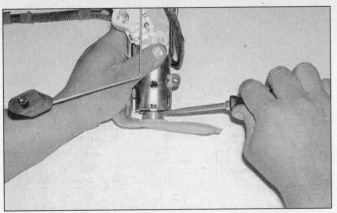

5.6 Remove the strainer from the base of the fuel pump

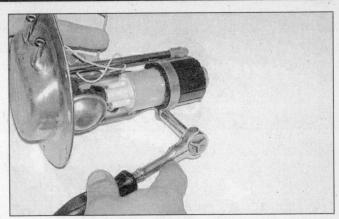

5.7 Loosen the fuel pump mounting clamp, then slide it down past the end of the mounting bracket

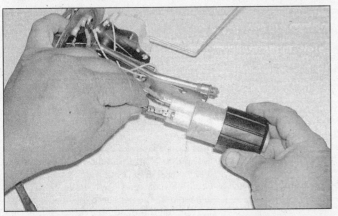

5.8 Disconnect the fuel pump electrical connector and the fuel line hose from the fuel pump

5.9 Carefully maneuver the fuel level sending unit assembly out of the tank as far as possible

sending unit are delicate. Don't bump or bend them during removal or the accuracy of the sending unit may be affected.

6 Carefully pry the fuel strainer off the fuel pump **(see illustration)**. Check the strainer for contamination and replace it, if necessary.

7 Loosen the fuel pump mounting clamp bolt **(see illustration)**.

8 Remove the fuel hose lower clamp, disconnect the electrical connector from the fuel pump and separate the fuel pump from the assembly **(see illustration)**.

1995 through 1999 models

9 Carefully maneuver the fuel level sending unit assembly out of the tank as far as possible **(see illustration)**.

10 Slide the hose clamp down the hose and remove the hose from the fitting **(see illustration)**.

11 Disconnect the fuel pump electrical connector from the cover **(see illustration)**.

12 Set the fuel level sending unit and cover assembly aside. **Caution:** *The fuel level float and sending unit are delicate. Don't bump or bend them during removal or the accuracy of the sending unit may be affected.*

13 Reach inside the fuel tank, depress the lock on the fuel pump housing and withdraw the fuel pump and housing assembly from

5.10 Slide the hose clamp down the hose and remove the hose from the fitting

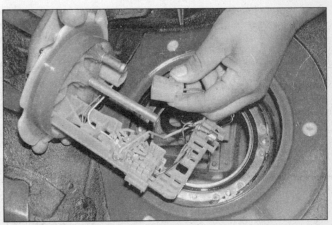

5.11 Disconnect the fuel pump electrical connector from the cover and set the fuel level sending unit and cover assembly aside

Chapter 4 Fuel and exhaust systems

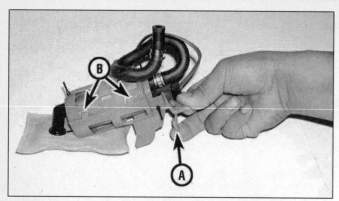

5.13 Reach inside the fuel tank, depress the lock on the fuel pump housing (A) and withdraw the fuel pump and housing assembly from the fuel tank (fuel pump removed from tank for clarity); then depress the two locking tabs (B) and separate the housing

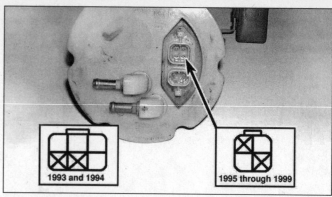

6.2 Connect an ohmmeter to the indicated terminals of the fuel level sending unit connector and check the resistance of the sending unit with the float positioned down (empty), then move the float up (full) and note the resistance - check for a smooth change in resistance as the float is moved

the fuel tank **(see illustration)**.

14 Carefully pry the fuel strainer off the fuel pump. Check the strainer for contamination and replace it, if necessary. Remove the fuel hose from the fuel pump outlet fitting.

15 Depress the locking tabs and separate the two halves of the fuel pump housing **(see illustration 5.13)**. Remove the fuel pump from the fuel pump housing.

2000 and later models

16 Relieve the fuel pressure (see Section 2).
17 Remove the rear seat (see Chapter 11).
18 Remove the fuel pump/fuel level sending unit access cover. Disconnect the electrical connector from the fuel pump/fuel level sending unit assembly and set the cover and wiring harness aside.
19 **Caution:** *Do not twist the quick disconnect connector nor use any tool to release the tabs.* Hold the sides of the fuel line quick disconnect connector, completely depress the connector tabs and pull out the tube inserted in the retainer. Disconnect both fuel lines.
20 Remove the fuel pump/fuel level sending unit six mounting screws.
21 Carefully maneuver the fuel pump/fuel level sending unit assembly out of the fuel tank. **Caution:** *The fuel level float and sending unit are delicate. Don't bump or bend them during removal or the accuracy of the sending unit may be affected.*
22 Using a screwdriver, release the locking tabs on the chamber flange and remove it from the fuel pump/fuel level sending unit chamber.
23 Carefully pull the fuel tank temperature sending unit wiring harness part way out of the fuel pump/fuel level sending unit. Do not try to disconnect it.
24 Using a screwdriver, release the snap fit portion and pull up on the fuel level sending unit and remove it.
25 Disconnect the fuel pump electrical connector.
26 **Caution:** *Do not twist the connector nor use any tool to release the tabs.* Hold the sides of the fuel pump line quick disconnect connector, completely depress the connector

tabs and pull out the tube inserted in the retainer. Disconnect both fuel pump fuel lines from the fuel level sending unit assembly.
27 Carefully pull the fuel level sending unit assembly part way straight up out of the chamber.
28 Using a screwdriver, release the two locking tabs and remove the fuel filter from the chamber. Remove the fuel pump bracket and pull out the fuel filter.
29 Pull down and remove the fuel pump from the fuel filter housing.
30 Remove the fuel pump support rubber from the fuel pump.

Installation

31 Installation of the fuel pump to the sending unit assembly is the reverse of removal.
32 Clean the fuel pump mounting flange and the tank mounting surface and seal ring groove.
33 Position a new O-ring around the opening in the fuel tank and guide the fuel pump/sending unit assembly into the tank.
34 Make sure the fuel lines are facing in their original position, then tighten the fuel pump/sending unit mounting bolts securely.
35 Connect the electrical connector, install the access cover and the rear seat.

6 Fuel level sending unit - check and replacement

Refer to illustration 6.2
Warning: *Gasoline is extremely flammable, so take extra precautions when you work on any part of the fuel system. See the* **Warning** *in Section 2.*
Note: *There are no service procedures for the 2000 and later models.*

1 Remove the fuel pump/fuel level sending unit assembly (see Sections 5).
2 Connect the probes of an ohmmeter to the two indicated terminals of the fuel sending unit electrical connector **(see illustration)**.
3 Position the float in the down (empty) position. Measure the resistance and compare it to the value listed in this Chapter's

Specifications.
4 Move the float up to the full position. Measure the resistance and compare it to the value listed in this Chapter's Specifications.
5 If the fuel level sending unit resistance is incorrect or if the resistance does not change smoothly as the float travels from empty to full, the fuel level sending unit assembly is defective.
6 On 1993 and 1994 models, remove the fuel pump from the sending unit assembly and install it onto the new unit (see Section 5). On 1995 through 1999 models, transfer the fuel temperature sensor to the new unit.
7 Install the assembly into the fuel tank with a new O-ring (see Section 5).
8 The remainder of installation is the reverse of removal.

7 Fuel tank - removal and installation

Refer to illustrations 7.7, 7.8 and 7.10
Warning: *Gasoline is extremely flammable, so take extra precautions when you work on any part of the fuel system. See the* **Warning** *in Section 2.*

1 Relieve the fuel pressure (see Section 2).
2 Disconnect the cable from the negative terminal of the battery.
3 Siphon the fuel from the fuel tank before removing the tank from the vehicle. **Warning:** *DO NOT start the siphoning action by mouth! Use a siphoning kit (available at most auto parts stores).*
4 Remove the rear seat and the fuel pump/fuel level sending unit access cover. Disconnect the electrical connectors for the fuel pump and fuel level sending unit (see Section 6).
5 On 2000 and later models, at the top of the fuel pump/fuel level sending unit, hold the sides of the fuel line quick disconnect connector, completely depress the connector tabs and pull out the tube inserted in the retainer. Disconnect both fuel lines. **Caution:** *Do not twist the connector nor use any tool to release the tabs.*

Chapter 4 Fuel and exhaust systems

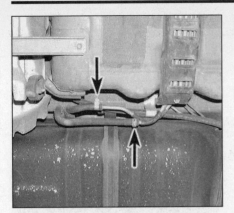

7.7 Disconnect the fuel feed line and the fuel return line (arrows)

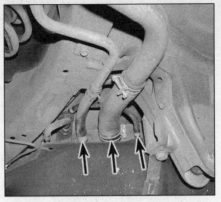

7.8 Disconnect the fuel filler hose, vent hose and EVAP hose (arrows) from the fuel tank

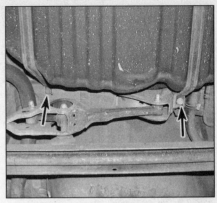

7.10 Remove the fuel tank strap bolts (arrows) and swing the straps out of the way

6   Raise the vehicle and support is securely on jackstands.
7   On 1993 through 1999 models, disconnect the fuel feed and return lines from the fuel tank **(see illustration)**.
8   Loosen the hose clamps and detach the fuel tank filler hose and vapor hose from the fuel filler neck and the fuel tank **(see illustration)**. Disconnect the evaporative emissions canister hoses.
9   Place a floor jack under the tank and position a wood plank between the jack pad and the tank. Raise the jack until it's supporting the tank.
10   Remove the bolts that retain the fuel tank mounting straps **(see illustration)**. The straps are hinged at the other end so you can swing them out of the way.
11   Slowly lower the jack while guiding the fuel tank past the exhaust pipe. Remove the tank from the vehicle.
12   If you're replacing the tank, or having it cleaned or repaired, refer to Section 8.
13   Refer to Section 5 to remove and install the fuel pump/sending unit assembly, if necessary.
14   Installation is the reverse of removal. Clean engine oil can be used as an assembly aid when pushing the fuel filler hose back onto the fuel tank.

8 Fuel tank - cleaning and repair

1   The fuel tank installed in the vehicles covered by this manual is not repairable. If the tank is damaged in any way it must be replaced. If cleaning is required, due to fuel contamination, the process should be carried out by a professional who has experience in this critical and potentially dangerous operation. Even after cleaning and flushing, explosive fumes can remain and ignite.

2   If the fuel tank is removed from the vehicle, it should not be placed in an area where sparks or open flames could ignite the fumes coming out of the tank. Be especially careful inside a garage where a natural gas-type appliance is located, because the pilot light could cause an explosion.

9 Air filter assembly - removal and installation

Refer to illustrations 9.2a, 9.2b and 9.5

1   Disconnect the cable from the negative terminal of the battery.
2   Disconnect the electrical connector from the mass airflow sensor and intake air temperature sensor **(see illustrations)**.
3   Loosen the retaining clamp securing the mass airflow sensor to the air intake duct or resonator.

9.2a Air intake duct and air filter housing installation details - 1993 and 1994 models

1   IAC valve air bypass hose
2   Mass airflow sensor
3   Air filter housing (upper)
4   Air filter housing retaining clips
5   Air filter housing (lower)
6   Air Intake duct
7   Hose clamps

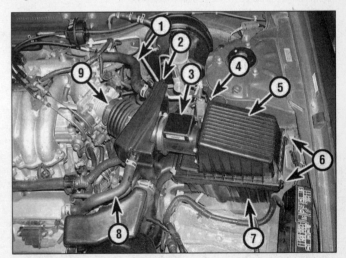

9.2b Air intake duct and air filter housing installation details - 1995 and later models

1   Crankcase ventilation hose
2   Air intake resonator
3   Mass airflow sensor
4   Intake air temperature sensor
5   Air cleaner housing (upper)
6   Air cleaner housing retaining clips
7   Air cleaner housing (lower)
8   Crankcase ventilation tube
9   Hose clamp

Chapter 4 Fuel and exhaust systems

9.5 Remove the bolts (arrows) and the lower air filter housing

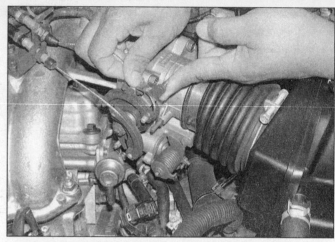

10.1 Rotate the throttle lever until the slot in the throttle lever aligns with the cable, then pass the cable through the slot

4   Detach the spring clips and remove the air filter upper housing and mass airflow sensor as an assembly.
5   Remove the lower air filter housing mounting bolts and remove it from the engine compartment (see illustration).
6   Installation is the reverse of removal.

10 Accelerator cable - removal, installation and adjustment

Note: *The adjustment procedure for the accelerator cable and the cruise control cable is similar except where noted below.*

Removal

Refer to illustrations 10.1, 10.2 and 10.3

1   Detach the accelerator cable and the cruise control cable (if equipped) from the throttle lever (see illustration).
2   Loosen the cable locknut and adjusting nut, then separate the accelerator cable from the cable bracket (see illustration).
3   From inside the passenger compartment, pull the cable end out from the accelerator pedal arm, then pass the cable through the slot in the arm. Remove the bolts securing the accelerator cable to the firewall (see illustration).
4   Disconnect any remaining cable clips.
5   Remove the cable through the firewall from the engine compartment side.

Installation

6   Installation is the reverse of removal. Be sure the cable is routed correctly and to fasten all the cable retaining clips.
7   If necessary, at the engine compartment side of the firewall, apply sealant to the accelerator cable bracket where it mates to the firewall to prevent water from entering the passenger compartment.

Adjustment

8   To adjust the accelerator cable:
   a) *Lift up on the cable to remove any slack.*
   b) *Turn the adjusting nut until the throttle lever just starts to move.*
   c) *Back off the adjusting nut 1-1/2 to 2 turns.*
   d) *Tighten the locknut.*
   e) *Verify that the throttle valve opens all the way when you depress the accelerator pedal to the floor and that it returns to the idle position when you release the accelerator. Verify the cable operates smoothly. It must not bind or stick.*

9   To adjust the cruise control cable:
   a) *Check the accelerator cable for proper adjustment.*
   b) *Turn the adjusting nut until the throttle lever just starts to move.*
   c) *Back off the adjusting nut 1/2 to 1 turn.*
   d) *Tighten the locknut and check for proper operation of the cruise control system.*

11 Fuel injection system - general information

All models are equipped with a multi-port fuel injection system. Fuel is delivered into each intake port in sequence with the engine firing order in accordance with engine demand through injectors (one per cylinder) mounted on the intake manifold. The intake

10.2 Loosen the accelerator cable locknut (B) and adjusting nut (A)

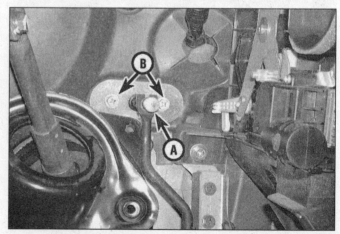

10.3 Working under the dash, pull the cable end (A) from the accelerator pedal recess and lift it through the slot, then remove the cable mounting bracket bolts (B)

Chapter 4 Fuel and exhaust systems

12.7 Use an automotive stethoscope to determine if the injectors are working properly

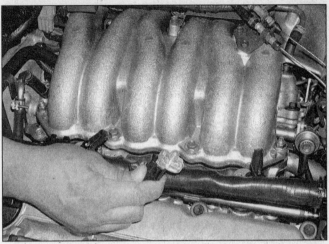

12.8 Install the fuel injector test light or "noid light" into the fuel injector electrical connector and confirm that it blinks when the engine is cranked or running

manifold incorporates an air intake plenum (upper manifold) to aid in air flow and distribution with a removable throttle body. The air intake plenum bolts to the lower intake manifold, which sits directly in the middle of the engine block.

The multi-port fuel injection system incorporates an on-board electronic engine control computer (known as the Powertrain Control Module - PCM) that accepts inputs from various engine sensors to compute the required fuel flow rate necessary to maintain a prescribed air/fuel ratio throughout the entire engine operational range. The computer then outputs a command to the fuel injectors to meter the approximate quantity of fuel. The system automatically senses and compensates for changes in altitude, load and speed.

The fuel delivery systems include an electric in-tank fuel pump which forces pressurized fuel through a series of metal and rubber lines and an inline fuel filter to the fuel rail assembly. The multi-port fuel injection system uses a single high-pressure pump mounted inside the tank.

The fuel rail assembly incorporates an electrically actuated fuel injector directly above each intake port. When energized, the injectors spray a metered quantity of fuel into the intake air stream.

A constant fuel supply is delivered to the injectors by the fuel rail. The fuel pressure regulator, positioned at the end of the fuel rail, maintains the system fuel pressure. Excess fuel passes through the regulator and returns to the fuel tank through a fuel return line.

Each injector is energized once every other crankshaft revolution in sequence with engine firing order. The period of time that the injectors are energized (known as "on time" or "pulse width") is controlled by the PCM. Air entering the engine is sensed by mass airflow and temperature sensors. The outputs of these, and other, sensors are processed by the PCM. The computer determines the needed injector pulse width and outputs a command to the injector to meter the exact quantity of fuel.

12 Fuel injection system - check

Refer to illustrations 12.7, 12.8 and 12.9

Warning: *Gasoline is extremely flammable, so take extra precautions when you work on any part of the fuel system. See the Warning in Section 2.*

Note: *The following procedure is based on the assumption that the fuel pump is working and the fuel pressure is adequate (see Section 3).*

1 Check all electrical connectors that are related to the system. Loose electrical connectors and poor grounds can cause many problems that resemble more serious malfunctions.

2 Check to see that the battery is fully charged, as the control unit and sensors depend on an accurate supply voltage in order to properly meter the fuel.

3 Check the air filter element - a dirty or partially blocked filter will severely impede performance and economy (see Chapter 1).

4 Check the fuses. If a blown fuse is found, replace it and see if it blows again. If it does, search for a wire shorted to ground in the fuel injection system wiring harness (see Chapter 12 and the wiring diagrams).

5 Check the condition of the vacuum hoses connected to the intake manifold.

6 Remove the air intake duct from the throttle body and check for dirt, carbon or other residue build-up in the throttle body, particularly around the throttle plate. **Caution:** *The throttle body on these models is coated with a sludge-resistant material designed to protect the bore and throttle plate. Do not attempt to clean the interior of the throttle body with carburetor or other spray cleaners. This throttle body is designed to resist sludge accumulation and cleaning may impair the performance of the engine.*

7 With the engine running, place an automotive stethoscope against each injector, one at a time, and listen for a clicking sound, indicating operation **(see illustration)**. If you don't have a stethoscope, you can place the tip of a long screwdriver against the injector and listen through the handle.

8 If an injector isn't functioning (not clicking), purchase a special injector test light (sometimes called a "noid" light) and install it into the injector electrical connector **(see illustration)**. Start the engine and check to see if the noid light flashes. If it does, the injector is receiving proper voltage. If it doesn't flash, further diagnosis should be performed by a dealer service department or other properly equipped repair facility.

9 With the engine OFF and the fuel injector electrical connectors disconnected, measure the resistance of each injector **(see illustration)**. Check the Specifications listed in this Chapter for the correct injector resistance.

10 The remainder of the system checks can be found in Chapter 6.

12.9 Measure the resistance across the two terminals of each injector (arrows) - resistance should be within Specifications

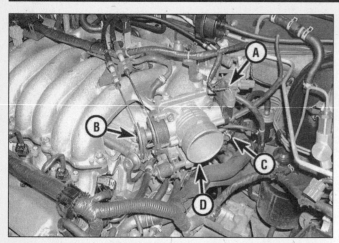

13.5 Disconnect the TPS connectors (A), the accelerator and cruise control cables (B), the coolant hoses (C) and the vacuum hoses (D)

13.6 Remove the throttle body mounting bolts (arrows) (1995 and later model shown)

13 Throttle body - removal and installation

Refer to illustrations 13.5 and 13.6

Warning: *Wait until the engine is completely cool before beginning this procedure.*

Note: *This procedure is shown on a 1993 through 1999 model. The throttle body on 2000 and later models is slightly different but this procedure pertains to all models.*

1 Disconnect the cable from the negative battery terminal.
2 Remove the air filter outlet tube from the intake duct or throttle body.
3 Disconnect the throttle position sensor connectors from the throttle body (see Chapter 6). Also label and detach all vacuum hoses from the throttle body.
4 Detach the accelerator cable (see Section 10) and if equipped, the cruise control cable.
5 Detach the coolant hoses from the throttle body **(see illustration)**. Plug the lines to prevent coolant loss.
6 Remove the throttle body mounting bolts **(see illustration)**. On 1993 and 1994 models, remove the intake duct. Remove the throttle body and gasket. Remove all traces of old gasket material from the throttle body and air intake plenum.
7 Installation is the reverse of removal. Be sure to use a new gasket. Adjust the accelerator cable and the cruise control cable (see Section 10). Check the coolant level and add, if necessary (see Chapter 1).

14 Fuel pressure regulator - removal and installation

Refer to illustrations 14.3 and 14.4

Warning: *Gasoline is extremely flammable, so take extra precautions when you work on any part of the fuel system. See the **Warning** in Section 2.*

1 Relieve the fuel pressure (see Section 2). Disconnect the cable from the negative terminal of the battery.
2 Remove the air intake duct and resonator from the throttle body and air filter assembly.
3 Clean any dirt from around the fuel pressure regulator. Detach the vacuum hose and the fuel return hose from the fuel pressure regulator **(see illustration)**.
4 Remove the two screws retaining the fuel pressure regulator **(see illustration)** and detach the regulator from the fuel rail.
5 Install new O-rings on the pressure regulator and lubricate them with a light coat of oil.
6 Installation is the reverse of removal. Tighten the pressure regulator mounting screws securely.

15 Fuel rail and injectors - removal and installation

Refer to illustrations 15.4, 15.5a, 15.5b, 15.6 and 15.8

Warning: *Gasoline is extremely flammable, so take extra precautions when you work on any part of the fuel system. See the **Warning** in Section 2.*

14.3 Disconnect the vacuum hose, loosen the hose clamp and remove the fuel return line from the fuel pressure regulator (arrow)

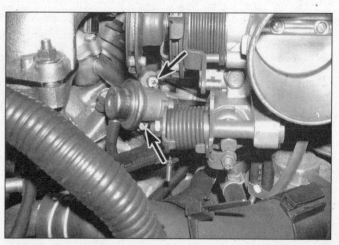

14.4 Remove the fuel pressure regulator screws (arrows)

Chapter 4 Fuel and exhaust systems

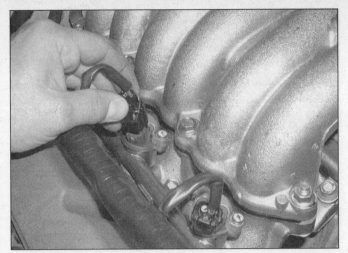

15.4 Disconnect the electrical connectors from the fuel injectors

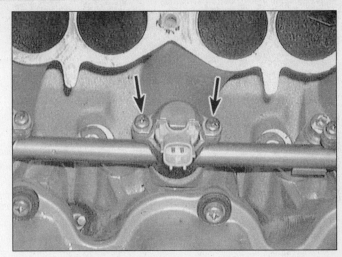

15.5a If fuel injector service is necessary, remove the cap screws (arrows) and pull the injector(s) from the fuel rail

Note: *On 1995 and later models, the fuel injectors on the front bank can be serviced without removing of the upper intake manifold. However servicing the fuel injectors on the rear bank requires removal of the upper intake manifold. It is not necessary to remove the fuel rail if you're only servicing a fuel injector. Individual fuel injectors can be removed without detaching the fuel rail.*

1 Relieve the fuel pressure (see Section 2).
2 Disconnect the cable from the negative terminal of the battery.
3 Remove the air intake plenum (upper intake manifold) from the lower intake manifold (see Chapter 2A or 2B).
4 Disconnect the fuel injector electrical connectors **(see illustration)**.
5 If servicing of the fuel injector(s) is necessary, remove the injector cap screws and cap, then pull the injector(s) from the fuel rail cup **(see illustrations)**. Inspect the injector O-rings (three per injector) for signs of deterioration. Replace as required. Lubricate the new O-rings with light grade oil. **Caution:** *Do not use silicone grease. It will clog the injectors.* Using a light twisting motion, install the injector(s) into the fuel rail cup. Ensure that the injector caps are clean and free of contamination and tighten the cap screws to the torque listed in this Chapter's Specifications.
6 If removal of the fuel rail assembly is necessary, disconnect the fuel feed and return lines then remove the fuel rail retaining bolts (two on each side) **(see illustration)**.
7 Using a rocking, side to side motion, carefully lift the fuel rail and the fuel injectors as an assembly from the lower intake manifold.
8 Installation is the reverse of removal with the following exceptions **(see illustration)**:

  a) Inspect the fuel rail insulators (one per injector) for signs of deterioration and replace as required.
  b) Install the fuel rail insulators into the lower intake manifold and lubricate them with a light coat of oil before installing the fuel rail assembly onto the lower intake manifold.

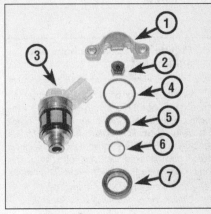

15.5b Fuel injector components

1 Cap
2 Insulator
3 Fuel injector
4 Top O-ring
5 Center O-ring
6 Bottom O-ring
7 Fuel rail insulator

15.6 Fuel rail mounting bolt locations (arrows)

15.8 Install the fuel rail insulators into the lower intake manifold and lubricate them with a light coat of oil before installing the fuel rail assembly

16.2 1995 and later models are not equipped with a distributor; therefore, the manufacturer has provided a special terminal (arrow) for connecting a tachometer to the system

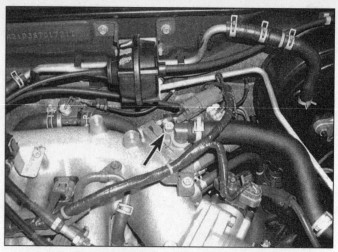

16.8 The idle adjustment screw (arrow) is located on top of the IAC valve

16 Engine idle speed - check and adjustment

1 Engine idle speed is the speed at which the engine operates when no accelerator pedal pressure is applied, as when stopped at a traffic light. This speed is critical to the performance of the engine itself, as well as many subsystems. Before checking the engine idle speed, check the following items:

a) Check the air filter for restriction.
b) Check the air intake system for leaks.
c) Check the vacuum hoses for leaks.
d) Check the battery and ignition system, including the ignition timing (see Chapter 5).
e) Check the fuel pressure (see Section 3).
f) Check the engine compression (see Chapter 2C).
g) Check the EGR and EVAP systems (see Chapter 6).
h) Check the TPS adjustment (see Chapter 6).
i) Check the On Board Diagnostic system for trouble codes (see Chapter 6).

Check

Refer to illustration 16.2

2 Connect a hand-held tachometer in accordance with the tool manufacturer's instructions. **Note:** *On 1995 and later models, a connector taped to the wiring harness on the right (passenger side) of the vehicle is provided for connecting a voltage-type tachometer* **(see illustration).**

3 Set the parking brake firmly and block the wheels to prevent the vehicle from rolling. Place the transaxle in Park or Neutral. Make sure all accessories are turned Off. On 1993 and 1994 models, remove the access panel to the PCM (see Chapter 6).

4 Start the engine, allow it to warm-up to normal operating temperature then shut it off.

5 PCM control of the idle system must be disabled before the idle speed can be checked or adjusted. Refer to Chapter 6, Section 2 and place the engine in the self-diagnosis mode *(Obtaining diagnostic system trouble codes)*. If any trouble codes are present repair the problem before proceeding. If no codes are present the Check Engine light will flash code 55 (or 0505).

6 With the system in the self-diagnosis mode, start the engine. Rev the engine to approximately 2000 rpm several times, then return the engine to idle. Verify that the ignition timing is correct (see Chapter 5).

7 Note the idle speed rpm on the tachometer and compare it to that listed on the VECI label or in this Chapter's Specifications. **Note:** *If the idle speed listed on the VECI label is different than that listed in this Chapter's Specifications, use the specification shown on the VECI label.*

Adjustment

Refer to illustration 16.8

8 If the idle speed is too low or too high, turn the idle speed adjustment screw to obtain the specified idle speed **(see illustration)**. **Caution:** *Do not attempt to adjust the idle speed with the idle stop screw on the throttle body. The idle stop screw is preset at the factory and should not be tampered with.*

9 Turn the engine off and disconnect the tachometer.

17 Exhaust system servicing - general information

Refer to illustrations 17.4a, 17.4b, 17.4c and 17.4d

Warning 1: *Inspection and repair of exhaust system components should be done only after enough time has elapsed after driving the vehicle to allow the system components to cool completely. Also, when working under the vehicle, make sure it is securely supported on jackstands.*

Warning 2: *All models covered by this manual are equipped with an exhaust system flex tube which is extremely sensitive to sharp bends. Do not allow the flex tube to hang downward during servicing or damage will occur.*

1 The exhaust system consists of the exhaust manifolds, the catalytic converter, the muffler, the tailpipe and all connecting pipes, brackets, hangers and clamps. The exhaust system is attached to the body with mounting brackets and rubber hangers. If any of the parts are improperly installed, excessive noise and vibration will be transmitted to the body.

2 Conduct regular inspections of the exhaust system to keep it safe and quiet. Look for any damaged or bent parts, open seams, holes, loose connections, excessive corrosion or other defects which could allow exhaust fumes to enter the vehicle. Deteriorated exhaust system components should not be repaired; they should be replaced with new parts.

3 If the exhaust system components are extremely corroded or rusted together, welding equipment will probably be required to remove them. The convenient way to accomplish this is to have a muffler repair shop remove the corroded sections with a cutting torch. If, however, you want to save money by doing it yourself (and you don't have a welding outfit with a cutting torch), simply cut off the old components with a hacksaw. If you have compressed air, special pneumatic cutting chisels can also be used. If you do decide to tackle the job at home, be sure to wear safety goggles to protect your eyes from metal chips and work gloves to protect your hands.

4 Here are some simple guidelines to fol-

Chapter 4 Fuel and exhaust systems

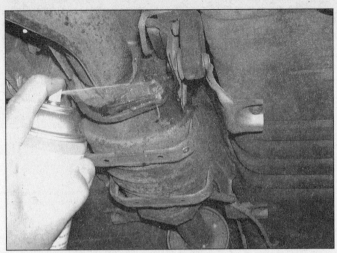

17.4a Be sure to apply penetrating lubricant to the exhaust system fasteners before attempting to remove them

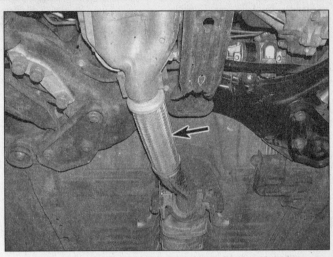

17.4b Do not allow the flex tube (arrow) to bend during servicing or damage will occur

low when repairing the exhaust system **(see illustrations)**:

a) Work from the back to the front when removing exhaust system components.
b) Apply penetrating oil to the exhaust system component fasteners to make them easier to remove.
c) Use new gaskets, hangers and clamps when installing exhaust systems components.
d) Apply anti-seize compound to the threads of all exhaust system fasteners during reassembly.
e) Be sure to allow sufficient clearance between newly installed parts and all points on the underbody to avoid overheating the floor pan and possibly damaging the interior carpet and insulation. Pay particularly close attention to the catalytic converter and heat shield.
f) Always remove oxygen sensors and connectors before servicing exhaust system components (see Chapter 6).

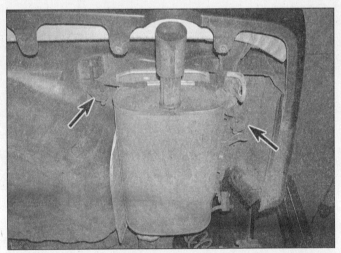

17.4c Check the condition of the rubber hangers (arrows) supporting the exhaust system

17.4d Be sure to allow sufficient clearance between the exhaust system and underbody, frame or suspension components

Notes

Chapter 5
Engine electrical systems

Contents

	Section
Alternator - removal and installation	13
Battery cables - check and replacement	4
Battery - check and replacement	3
Battery - emergency jump starting	2
Battery – maintenance and charging	See Chapter 1
Charging system - check	12
Charging system - general information and precautions	11
Distributor (1993 and 1994 models) - removal and installation	9
Drivebelt check, adjustment and replacement	See Chapter 1
General information	1
Ignition coil - check and replacement	7

	Section
Ignition system - check	6
Ignition system - general information	5
Ignition timing - check and adjustment	10
Spark plug replacement	See Chapter 1
Power transistor (1993 and 1994 models) - check and replacement	8
Starter motor and circuit - in-vehicle check	15
Starter motor - removal and installation	16
Starter solenoid - replacement	17
Starting system - general information and precautions	14

Specifications

Battery voltage	
Engine off	12 volts
Engine running	14 to 15 volts
Firing order	1-2-3-4-5-6
Ignition coil resistance (at 68-degrees F)	
1993 and 1994	
Primary resistance	1.0 ohm
Secondary resistance	8 to 12 K-ohms
1995 and later	Not applicable
Ignition timing	13 to 17 degrees BTDC
Spark plug wire resistance	5 to 9 K-ohms per foot (25 K-ohms maximum)

Chapter 5 Engine electrical systems

1.1a Starting, charging and ignition system components - 1993 and 1994 models

1. Spark plug
2. Battery
3. Battery cable
4. Spark plug wires
5. Ignition coil and power transistor
6. Distributor

1 General information

Refer to illustrations 1.1a and 1.1b

The engine electrical systems include all ignition, charging and starting components **(see illustrations)**. Because of their engine-related functions, these components are considered separately from chassis electrical devices like the lights, instruments, etc.

Be very careful when working on the engine electrical components. They are easily damaged if checked, connected or handled improperly. The alternator is driven by an engine drivebelt which could cause serious injury if your hands, hair or clothes become entangled in it with the engine running. Both the starter and alternator are connected directly to the battery and could arc or even cause a fire if mishandled, overloaded or shorted out.

1.1b Starting, charging and ignition system components - 1995 and later models

1. Starter motor (not visible)
2. Battery cable
3. Battery
4. Ignition coil (under cover)
5. Alternator (not visible)

Chapter 5 Engine electrical systems

3.1a Use a battery hydrometer to draw electrolyte from the battery cell - this hydrometer is equipped with a thermometer to make temperature corrections

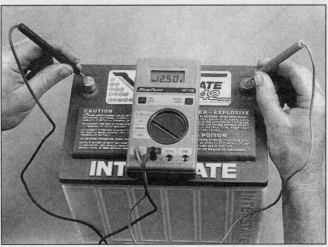

3.1b To test the open circuit voltage of the battery, connect the black probe of the voltmeter to the negative terminal and the red probe to the positive terminal of the battery - a fully charged battery should indicate approximately 12.5 volts depending on the outside air temperature

Never leave the ignition switch on for long periods of time with the engine off. Don't disconnect the battery cables while the engine is running. Correct polarity must be maintained when connecting battery cables from another source, such as another vehicle, during jump starting. Always disconnect the negative cable first and hook it up last or the battery may be shorted by the tool being used to loosen the cable clamps.

Additional safety related information on the engine electrical systems can be found in *Safety first* near the front of this manual. It should be referred to before beginning any operation included in this Chapter.

2 Battery - emergency jump starting

Refer to the *Booster battery (jump) starting* procedure at the front of this manual.

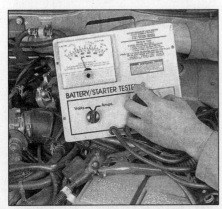

3.1c Some battery load testers are equipped with an ammeter which enables the battery load to be precisely dialed in, as shown – less expensive testers have a load switch and a voltmeter only

3 Battery - check and replacement

Note: *Anytime the battery is disconnected stored operating parameters may be lost from the PCM causing the engine to run rough for sometime while the PCM relearns the information.*

Check

Refer to illustrations 3.1a, 3.1b, 3.1c and 3.1d

1 A battery cannot be accurately tested until it is at or near a fully charged state. Disconnect the negative battery cable, then the positive cable from the battery and perform the following tests:

a) **Battery state of charge test** - Visually inspect the indicator eye (if equipped) on the top of the battery, if the indicator eye is dark in color, charge the battery as described in Chapter 1. If the battery is equipped with removable caps, check the battery electrolyte. The electrolyte level should be above the upper edge of the plates. If the level is low, add distilled water. DO NOT OVERFILL. The excess electrolyte may spill over during periods of heavy charging. Test the specific gravity of the electrolyte using a hydrometer **(see illustration)**. Remove the caps and extract a sample of the electrolyte and observe the float inside the barrel of the hydrometer. Follow the instructions from the tool manufacturer and determine the specific gravity of the electrolyte for each cell. A fully charged battery will indicate approximately 1.270 (green zone). If the specific gravity of the electrolyte is low (red zone), charge the battery as described in Chapter 1.

b) **Open voltage circuit test** - Using a digital voltmeter, perform an open voltage circuit test **(see illustration)**. **Note:** *The battery's surface charge must be removed before accurate voltage measurements can be made. Turn On the high beams for ten seconds, then turn them Off, let the vehicle stand for two minutes.* With the engine and all accessories Off, connect the negative probe of the voltmeter to the negative terminal of the battery and the positive probe to the positive terminal of the battery. The battery voltage should be approximately 12.5 volts. If the battery is less than the specified voltage, charge the battery before proceeding to the next test. Do not proceed with the battery load test until the battery is fully charged.

c) **Battery load test** - An accurate check of the battery condition can only be performed with a load tester (available at most auto parts stores). This test evaluates the ability of the battery to operate the starter and other accessories during periods of heavy amperage draw (load). Install a special battery load testing tool onto the battery terminals **(see illustration)**. Load test the battery according to the tool manufacturer's instructions. This tool utilizes a carbon pile to increase the load demand (amperage draw) on the battery. Maintain the load on the battery for 15 seconds and observe that the battery voltage does not drop below 9.6 volts. If the battery condition is weak or defective, the tool will indicate this condition immediately. **Note:** *Cold temperatures will cause the minimum voltage requirements to drop slightly. Follow the chart given in the tool manufacturer's instructions to compensate for cold climates. Minimum load voltage for freezing temperatures (32 degrees F) should be approximately 9.1 volts.*

Chapter 5 Engine electrical systems

3.1d To find out whether there's a drain on the battery, detach the negative cable and connect a test light between the battery post and the cable clamp

d) **Battery drain test** - *This test will indicate whether there's a constant drain on the vehicle's electrical system that can cause the battery to discharge. Make sure all accessories are turned Off. If the vehicle has an underhood light, verify it's working properly, then disconnect it. Disconnect the cable from the negative*

4.2 Typical battery cable problems

3.3 Remove the nuts (arrows) from the battery hold-down clamp

terminal of the battery and attach one lead of a test light to the negative battery cable and the other end to the negative battery terminal **(see illustration)**. *The test light should not glow. If the test light glows, it indicates a constant drain on the battery which could cause the battery to discharge.* **Note:** *On vehicles equipped with engine control computers, digital clocks, digital radios, power seats with memory and/or other components which normally cause a key-off battery drain, it's normal for the test light to glow dimly. If you suspect the drain is excessive, install an ammeter in place of the test light. The reading should not exceed 0.05 amps (50 milliamps).*

Replacement

Refer to illustrations 3.3 and 3.6
Caution: *Always disconnect the negative cable first and hook it up last or the battery may be shorted by the tool being used to loosen the cable clamps.*
2 Disconnect the negative battery cable, then the positive cable from the battery.
3 Remove the battery hold-down clamp **(see illustration)**.
4 Lift out the battery. Be careful - it's heavy. **Note:** *Battery straps and handlers are available at most auto parts stores for a rea-*

4.4a Detach any battery cable ties or retaining clips (arrow)

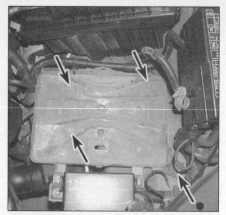

3.6 Battery tray mounting bolts (arrows)

sonable price. They make it easier to remove and carry the battery.
5 While the battery is out, inspect the battery tray for corrosion.
6 If corrosion exists on the battery tray, detach the bolts and remove the tray from the engine compartment **(see illustration)**. Clean the deposits from the metal underneath the tray to prevent further corrosion.
7 If you are replacing the battery, make sure you replace it with a battery with the identical dimensions, amperage rating, cold cranking rating, etc.
8 Installation is the reverse of removal.

4 Battery cables - check and replacement

Refer to illustrations 4.2, 4.4a, 4.4b, 4.4c and 4.4d
1 Periodically inspect the entire length of each battery cable for damage, cracked or burned insulation and corrosion. Poor battery cable connections can cause starting problems and decreased engine performance.
2 Check the cable-to-terminal connections at the ends of the cables for cracks, loose wire strands and corrosion **(see illustration)**. The presence of white, fluffy deposits under the insulation at the cable terminal connection is a sign that the cable is corroded and should be replaced. Check the terminals for distortion, missing mounting bolts and corrosion.
3 When replacing the cables, always disconnect the negative cable first and hook it up last or the battery may be shorted by the tool used to loosen the cable clamps. Even if only the positive cable is being replaced, be sure to disconnect the negative cable from the battery first.
4 Disconnect and remove the cable **(see illustrations)**. Make sure the replacement cable is the same length and diameter.
5 Clean the threads of the relay or ground connection with a wire brush to remove rust and corrosion. Apply a light coat of petroleum jelly to the threads to prevent future corrosion.

Chapter 5 Engine electrical systems

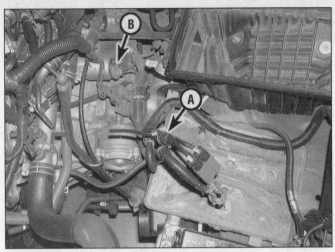

4.4b The positive cable is fastened to a fusible link connector (A) and at the starter solenoid (B)

4.4c The negative cable is fastened to the body (arrow) . . .

6 Attach the cable to the relay or ground connection and tighten the mounting nut/bolt securely.
7 Before connecting the new cable to the battery, make sure that it reaches the battery post without having to be stretched. Clean the battery posts thoroughly and apply a light coat of petroleum jelly to prevent corrosion (see Chapter 1).
8 Connect the positive cable first, followed by the negative cable.

5 Ignition system - general information

1 The ignition system is designed to ignite the fuel/air charge entering each cylinder at just the right moment. It does this by producing a high voltage spark between the electrodes of each spark plug.

1993 and 1994 models

2 The 1993 and 1994 SOHC models are equipped with an electronic ignition system which consists of the distributor, camshaft position sensor (located in the distributor), the power transistor, the ignition coil, an ignition circuit resistor/condenser and the primary and secondary wiring.
3 The camshaft position sensor is the basis of this computer controlled ignition system. It monitors engine speed and piston position and relays this data to the PCM which in turn controls the fuel injection duration (fuel injector on/off time) and ignition timing. The camshaft position sensor consists of a rotor plate, Light Emitting Diodes (LED) and photo diodes, which produce a wave forming circuit. This signal is then sent to the PCM, which produces an ignition signal. The power transistor amplifies the ignition signal from the PCM and intermittently grounds the primary circuit to the ignition coil which generates high voltage in the secondary circuit, thus sending spark from the ignition coil to the distributor, through the spark plug wires

and to the spark plugs.
4 The Powertrain Control Module (PCM) receives input signals from various sensors and switches and controls all spark timing advance and retard functions through the power transistor and ignition coil (see Chapter 6 for more information).
5 The ignition system is also integrated with a spark control system which uses a knock sensor in connection with the Powertrain Control Module to retard spark timing. The knock sensor system allows the engine to have maximum spark advance without spark knock which also improves driveability and fuel economy.
6 The secondary (spark plug) wires are a carbon-impregnated cord conductor encased in a rubber jacket with an outer silicone jacket. This type of wire will withstand very high temperatures and provides an excellent insulator for the high secondary ignition voltage. Silicone spark plug boots form a tight seal on the plug. The boot should be twisted 1/2-turn before removing (for more information on the spark plug wires refer to Chapter 1).

1995 and later models

7 1995 and later models are equipped with a distributorless ignition system. The system consists of six individual ignition coils located above each spark plug and connected directly to the spark plug. The PCM controls the operation of the ignition coils, firing each coil in sequence. The PCM uses information primarily supplied by the crankshaft position sensor and camshaft position sensor to determine the firing order sequence. In addition to the crankshaft and camshaft sensor signals, the PCM looks at the input from various other sensor to determine the optimum ignition timing.
8 The system is equipped with a knock sensor to detect detonation, or spark knock (usually caused by the use of sub-standard fuel). If a knock signal is received, the PCM will retard the timing until the knock is eliminated.

4.4d . . . and at the cylinder head (arrow)

6 Ignition system - check

Warning: *Because of the high voltage generated by the ignition system, extreme care should be taken whenever an operation is performed involving ignition components. This not only includes the power transistor, coil, distributor and spark plug wires, but related components such as plug connectors, tachometer and other test equipment.*

1 If a malfunction occurs and the vehicle won't start, do not immediately assume that the ignition system is causing the problem. First, check the following items:

a) Make sure the battery cable clamps, where they connect to the battery, are clean and tight.
b) Test the condition of the battery (see Section 3). If it does not pass all the tests, replace it with a new battery.
c) Check the external distributor and ignition coil wiring and connections.
d) Check the fusible links (if equipped) inside the engine compartment fuse box (see Chapter 12). If they're burned, determine the cause and repair the circuit

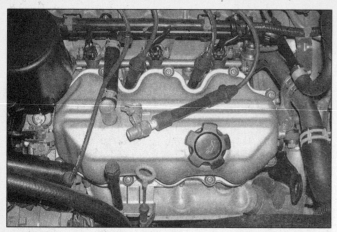

6.2 To use a calibrated ignition tester, disconnect a spark plug wire, clip the tester to a convenient ground (like a valve cover bolt) and operate the starter - if there is enough power to fire the plug, spark will be visible between the electrode tip and the tester body

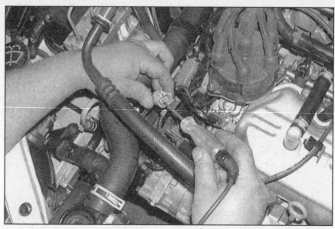

6.7 Disconnect the electrical connector from the ignition coil and check for battery voltage to the coil at the white/green wire terminal with the ignition key On, then connect an LED test light to the positive battery terminal, connect the test light probe to the light green wire terminal of the coil harness connector and watch for a blinking light when the engine is cranked

1993 and 1994 models

Refer to illustrations 6.2, 6.7 and 6.9

2   If the engine turns over but won't start, make sure there is sufficient secondary ignition voltage to fire the spark plug. Disconnect the spark plug wire from any spark plug and attach it to a calibrated ignition system tester (available at most auto parts stores). Connect the clip on the tester to a bolt or metal bracket on the engine **(see illustration)**. Crank the engine and watch the end of the tester to see if a bright blue, well-defined spark occurs (weak spark or intermittent spark is the same as no spark).

3   If spark occurs, sufficient voltage is reaching the plug to fire it (repeat the check at the remaining plug wires to verify that the distributor cap and rotor are OK). However, the plugs themselves may be fouled, so remove and check them as described in Chapter 1.

4   If no spark or intermittent spark occurs, check the cap, rotor and spark plug wires for damage and corrosion as described in Chapter 1. If moisture is present, dry out the cap and rotor, then reinstall the cap and repeat the spark test.

5   If there is no spark, detach the coil secondary wire from the distributor cap and hook it up to the tester (reattach the plug wire to the spark plug), then repeat the spark check. If spark now occurs, the distributor cap, rotor or plug wire(s) may be defective.

6   If no spark occurs, remove the coil-to-cap wire and check the terminals for damage. Using an ohmmeter, check the wire for an open or high resistance. If in doubt about its condition, substitute a known good wire and retest for spark.

7   If no spark occurs, check for battery voltage to the ignition coil from the ignition switch with the ignition key On (engine not running). Attach a 12 volt test light to the battery negative (-) terminal or other good ground. Disconnect the coil electrical connector and check for power at the white/green wire terminal **(see illustration)**. Battery voltage should be available. If there is no battery voltage, check the wiring and/or circuit between the coil and ignition switch. **Note:** *Refer to the wiring diagrams at the end of Chapter 12 for wire color identification for testing and additional information on the circuits.*

8   If battery voltage is available to the ignition coil, attach an LED test light to the battery positive (+) terminal. Connect the test light probe to the light green wire terminal of the coil harness connector (vehicle harness side) **(see illustration 6.7)**, then crank the engine. Confirm that the test light flashes. This test checks for the trigger signal (ground) from the computer and the power

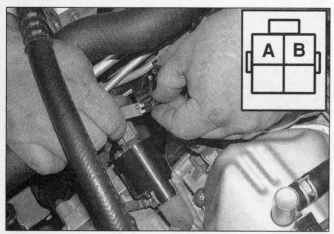

6.9 Remove the resistor from its square case and check the resistance between terminals A and B. There should be approximately 2.2 K-ohms

6.12 To check for spark on a 1995 and later model, remove an ignition coil and insert a calibrated ignition tester into the spark plug boot, clip the tester to a convenient ground (like a valve cover bolt) and operate the starter - if there is enough power to fire the plug, spark will be visible between the electrode tip and the tester body

Chapter 5 Engine electrical systems

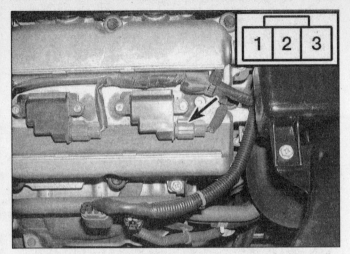

6.14 Disconnect the electrical connector from the ignition coil and check for battery voltage to the coil at terminal no. 1 with the ignition key On

6.16 Disconnect the condenser (arrow) from the harness connector and measure the resistance across the two terminals of the condenser - resistance should be very high (1 M-ohm or greater)

transistor. If a trigger signal is present at the coil, the computer, camshaft position sensor and or the power transistor are functioning properly. **Caution:** *Use only an LED test light to avoid damaging the PCM.*

9   If a trigger signal is not present at the ignition coil, check the ignition circuit resistor/condenser **(see illustration)** and the power transistor (see Section 8). **Note:** *The resistor/condenser and connector is taped to the wiring harness just to the left of the distributor.* If the power transistor and the resistor are good, refer to Chapter 6 and check the camshaft position sensor.

10 If battery voltage and a trigger signal exist at the ignition coil and there is no spark, check the primary and secondary resistance of the ignition coil (see Section 7). If an open is found (verified by an infinite reading), replace the coil.

11 If all the components are good and there is no spark, have the PCM checked by a dealer service department or other qualified repair shop.

1995 and later models

Refer to illustrations 6.12, 6.14 and 6.16

12 If the engine turns over but won't start, make sure there is sufficient secondary ignition voltage to fire the spark plug. Remove an ignition coil (see Section 7) and attach a calibrated ignition system tester (available at most auto parts stores) to the spark plug boot (be sure to reconnect the electrical connector to the coil). Connect the clip on the tester to a bolt or metal bracket on the engine **(see illustration)**. Crank the engine and watch the end of the tester to see if a bright blue, well-defined spark occurs (weak spark or intermittent spark is the same as no spark).

13 If spark occurs, sufficient voltage is reaching the plug to fire it (repeat the check at the remaining ignition coils to verify that the ignition coils are good). However, the plugs themselves may be fouled, so remove and check them as described in Chapter 1.

14 If no spark occurs, remove the spark plug boot and check the terminals for damage. Using an ohmmeter, check the boot for an open or high resistance. Check for battery voltage to the ignition coil from the ignition switch with the ignition key On (engine not running). Attach a 12 volt test light to the battery negative (-) terminal or other good ground. Disconnect the coil electrical connector and check for power at terminal no. 1 (red wire) **(see illustration)**. Battery voltage should be available. If there is no battery voltage, check the wiring and/or circuit between the coil, ECCS relay and ignition switch. Also check the ground circuit for continuity. **Note:** *Refer to the wiring diagrams at the end of Chapter 12 for wire color identification for testing and additional information on the circuits.*

15 If battery voltage is available to the ignition coil, connect the positive probe of a voltmeter to terminal no. 3 of the coil harness connector (vehicle harness side) **(see illustration 6.14)**, connect the negative probe to a good engine ground point then crank the engine. Approximately 0.2-volt (200 millivolts) should be indicated on the meter. This test checks for the trigger signal from the computer. If a trigger signal is present at the coil, the computer and crankshaft position sensors are functioning properly. **Note:** *An accurate check of the ignition coil trigger signal must be performed with an oscilloscope. The trigger signal should pulse from zero to 0.2 volts (average) several times per second as the engine is cranked.*

16 If a trigger signal is not present at the ignition coil, check the ignition circuit condenser **(see illustration)**. **Note:** *The condenser and connector are taped to the wiring harness at the transaxle end of the engine.* If the condenser is good, refer to Chapter 6 and check the camshaft position sensor and crankshaft position sensors.

17 If battery voltage, continuity to ground and a trigger signal exist at the ignition coil and there is no spark, check the ignition coil (see Section 7). If the ignition coil is defective, replace the coil.

18 If all the components are good and there is no spark, have the PCM checked by a dealer service department or other qualified repair shop.

7 Ignition coil - check and replacement

1993 and 1994 models

Refer to illustrations 7.1, 7.2 and 7.6

Check

1   With the ignition off, disconnect the electrical connector from the coil. To check the coil primary resistance, connect an ohmmeter across the two terminals of the coil connector **(see illustration)**. The resistance should be as listed in this Chapter's Specifications for primary resistance. If not, replace the coil.

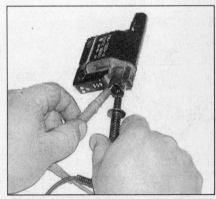

7.1 To check the primary resistance of the ignition coil, connect the probes of an ohmmeter to the two terminals of the coil connector

Chapter 5 Engine electrical systems

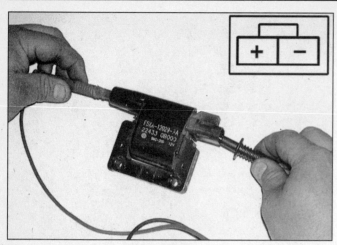

7.2 To check the coil secondary resistance, connect the ohmmeter to the positive (+) primary terminal and the coil secondary terminal

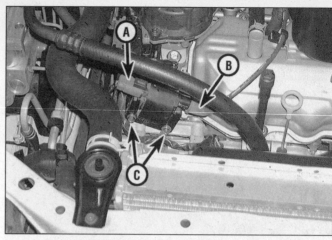

7.6 Detach the electrical connector (A), the secondary high tension lead (B), the coil mounting bolts (C) and remove the ignition coil from the engine

2   Connect an ohmmeter across the secondary terminal (the one that the coil-to-cap wire connects to) and to the positive primary terminal **(see illustration)**. The resistance should be as listed in this Chapter's Specifications for secondary resistance. If not, replace the coil.

Replacement

3   Disconnect the cable from the negative terminal of the battery.
4   Disconnect the ignition coil electrical connector.
5   Disconnect the secondary high tension (coil-to-cap) wire from the coil.
6   Remove the coil mounting bolts and detach it from the engine **(see illustration)**.
7   Installation is the reverse of the removal procedure. Before installing the spark plug wire connector into the ignition coil, coat the interior of the boot with silicone dielectric compound.

1995 and later models

Refer to illustrations 7.8 and 7.13

Check

8   With the ignition off, disconnect the electrical connector from the ignition coil. Connect the positive (+) probe of an ohmmeter to terminal no. 1 and the negative (-) probe to terminal no. 2 of the coil connector **(see illustration)** - the meter should indicate infinite resistance. Reverse the meter leads (negative probe to terminal no. 1 and positive probe to terminal no. 2) - continuity should be indicated, but not zero ohms. If the results are not as specified, replace the coil.
9   Connect an ohmmeter between the secondary terminal (the one that the spark plug connects to) and terminal no. 1 **(see illustration 7.8)** - the meter should indicate infinite resistance. If not, replace the coil.
10   Check each coil in the same manner.

Replacement

11   Disconnect the cable from the negative terminal of the battery.
12   Disconnect the ignition coil electrical connector.
13   Remove the coil mounting screws and pull the coil up with a twisting motion **(see illustration)**.
14   Installation is the reverse of the removal procedure. Before installing the ignition coil, coat the interior of the boot with silicone dielectric compound.

8   Power transistor (1993 and 1994 models) - check and replacement

Caution: *The power transistor is a delicate and relatively expensive electrical component. Failure to follow the step-by-step procedures could result in damage to the module and/or other electronic devices, including the PCM. Additionally, all PCM controlled devices are protected by a Federally mandated emissions warranty. Check with the dealer concerning this warranty before attempting to diagnose and replace this unit yourself.*

Check

Refer to illustration 8.3

1   Disconnect the cable from the negative terminal of the battery.
2   Disconnect the electrical connector from the power transistor.
3   Using an ohmmeter, connect the positive probe of the ohmmeter to terminal A and the negative probe of the ohmmeter to terminal B **(see illustration)**. The meter should indicate continuity. Connect the negative probe of the ohmmeter to terminal A and the positive probe of the ohmmeter to terminal B. The meter

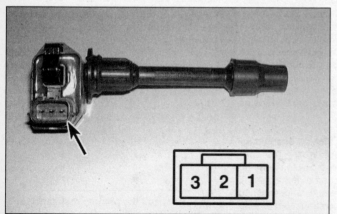

7.8 Ignition coil terminal guide - 1995 and later models

7.13 Disconnect the electrical connector, remove the mounting screws (arrows) and pull the coil up with a twisting motion

Chapter 5 Engine electrical systems

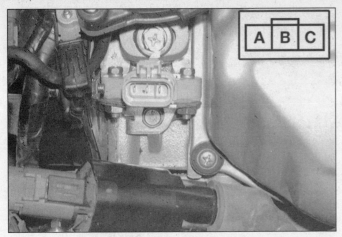

8.3 Power transistor terminal guide

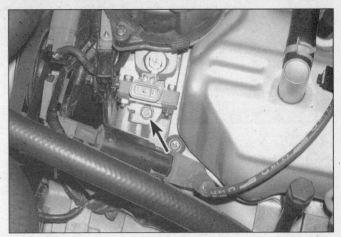

8.7 Remove the power transistor mounting bolt (arrow) and the power transistor

should indicate no continuity (infinity).
4 If the test results are incorrect, replace the power transistor with a new part.

Replacement

Refer to illustration 8.7

5 Disconnect the cable from the negative terminal of the battery.
6 Disconnect the electrical connector from the power transistor.
7 Remove the bolt that secures the power transistor and retaining bracket to the cylinder head **(see illustration)**.
8 Remove two retaining bracket bolts. Install a new power transistor onto the retaining bracket.
9 The remainder of the installation is the reverse of removal.

9 Distributor (1993 and 1994 models) - removal and installation

Removal

Refer to illustrations 9.5a and 9.5b

1 Disconnect the cable from the negative terminal of the battery.

2 Remove the distributor cap cover (if equipped) and note the position of the raised "1" on the distributor cap. This marks the location for the number one cylinder spark plug wire terminal. **Note:** *Some distributor caps may not be marked with the number 1 terminal position.*
3 Disconnect the main electrical connector and ground connector from the distributor. Follow the wires as they exit the distributor to find the connector if necessary.
4 Remove the distributor cap (see Chapter 1). Using a socket and breaker bar on the crankshaft pulley bolt, rotate the engine until the rotor is pointing toward the number one spark plug terminal (see the TDC locating procedure in Chapter 2A).
5 Make a mark on the edge of the distributor base directly below the rotor tip and inline with it **(see illustration)**. Also, mark the distributor base and cylinder head to ensure that the distributor can be reinstalled correctly **(see illustration)**.
6 Remove the distributor hold-down bolt, then pull the distributor straight out to remove it. **Caution:** *DO NOT turn the engine while the distributor is removed, or the alignment marks will be useless.*

Installation

7 Insert the distributor into the engine in exactly the same relationship to the block that it was in when removed.
8 To mesh the helical gears on the camshaft and the distributor, it may be necessary to turn the rotor slightly. Make sure the distributor is seated completely and the alignment marks made previously are aligned. If not, remove the distributor and reposition it. **Note:** *If the crankshaft has been moved while the distributor is out, locate Top Dead Center (TDC) for the number one piston (see Chapter 2A) and position the distributor and rotor accordingly.*
9 Loosely install the hold-down bolt.
10 Install the distributor cap and tighten the screws securely.
11 Plug in the electrical connectors.
12 Reattach the spark plug wires to the plugs (if removed).
13 Connect the cable to the negative terminal of the battery.
14 Check and, if necessary, adjust the ignition timing (see Section 10) and tighten the distributor hold-down bolt securely. Reinstall the distributor cap cover (if equipped).

9.5a Apply an alignment mark on the perimeter of the distributor body in line with the rotor tip (arrows)

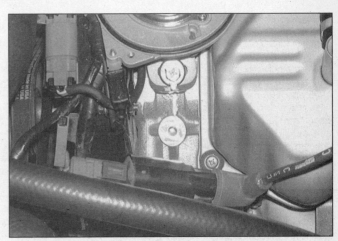

9.5b Mark the base of the distributor body and the cylinder head to clearly define the position of the distributor

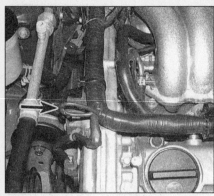

10.1 On 1995 and later models, a loop (arrow) is provided for connecting a timing light inductive pick-up

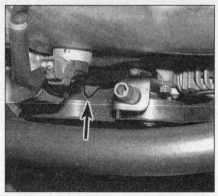

10.3a Locate the timing pointer (arrow) and notches on the crankshaft pulley - the TDC notch is marked with yellow paint

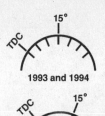

10.3b On 1993 and 1994 models, each notch represents 5-degree increments - on 1995 and later models, the first notch clockwise from the TDC mark is the 15-degree BTDC mark

10 Ignition timing - check and adjustment

Refer to illustrations 10.1, 10.3a and 10.3b
Note: *Timing is adjustable on 1993 and 1994 models only. On 1995 and later models, timing is not adjustable, but the check can be performed to verify proper operation of the ignition control system.*

1 With the ignition switch off, connect a timing light in accordance with the tool manufacturer's instructions. On 1993 and 1994 models, install the inductive pick-up onto the number one cylinder spark plug wire. On 1995 and later models install the inductive pick-up onto the loop in the ignition coil wiring harness **(see Illustration)**. **Note:** *If your timing light is unable to pick-up a signal from the wiring harness loop, remove the number one cylinder ignition coil, reconnect the electrical connector, install a spark plug wire between the ignition coil and the spark plug and connect the timing light to the spark plug wire.*
2 Set the parking brake firmly and block the wheels to prevent the vehicle from rolling. Place the transaxle in Park or Neutral.
3 Locate the timing notches on the crankshaft pulley and the pointer on the timing cover **(see illustrations)**. The Top Dead Center (TDC) notch on the crankshaft pulley is marked with yellow paint. On 1993 and 1994 models, notches on the pulley are spaced in 5 degree increments, clockwise from the TDC notch. Locate the specified notch (three notches clockwise from the yellow TDC mark indicates 15-degrees BTDC, for instance). On 1995 and later models, the first notch clockwise from the TDC notch indicates 15-degrees BTDC. Clean the marks, if necessary, so they will be easy to see.
4 Start the engine, allow it to warm-up to normal operating temperature then shut it off. PCM control of the timing must be disabled before the timing can be checked or adjusted. Refer to Chapter 6, Section 2 and place the engine in the self-diagnosis mode *(Obtaining diagnostic system trouble codes)*. If any trouble codes are present repair the problem before proceeding. If no codes are present the Check Engine light will flash code 55 (or 0505).
5 With the system in the self-diagnosis mode, start the engine. Rev the engine to appoximately 2000 rpm several times, then return the engine to idle. Verify that the engine idle speed is correct (see Chapter 4).
6 Aim the timing light at the timing marks on the front of the engine and check the ignition timing. The specified notch on the pulley will appear stationary and be aligned with the pointer if the timing is correct.
7 On 1993 and 1994 models, if an adjustment is required, loosen the distributor hold-down bolt and rotate the distributor slightly until the timing is correct. Tighten the bolt and recheck the timing.
8 On 1995 and later models, if the timing is not as specified, check the crankshaft and camshaft position sensors (see Chapter 6). If the sensors are good, have the PCM diagnosed by a dealer service department or other qualified repair shop.
9 Shut the engine off and disconnect the timing light.

11 Charging system - general information and precautions

The charging system includes the alternator, a voltage regulator (mounted on the backside of the alternator), a charge indicator or warning light, the battery, a large fusible link and the wiring between all the components. The charging system supplies electrical power for the ignition system, the lights, the radio, etc. The alternator is driven by a drivebelt at the front of the engine.
The purpose of the voltage regulator is to limit the alternator's voltage to a preset value. This prevents power surges, circuit overloads, etc., during peak voltage output. All models are equipped with integral type voltage regulator, If a voltage regulator malfunctions, it will be necessary to replace the entire alternator.
The charging system is protected by a large fusible link which is located in the engine compartment fuse box. In the event of charging system problems, check the fusible link for damage or broken contacts.
The charging system doesn't ordinarily require periodic maintenance. However, the drivebelt, battery and wires and connections should be inspected at the intervals outlined in Chapter 1.
Be very careful when making electrical circuit connections to a vehicle equipped with an alternator and note the following:

a) When reconnecting wires to the alternator from the battery, be sure to note the polarity.
b) Before using arc welding equipment to repair any part of the vehicle, disconnect the wires from the alternator and the battery terminals.
c) Never start the engine with a battery charger connected.
d) Always disconnect both battery cables before using a battery charger (always disconnect negative cable first, positive cable last).

12 Charging system - check

Refer to illustrations 12.3 and 12.7

1 If a malfunction occurs in the charging circuit, do not immediately assume that the alternator is causing the problem. First, check the following items:

a) The battery cables where they connect to the battery. Make sure the connections are clean and tight.
b) Check the battery state of charge (see Section 3).
c) Check the external alternator wiring and connections.
d) Check the drivebelt condition and tension (see Chapter 1).
e) Check the alternator mounting bolts for tightness.
f) Run the engine and check the alternator for abnormal noise.
g) Check the fusible links in the engine compartment fuse box (see Chapter 12). If they're burned, determine the cause and repair the circuit.
h) Refer to wiring diagrams in Chapter 12 and check all the fuses in series with the charging system. The location of the fuses may vary from year to year but the designations are the same.

Chapter 5 Engine electrical systems

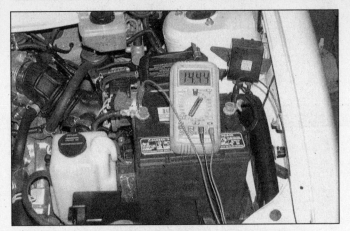

12.3 To measure charging voltage, attach the voltmeter leads to the battery terminals, start the engine and record the voltage reading

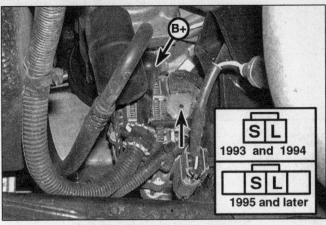

12.7 Alternator terminal identification

2 Using a voltmeter, check the battery voltage with the engine off. It should be approximately 12-volts.

3 Start the engine and check the battery voltage again. The voltage should be greater than the voltage measured in Step 2, but not more than 15-volts (see illustration).

4 Turn ON the headlights. The voltage may drop slightly but should be greater than the voltage measured in Step 2, if the charging system is working properly.

5 If the charging voltage is greater than 15-volts, the voltage regulator is defective. Replace the alternator (see Section 13).

6 If the charging voltage is less than the voltage measured in Step 2, check the alternator as follows:

7 Using a voltmeter and working on the backside of the alternator, backprobe the "B+" terminal. There should be 12 volts present with the ignition key Off (see illustration).

8 With the ignition key On (engine not running), backprobe each terminal. There should be 12 volts at the "S" terminal, 1.5 to 2.0 volts at the "L" terminal and 12 volts at the "B+" terminal.

9 Start the engine, then raise the engine speed to 2000 rpm and backprobe each terminal again. There should be 14.0 to 14.7 volts at the "S" terminal and "B+" terminal and 13.0 to 14.0 volts at the "L" terminal. Warning: *Make sure the meter leads, loose clothing, long hair, etc. are away from the moving parts of the engine (drivebelt, cooling fan, etc.) before starting the engine.*

10 If the voltages are not as specified, check the wiring harness. If the wiring harness is not defective, replace the alternator.

11 If you suspect that there is a voltage drain on the battery while the vehicle is sitting in the driveway, see Section 3 and perform a battery drain test.

12 If a drain is indicated, carefully remove the fuses one-by-one that govern accessories such as radio, blower motor, trunk lights, etc. until the test light goes out. Trace the short circuit in the particular fused circuit and repair the problem. Recheck the electrical system as described.

13 If all the fuses are pulled out and the test light remains lit, remove the alternator output cable at the rear of the alternator then unplug all the connectors from the backside of the alternator. If the test light goes out, then there is an internal drain in the alternator or voltage regulator. Replace the alternator.

13 Alternator - removal and installation

1993 and 1994 models

Refer to illustrations 13.5a and 13.5b

1 Disconnect the cable from the negative terminal of the battery.

2 Raise the vehicle and support it securely on jackstands. Remove the lower splash shield (if equipped) from beneath the alternator.

3 Remove the drivebelt (see Chapter 1).

4 Disconnect the electrical connector and the alternator output wire from the alternator.

5 Remove the bolts and separate the alternator from the engine (see illustrations).

6 Installation is the reverse of removal.

7 Install the drivebelt and reconnect the cable to the negative terminal of the battery. Adjust the drivebelt following the procedure in Chapter 1.

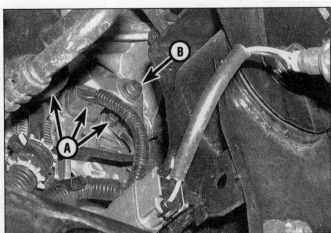

13.5a Working at the rear of the alternator, disconnect the wiring (A) and the rear retaining bolt (B)

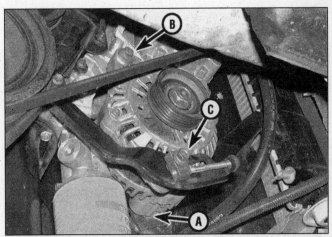

13.5b Working at the front of the alternator, detach the wiring harness retaining bracket (A) the upper bolt (B) and the adjuster lock bolt (C)

Chapter 5 Engine electrical systems

13.10 Disconnect the alternator electrical connector and output wire

13.13 Remove the mounting bolts and remove the alternator

1995 and later models

Refer to illustrations 13.10 and 13.13

8   Disconnect the cable from the negative terminal of the battery.
   Raise the vehicle and support it securely on jackstands. Remove the lower splash shield beneath the alternator and the right-side inner fender splash shield.
9   Remove the drivebelt (see Chapter 1).
10 Disconnect the electrical connector and the alternator output wire from the alternator **(see Illustration)**.
11 Remove the engine cooling fan and shroud (see Chapter 3).
12 Remove the air conditioning compressor mounting bolts (without disconnecting the refrigerant lines) and slide the compressor forward.
13 Remove the alternator mounting bolts and remove the alternator from the engine **(see Illustration)**.
14 Installation is the reverse of removal.
15 Install the drivebelt and reconnect the cable to the negative terminal of the battery. Adjust the drivebelt following the procedure in Chapter 1.

14 Starting system - general information and precautions

The sole purpose of the starting system is to crank the engine over fast enough to allow it to start. The starting system is composed of the starter motor, starter inhibitor or interlock relay, battery, ignition switch and connecting wiring.
   The starter circuit is equipped with an inhibitor or interlock relay. Turning the ignition key to the Start position actuates the starter relay through the starter control circuit. The starter relay then connects battery power to the starter solenoid. The starter solenoid connects battery power to the starter motor and the starter motor turns. The relay is located in the engine compartment fuse/relay box (see Chapter 12). The starter/solenoid assembly is mounted to the transmission bellhousing.
   Automatic transaxle models are equipped with a Park/Neutral position switch in the starter control circuit, which prevents operation of the starter unless the shift lever is in Neutral or Park. Manual transaxle models are equipped with a clutch interlock switch in the starter control circuit, which prevents starter operation unless the clutch pedal is depressed.

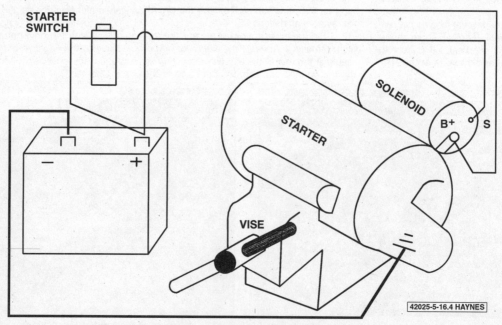

15.7 Starter motor bench testing details

Chapter 5 Engine electrical systems

Never operate the starter motor for more than 15 seconds at a time without pausing to allow it to cool for at least two minutes. Excessive cranking can cause overheating, which can seriously damage the starter.

15 Starter motor and circuit - in-vehicle check

Refer to illustration 15.7

1 If a malfunction occurs in the starting circuit, do not immediately assume that the starter is causing the problem. First, check the following items:
   a) Make sure the battery cable clamps, where they connect to the battery, are clean and tight.
   b) Check the condition of the battery cables (see Section 4). Replace any defective battery cables with new parts.
   c) Test the condition of the battery (see Section 3). If it does not pass all the tests, replace it with a new battery.
   d) Check the starter solenoid wiring and connections.
   e) Check the starter mounting bolts for tightness.

2 If the starter does not activate when the ignition switch is turned to the start position, check for battery voltage to the solenoid with the ignition switch Off. There should be battery voltage at the positive battery cable on the solenoid if the battery and/or cables are in good working order.

3 Backprobe the S terminal on the starter solenoid and check for voltage as the ignition switch is turned to the start position. This will determine if the solenoid is receiving the correct voltage signal from the ignition switch. If voltage is not available, check the fusible links in the engine compartment fuse box (see Chapter 12). If they're burned, determine the cause and repair the circuit. Also, check the related fuses in the passenger compartment fuse panel (see Chapter 12). If the fuses and fusible links are OK, check the starter inhibitor or interlock relay and circuit for

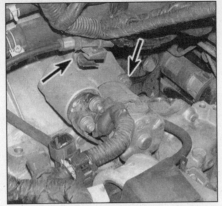

16.4 Remove the electrical connector, battery cable and the starter mounting bolts (arrows) and separate the assembly from the transaxle bellhousing

proper operation. Refer to Chapter 12 for the relay locations, wiring diagrams and the relay checking procedure.

4 If the starter relay circuit is not functioning, check the operation of the Park/Neutral position switch or clutch interlock switch (see Chapter 7 or 8). Make sure the shift lever is in PARK or NEUTRAL or the clutch pedal is fully depressed when attempting to start the engine.

5 If the vehicle is equipped with an anti-theft alarm, check the circuit and the control module for shorts or damaged components.

6 If the starter is receiving voltage but does not activate, most likely the solenoid is defective, but in some rare cases, the engine may be seized. Verify the engine is not seized by rotating the crankshaft pulley (see Chapter 2A) before proceeding.

7 If voltage is available at the starter solenoid and there is no movement from the starter motor, remove the starter from the engine (see Section 16) and bench test the starter. Mount the starter/solenoid assembly in a large vise on a sturdy bench. Install one jumper cable from the negative terminal (-) of a fully charged 12-volt automotive battery to the body of the starter **(see illustration)**.

Install another jumper cable from the positive terminal (+) of the battery to the battery terminal on the starter. Install a starter switch between the positive terminal of the battery and the starter solenoid terminal. Apply battery voltage to the solenoid terminal (for 10 seconds or less) and observe the solenoid plunger, shift lever and overrunning clutch extend and rotate the pinion drive. If the pinion drive extends but does not rotate, the solenoid is operating but the starter motor is defective. If there is no movement but the solenoid clicks, the solenoid and/or the starter motor is defective. If the solenoid plunger extends and rotates the pinion drive at approximately 3,000 rpm, the starter/solenoid assembly is working properly.

16 Starter motor - removal and installation

Refer to illustration 16.4

1 Disconnect the cable from the negative terminal of the battery.

2 Remove the air intake duct and the air cleaner housing assembly from the engine compartment (see Chapter 4).

3 Disconnect the battery cable and the solenoid terminal connection from the starter solenoid.

4 Remove the starter motor mounting bolts **(see illustration)** and detach the starter from the engine.

5 Installation is the reverse of removal.

17 Starter solenoid - replacement

Refer to illustrations 17.3a and 17.3b

1 Remove the starter assembly from the engine compartment (see Section 16).

2 Remove the electrical connector from the solenoid lower terminal.

3 Remove the solenoid mounting screws and separate the solenoid from the starter body **(see illustrations)**.

4 Installation is the reverse of removal.

17.3a Remove the solenoid mounting screws (arrows)

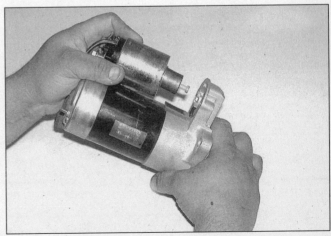

17.3b Separate the solenoid from the starter

Notes

Chapter 6
Emissions and engine control systems

Contents

	Section		Section
Camshaft position sensor - check and replacement	10	Manifold absolute pressure sensor and solenoid valve - check and replacement	6
Catalytic converter	21	Mass airflow sensor - check and replacement	5
Crankshaft position sensor - check and replacement	9	On-Board Diagnosis system and trouble codes	2
Engine coolant temperature sensor - check and replacement	8	Oxygen sensor - check and replacement	12
Evaporative emissions control system	20	Positive crankcase ventilation system	18
Exhaust gas recirculation system	19	Power steering pressure switch - check and replacement	11
Fuel temperature sensor - check and replacement	14	Powertrain Control Module - removal and installation	3
General information	1	Power valve control system	17
Idle air control system	16	Throttle position sensor - check, replacement and adjustment	4
Intake air temperature sensor - check and replacement	7	Vehicle speed sensor - check and replacement	15
Knock sensor - general information	13		

1 General information

Refer to illustrations 1.1a, 1.1b, 1.7a and 1.7b

To prevent pollution of the atmosphere from incompletely burned and evaporating gases, and to maintain good driveability and fuel economy, a number of emission control systems are incorporated **(see illustrations)**. They include the:

Electronic engine control system
Evaporative emission controls system (EVAP)
Positive Crankcase Ventilation (PCV) system
Exhaust Gas Recirculation (EGR) system
Catalytic converter

All of these systems are linked, directly or indirectly, to the emission control system. The Sections in this Chapter include general

1.1a Typical emission and engine control system components - 1993 and 1994 models

1. Power steering pressure switch
2. Engine coolant temperature sensor
3. PCV valve
4. Power valve actuator
5. Mass airflow sensor
6. Idle air control valve
7. EGR valve
8. Knock sensor (below exhaust manifold)
9. Throttle position sensor
10. Camshaft position sensor (inside distributor)

6-2　Chapter 6　Emissions and engine control systems

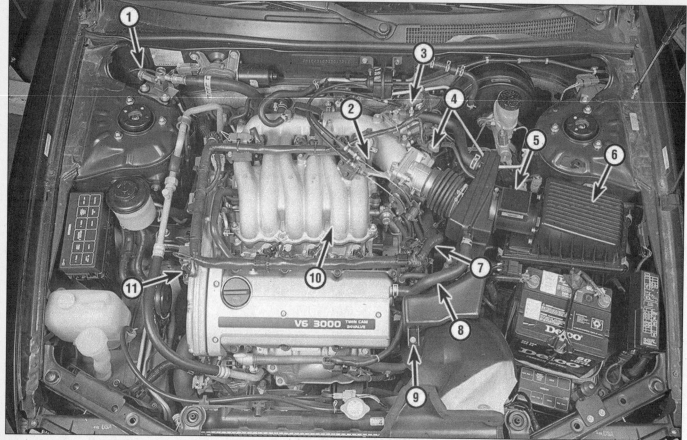

1.1b Typical emission and engine control system components - 1995 and later models (typical)

1. Power steering pressure switch
2. PCV valve
3. Idle air control valve and fast idle control solenoids
4. Throttle position sensor
5. Mass airflow sensor
6. Intake air temperature sensor
7. EGR and EVAP control solenoids
8. Engine coolant temperature sensor
9. Manifold absolute pressure sensor (location)
10. Knock sensor (beneath intake manifold)
11. Camshaft position sensor

descriptions, checking procedures within the scope of the home mechanic and component replacement procedures (when possible) for each of the systems listed above.

Before assuming that an emissions control system is malfunctioning, check the fuel and ignition systems carefully. The diagnosis of some emission control devices requires specialized tools, equipment and training. If checking and servicing become too difficult or if a procedure is beyond your ability, consult a dealer service department. Remember, the most frequent cause of emissions problems is simply a loose or broken vacuum hose or wire, so always check the hose and wiring connections first.

This doesn't mean, however, that emission control systems are particularly difficult

1.7a The Vehicle Emission Control Information (VECI) label is located in the engine compartment and contains information on the emission devices on your vehicle

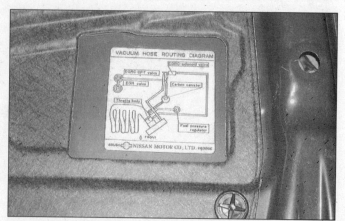

1.7b A vacuum hose routing diagram provides the specific vacuum hose routing information for the vehicle

Chapter 6 Emissions and engine control systems

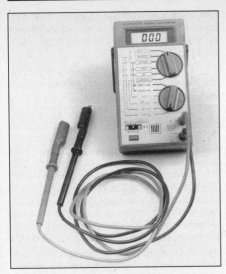

2.1 Digital multimeters can be used for testing all types of circuits; because of their high impedance, they are much more accurate than analog meters for measuring low-voltage computer circuits

2.2 Scanners like the Actron Scantool and the AutoXray XP240 are powerful diagnostic aids - programmed with comprehensive diagnostic information, they can tell you just about anything you want to know about your engine management system

2.4 Trouble code tools simplify the task of extracting the trouble codes on OBD-I systems

to maintain and repair. You can quickly and easily perform many checks and do most of the regular maintenance at home with common tune-up and hand tools. **Note:** *Because of a Federally mandated warranty which covers the emission control system components, check with your dealer about warranty coverage before working on any emissions-related systems. Once the warranty has expired, you may wish to perform some of the component checks and/or replacement procedures in this Chapter to save money.*

Pay close attention to any special precautions outlined in this Chapter. It should be noted that the illustrations of the various systems may not exactly match the system installed on the vehicle you're working on because of changes made by the manufacturer during production or from year-to-year.

A Vehicle Emissions Control Information (VECI) label is located in the engine compartment **(see illustrations)**. This label contains important emissions specifications and adjustment information, as well as a vacuum hose schematic with emissions components identified. When servicing the engine or emissions systems, the VECI label in your particular vehicle should always be checked for up-to-date information.

2 On-Board Diagnosis (OBD) system and trouble codes

Diagnostic tool information

Refer to illustrations 2.1, 2.2 and 2.4

1 A digital multimeter is necessary for checking fuel injection and emission related components **(see illustration)**. A digital volt-ohmmeter is preferred over the older style analog multimeter for several reasons. The analog multimeter cannot display the volts-ohms or amps measurement in hundredths and thousandths increments. When working with electronic circuits which are often very low voltage, this accurate reading is most important. Another good reason for the digital multimeter is the high impedance circuit. The digital multimeter is equipped with a high resistance internal circuitry (10 million ohms). Because a voltmeter is hooked up in parallel with the circuit when testing, it is vital that none of the voltage being measured should be allowed to travel the parallel path set up by the meter itself. This dilemma does not show itself when measuring larger amounts of voltage (9 to 12 volt circuits) but if you are measuring a low voltage circuit such as the oxygen sensor signal voltage, a fraction of a volt may be a significant amount when diagnosing a problem. Obtaining the diagnostic trouble codes is one exception where using an analog voltmeter is necessary.

2 Hand-held scanners are the most powerful and versatile tools for analyzing engine management systems used on later model vehicles **(see illustration)**. Each brand scan tool must be examined carefully to match the year, make and model of the vehicle you are working on. Often interchangeable cartridges are available to access the particular manufacturer (Ford, GM, Chrysler, etc.). Some manufacturers will specify by continent (Asia, Europe, USA, etc.).

3 With the arrival of the second generation Federally mandated emission control system (OBD-II), a specially designed scanner has been developed. Several tool manufacturers have released OBD-II scan tools for the home mechanic. Ask the parts salesman at a local auto parts store for additional information concerning dates and costs.

4 Another type of code reader is available at parts stores **(see illustration)**. These tools simplify the procedure for extracting codes from the engine management computer on some models by simply "plugging in" to the diagnostic connector on the vehicle wiring harness.

On-Board Diagnostic system general description

5 All models are equipped with an On-Board Diagnostic (OBD) system. 1993 and 1994 models are equipped with the first generation OBD-I system. 1995 and later are equipped with the second generation OBD-II system. Both systems consist of an onboard computer, known as the Powertrain Control Module (PCM), and information sensors, which monitor various functions of the engine and send data to the PCM. Based on the data and the information programmed into the computer's memory, the PCM generates output signals to control various engine functions via control relays, solenoids and other output actuators.

6 The PCM is the "brain" of the electronic engine control system. It receives data from a number of sensors and other electronic components (switches, relays, etc.). Based on the information it receives, the PCM generates

output signals to control various relays, solenoids and other actuators. The PCM is specifically calibrated to optimize the emissions, fuel economy and driveability of the vehicle.

7   Because of a Federally mandated warranty which covers the emissions system components and because any owner-induced damage to the PCM, the sensors and/or the control devices may void the warranty, it isn't a good idea to attempt diagnosis or replacement of the PCM at home while the vehicle is under warranty. Take the vehicle to a dealer service department if the PCM or a system component malfunctions.

Information sensors

8   **Camshaft position sensor** - The camshaft position sensor provides information on camshaft position and the engine speed signal to the PCM. The PCM uses this signal to control ignition timing and fuel injection.

9   **Crankshaft position sensors** - 1995 and later models use two crankshaft position sensors, while earlier models have only one. The crankshaft position sensor (REF) is used to detect engine speed. Crankshaft position sensor (POS) is used to detect TDC for each cylinder. The PCM uses the signals to control ignition timing and fuel injection. They are also used to detect engine misfires on OBD-II systems.

10 **Engine coolant temperature sensor** - The engine coolant temperature sensor monitors engine coolant temperature and sends the PCM a voltage signal that affects PCM control of the fuel mixture, ignition timing, and EGR operation.

11 **Exhaust gas recirculation temperature sensor** - The EGR temperature sensor is used to monitor the rate and flow of exhaust gas recirculation into the intake system.

12 **Fuel temperature sensor** - The fuel temperature sensor provides the PCM with fuel temperature information. The PCM uses this input signal for diagnostic purposes only.

13 **Intake air temperature sensor** - The intake air temperature sensor provides the PCM with intake air temperature information. The PCM uses this information to control fuel injection, ignition timing, and EGR system operation.

14 **Knock sensor** - The knock sensor is a piezoelectric element that detects the sound of engine detonation, or "pinging". The PCM uses the input signal from the knock sensor to recognize detonation and retard spark advance to avoid engine damage.

15 **Manifold absolute pressure sensor** - This sensor is used on 1996 and later models only. The sensor is used in conjunction with a MAP sensor solenoid valve to monitor intake manifold pressure and ambient barometric pressure. The PCM uses this input signal for diagnostic purposes only.

16 **Mass airflow sensor** - The mass airflow sensor measures the molecular mass of the intake airflow entering the engine. The mass airflow sensor, along with the intake air temperature sensor, provide mass airflow and air temperature information for the most precise fuel metering.

17 **Oxygen sensor** - The oxygen sensor generates a voltage signal that varies with the difference between the oxygen content of the exhaust and the oxygen in the surrounding air. The PCM uses this information to determine if the fuel system is running rich or lean.

18 **Power steering pressure switch** - The power steering pressure switch is used to detect excessive line pressure in the power steering system. The PCM uses this input signal to adjust the idle speed under increased engine loads during low-speed vehicle maneuvers.

19 **Throttle position sensor** - The throttle position sensor senses throttle movement and position, then transmits a voltage signal to the PCM. This signal enables the PCM to determine when the throttle is closed, in a cruise position, or wide open.

20 **Vehicle speed sensor** - The vehicle speed sensor provides information to the PCM to indicate vehicle speed.

21 **Miscellaneous PCM inputs** - In addition to the various sensors, the PCM monitors various switches, circuits and systems to determine vehicle operating conditions. The switches, circuits and systems include:
   a) Air conditioning system
   b) Antilock brake system
   c) Battery voltage
   d) EVAP system
   e) Ignition switch
   f) Park/neutral position switch
   g) Sensor signal and ground circuits
   h) Transaxle control system

Output actuators

22 **Air conditioning clutch relay** - The PCM will de-energize the air conditioning compressor relay during periods of heavy acceleration.

23 **Check Engine light** - The PCM will illuminate the Check Engine light if a malfunction in the electronic engine control system occurs.

24 **Cooling fan control relays** - The PCM controls the operation of the cooling fans according to information received from the engine coolant temperature sensor.

25 **EGR vacuum control solenoids** - On 1998 and earlier models, the EGR vacuum solenoid is controlled by the PCM to regulate the opening of the vacuum-operated EGR valve.

26 **EGR valve** - On 1999 models, EGR flow is controlled by the PCM with an electronic EGR valve.

27 **EVAP canister purge valve** - The evaporative emission canister purge valve is a solenoid valve, operated by the PCM to purge the fuel vapor canister and route fuel vapor to the intake manifold for combustion.

28 **Fast idle control solenoid** - The fast idle control solenoid is used on 1995 and later models to increase idle speed during cold operation. It functions similarly to a choke on a carbureted vehicle.

29 **Front engine mount** - On some models, the PCM controls the operation of the front engine mount according to engine speed. Two engine mount settings are used to control engine vibrations.

30 **Fuel injectors** - The PCM opens the fuel injectors individually in firing order sequence. The PCM also controls the time the injector is open, called the "pulse width." The pulse width of the injector (measured in milliseconds) determines the amount of fuel delivered. For more information on the fuel delivery system and the fuel injectors, including injector replacement, refer to Chapter 4.

31 **Fuel pump relay** - The fuel pump relay is activated by the PCM with the ignition switch in the Start or Run position. When the ignition switch is turned on, the relay is activated to supply initial line pressure to the system. Refer to Chapter 12 or your owner's manual for more information on relay location. For more information on fuel pump check and replacement, refer to Chapter 4.

32 **Idle air control valve** - The IAC valve controls the amount of air to bypass the throttle plate when the throttle valve is closed or at idle position. The IAC valve opening and the resulting airflow is controlled by the PCM.

33 **Oxygen sensor heater** - The PCM controls the operation of the oxygen sensor heater. The oxygen sensor heater allows the oxygen sensor to reach operating temperature quickly.

34 **Power transistor** - The power transistor amplifies the ignition signal from the PCM and intermittently grounds the primary circuit to the ignition coils which generates high voltage in the secondary circuit, thus sending spark from the ignition coil to the distributor or directly to the spark plug. Refer to Chapter 5 for more information on the power transistor and ignition coil(s).

35 **Power valve control solenoid** - On 1993 and 1994 models, the power valve is controlled by the PCM with the power valve control solenoid.

36 **Transmission Control Module (TCM)** - The TCM receives input signals from various sensors and switches such as the vehicle speed sensor, Park/Neutral position switch, turbine shaft speed sensor, throttle position sensor and the camshaft position sensor to determine shifting points, required line pressure and torque converter lock-up operations of the transaxle. The TCM is a separate control module from the PCM although both control modules are used to determine operational characteristics of the transaxle.

Obtaining diagnostic system trouble codes

Refer to illustrations 2.38 and 2.39

37 The PCM will illuminate the CHECK ENGINE light (also known as the Malfunction

Chapter 6 Emissions and engine control systems

2.38 To read the trouble codes on either the inspection lamps or Check Engine light, use the mode selector on the side of the PCM

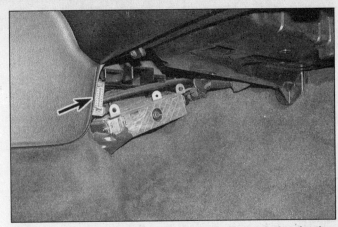

2.39 The Data Link Connector (DLC), typically located under the instrument panel, is used to access the On-Board Diagnostic system with a generic Scan tool

Indicator Lamp) on the dash if it recognizes a component fault. It will continue to set the light until the codes are cleared or the PCM does not detect any malfunction for several consecutive drive cycles.

38 The diagnostic codes can be extracted from the PCM using two methods. The first method requires access to the PCM to turn the diagnostic mode selector and flash the trouble codes on either the inspection lamps or the CHECK ENGINE light. To extract the diagnostic trouble codes using this method, remove the PCM from its mounting bracket (without disconnecting the electrical connectors) and proceed as follows (see Illustration):

1993 and 1994 models

a) *Turn the ignition key ON (engine not running). The CHECK ENGINE light on the dash should remain ON. This indicates that the PCM is receiving power and the CHECK ENGINE light bulb is not defective.* **Note:** *Failure to follow this procedure exactly as described may erase stored trouble codes from the PCM memory.*

b) *Using a screwdriver, turn the mode selector on the side of the PCM fully clockwise. The inspection lamps will begin to flash; after three flashes, turn the mode selector fully counterclockwise.*

c) *Carefully observe the inspection lamps. The red LED will flash the first digit of the trouble code and then the green LED will flash the second digit. Record each trouble code number displayed onto paper. For example, code 43 (throttle position sensor) is indicated by four flashes of the red LED followed by three flashes of the green LED. If everything in the self diagnosis system is functioning properly, the LEDs will flash a code 55. Refer to the accompanying trouble code charts for trouble code identification.*

d) *If the ignition key is turned OFF during the code extraction process and possibly turned back ON, the self diagnostic system will automatically invalidate the procedure. Restart the procedure to extract the codes.* **Note:** *The self diagnostic system cannot be accessed if the engine is running.*

1995 and later models

a) *Turn the ignition key ON (engine not running). The CHECK ENGINE light on the dash should remain ON. This indicates that the PCM is receiving power and the CHECK ENGINE light bulb is not defective.* **Note:** *Failure to follow this procedure exactly as described may erase stored trouble codes from the PCM memory.*

b) *Using a screwdriver, turn the mode selector on the side of the PCM fully clockwise, wait at least two seconds and turn the mode selector fully counterclockwise.*

c) *Carefully observe the CHECK ENGINE light on the instrument panel. The CHECK ENGINE light will flash the first two digits of the trouble code with long (approximately 0.6 second) flashes, pause approximately 2.0 seconds and then flash the second two digits with short (approximately 0.3 second) flashes. Record each trouble code number displayed onto paper. For example, code 0403 (throttle position sensor) is indicated by four long flashes, pause, followed by three short flashes. If everything in the self diagnosis system is functioning properly, the CHECK ENGINE light will flash a code 0505. Refer to the accompanying trouble code charts for trouble code identification.*

d) *If the ignition key is turned OFF during the code extraction process and possibly turned back ON, the self diagnostic system will automatically invalidate the procedure. Restart the procedure to extract the codes.* **Note:** *The self diagnostic system cannot be accessed if the engine is running.*

39 The second method is used on 1995 and later (OBD-II equipped) vehicles only. It uses a special SCAN tool that is programmed to interface with the OBD-II system by plugging into the Data Link Connector (DLC) **(see illustration)**. When used, the SCAN tool has the ability to diagnose in-depth driveability problems and it allows freeze frame data to be retrieved from the PCM stored memory. Freeze frame data is an OBD-II feature that records all related sensor and actuator activity on the PCM data stream whenever an engine control or emissions fault is detected and a DTC is set. This ability to look at the circuit conditions and values when the malfunction occurs provides a valuable tool when trying to diagnose intermittent driveability problems. **Note:** *OBD-II SCAN tools use different trouble code number designations (referred to as P0 or P1 codes) than the CHECK ENGINE light/ MIL lamp codes described above. Refer to the SCAN tool column of the OBD-II trouble code chart below for trouble code identification.* If the tool is not available and intermittent driveability problems exist, have the vehicle checked at a dealer service department or other qualified repair shop.

Clearing codes

40 After the system has been repaired, the codes can be cleared from the PCM memory. Clear the codes as follows: **Caution:** *Do not disconnect the battery from the vehicle to clear the codes. This will erase stored operating parameters from the memory and cause the engine to run rough for a period of time while the computer relearns the information.* **Note:** *If using a OBD-II SCAN tool, scroll the menu for the function that describes "CLEARING CODES" and follow the prescribed method for that particular SCAN tool.*

1993 and 1994 models

a) *Obtain the trouble codes as described previously.*

b) *Wait at least two seconds, then turn the mode selector clockwise. The LEDs will begin to flash.*

c) *After the LEDs flash four times, turn the mode selector counterclockwise.*

d) *Turn the ignition key Off.*

Chapter 6 Emissions and engine control systems

1995 and later models
a) Obtain the trouble codes as described previously.
b) Wait at least two seconds, then turn the mode selector clockwise.
c) Wait at least two seconds, then turn the mode selector counterclockwise.
d) Turn the ignition key Off.

41 Always clear the codes from the PCM before starting the engine for the first time after installing a new electronic emission control component. The PCM will often store trouble codes during sensor malfunctions. The PCM will also record new trouble codes if a new sensor is allowed to operate before the parameters from the old sensor have been erased. Clearing the codes will allow the computer to relearn the new operating parameters relayed by the new component. During the computer relearning process, the engine may experience a rough idle or slight driveability changes. This period of time, however, should last no longer than 15 to 20 minutes.

Diagnostic trouble code identification

42 The accompanying list of diagnostic trouble codes is a compilation of all the codes that may be encountered. Not all codes pertain to all models and not all codes will illuminate the Check Engine light when set. The codes listed under the "Check Engine Light Flash Code" column are codes that may be displayed by the Check Engine light on 1995 and later models.

OBD-I trouble codes - 1993 and 1994 models

Trouble code	Code identification
Code 11	Camshaft position sensor or circuit fault
Code 12	Mass airflow sensor or circuit fault
Code 13	Engine coolant temperature sensor or circuit fault
Code 14	Vehicle speed sensor or circuit fault
Code 21	Ignition signal circuit fault
Code 31	Powertrain Control Module fault
Code 32	EGR control solenoid or circuit fault
Code 33	Oxygen sensor or circuit fault
Code 34	Knock sensor or circuit fault
Code 35	EGR temperature sensor or circuit fault
Code 43	Throttle position sensor or circuit fault
Code 45	Fuel injector leak
Code 51	Fuel injector signal circuit fault
Code 55	No codes identified

OBD-II trouble codes - 1995 and later models

Scan tool trouble code	CHECK ENGINE light flash code	Code identification
P0000	0505	No codes identified
P0100	0102	Mass airflow sensor or circuit fault
P0105	0803	Manifold absolute pressure sensor or circuit fault
P0110	0401	Intake air temperature sensor or circuit fault
P0115	0103	Engine coolant temperature sensor or circuit fault
P0120	0403	Throttle position sensor or circuit fault
P0125	0908	Engine coolant temperature sensor or circuit fault
P0130	0307	Unable to obtain closed loop operation (right bank)
P0130	0503	Upstream oxygen sensor or circuit fault (right bank)
P0131	0411	Upstream oxygen sensor lean shift monitor fault (right bank)
P0132	0410	Upstream oxygen sensor rich shift monitor fault (right bank)
P0133	0409	Upstream oxygen sensor circuit slow response fault (right bank)
P0134	0412	Upstream oxygen sensor high voltage fault (right bank)
P0135	0901	Upstream oxygen sensor heater fault (right bank)
P0136	0707	Downstream oxygen sensor or circuit fault
P0137	0511	Downstream oxygen sensor minimum voltage monitor fault
P0138	0510	Downstream oxygen sensor maximum voltage monitor fault
P0139	0707	Downstream oxygen sensor circuit slow response fault
P0140	0512	Downstream oxygen sensor high voltage fault
P0141	0902	Downstream oxygen sensor heater or circuit fault
P0150	0303	Upstream oxygen sensor or circuit fault (left bank)
P0150	0308	Unable to obtain closed loop operation (left bank)
P0151	0415	Upstream oxygen sensor lean shift monitor fault (left bank)
P0152	0414	Upstream oxygen sensor rich shift monitor fault (left bank)
P0153	0413	Upstream oxygen sensor circuit slow response fault (left bank)
P0154	0509	Upstream oxygen sensor high voltage fault (left bank)
P0155	0101	Upstream oxygen sensor heater fault (left bank)

Chapter 6 Emissions and engine control systems

Scan tool trouble code	CHECK ENGINE light flash code	Code identification
P0157	0314	Downstream oxygen sensor minimum voltage monitor fault
P0158	0313	Downstream oxygen sensor maximum voltage monitor fault
P0159	0708	Downstream oxygen sensor circuit slow response fault
P0160	0315	Downstream oxygen sensor high voltage fault
P0161	1002	Downstream oxygen sensor heater or circuit fault
P0170	0706	Fuel injection system lean or rich (right bank)
P0171	0115	Fuel injection system lean (right bank)
P0172	0114	Fuel injection system rich (right bank)
P0173	0806	Fuel injection system lean or rich (left bank)
P0174	0210	Fuel injection system lean (left bank)
P0175	0209	Fuel injection system rich (left bank)
P0180	0402	Fuel tank temperature sensor or circuit fault
P0300	0701	Multiple cylinder misfire detected
P0301	0608	Cylinder no. 1 misfire detected
P0302	0607	Cylinder no. 2 misfire detected
P0303	0606	Cylinder no. 3 misfire detected
P0304	0605	Cylinder no. 4 misfire detected
P0305	0604	Cylinder no. 5 misfire detected
P0306	0603	Cylinder no. 6 misfire detected
P0325	0304	Knock sensor or circuit fault
P0335	0802	Crankshaft position sensor (POS) or circuit fault
P0340	0101	Camshaft position sensor or circuit fault
P0400	0302	EGR insufficient or excessive flow detected
P0402	0306	EGR backpressure transducer or circuit fault
P0403	0515	EGR valve or circuit
P0420	0702	Catalyst system fault (right bank)
P0430	0703	Catalyst system fault (left bank)
P0440	0705	EVAP system leak
P0443	0807	EVAP canister purge control valve circuit fault
P0443	1008	EVAP canister purge control valve circuit fault
P0446	0903	EVAP canister vent control valve circuit fault
P0450	0704	EVAP system pressure sensor or circuit fault
P0500	0104	Vehicle speed sensor or circuit fault
P0505	0205	Idle air control valve or circuit fault
P0510	0203	Closed throttle position switch or circuit fault
P0600	0504	A/T control unit or circuit fault
P0605	0301	PCM or circuit fault
P0705	1003	Park/Neutral position switch or circuit fault
P0705	1101	Park/Neutral position switch or circuit fault
P0710	1208	Transaxle temperature sensor or circuit fault
P0720	1102	Vehicle speed sensor or circuit fault
P0725	1207	Engine speed signal or circuit fault
P0731	1103	A/T first gear signal fault
P0732	1104	A/T second gear signal fault
P0733	1105	A/T third gear signal fault
P0734	1106	A/T fourth gear signal fault
P0740	1204	TCC solenoid or circuit fault
P0744	1107	A/T TCC solenoid valve fault
P0745	1205	Line pressure solenoid or circuit fault
P0750	1108	Shift solenoid A or circuit fault
P0755	1201	Shift solenoid B or circuit fault
P1105	1302	Manifold absolute pressure solenoid valve or circuit fault
P1148	0307	Closed loop control fault (right bank)
P1168	0308	Closed loop control fault (left bank)
P1220	1305	Fuel pump control module or circuit fault
P1320	0201	Ignition signal primary circuit fault
P1335	0407	Crankshaft position sensor (REF) or circuit fault

OBD-II trouble codes - 1995 and later models (continued)

Scan tool trouble code	CHECK ENGINE light flash code	Code identification
P1336	0905	Crankshaft position sensor (POS) fault or signal plate damage
P1400	1005	EGR solenoid valve or circuit fault
P1401	0305	EGR temperature sensor or circuit fault
P1402	0514	EGR high flow detected
P1440	0213	EVAP system small leak
P1441	0801	Vacuum cut bypass valve or circuit fault
P1444	0214	EVAP canister purge volume control valve or circuit fault
P1445	1008	EVAP canister purge volume control valve or circuit fault
P1446	0215	EVAP canister vent control valve closed
P1447	0111	EVAP purge flow monitor fault
P1448	0309	EVAP canister vent control valve open
P1490	0801	Vacuum cut bypass valve or circuit fault
P1491	0311	Vacuum cut bypass valve or circuit fault
P1492	0807	EVAP canister purge control valve solenoid valve or circuit fault
P1493	0312	EVAP canister purge control valve solenoid valve or circuit fault
P1605	0804	A/T control unit or circuit fault
P1705	1206	Throttle position sensor to A/T signal fault
P1706	1003	Park/Neutral position switch or circuit fault
P1760	1203	Overrun clutch solenoid or circuit fault
P1900	0208	Engine cooling fan or circuit fault
P1900	1308	Engine cooling fan or circuit fault

3 Powertrain Control Module - removal and installation

Refer to illustrations 3.4 and 3.5

Warning: *On models equipped with a Supplemental Restraint System (SRS) (more commonly known as airbags), always disable the airbag system before working in the vicinity of the airbag system components to avoid the possibility of accidental deployment of the airbag, which could cause personal injury (see Chapter 12).*

Caution: *Static electricity can damage the PCM. Be sure to ground yourself to the vehicle body before touching the PCM or the electrical connector. Store the PCM on an anti-static pad once it is removed.*

Note: *Anytime the battery is disconnected stored operating parameters may be lost from the PCM causing the engine to run rough for sometime while the PCM relearns the information.*

1 The Powertrain Control Module (PCM) is located in front of the center console, to the right of the accelerator pedal.
2 Disconnect the cable from the negative terminal of the battery.
3 Remove the side trim panels in front of the console.
4 Disconnect the harness connector from the PCM **(see illustration)**.
5 Remove the retaining bolts from the PCM mounting bracket **(see illustration)**.
6 Carefully slide the PCM from under the instrument panel.
7 Installation is the reverse of removal.

3.4 Remove the bolt (arrow) and carefully detach the electrical connector from the PCM

3.5 Remove the PCM bracket mounting bolts (arrow)

4 Throttle position sensor - check, replacement and adjustment

1 The Throttle Position Sensor (TPS) is a variable-resistance potentiometer, mounted on the side of the throttle body and connected to the throttle shaft. By monitoring the output voltage from the TPS, the PCM can determine fuel delivery based on throttle valve angle (driver demand). A broken or loose TPS can cause intermittent bursts of fuel from the injector and an unstable idle because the PCM thinks the throttle is moving. A problem with the TPS or circuit will set a diagnostic trouble code. On most models, a wide open/closed throttle switch is incorporated into the throttle position sensor. The wide open/closed throttle switch closes the particular throttle circuit in response to the position of the throttle valve (wide open or closed).

Check

Refer to illustrations 4.2 and 4.3

Note: *Performing the following test may set a diagnostic trouble code and illuminate the Check Engine light. Clear the diagnostic trouble code after performing the tests and making the necessary repairs (see Section 2).*

2 Before checking the TPS, check the voltage supply and ground circuits from the PCM to the TPS. Disconnect the TPS electri-

Chapter 6 Emissions and engine control systems

4.2 The TPS (A) and the wide open/closed throttle switch (B) are mounted to the side of the throttle body (1995 and later model shown)

4.3 TPS and wide open/closed throttle switch terminal identification (1995 and later model shown)

cal connector **(see illustration)** and connect the positive probe of a voltmeter to the green/black (1993 and 1994) or red (1995 and later) wire terminal at the harness connector. Connect the negative probe to the black wire terminal. Turn the ignition key ON (engine not running), the voltmeter should indicate approximately 5.0 volts. If the specified voltage is not present, check the circuits from the connector to the PCM. If the circuits are good, have the PCM diagnosed by a dealer service department or other qualified repair shop.

3   To check the operation of the TPS, connect an ohmmeter to terminal no. 2 and terminal no. 3 on the TPS **(see illustration)**. Check the resistance of TPS with the throttle fully closed, then gradually open the throttle to full throttle. On 1993 and 1994 models, there should be 1.0 K-ohms of resistance with the throttle closed and 9.0 K-ohms at full throttle. On 1995 and later models, there should be 0.5 K-ohms of resistance with the throttle closed and 4.0 K-ohms at full throttle. The resistance should increase smoothly as the throttle is opened.

4   Disconnect the wide open/closed throttle position switch connector. Connect an ohmmeter to terminal no. 5 and terminal no. 6 of the switch **(see illustration 4.3)**. Check for continuity at the switch with the throttle fully closed, then slightly open the throttle. There should be continuity with the throttle completely closed and NO continuity when the throttle is opened. If the readings are incorrect, replace the TPS.

Replacement

5   Disconnect the cable from the negative terminal of the battery.
6   Remove the air intake duct (see Chapter 4) and disconnect the TPS and wide open/closed throttle switch electrical connectors.
7   Remove the two retaining screws and separate the TPS from the throttle body.

8   Install the new TPS leaving the mounting screws loose.
9   Adjust the TPS as described and tighten the screws securely.

Adjustment

10   Remove the air intake duct (see Chapter 4).
11   Loosen the TPS mounting screws.
12   On models with a wide open/closed throttle switch, disconnect the wide open/closed throttle switch electrical connector and connect an ohmmeter to the closed throttle switch as described in Step 4. Insert a 0.012-inch (0.3-mm) feeler gauge between the throttle lever and the throttle stop screw - continuity should be indicated on the meter. Remove the 0.012-inch (0.3-mm) feeler gauge and insert a 0.016-inch (0.4-mm) feeler gauge - no continuity should be indicated. Rotate the TPS until the desired results are achieved and tighten the mounting screws. **Caution:** *Do not alter the position of the throttle stop screw. The throttle stop screw is preset at the factory and does not require adjustment.*
13   On models without a wide open/closed throttle position switch, backprobe the white (+) and black (-) wire terminals without disconnecting the electrical connector and connect a voltmeter to the probes. Use suitable probes and be very careful not to damage the wiring harness (see Chapter 12 for additional information on how to backprobe a connector). With the ignition key ON (engine not running) and the throttle fully closed, the voltage should read approximately 0.4 to 0.5 volts. If the readings are incorrect, rotate the TPS body until the correct voltage is attained and tighten the mounting screws.

5   Mass airflow sensor - check and replacement

1   The Mass Airflow (MAF) sensor is installed in the air intake duct. This sensor uses a hot-wire sensing element to measure the molecular mass (or weight) of air entering the engine. The air passing over the hot wire causes it to cool, and the sensor converts this temperature change into a voltage signal to the PCM. The PCM in turn calculates the required fuel injector pulse width to obtain the necessary air/fuel ratio. A defective MAF sensor can cause surging, stalling, rough idle and other driveability problems. The electronic engine control system can detect several different MAF sensor problems and set trouble codes to indicate the specific fault.

Check

Refer to illustration 5.5
Note: *Performing the following test may set a diagnostic trouble code and illuminate the Check Engine light. Clear the diagnostic trouble code after performing the tests and making the necessary repairs (see Section 2).*

2   Before checking the MAF sensor operation, check the power and ground circuits to the MAF sensor.
3   Disconnect the MAF sensor electrical connector and connect the positive lead of a voltmeter to the white/black (1993 and 1994) or red (1995 and later) wire terminal of the harness connector and the negative lead to a good engine ground point. Turn the ignition On but do not start the engine. The meter should indicate approximately battery voltage. If battery voltage is not present, check the ECCS relay and related circuits (see Chapter 12 and the wiring diagrams).
4   On 1993 and 1994 models, check for continuity to ground on the harness connector black wire terminal. If continuity is not indicated, check the ground circuit.
5   Reconnect the electrical connector to the MAF sensor. Using a suitable probe, backprobe the MAF sensor white wire terminal **(see Illustration)**. See Chapter 12 for additional information on how to backprobe a connector. Connect the positive lead of a

5.5 Backprobe the white wire terminal of the MAF sensor (arrow) and check the sensor output voltage

5.9 Loosen the retaining clamp, disconnect the air inlet duct, remove the screws and separate the MAF sensor from the air filter housing

voltmeter to the probe and connect the negative lead to a good engine ground point. Turn the ignition key On. The meter should indicate less than 0.5 volt (1993 and 1994) or less than 1.0 volt (1995 and later).

6 Start the engine and check the voltage; it should be 1.0 to 1.7 volts at idle. Increase the engine rpm. The MAF signal voltage should increase from 1.7 to 3.0 volts. It is impossible to simulate driving conditions in the driveway, but it is necessary to watch the voltmeter for an increase in signal voltage as the engine speed is raised. The engine is not under load, but signal voltage should vary slightly.

7 If the voltage readings are correct the MAF sensor is operating properly. Refer to the wiring diagrams and check the wiring harness for open circuits or a damaged harness. If the circuits are good, have the PCM diagnosed by a dealer service department or other qualified repair shop. **Note:** *If MAF related driveability problems continue but these general tests don't indicate a MAF fault, have the sensor tested by a dealer service department or other qualified repair shop. A MAF sensor can develop voltage signal problems that can't be seen on a voltmeter. The PCM can see such signal faults and a driveability problem will result.*

Replacement

Refer to illustration 5.9

Note: *The plastic MAF sensor body and the metal air duct on which it is mounted are an assembly that must be replaced as a unit. Do not try to separate the sensor body from the metal duct.*

8 Disconnect the electrical connector from the MAF sensor.

9 Remove the clamp that secures the sensor to the intake air duct. Remove the four fasteners securing the sensor to the air filter housing **(see illustration)**.

10 Install and connect the new sensor.

6 Manifold absolute pressure sensor and solenoid valve - check and replacement

1 1996 and later models use a Manifold Absolute Pressure (MAP) sensor in conjunction with a MAP sensor solenoid valve to monitor the intake manifold pressure and ambient barometric pressure changes resulting from changes in engine load and speed. The PCM uses the MAP sensor signal for diagnostic purposes. When the PCM applies voltage to the solenoid valve (ON) it switches the vacuum signal and allows the MAP sensor to monitor ambient barometric pressure. When voltage is not supplied to the solenoid valve (OFF) it switches the vacuum signal again and allows the MAP sensor to monitor intake manifold pressure. As the vacuum signal changes the MAP sensor converts this information into a voltage output signal and sends it to the PCM. The voltage will vary from 0.5 volts at closed throttle (high vacuum) and approximately 5.0 volts at wide open throttle (low vacuum). The voltage range values will vary slightly according to changes in altitude. The MAP sensor and the solenoid valve are mounted at the rear (transaxle end) of the left-bank (front) cylinder head. The PCM can detect several different MAP sensor problems and set trouble codes to indicate the specific fault.

Check

Note: *Performing the following test may set a diagnostic trouble code and illuminate the Check Engine light. Clear the diagnostic trouble code after performing the tests and making the necessary repairs (see Section 2).*

MAP sensor

2 Disconnect the vacuum hose from the MAP sensor and attach a vacuum gauge to the hose. Start the engine and allow it to idle, after approximately 5 seconds vacuum should be indicated on the gauge. If vacuum is not present, check the vacuum hose from the MAP sensor to the solenoid valve for cracks and clogging. If the hose is good, proceed with the solenoid valve check. If vacuum is present, proceed with the MAP sensor check.

3 Disconnect the electrical connector from the MAP sensor. Connect a voltmeter to the red wire terminal (+) and the black wire terminal (-). With the ignition key ON (engine not running), the meter should indicate approximately 5.0 volts. If the specified voltage is not present, check the circuits from the connector to the PCM. If the circuits are good, have the PCM diagnosed by a dealer service department or other qualified repair shop.

4 Turn the ignition key Off and connect the electrical connector to the MAP sensor. Using a suitable probe, backprobe the white wire terminal of the connector (see Chapter 12 for additional information on how to back-probe a connector). Connect the positive lead of a voltmeter to the probe and connect the negative lead to a good engine ground point. Turn the ignition key ON (engine not running). There should be approximately 3.2 to 4.8 volts indicated on the meter.

5 Disconnect the vacuum hose from the sensor. Connect a hand-held vacuum pump the sensor and apply approximately 8 in-Hg of vacuum. The MAP signal voltage should decrease to approximately 1.0 to 1.4 volts as vacuum is applied to the sensor. If the MAP sensor does not respond as described, replace the sensor.

Solenoid valve

Refer to illustration 6.8

6 Disconnect the intake manifold vacuum source hose from the solenoid valve and attach a vacuum gauge to the hose. Start the engine and allow it to idle, intake manifold vacuum should be indicated on the gauge. If vacuum is not present, check the hose from the vacuum source to the solenoid valve for cracks and clogging. If vacuum is present, proceed with the solenoid valve check.

7 Disconnect the electrical connector from the solenoid valve and check for battery voltage between the two terminals of the harness connector with the ignition key ON (engine not running). If voltage is not present, refer to the wiring diagrams and check the wiring harness for open circuit or a damaged harness from the fuse box to the connector (don't forget to check the fuses first).

8 If battery voltage is present, remove the solenoid valve and test it off the vehicle. Using a pair of fused jumper wires, connect a 12 volt battery source and ground to the two terminals of the solenoid valve. The solenoid should click as voltage is applied. With battery voltage applied, air should pass between ports A and B **(see Illustration)**. With no voltage applied, air should pass between ports A and C. If the solenoid does not operate as described, replace the solenoid.

Chapter 6 Emissions and engine control systems   6-11

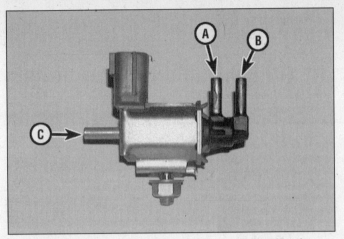

6.8 MAP sensor solenoid valve vacuum port identification
A To MAP sensor
B To intake air duct
C To intake manifold vacuum source

7.3a Disconnect the electrical connector from the intake air temperature sensor (arrow) and measure the resistance across the two terminals of the sensor connector

Replacement

9 Disconnect the cable from the negative terminal of the battery.
10 Disconnect the electrical connector from the MAP sensor and/or solenoid valve.
11 Remove the MAP sensor and/or solenoid valve mounting screws. Detach the vacuum hose(s) and remove the MAP sensor and/or solenoid valve.
12 Installation is the reverse of removal.

7 Intake air temperature sensor - check and replacement

1 The IAT sensor is a thermistor that changes resistance as temperature changes. The sensor is installed in the intake air duct to sense air temperature. As temperature increases, sensor resistance decreases and vice versa. The PCM uses this information to compute the intake temperature and fine tune fuel metering. A problem in the IAT sensor circuit will set a trouble code. The fault may be in the circuit wiring or connections or in the sensor itself.

Check

Refer to illustrations 7.3a and 7.3b
Note: *Performing the following test will set a diagnostic trouble code and illuminate the Check Engine light. Clear the diagnostic trouble code after performing the tests and making the necessary repairs (see Section 2).*
2 Before checking the intake air temperature sensor, check the voltage supply and ground circuits from the PCM. Disconnect the electrical connector from the intake air temperature sensor and connect a voltmeter to the two terminals of the harness connector. Turn the ignition key On - the voltage should read approximately 5.0 volts. If the voltage is incorrect, check the wiring from the intake air temperature sensor to the PCM. If the circuits are good, have the PCM checked

at a dealer service department or other properly equipped repair facility.
3 With the ignition switch OFF, disconnect the electrical connector from the intake air temperature sensor. Using an ohmmeter, measure the resistance between the two terminals on the sensor with the engine cool. Reconnect the electrical connector to the sensor, start the engine and warm it up until it reaches operating temperature, disconnect the connector and check the resistance again. Compare your measurements to the resistance chart **(see Illustrations)**. If the sensor resistance test results are incorrect, replace the intake air temperature sensor.
Note: *A more accurate check may be performed by removing the sensor and suspending the tip of the sensor in a container of water. Heat the water on the stove while you monitor the resistance of the sensor.*

Replacement

4 Disconnect the electrical connector, then carefully remove the IAT sensor from the air intake duct. Be careful not to damage any of the plastic parts.
5 Install and connect the new sensor.

8 Engine coolant temperature sensor - check and replacement

1 Like the intake air temperature sensor, the engine coolant temperature sensor is a thermistor, which is a variable resistor that changes its resistance as temperature changes. On 1993 and 1994 models, the engine coolant temperature sensor is located in the coolant inlet pipe behind the distributor. On 1995 and later models, the sensor is located in the coolant outlet pipe behind the left (front) cylinder head. The engine coolant temperature sensor senses coolant temperature. As coolant temperature increases, sensor resistance decreases and vice versa. The

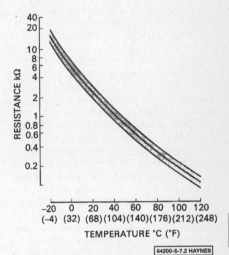

7.3b Intake air temperature sensor and engine coolant temperature sensor approximate temperature vs. resistance values

PCM uses this information to compute the engine operating temperature. A problem in the engine coolant temperature sensor circuit will set a trouble code. The fault may be in the circuit wiring or connections or in the sensor itself.

Check

Refer to illustration 8.2a and 8.2b
Note: *Performing the following test will set a diagnostic trouble code and illuminate the Check Engine light. Be prepared to clear the diagnostic trouble code after performing the tests and making the necessary repairs (see Section 2).*
2 Before checking the engine coolant temperature sensor, check the voltage supply and ground circuits from the PCM. Disconnect the electrical connector from the

6-12 Chapter 6 Emissions and engine control systems

8.2a On 1993 and 1994 models, the engine coolant temperature sensor is located in the coolant inlet pipe behind the distributor

8.2b On 1995 and later models, the engine coolant temperature sensor is located in the coolant outlet pipe at the rear of the left (front) cylinder head

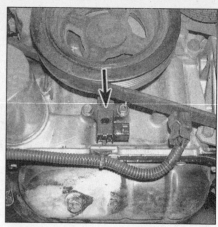

9.2 The crankshaft position sensor (REF) (arrow) is located on the aluminum oil pan below the crankshaft pulley

engine coolant temperature sensor and connect a voltmeter to the two terminals of the harness connector **(see illustrations)**. Turn the ignition key On - the voltage should read approximately 5.0 volts. If the voltage is incorrect, check the wiring from the engine coolant temperature sensor to the PCM. If the circuits are good, have the PCM checked at a dealer service department or other properly equipped repair facility.

3 Using an ohmmeter, measure the resistance between the two terminals on the sensor with the engine cool. Reconnect the electrical connector to the sensor, start the engine and warm it up until it reaches operating temperature, disconnect the connector and check the resistance again. Compare your measurements to the resistance chart **(see Illustration 7.3b)**. If the sensor resistance test results are incorrect, replace the engine coolant temperature sensor. *Note: A more accurate check may be performed by removing the sensor and suspending the tip of the sensor in a container of water. Heat the water on the stove while you monitor the resistance of the sensor.*

Replacement

Warning: *Wait until the engine is completely cool before performing this procedure.*

4 Before installing the new sensor, wrap the threads with Teflon sealing tape to prevent leakage and thread corrosion.

5 Disconnect the electrical connector and remove the engine coolant temperature sensor from the coolant pipe. Install the new sensor as quickly as possible to minimize coolant loss. Tighten the sensor securely and reconnect the electrical connector.

6 Check the coolant level as described in Chapter 1, adding coolant, if necessary. Start the engine and allow it to reach normal operating temperature, then check for coolant leaks. Check the coolant level in the expansion tank after the engine has warmed up and then cooled down again.

9 Crankshaft position sensor - check and replacement

1 Two crankshaft position sensors are used on 1995 and later models. Crankshaft position sensor (REF) is mounted at the front of the engine, on the aluminum oil pan section below the crankshaft pulley. It detects gaps in the crankshaft pulley sensor ring corresponding to TDC of each cylinder (120-degree signal). Crankshaft position sensor (POS) is mounted on the aluminum oil pan section at the rear of the engine near the transaxle bellhousing. It detects gaps in the flywheel/driveplate gear teeth to determine crankshaft speed (one degree signal). Both sensors use a permanent magnet, core and coil. The changing gap causes the magnetic field near the sensor to change, this in turn varies the voltage signal to the PCM. The sensor signals are used by the PCM for ignition timing, fuel synchronization and detecting engine misfire.

Check

Refer to illustrations 9.2 and 9.3

Note: *Performing the following test will set a diagnostic trouble code and illuminate the Check Engine light. Be prepared to clear the diagnostic trouble code after performing the tests and making the necessary repairs (see Section 2).*

Crankshaft position sensor (REF)

2 Disconnect the sensor electrical connector **(see Illustration)**. Using an ohmmeter, measure the resistance across the two crankshaft sensor terminals. Resistance should be 470 to 570 ohms at 68-degrees F (20-degrees C). If the resistance values are incorrect, replace the sensor. If the resistance values are correct, refer to the wiring diagrams and check the wiring harness for an open circuit to the PCM or a damaged harness. Check for continuity to ground on the black wire of the harness connector. If the sensor and the wiring harness are both good, have the PCM diagnosed by a dealer service department or other qualified repair shop.

Crankshaft position sensor (POS)

3 Disconnect the sensor electrical connector **(see Illustration)**. Turn the ignition key On. Using a voltmeter check for battery voltage at the Red wire terminal of the harness connector. If voltage is not present, check the circuit from the battery to the ECCS relay (don't forget to check the fuses), check the ECCS relay and the circuit from the relay to the sensor connector (see Chapter 12 and the wiring diagrams). Check for 5.0 volts at the white wire terminal. If voltage is not present, check the circuit from the PCM to the sensor connector. If the circuit is good, have the PCM diagnosed by a dealer service department or other qualified repair shop. Check for continuity to ground at the black wire terminal of the sensor connector.

4 Remove the sensor. Turn the ignition key Off and connect the electrical connector

9.3 The crankshaft position sensor (POS) (arrow) is located on the transaxle bellhousing

Chapter 6 Emissions and engine control systems

10.1 On 1993 and 1994 models, the camshaft position sensor (arrow) is an integral part of the distributor

10.6 On 1995 and later models, the camshaft position sensor (arrow) is located on the timing chain cover

to the sensor. Using a suitable probe, backprobe the white wire terminal of the sensor connector (see Chapter 12 for additional information on how to backprobe a connector). Connect the positive lead of a voltmeter to the probe and connect the negative lead to a good engine ground point. Turn the ignition key On. Touch the tip of the sensor with a metal object (such as a screwdriver) and quickly pull it away. The voltmeter should indicate 5.0 volts when contacted and quickly drop to zero volts when the metal object is pulled away. If the sensor does not react as described, replace the sensor. **Note:** *If in the rare case that the wiring, PCM and sensor are all good and the system continues to set a diagnostic trouble code, check the flywheel/driveplate for broken or stripped gear teeth.*

Replacement

5   Disconnect the electrical connector from the sensor.
6   Remove the crankshaft sensor retaining bolt and remove the sensor.
7   Installation is the reverse of removal.

10 Camshaft position sensor - check and replacement

Note: *Performing the following test will set a diagnostic trouble code and illuminate the Check Engine light. Be prepared to clear the diagnostic trouble code after performing the tests and making the necessary repairs (see Section 2).*

1993 and 1994 models

Refer to illustration 10.1

1   The camshaft position sensor monitors engine speed and piston position and relays this data to the computer which in turn controls the fuel injection duration (fuel injector on/off time) and ignition timing. The camshaft position sensor consists of a rotor plate and a wave forming circuit. The rotor plate has 360 slits for each degree, or one percent signal (engine speed signal) and six slits for the 120-degree camshaft position signal. Light Emitting Diodes (LED) and photo diodes are built into the wave forming circuit. When the rotor plate passes the space between the LED and the photo diode, the slits on the rotor plate continually cut the beam of light sent to the photo diode from the LED. They are then converted into on-off pulses by the wave forming circuit and then sent to the ECM. The camshaft position sensor is an integral part of the distributor **(see Illustration)**.
2   Disconnect the electrical connector from the wiring harness leading to the distributor. Turn the ignition key On. Using a voltmeter, check for battery voltage at the white/black wire terminal of the harness connector. If voltage is not present, check the circuit from the battery to the ECCS relay (don't forget to check the fuses), check the ECCS relay and the circuit from the relay to the distributor connector (see Chapter 12 and the wiring diagrams). Using an ohmmeter, check for continuity to ground at the black wire terminal.
3   Turn the ignition OFF and remove the distributor from the engine (see Chapter 5). Reconnect the distributor connector. Using a suitable probe, backprobe the green/black wire terminal of the distributor connector (see Chapter 12 for additional information on how to backprobe a connector). Connect the positive lead of a voltmeter to the probe and connect the negative lead to a good engine ground point. With the ignition ON, slowly rotate the distributor shaft and check the voltage, the meter should fluctuate between zero volts and 5.0 volts six times per revolution of the distributor shaft . This tests camshaft position sensor 120-degree signal.
4   Turn the ignition OFF and backprobe the green/yellow wire terminals with the voltmeter. With the ignition ON, slowly rotate the distributor shaft, the meter should fluctuate between zero volts and 5.0 volts 360 times per revolution of the distributor shaft. This tests camshaft position sensor 1-degree signal.
5   If the camshaft position sensor doesn't operate as described, the distributor must be replaced (see Chapter 5). The camshaft position sensor is an integral part of the distributor and not serviced separately.

1995 and later models

Refer to illustration 10.6

6   The camshaft position sensor is located on the timing chain cover at the front of the engine **(see Illustration)**. The sensor uses a permanent magnet, core and coil to detect a gap in the camshaft sprocket. The changing gap causes the magnetic field near the sensor to change, this in turn varies the voltage signal to the PCM. The camshaft position sensor provides information on cylinder TDC position to the PCM.
7   Disconnect the electrical connector from the sensor. Using an ohmmeter, measure the resistance across the two camshaft sensor terminals. Resistance should be 1,440 to 1,760 ohms at 68-degrees F (20-degrees C) for a Hitachi made sensor or 2,090 to 2,550 ohms at 68-degrees F (20-degrees C) for a Mitsubishi made sensor. If the resistance values are incorrect, replace the sensor.
8   If the resistance values are correct, refer to the wiring diagrams and check the wiring harness for an open circuit to the PCM or a damaged harness. Check for continuity to ground on the black wire of the harness connector. If the sensor and the wiring harness are both good, have the PCM diagnosed by a dealer service department or other qualified repair shop.

11 Power steering pressure switch - check and replacement

1   The power steering pressure switch is a normally open switch, mounted in the pressure line between the steering gear and the power steering pump. When steering system pressure reaches a high-pressure setpoint, the power steering pressure switch closes and sends a signal to the PCM that the PCM uses to maintain engine idle speed during parking maneuvers. The PCM can detect switch problems and set trouble codes to

11.2 The power steering pressure switch (arrow) is located in the right rear corner of the engine compartment

12.6a The upstream oxygen sensor is located in the exhaust pipe between the exhaust manifold and the catalytic converter

indicate specific faults. Check the operation of the power steering pressure switch if the engine stalls during parking or if the engine idles continuously at high rpm.

Check

Refer to illustration 11.2
Note: *Performing the following test will set a diagnostic trouble code and illuminate the Check Engine light. Be prepared to clear the diagnostic trouble code after performing the tests and making the necessary repairs (see Section 2).*

2 Disconnect the power steering pressure switch connector and connect an ohmmeter to the terminals on the switch body **(see illustration)**.
3 Start the engine and let it idle. **Warning:** *Make sure the meter leads, loose clothing, long hair, etc. are away from the moving parts of the engine (drivebelt, cooling fan, etc.) before starting the engine.*
4 Turn the steering wheel to point the front wheels straight ahead and read the ohmmeter. It should indicate an open circuit (infinite resistance).
5 Turn the steering wheel to either side and watch the ohmmeter. The power steering pressure switch should close as the wheel nears the steering stop on either side, and the meter should indicate continuity of close to zero ohms.
6 If the switch fails either test, replace it. If the switch is OK, troubleshoot the engine idle control operation if high idle speed or stalling problems continue. Also check the wiring harness for an open circuit to the PCM or a damaged harness and check for continuity to ground on the black wire of the harness connector.

Replacement

7 Disconnect the electrical connector from the switch.
8 Using a back-up wrench on the hex fitting, unscrew the switch from the junction block.
9 Install and connect the new switch.

Refer to Chapter 10 and bleed air from the power steering system. Add fluid as required (see Chapter 1).

12 Oxygen sensor - check and replacement

1 The oxygen in the exhaust reacts with the oxygen sensor to produce a voltage output that varies from 0.1 volt (high oxygen, lean mixture) to 0.9 volt (low oxygen, rich mixture). The upstream oxygen sensor in the exhaust system provides a feedback signal to the PCM that indicates the amount of leftover oxygen in the exhaust. The PCM monitors this variable voltage continuously to determine the required fuel injector pulse width and to control the engine air/fuel ratio. A mixture ratio of 14.7 parts air to 1 part fuel is the ideal ratio for minimum exhaust emissions, as well as the best combination of fuel economy and engine performance. Based on oxygen sensor signals, the PCM tries to maintain this air/fuel ratio of 14.7:1 at all times.
2 The downstream oxygen sensor in the exhaust system has no effect on PCM control of the air/fuel ratio. This sensor is identical to the upstream sensor and operates in the same way. The PCM uses the downstream signal, however, for the catalyst monitor system. A downstream oxygen sensor will produce a slower fluctuating voltage signal that reflects the lower oxygen content in the post-catalyst exhaust. **Note:** *1993 and 1994 models are equipped with one (upstream) oxygen sensor, while 1995 and later models are equipped with two upstream (one per cylinder bank) and one downstream oxygen sensor.*
3 An oxygen sensor produces no voltage when it is below its normal operating temperature of about 600-degrees F (318-degrees C). During this warm-up period, the PCM operates in an open-loop fuel control mode. It does not use the oxygen sensor signal as a feedback indication of residual oxygen in the exhaust. Instead, the PCM controls fuel metering based on the inputs of other sensors and its own programs.
4 Proper operation of an oxygen sensor depends on four conditions:

a) **Electrical** - The low voltages generated by the sensor require good, clean connections which should be checked whenever a sensor problem is suspected or indicated.
b) **Outside air supply** - The sensor needs air circulation to the internal portion of the sensor. Whenever the sensor is installed, make sure the air passages are not restricted.
c) **Proper operating temperature** - The PCM will not react to the sensor signal until the sensor reaches approximately 600-degrees F (318-degrees C). This factor must be considered when evaluating the performance of the sensor.
d) **Unleaded fuel** - Unleaded fuel is essential for proper operation of the sensor.

5 The PCM can detect several different oxygen sensor problems and set DTC's to indicate the specific fault. When an oxygen sensor fault occurs that sets a DTC, the PCM will disregard the oxygen sensor signal voltage and revert to open-loop fuel control as described previously.

Check

Refer to illustrations 12.6a, 12.6b and 12.6c
Caution: *The oxygen sensor is very sensitive to excessive circuit loads and circuit damage of any kind. For safest testing, install jumper wires in the oxygen sensor connector to connect your voltmeter. If jumper wires aren't available, carefully backprobe the wires in the connector shell with straight pins or similar devices. Do not puncture the oxygen sensor wires or try to backprobe the sensor itself. Use only a digital voltmeter to test an oxygen sensor. Refer to Chapter 12 for additional information on how to backprobe an electrical connector.*
Note: *Performing the following test will set a*

Chapter 6 Emissions and engine control systems

12.6b Follow the wiring harness up from the oxygen sensor to locate the electrical connector (arrow)

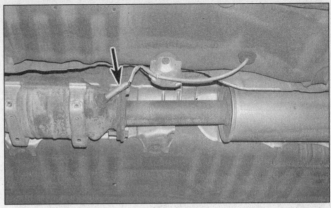

12.6c The downstream oxygen sensor is located at the catalytic converter outlet (1995 and later models) - note that the wiring harness goes through the floorboard; pull the grommet out to locate the electrical connector

diagnostic trouble code and illuminate the Check Engine light. Clear the diagnostic trouble code after performing the tests and making the necessary repairs (see Section 2).

6 Backprobe the white wire terminal of the oxygen sensor connector **(see Illustrations)**. Connect the positive lead of the voltmeter to the probe and the negative lead to a good engine ground point. Start the engine and warm up to normal operating temperature; check the oxygen sensor signal voltage.

a) Voltage from an upstream sensor should range from 100 to 900 millivolts (0.1 to 0.9 volt) and switch actively between high and low readings.

b Voltage from a downstream sensor should also read between 100 to 900 millivolts (0.1 to 0.9 volt) but it should not switch actively. The downstream oxygen sensor voltage may stay toward the center of its range (about 400 millivolts) or stay for relatively longer periods of time at the upper or lower limits of the range.

7 Check the battery voltage supply and ground circuits to the oxygen sensor heater. Disconnect the electrical connector and connect the positive lead of the voltmeter to the green/black (1993 and 1994) or red/black (1995 and later) terminal of the sensor connector. Connect the negative lead to the black (1993 and 1994) or blue, blue/yellow (1995 and later) wire terminal. With the ignition ON, the meter should indicate approximately battery voltage. Refer to the wiring diagrams in Chapter 12 for more information on the oxygen sensor circuits.

8 Check the resistance of the oxygen sensor heater. With the connector disconnected. Connect an ohmmeter to the two oxygen sensor heater terminals of the connector (oxygen sensor side). The oxygen sensor pigtail is generally not color coded. The oxygen sensor heater resistance should be as follows:

a) 1993 and 1994 models - 3.0 to 1,000 ohms

b) 1995 and 1996 models - upstream sensors, 2.3 to 4.3 ohms; downstream sensor, 5.2 to 8.2 ohms

c) 1997 and later models - 2.3 to 4.3 ohms

9 If an open circuit or excessive resistance is indicated, replace the oxygen sensor. **Note:** *If the tests indicate that a sensor is good, and not the cause of a driveability problem or diagnostic trouble code, check the wiring harness and connectors between the sensor and the PCM for an open or short circuit. If no problems are found, have the vehicle checked by a dealer service department or other qualified repair shop.*

Replacement

Refer to illustration 12.13

10 The exhaust pipe contracts when cool, and the oxygen sensor may be hard to loosen when the engine is cold. To make sensor removal easier, start and run the engine for a minute or two; then shut it off. Be careful not to burn yourself during the following procedure. Also observe these guidelines when replacing an oxygen sensor.

a) The sensor has a permanently attached pigtail and electrical connector which should not be removed from the sensor. Damage or removal of the pigtail or electrical connector can harm operation of the sensor.

b) Keep grease, dirt and other contaminants away from the electrical connector and the louvered end of the sensor.

c) Do not use cleaning solvents of any kind on the oxygen sensor.

d) Do not drop or roughly handle the sensor.

11 Raise the vehicle and place it securely on jackstands.

12 Disconnect the electrical connector from the sensor.

13 Using a suitable wrench or special oxygen sensor socket, unscrew the sensor from the exhaust manifold **(see illustration)**.

14 Anti-seize compound must be used on the threads of the sensor to aid future removal. The threads of most new sensors will be coated with this compound. If not, be

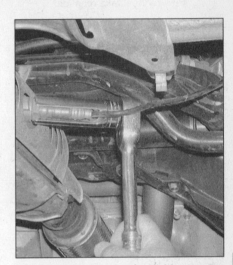

12.13 A special socket (available at many auto parts stores) that allows clearance for the wiring harness may be required for oxygen sensor removal

sure to apply anti-seize compound before installing the sensor.

15 Install the sensor and tighten it securely.

16 Lower the vehicle and reconnect the electrical connector for the sensor.

13 Knock sensor - check and replacement

1 The knock sensor detects abnormal vibration (spark knock or pinging) in the engine. The knock control system is designed to reduce spark knock during periods of heavy detonation. This allows the engine to use maximum spark advance to improve driveability. Knock sensors produce AC output voltage which increases with the severity of the knock. The signal is fed into the PCM and the timing is retarded to compensate for the severe detonation. On 1993 and 1994 models, the knock sensor is located on the right (rear) side of the engine

6-16   Chapter 6 Emissions and engine control systems

13.2 On 1995 and later models, the knock sensor sub-harness connector is located next to the intake manifold

13.10 On 1995 and later models, the knock sensor is located under the intake manifold

14.4 Remove the fuel pump/fuel level sending unit access cover and measure the resistance across the two fuel temperature sensor terminals (arrows)

block below the exhaust manifold. On 1995 and later models, the knock sensor is located below the intake manifold.

Check

Refer to illustrations 13.2

2   Disconnect the knock sensor harness connector **(see illustration)**. Using an ohmmeter, check for continuity between the two terminals of the knock sensor connector. Continuity should be indicated. If an open circuit is indicated, replace the knock sensor. **Note:** *The knock sensor resistance is very high, use an ohmmeter cable of measuring at least 10 M-ohms.*

Replacement

Refer to illustration 13.10

1993 and 1994 models

Warning: *Wait for the engine to cool completely before performing this procedure.*
3   Drain the cooling system (see Chapter 1).
4   Raise the vehicle and support it securely on jackstands.
5   Disconnect the knock sensor harness connector and detach any retaining clips.
6   Remove the knock sensor from the engine block.
7   Installation is the reverse of removal.
8   Refill the cooling system (see Chapter 1).

1995 and later models

9   Remove the upper and lower intake manifold (see Chapter 2A).
10 Disconnect the electrical connector from the knock sensor.
11 Remove the sensor retaining bolt and the sensor from the engine block **(see illustration)**.
12 Installation is the reverse of removal.
13 Install the intake manifold (see Chapter 2A).

14 Fuel temperature sensor - check and replacement

1   Some 1995 and later models are equipped with a fuel temperature sensor. The fuel temperature sensor senses the fuel temperature of the fuel inside the tank. The sensor is a thermistor, which is a variable resistor that changes its resistance as temperature changes. As fuel temperature increases, sensor resistance decreases and vice versa. The sensor is located in the fuel tank on the fuel level sending unit. The PCM uses this information for diagnostic purposes. A problem in the fuel temperature sensor circuit will set a trouble code.

Check

Refer to illustration 14.4

Note: *Performing the following test will set a diagnostic trouble code and illuminate the Check Engine light. Clear the diagnostic trouble code after performing the tests and making the necessary repairs (see Section 2).*
2   Remove the rear seat and the fuel pump/fuel level sending unit access cover (see Chapter 4).
3   Before checking the fuel temperature sensor, check the voltage supply and ground circuits from the PCM. Disconnect the electrical connector from the fuel level sending unit/fuel temperature sensor. Connect a voltmeter to the pink/blue (+) and black (-) wire terminals of the harness connector. Turn the ignition key On - the voltage should read approximately 5.0 volts. If the voltage is incorrect, check the wiring from the sensor connector to the PCM and from the connector to the engine ground point. If the circuits are good, have the PCM checked at a dealer service department or other properly equipped repair facility.
4   Using an ohmmeter, measure the resistance between the two fuel temperature sensor terminals **(see Illustration)**. With the system at room temperature (68-degrees F, 20-degrees C), sensor resistance should be 2,300 to 2,700 ohms. A more accurate check may be performed by removing the sensor and suspending the tip of the sensor in a container of water. Heat the water on the stove while you monitor the resistance of the sensor. Compare your measurements to the intake air temperature sensor resistance chart **(see Illustration 7.3)**. If the sensor resistance test results are incorrect, replace the fuel temperature sensor.

Replacement

5   Remove the fuel level sending unit (see Chapter 4).
6   Remove the fuel temperature sensor from the fuel level sending unit.
7   Installation is the reverse of removal.

15 Vehicle speed sensor - check and replacement

1   The vehicle speed sensor is a permanent magnet generator mounted on the transaxle case. It produces an AC voltage sine wave, the frequency of which is proportional to vehicle speed. The PCM uses the sensor input signal for several different engine and transmission control functions. The vehicle speed sensor signal also drives the speedometer on the instrument panel. A defective vehicle speed sensor can cause various driveability and transmission problems. The PCM can detect sensor problems and set trouble codes to indicate specific faults.

Check

Refer to illustration 15.3

2   Remove the vehicle speed sensor from the transaxle as described below.
3   Connect a voltmeter to the two terminals of the vehicle speed sensor, set the meter on the AC scale and spin the sensor

Chapter 6 Emissions and engine control systems

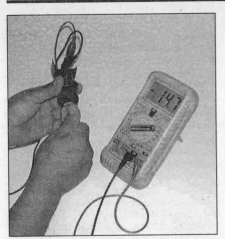

15.3 Remove the vehicle speed sensor and check for a pulsing AC voltage signal as the vehicle speed sensor gear is turned

15.8 Location of the vehicle speed sensor (arrow)

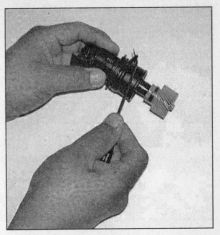

15.9 Inspect the sensor O-ring – install a new one if damaged or if you're installing a new sensor

drive gear by hand **(see illustration)**. The sensor should generate approximately 0.5 volts AC.

4 If no AC voltage signal is produced, replace the sensor.

5 If the vehicle speed sensor is good, check for continuity between the sensor connector and the instrument cluster (refer to the wiring diagrams). If the wiring is good, have the instrument cluster and PCM diagnosed by a dealer service department or other qualified repair shop.

Replacement

Refer to illustrations 15.8 and 15.9

6 Raise the vehicle and support it securely on jackstands.

7 Disconnect the electrical connector from the vehicle speed sensor.

8 Remove the hold-down bolt and clamp and remove the vehicle speed sensor from the transaxle **(see illustration)**.

9 Inspect the O-ring on the sensor **(see illustration)** and replace it if damaged. If you are installing a new sensor, use a new O-ring.

10 Installation is the reverse of removal.

16 Idle air control system

Idle air control valve

Refer to illustration 16.3

1 The Idle Air Control (IAC) valve controls the amount of air that bypasses the throttle valve, which controls the engine idle speed. The IAC valve is a stepper motor type actuator mounted on the upper intake plenum (manifold) and controlled by voltage pulses from the PCM. The IAC valve pintle moves in or out, allowing more or less intake air into the system. To increase idle speed, the PCM commands the stepper motor to pull the IAC valve pintle from the seat, allowing more air to bypass the throttle bore. To decrease idle speed, the PCM commands the IAC valve pintle towards the seat, reducing the air flow.

Check

Note: *Performing the following test will set a diagnostic trouble code and illuminate the Check Engine light. Clear the diagnostic trouble code after performing the tests and making the necessary repairs (see Section 2).*

2 Before checking the IAC valve, check the power supply to the valve. Disconnect the electrical connector from the IAC valve. Turn the ignition key On. Connect the negative probe of a voltmeter to a good engine ground point and probe each of the two red wire terminals of the IAC valve electrical connector (harness side) in turn. Battery voltage should be present at each of the red wire terminals. If battery voltage is not present, check the circuit from the battery to the ECCS relay (don't forget to check the fuses), the ECCS relay and the circuit from the ECCS relay to the IAC valve (see Chapter 12 and the wiring diagrams).

3 To check the IAC valve, use an ohmmeter to measure the resistance across the indicated terminals on the IAC valve **(see illustration)**. There should be approximately 30 ohms resistance at 68-degrees F (20-degrees C) across each pair. If the resistance is incorrect, replace the IAC valve. If the IAC valve is good, have the PCM diagnosed by a dealer service department or other qualified repair shop.

Replacement

4 Disconnect the electrical connector from the IAC valve.

5 On 1993 and 1994 models, remove the hose from the IAC valve housing. Remove the four mounting bolts and separate the IAC valve from the intake manifold.

6 On 1995 and later models, remove the two IAC valve attaching screws and withdraw the valve from the IAC housing.

7 Clean the sealing surface and the bore of the IAC housing with a shop rag or soft cloth to ensure a good seal. **Caution:** *The IAC valve itself is an electrical component and must not be soaked in any liquid cleaner, as damage may result.*

8 Position a new O-ring on the housing. Lubricate the O-ring with a light film of engine oil.

9 Install the IAC valve and tighten the screws securely.

10 The remainder of the installation is the reverse of the removal.

Fast idle control solenoid

Refer to illustration 16.12

11 1995 and later models are equipped with two fast idle control solenoids. The solenoids are used to maintain a constant idle speed due to increased engine loads.

16.3 To check the IAC valve, measure the resistance across terminals 1 and 2; 2 and 3; 4 and 5; 5 and 6; the resistance should be equal across each pair (approximately 30 ohms)

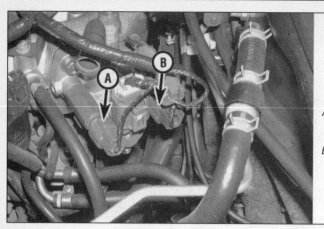

16.12 The fast idle control solenoids are located on the IAC valve housing

A Fast idle control solenoid for power steering
B Fast idle control solenoid for air conditioning

16.21 When checking the fast idle cam adjustment, the mark on the fast idle cam (A) should align with the center of the cam follower pin (B)

One solenoid is energized when the power steering is operated and the other is energized when the air conditioning is activated. When either fast idle control solenoid is energized it allows additional air to enter the intake manifold thus raising the idle speed. A fault in this system will not be detected by the OBD system.

Check

12 To check the system, start the engine and turn the steering wheel to its full stop one way, then the other and return to center, the idle speed should remain steady. Switch the air conditioning system On, the idle speed should raise slightly. If the idle speed drops during one of the situations, test that particular solenoid **(see Illustration)**.
13 Disconnect the electrical connector from the solenoid. Connect the positive probe of a voltmeter to the red/yellow or yellow/blue terminal of the harness connector, connect the negative lead to a good engine ground point. Turn the ignition key On and if testing the air conditioning fast idle control solenoid, switch the air conditioning system On. Battery voltage should be indicated on the meter. If battery voltage is not present, check the circuit from the fuse box to the solenoid (don't forget to check the fuses). If the air conditioning clutch is not energized when the system is switched On, check the air conditioning relay as well. Also check the power steering pressure switch (see Section 11) and the ground circuits for continuity from the connector to ground (see the wiring diagrams).
14 Turn the ignition key OFF. Using a pair of fused jumper wires apply battery voltage and ground to the solenoid terminals. The solenoid should make a distinct clicking sound. If the solenoid does not click, replace the solenoid. Remove the solenoid (proceed to Step 5) and check the plunger for free movement. If the plunger is seized or the detent spring is broken replace the solenoid.

Replacement

15 Disconnect the electrical connector from the fast idle control solenoid.
16 Using a wrench remove the solenoid from the IAC valve housing **(see illustration 16.12)**.
17 Install a new copper washer onto the solenoid and tighten the solenoid securely.
18 The remainder of the installation is the reverse of removal.

Fast idle cam

Refer to illustration 16.21

19 A fast idle cam and thermo element is used on 1995 and later models to increase idle speed during cold operating conditions. When the engine temperature is cold the thermo element plunger is in its normal retracted position and the fast idle cam acts on the cam follower lever to lift the throttle lever slightly off the throttle stop, increasing the idle speed. As the engine warms-up, the thermo element plunger extends, rotating the fast idle cam and allowing the throttle lever to return to its base setting.

Check and adjustment

Note: *Performing the following procedure will set a diagnostic trouble code and illuminate the Check Engine light. Clear the diagnostic trouble code after performing the procedure (see Section 2).*

20 Disconnect the electrical connector from the engine coolant temperature sensor and connect an ohmmeter to the two terminals of the sensor (see Section 8).
21 Start the engine and monitor the resistance of the coolant temperature sensor. When the sensor resistance is 1,650 to 2,400 ohms, the long mark on the fast idle cam should align with the center of the cam follower lever pin **(see Illustration)**.
22 Continue to run the engine. When the coolant temperature resistance is 260 to 390 ohms, the short mark on the fast idle cam should align with the center of the cam follower lever pin. If it doesn't, loosen the locknut and turn the adjustment screw until the mark is aligned.
23 If the fast idle cam does not operate as described, or cannot be adjusted properly, replace the throttle body (see Chapter 4).

17 Power valve control system

1 1993 and 1994 models are equipped with a power valve system. The power valve system consists of the power valve (located inside the intake manifold), the power valve actuator, the power valve control solenoid and the PCM. The power valve diverts the path of the incoming air through the intake manifold, one path being longer than the other. At low engine speeds vacuum is applied to the power valve actuator, the power valve is closed and air is diverted through the longer path to enhance maximum torque at low speed. At a preset engine speed the PCM de-energizes the solenoid, vacuum is vented from the actuator and the power valve opens. Air is then drawn through the shorter path enhancing high speed power.
2 To check the system, start the engine and rev the engine to 3,000 rpm several times while watching the power valve actuator. The actuator rod should move in-and-out correspondingly to engine speed. If the actuator is not responding, stop the engine and continue testing.
3 Check the vacuum hoses from the vacuum source to the actuator for leaks or damage. Check the vacuum tank to make sure it holds vacuum. Check the one-way valve - it should allow air to pass one way but not the other.
4 Disconnect the electrical connector from the power valve control solenoid. Turn the ignition key on and using a voltmeter, check for battery voltage at the white/blue wire terminal of the harness connector. If battery voltage is not present, check the circuit from the connector to the fuse box. Check for continuity from the connector to the PCM on the orange wire.
5 Remove the vacuum lines from the power valve control solenoid ports. Using a pair of fused jumper wires, connect battery voltage and ground to the two terminals of the power valve control solenoid. The solenoid should click and air should be allowed to pass between the two ports. Remove the jumpers, no air should pass between the two ports. If the solenoid does

Chapter 6 Emissions and engine control systems

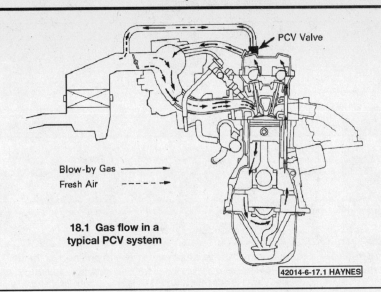

18.1 Gas flow in a typical PCV system

not operate as described, replace the power valve control solenoid.

6   Remove the vacuum hose from the power valve actuator. Connect a hand-held vacuum pump to the actuator and apply vacuum. The rod should move and the actuator should hold vacuum. If it doesn't, replace the actuator.

7   If all the above tests are good, have the PCM diagnosed by a dealer service department or other qualified repair shop.

18 Positive crankcase ventilation system

Refer to illustration 18.1

1   The Positive Crankcase Ventilation (PCV) system reduces hydrocarbon emissions by scavenging crankcase vapors. It does this by circulating fresh air from the air filter through the crankcase, where it mixes with blow-by gases and is then rerouted through a PCV valve to the intake manifold **(see illustration)**.

2   The PCV system consists of a replaceable PCV valve and the crankcase ventilation hoses.

3   To maintain idle quality, the PCV valve restricts the flow when the intake manifold vacuum is high. If abnormal operating conditions arise, the system is designed to allow excessive amounts of blow-by gases to flow back through the crankcase vent tube into the air filter to be consumed by normal combustion.

4   Checking and replacement of the PCV valve and filter is covered in Chapter 1.

19 Exhaust gas recirculation system

1   The Exhaust Gas Recirculation (EGR) system is used to lower NOx (oxides of nitrogen) emission levels caused by high combustion temperatures. The EGR valve recirculates a small amount of exhaust gases into the intake manifold. The additional mixture lowers the temperature of combustion thereby reducing the formation of NOx compounds.

1998 and earlier models

Refer to illustration 19.2

2   On 1998 and earlier models, the EGR system is equipped with an EGR valve, an EGR control solenoid valve which receives ported and manifold vacuum, an EGR temperature sensor and a EGR control backpressure transducer valve **(see illustration)**. The operation of the system is controlled by the PCM which operates the EGR control solenoid. The manifold vacuum system utilizes a vacuum tap in the air intake system positioned after the throttle valve. The ported vacuum control system uses a vacuum tap in the throttle body which is exposed to an increasing percentage of manifold vacuum as the throttle valve is opened during acceleration.

3   The backpressure transducer valve monitors the exhaust backpressure as the engine rpm increases or decreases to aid in controlling the amount of the EGR vacuum signal. The EGR temperature sensor is used to inform the PCM of temperature changes in the EGR passage way. This helps the PCM determine the EGR On/Off time.

System check

Refer to illustration 19.5

4   Check all hoses for cracks, kinks, broken sections and proper connection. Inspect all system connections for damage, cracks and leaks.

5   To check the EGR system operation, bring the engine up to operating temperature and, with the transmission in Neutral (parking brake set and tires blocked to prevent movement), allow it to idle. Open the throttle so the engine speed is between 2,000 and 4,000 rpm and then allow it to close. The EGR valve stem should move if the control system is working properly **(see illustration)**. The test

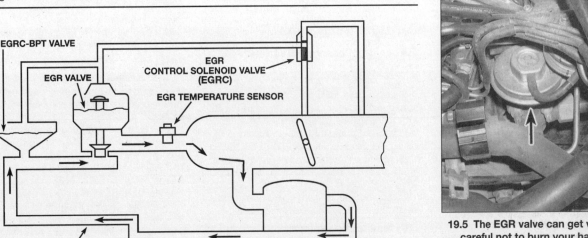

19.2 Schematic of a typical backpressure transducer EGR system

19.5 The EGR valve can get very hot; be careful not to burn your hand when checking for EGR valve diaphragm movement - the diaphragm and rod beneath the top of the valve (arrow) should move when the engine is revved

19.11 The EGR control solenoid (arrow) is mounted at the rear of the left (front) cylinder head (1995 and later model shown)

19.14 Connect the vacuum pump hose to one port of the EGR control backpressure transducer valve and apply vacuum - there should be a moderate amount of leakage with no pressure applied to the bottom port

should be repeated several times. Movement of the stem indicates the control system is functioning correctly. If the EGR valve stem does not move, Check the vacuum signal to the EGR valve. Remove the hose from the valve, place your finger over the end of the hose and perform the procedure again. If no vacuum is present at the end of the hose check the hose connections to make sure they are not leaking or clogged, then check the EGR control solenoid (see Step 11).

Component checks

EGR Valve

6   With the engine Off, disconnect the vacuum hose and apply ten inches of vacuum with a hand-held vacuum pump to the EGR valve. If the valve opens, measure the valve travel to make sure it is approximately 1/8-inch. If the stem does not move, replace the EGR valve with a new one.

7   Next apply vacuum with the pump and then clamp the hose shut. The valve should stay open for 30 seconds or longer. If it does not, the diaphragm is leaking and the valve should be replaced with a new one.

8   Start the engine and apply vacuum to the valve. The engine should idle roughly when the valve is open. If it doesn't, the passages in the manifold are probably clogged. If there's no difference in idle quality with the valve open or closed the EGR valve is probably not closing all the way. Remove the EGR valve and inspect the poppet and seat area for deposits.

9   If the deposits are more than a thin film of carbon, the valve should be cleaned. To clean the valve, apply solvent and allow it to penetrate and soften the deposits, making sure that none gets on the valve diaphragm, as it could be damaged.

10   Use a vacuum pump to hold the valve open and carefully scrape the deposits from the seat and poppet area with a tool. Inspect the poppet and stem for wear and replace the valve with a new one if wear is found.

EGR control solenoid

Refer to illustration 19.11

11   First check for a vacuum signal to the solenoid with the engine running, If no vacuum is present check the hose to and from the solenoid for cracks and clogging. If the hose is OK, check the throttle body port for clogging. If vacuum is present at the solenoid, disconnect the solenoid valve harness connector and check for battery voltage between the terminals with the ignition key ON (engine not running) **(see illustration)**. Battery voltage should be present. If voltage is not present at the solenoid valve refer to the wiring diagrams and check the wiring harness for open circuits or a damaged harness.

12   To check the operation of this EGR control solenoid, disconnect the electrical connector and connect a ground wire to one of its terminals and fused battery voltage to the other terminal.

13   It should be possible to blow air through the two vacuum ports that are adjacent to one another when there is battery power applied, but impossible to do so when it is not energized with battery voltage.

EGR control backpressure transducer valve

Refer to illustration 19.14

14   Locate the EGR control backpressure transducer valve and plug one of the ports with a finger **(see illustration)**. Use a hand-held vacuum pump and apply vacuum to the valve. The valve should leak. **Note:** *It should stop leaking when pressure greater than 0.145 psi (1.0 kPa) is applied to the port under the valve (the valve must be removed from the engine to apply pressure at this point).* If the valve is defective replace it.

EGR temperature sensor

Refer to illustration 19.15

15   Remove the EGR temperature sensor **(see illustration)**. Submerge the tip of the sensor in a container of water. Heat the water on the stove while monitoring the resistance

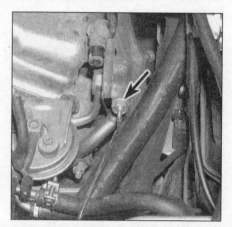

19.15 The EGR temperature sensor (arrow) is located in the pipe leading from the EGR valve to the intake manifold

of the sensor. Resistance should decrease as temperature increases and at 212-degrees F (100-degrees C) the sensor resistance should measure 80 to 100 K-ohms.

Component replacement

EGR valve

Refer to illustration 19.17

16   Remove the air intake duct from the throttle body and air filter housing (see Chapter 4).

17   Detach the vacuum line from the EGR valve. Disconnect the EGR pipe nut **(see illustration)**.

18   Remove the EGR valve mounting fasteners.

19   Remove the EGR valve and gasket from the manifold. Discard the gasket.

20   With a wire wheel, buff the exhaust deposits from the EGR valve mounting surface on the manifold and, if you plan to use the same valve, the mounting surface of the valve itself. Look for exhaust deposits in the valve outlet. Remove deposit build-up with a screwdriver. **Caution:** *Never wash the valve*

Chapter 6 Emissions and engine control systems

19.17 Loosen the EGR pipe flange nut (arrow), disconnect the pipe from the EGR valve and remove the two mounting fasteners (arrows)

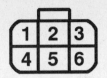

19.33 To check the EGR valve stepper motor coils, disconnect the electrical connector and measure the resistance across terminals 1 and 2; 2 and 3; 4 and 5; 5 and 6; of the EGR valve in turn - resistance should be approximately 22 ohms across each pair

in solvents or degreaser - both agents will permanently damage the diaphragm. Sandblasting is also not recommended because it will affect the operation of the valve.

21 If the EGR passage contains an excessive build-up of deposits, clean it out with a wire wheel. Make sure that all loose particles are completely removed to prevent them from clogging the EGR valve or from being ingested into the engine.

22 If there are large amounts of deposits within the EGR valve, remove the EGR pipe from the exhaust manifold and clean out the deposits inside the tube. **Note:** *Remove the EGR control backpressure transducer valve and pipe and clean any deposits from the passages.*

23 Installation is the reverse of removal.

EGR control solenoid

24 Unplug the electrical connector from the solenoid.

25 Clearly label and detach the vacuum hoses.

26 Remove the solenoid mounting nut and remove the solenoid.

27 Installation is the reverse of removal.

EGR temperature sensor

28 Disconnect the electrical connector and unscrew the sensor from the pipe **(see illustration 19.15).**

29 Apply anti-seize to the threads of the sensor before installing it.

30 Installation is the reverse of removal.

1999 and later models

31 On 1999 and later models, the EGR system consists of a stepper motor-type EGR valve, an EGR temperature sensor and the PCM. The PCM controls the EGR flow rate by reversing the direction of the stepper motor, opening or closing the EGR passage in small increments. This system allows for precise control of EGR flow, achieving optimum EGR flow depending on engine operating conditions.

Component checks

EGR valve

Refer to illustration 19.33

32 Before checking the EGR valve, check the power supply to the valve. Disconnect the electrical connector from the EGR valve. Turn the ignition key On. Connect the negative probe of a voltmeter to a good engine ground point and probe each of the two red wire terminals of the EGR valve electrical connector (harness side) in turn. Battery voltage should be present at each of the red wire terminals. If battery voltage is not present, check the circuit from the battery to the ECCS relay (don't forget to check the fuses), the ECCS relay and the circuit from the ECCS relay to the EGR valve (see Chapter 12 and the wiring diagrams).

33 To check the EGR valve, use an ohmmeter to measure the resistance across the indicated terminals on the EGR valve **(see Illustration).** There should be approximately 22 ohms resistance at 68-degrees F (20-degrees C) across each pair. If the resistance is incorrect, replace the EGR valve. If the EGR valve is good, have the PCM diagnosed by a dealer service department or other qualified repair shop.

EGR temperature sensor

34 Remove the EGR temperature sensor. Submerge the tip of the sensor in a container of water. Heat the water on the stove while monitoring the resistance of the sensor. Resistance should decrease as temperature increases and at 212-degrees F (100-degrees C) the sensor resistance should measure 10 to 30 K-ohms.

Component replacement

EGR valve

35 Remove the air intake duct from the throttle body and air filter housing.

36 Position the hoses and wiring harness aside to access the EGR valve.

37 Disconnect the EGR pipe.

38 Remove the EGR valve mounting fasteners.

39 Remove the EGR valve and gasket from the manifold. Discard the gasket.

40 Using a gasket scraper, clean the EGR valve mounting surface.

41 Installation is the reverse of removal.

EGR temperature sensor

42 Disconnect the electrical connector and unscrew the sensor from the manifold.

43 Apply anti-seize to the threads of the sensor before installing it.

44 Installation is the reverse of removal.

20 Evaporative emissions control system

General description

Refer to illustration 20.1

1 The Evaporative Emissions Control (EVAP) system absorbs fuel vapors and, during engine operation, releases them into the engine intake where they mix with the incoming air-fuel mixture. The EVAP system consists of a charcoal-filled canister and the lines connecting the canister to the fuel tank,

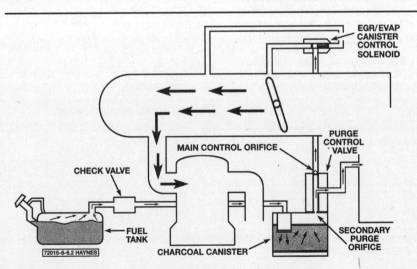

20.1 Schematic of the EVAP system

ported vacuum and intake manifold vacuum **(see illustration)**.

2   When the engine is not operating fuel vapors are transferred from the fuel tank, throttle body and intake manifold to the charcoal canister where they are stored. When the engine is running, the fuel vapors are purged from the canister by the purge control valve. The gasses are consumed in the normal combustion process.

3   On all 1993 and 1994 models and 1995 Federal emissions models, a vacuum operated purge control valve is mounted on top of the canister. On 1995 California emissions models and all 1996 and later models, an electronic purge control valve, controlled by the PCM is used. The electronic system is also equipped with a canister purge cut valve to close the purge line during deceleration and idling conditions.

Check

4   Poor idle, stalling and poor driveability can be caused by a defective canister purge control valve, a damaged canister, split or cracked hoses or hoses connected to the wrong tubes.

5   Evidence of fuel loss or fuel odor can be caused by the following items: fuel leaking from fuel lines or a cracked or damaged canister, an inoperative fuel tank check valve, an inoperative purge valve, disconnected, misrouted, kinked, deteriorated or damaged vapor or control hoses or an improperly seated air filter or air filter gasket.

6   Inspect each hose attached to the canister for kinks, leaks and breaks along its entire length. Repair or replace as necessary.

7   Inspect the canister. If it is cracked or damaged, replace it.

8   Look for fuel leaking from the bottom of the canister. If fuel is leaking, replace the canister and check the hoses and hose routing.

Canister replacement

All 1993 and 1994 models and 1995 Federal emissions model

Refer to illustration 20.10

9   Clearly label, then detach, all vacuum lines from the canister.

10   Remove the canister mounting bolts and pull the canister out of the mounting bracket **(see illustration)**.

11   Installation is the reverse of removal.

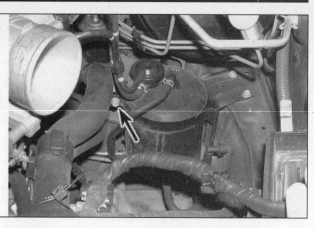

20.10 EVAP canister mounting bolt (arrow) - all 1993 and 1994 models and 1995 Federal emissions model

1995 California emissions model and all 1996 and later models

Note: On 1995 California emissions model and all 1996 and later models, the EVAP canister is located on the left side of the vehicle, behind the fuel tank.

12   Raise the vehicle and support it securely on jackstands.

13   Clearly label, then detach, all vacuum lines from the canister.

14   Remove the canister mounting bolts and remove the canister.

15   Installation is the reverse of removal.

21   Catalytic converter

Note: Because of a Federally mandated extended warranty which covers emissions-related components such as the catalytic converter, check with a dealer service department before replacing the converter at your own expense.

General description

1   The catalytic converter is an emission control device added to the exhaust system to reduce pollutants from the exhaust gas stream. A single-bed converter design is used in combination with a three-way (reduction) catalyst. The catalytic coating on the three-way catalyst contains platinum and rhodium, which lowers the levels of oxides of nitrogen (NOx) as well as hydrocarbons (HC) and carbon monoxide (CO).

Check

2   The test equipment for a catalytic converter is expensive and highly sophisticated. If you suspect that the converter on your vehicle is malfunctioning, take it to a dealer or authorized emissions inspection facility for diagnosis and repair.

3   Whenever the vehicle is raised for servicing of underbody components, check the converter for leaks, corrosion, dents and other damage. Check the welds/flange bolts that attach the front and rear ends of the converter to the exhaust system. If damage is discovered, the converter should be replaced.

4   Although catalytic converters don't break too often, they can become plugged. The easiest way to check for a restricted converter is to use a vacuum gauge to diagnose the effect of a blocked exhaust on intake vacuum.

  a) Connect a vacuum gauge to an intake manifold vacuum source.
  b) Warm the engine to operating temperature, place the transaxle in park and apply the parking brake.
  c) Note and record the vacuum reading at idle.
  d) Open the throttle until the engine speed is about 2000 rpm.
  e) Release the throttle quickly and record the vacuum reading.
  f) Perform the test three more times, recording the reading after each test.
  g) If the reading after the fourth test is more than one in-Hg lower than the reading recorded at idle, the catalytic converter, muffler or exhaust pipes may be plugged or restricted.

Component replacement

5   Refer to the exhaust system servicing section in Chapter 4.

Chapter 7 Part A
Manual transaxle

Contents

	Section		Section
Back-up light and neutral position switch, check and replacement	4	Oil seal replacement	3
General information	1	Shift linkage, removal and installation	2
Manual transaxle lubricant change	See Chapter 1	Transaxle, removal and installation	5
Manual transaxle lubricant level check	See Chapter 1	Transaxle overhaul, general information	6

Specifications

Transaxle fluid type .... See Chapter 1

Torque specifications	Nm	Ft-lbs (unless otherwise indicated)
Engine-to-transaxle bolts		
1993 and 1994 **(see illustration 5.24a)**		
Bolt A (25mm)	16 to 21	144 to 180 in-lbs
Bolt B (28mm)	30 to 40	22 to 30
Bolt C (57mm)	39 to 49	29 to 36
Bolt D (57mm)	39 to 49	29 to 36
Bolt E (64mm)	39 to 49	29 to 36
Bolt F (30mm)	21 to 27	15 to 20
Transaxle-to-engine gusset bolts	30 to 40	22 to 30
1995 and later		
Bolt A (52mm)	70 to 79	51 to 59
Bolt B (65mm)	70 to 79	51 to 59
Bolt C (124mm)	70 to 79	51 to 59
Bolt D (40mm)	35 to 47	26 to 35
Bolt E (40mm)	35 to 47	26 to 35

Chapter 7 Part A Manual transaxle

1 General information

The vehicles covered by this manual are equipped with either a 5-speed manual transaxle or a 4-speed automatic transaxle. Information on the manual transaxle is included in this Part of Chapter 7. Service procedures for the automatic transaxle are contained in Chapter 7, Part B.

The manual transaxle is a compact, two-piece, lightweight aluminum alloy housing containing both the transmission and differential assemblies.

Because of the complexity of the transaxle and the special tools needed to work on it, internal repair procedures for the manual transaxle are beyond the scope of this manual. The information in this Chapter is devoted to removal and installation procedures.

2 Shift linkage and lever - removal and installation

Refer to illustrations 2.3, 2.4 and 2.5

Warning: *Some models covered by this manual are equipped with a Supplemental Restraint System, more commonly known as airbags. Always disable the airbag system before working in the vicinity of any airbag system components (see Chapter 12).*

1   Disconnect the cable from the negative terminal of the battery. On airbag-equipped models also disconnect the positive cable, then wait three minutes before proceeding.
2   Raise the vehicle and support it securely on jackstands. Remove the catalytic converter heat shield.
3   Disconnect the support and control rod from the transaxle **(see illustration)**.
4   Remove the nut and bolt that attach the control rod to the shift lever **(see illustration)**. Disconnect the return spring, remove the nuts that attach the support rod to the shift lever socket and mass damper **(see illustration)** and remove the shift linkage assembly. Inspect the condition of all the bushings at both ends of the support and control rods. If they're cracked or worn, replace them.
5   Remove the console (see Chapter 11). Detach the shift lever dust boot from the floor **(see illustration)**, pull out the shift lever and slide off the dust boot, seat, insulator and shift lever socket. Inspect these parts for cracks and tears. Replace as necessary. The condition of the shift lever socket and the seat are especially important; if either of these parts is damaged, shifting will be difficult. Also inspect the condition of the large rubber shift lever boot. If it's damaged, replace it.
6   Installation is the reverse of removal. Make sure you lubricate all friction surfaces, the spherical bearing surface of the shift lever, the inside of the shift lever socket and the inside of the seat, with silicone grease. Also be sure to lubricate all support and control rod bushings and collars with silicone grease. Finally, make sure you tighten all fasteners securely.

3 Oil seal replacement

1   Oil leaks can occur as a result of worn seals or O-rings. Replacement of these seals or O-rings is relatively easy, since the repairs can usually be performed without removing the transaxle from the vehicle.

Driveaxle oil seals

Refer to illustrations 3.4 and 3.5

2   The driveaxle oil seals are located on the sides of the transaxle, where the inner ends of the driveaxles are splined into the differential side gears. If you suspect that a driveaxle oil seal is leaking, raise the vehicle and support it securely on jackstands. If the seal is leaking, you'll see lubricant on the side of the transaxle, below the seal.
3   Remove the driveaxle(s) (see Chapter 8).
4   Using a seal removal tool, screwdriver or prybar, carefully pry the seal out of the transaxle bore **(see illustration)**.
5   Using a large section of pipe or a large deep socket as a drift, install the new oil seal. Drive it into the bore squarely and make sure

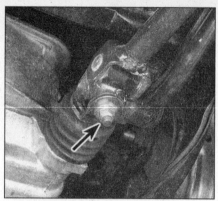

2.3 To disconnect the control rod from the transaxle, remove this nut and bolt (arrow); to disconnect the support rod, remove the fastener(s) (not visible in this photo) that connect the forward end of the support rod to the transaxle

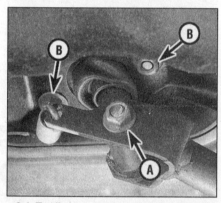

2.4 To disconnect the control rod from the shift lever, remove nut and bolt "A"; to disconnect the support rod from the shift lever socket, remove the two nuts "B"

2.5 To remove the large shift lever dust boot, remove the two front nuts (upper arrows) and the two rear nuts from below (not visible in this photo)

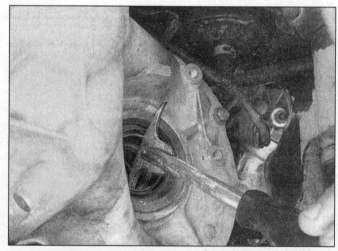

3.4 Carefully pry out the driveaxle oil seal with a seal removal tool or a screwdriver; make sure you don't damage the seal bore or the new seal may leak

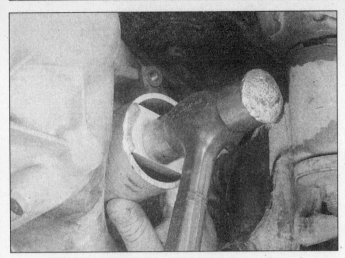

3.5 Use a seal installer, a large socket or a piece of pipe to install the new seal

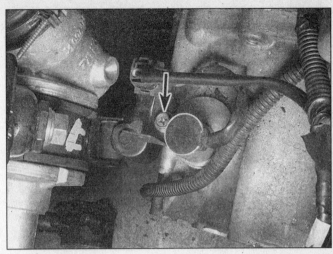

3.8 To remove the vehicle speed sensor, disconnect the electrical connector, remove the hold-down bolt (arrow) and remove the sensor assembly from the transaxle (later model shown)

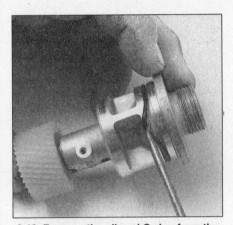

3.10 Remove the oil seal O-ring from the pinion gear assembly with a small hook tool or a small screwdriver; make sure you don't gouge the groove (typical)

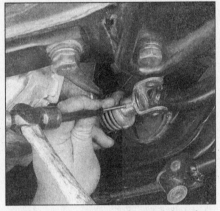

3.15a Drive out the yoke retaining pin with a hammer and punch . . .

it's fully seated **(see illustration)**.

6   Lubricate the lip of the new seal with multi-purpose grease.

7   Install the driveaxle(s). Be careful not to damage the lip(s) of the new seal(s).

Vehicle speed sensor (speedometer drive) O-ring

Refer to illustrations 3.8 and 3.10

8   The vehicle speed sensor **(see illustration)** is located on the transaxle housing. Look for lubricant around the housing to determine if the O-ring is leaking.

9   Disconnect the electrical connector, remove the hold-down bolt and remove the pinion assembly and vehicle speed sensor from the transaxle.

10   Using a scribe or a small screwdriver, remove the O-ring seal **(see illustration)**.

11   Install a new O-ring on the pinion gear housing. Smear some transmission lubricant on the O-ring before installing the sensor.

12   Installation is the reverse of removal.

Control rod seal

Refer to illustrations 3.15a, 3.15b, 3.15c, 3.16, 3.17 and 3.18

13   Raise the vehicle and place it securely on jackstands.

14   Disconnect the transaxle control rod from the yoke (see Section 2).

15   Remove the yoke retaining pin, the yoke and the dust boot **(see illustrations)**.

16   Remove the control rod seal **(see illustration)**.

17   Install a new seal **(see illustration)**.

3.15b . . . remove the yoke . . .

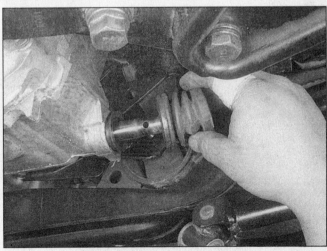

3.15c . . . and remove the dust boot

3.16 Pry out the control rod seal

3.17 Install the new control rod seal with a large deep socket

18 Install the yoke and the yoke retaining pin **(see illustration)**.
19 Install the control rod (see Section 2).
20 Remove the jackstands and lower the vehicle.

4 Back-up light and neutral position switch - check and replacement

Refer to illustrations 4.1 and 4.3

Check

1 Raise the vehicle and place it securely on jackstands. Locate the back-up light switch **(see illustration)**.
2 Disconnect the switch electrical connector.
3 Place the transaxle in Reverse and verify that there's continuity between connector terminals 2 and 4 (1998 and earlier models) or 1 and 3 (1999 models) **(see illustration)**. Put the transaxle in Neutral and verify that there's continuity between terminals 1 and 3 (1998 and earlier models) or 2 and 4 (1999 models). Verify that there's no continuity between those terminal pairs in any other gear.
4 If continuity isn't as specified, replace the back-up light switch.

Replacement

5 Drain the transaxle lubricant (see Chapter 1).
6 Disconnect the switch electrical connector.
7 Remove the switch hold-down bolt and pull the switch straight out of the transaxle.
8 Apply a light coat of clean oil to a new O-ring, install the new switch and O-ring, and tighten the switch hold-down bolt securely.
9 Plug in the electrical connector.
10 Check the switch as described above to ensure it's working properly.
11 Fill the transaxle with the specified lubricant (see Chapter 1).
12 Remove the jackstands and lower the vehicle.

3.18 Install the dust boot and yoke and secure the yoke with a *new* roll pin

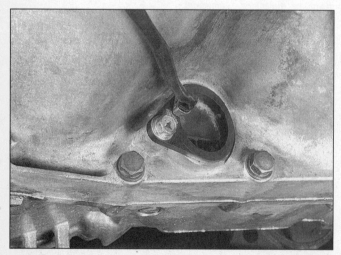

4.1 The back-up light and neutral position switch is located on the lower left side of the transaxle

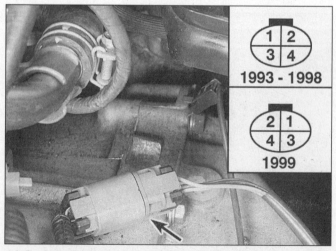

4.3 Back-up light and neutral position switch electrical connector location and terminal guide (terminal numbers indicate switch side of harness)

Chapter 7 Part A Manual transaxle

5.3 To detach the clutch hydraulic line from the transaxle, remove this clip and pull the line out of the bracket

5.7 Upper transaxle-to-engine bolts (arrows)

5.20 With the transaxle supported on a transmission jack or floor jack, carefully pull the transaxle away from the engine until the input shaft clears the clutch hub

5 Transaxle - removal and installation

Refer to illustrations 5.3, 5.7, 5.20, 5.24a and 5.24b

Removal

1 Remove the battery and the battery tray (see Chapter 5).
2 Remove the air cleaner housing, mass airflow sensor and air intake duct (see Chapter 4).
3 Remove the clutch release cylinder from the transaxle (see Chapter 8). Remove the clip **(see illustration)** and disconnect the clutch hydraulic line at the bracket on top of the transaxle. **Caution:** *Don't depress the clutch pedal while the release cylinder is removed.*
4 Disconnect the electrical connector for the vehicle speed sensor (see Section 3). Disconnect all ground wires.
5 On 1995 and later models, remove the crankshaft position sensor (see Chapter 6).
6 Remove the starter motor from the transaxle (see Chapter 5).
7 Remove the upper transaxle-to-engine bolts **(see illustration)**. Note: *The bolts securing the transaxle to the engine are different lengths. When removing the bolts, mark their positions or lay them out in order so they can be reinstalled in their original locations.*
8 Loosen the wheel lug nuts. Raise the vehicle and support it securely on jackstands. Remove the wheels.
9 Disconnect the electrical connector for the back-up light and neutral position switch (see Section 4).
10 Remove the section of exhaust pipe under the engine (see Chapter 4).
11 Drain the transaxle fluid (see Chapter 1).
12 Disconnect the support rod and the shift control rod from the transaxle (see Section 2).
13 Remove the driveaxles (see Chapter 8).
14 Support the engine. This can be done from above with an engine hoist or engine support fixture, or from underneath by placing a jack (with a wood block as an insulator) under the engine oil pan. The engine must be supported at all times while the transaxle is out of the vehicle.
15 Support the transaxle with a jack (preferably a special jack made for this purpose). Safety chains will help steady the transaxle on the jack.
16 Remove the center crossmember from under the engine/transaxle.
17 Raise the engine and transaxle slightly and disconnect the left transaxle mount and, on models so equipped, the oil pan-to-bellhousing bolts (see Chapter 2).
18 Remove the remaining bellhousing-to-engine bolts.
19 Make a final inspection of the transaxle for any wires and hoses that have been overlooked.
20 Lower the jacks supporting the engine and transaxle slightly, then move the transaxle toward the side of the vehicle **(see illustration)**. Once the input shaft is clear of the splines in the clutch hub, lower the transaxle and remove it from under the vehicle. Try to keep the transaxle as level as possible.
21 While the transaxle is removed, be sure to inspect the clutch components (see Chapter 8). In most cases, new clutch components should be routinely installed whenever the transaxle is removed.

Installation

22 If removed, install the clutch components (see Chapter 8).
23 With the transaxle secured to the jack as on removal, raise it into position and then carefully engage the input shaft with the splines in the clutch hub. Do not use excessive force to install the transaxle, if the input shaft does not slide into place, readjust the angle of the transaxle so it is in the same plane as the engine. If the engine and transaxle are in the same plane, but the input shaft still won't engage the clutch hub, turn the input shaft slightly and the splines on the shaft will engage properly with the splines in the clutch hub.
24 Install the transaxle-to-engine bolts **(see illustrations)**. Tighten the bolts to the torque values listed in this Chapter's Specifications.
25 Install the transaxle mount nuts and bolts. Tighten all nuts and bolts securely.
26 Install the transaxle mount and center crossmember.
27 Remove all transaxle and engine supports. Install the various items removed pre-

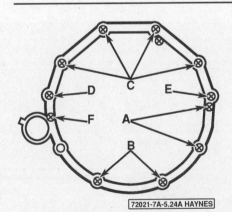

5.24a Transaxle bolt locations - 1993 and 1994 models

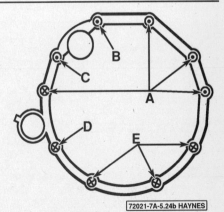

5.24b Transaxle bolt locations - 1995 and later models

viously. Refer to Chapter 8 for driveaxle installation, Chapter 4 for exhaust pipe installation, Chapter 5 for starter motor installation and Chapter 6 for crankshaft position sensor installation.

28 Make a final check that all electrical wiring has been connected and that the transaxle has been filled with the specified lubricant to the proper level (see Chapter 1). Lower the vehicle.

29 Connect the negative battery cable. Road test the vehicle to check for proper transaxle operation and check for leakage.

6 Transaxle overhaul - general information

1 Overhauling a manual transaxle is a difficult job for the do-it-yourselfer. It involves the disassembly and reassembly of many small parts. Numerous clearances must be precisely measured and, if necessary, changed with select fit spacers and snap-rings. As a result, if transaxle problems arise, it can be removed and installed by a competent do-it-yourselfer, but overhaul should be left to a transmission repair shop. Rebuilt transaxles may be available, check with your dealer parts department and auto parts stores. At any rate, the time and money involved in an overhaul is almost sure to exceed the cost of a rebuilt unit.

2 Nevertheless, it's not impossible for an inexperienced mechanic to rebuild a transaxle if the special tools are available and the job is done in a deliberate step-by-step manner so nothing is overlooked.

3 The tools necessary for an overhaul include internal and external snap-ring pliers, a bearing puller, a slide hammer, a set of pin punches, a dial indicator and possibly a hydraulic press. In addition, a large, sturdy workbench and a vise or transaxle stand will be required.

4 During disassembly of the transaxle, make careful notes of how each piece comes off, where it fits in relation to other pieces and what holds it in place. If you note how each part is installed before removing it, getting the transaxle back together again will be much easier.

5 Before taking the transaxle apart for repair, it will help if you have some idea what area of the transaxle is malfunctioning. Certain problems can be closely tied to specific areas in the transaxle, which can make component examination and replacement easier. Refer to the *Troubleshooting* section at the front of this manual for information regarding possible sources of trouble.

Chapter 7 Part B
Automatic transaxle

Contents

	Section		Section
Automatic transaxle - removal and installation	7	Park Neutral Position switch - check, adjustment and replacement	6
Automatic transaxle fluid and filter change	See Chapter 1	Shift cable - check, adjustment and replacement	3
Automatic transaxle fluid level check	See Chapter 1	Shift interlock system - description, check and component replacement	5
Diagnosis and trouble codes	2		
Driveaxle oil seals	See Chapter 7A	Shift lever - removal and installation	4
Engine mounts - check and replacement	See Chapter 2A		
General information	1		

Specifications

Transaxle fluid type and capacity .... See Chapter 1

Torque specifications

	Nm	Ft-lbs (unless otherwise indicated)
Driveplate-to-torque converter bolts	44 to 59	33 to 43
Transaxle-to-engine bolts		
1993 models **(see illustration 7.26a)**		
Bolt no. 1	30 to 40	22 to 30
Bolt no. 2	39 to 49	29 to 36
Bolt no. 3	30 to 40	22 to 30
1994 models **(see illustration 7.26b)**		
Bolt no. 1	30 to 40	22 to 30
Bolt no. 2	39 to 49	29 to 36
Bolt no. 3	30 to 40	22 to 30
Bolt no. 4	6 to 8	52 to 69 in-lbs
Bolt no. 5	30 to 40	22 to 30
1995 through 1997 models (all bolts)	70 to 79	51 to 59
1998 and later models **(see illustration 7.26d)**		
Bolt no. 1	39 to 49	29 to 36
Bolt no. 2	30 to 36	22 to 27
Bolt no. 3	30 to 36	22 to 27
Bolt no. 4	74 to 83	54 to 61
Bolt no. 5	30 to 36	22 to 27
Bolt no. 6	30 to 36	22 to 27

1 General information

All models covered by this manual are equipped with either a 5-speed manual transaxle or a 4-speed automatic transaxle. All information on the automatic transaxle is included in this Part of Chapter 7. Information for the manual transaxle can be found in Part A of this Chapter.

Because of the complexity of the automatic transaxle and the specialized equipment needed to service it, this Chapter contains only those procedures related to general diagnosis, routine maintenance, adjustment, and removal and installation.

If the transaxle requires major repair work, it should be taken to a dealer service department or an automotive or transmission repair shop. You can, however, save money by removing and installing the transaxle yourself, even if the repair work is done by a shop.

2 Diagnosis and trouble codes

Note: *Automatic transaxle malfunctions may be caused by five general conditions: poor engine performance, improper adjustments, hydraulic malfunctions, mechanical malfunctions or malfunctions in the computer or its signal network. Diagnosis of these problems should always begin with a check of the easily repaired items: fluid level and condition (see Chapter 1), shift linkage adjustment and throttle linkage adjustment. Next, perform a road test to determine if the problem has been corrected or if more diagnosis is necessary. If the problem persists after the preliminary tests and corrections are completed, additional diagnosis should be done by a dealer service department or transmission repair shop. Refer to the Troubleshooting section at the front of this manual for information on symptoms of transaxle problems.*

Preliminary checks

1 Drive the vehicle to warm the transaxle to normal operating temperature.
2 Check the fluid level as described in Chapter 1:

   a) If the fluid level is unusually low, add enough fluid to bring the level within the designated area of the dipstick, then check for external leaks (see below).
   b) If the fluid level is abnormally high, drain off the excess, then check the drained fluid for contamination by coolant. The presence of engine coolant in the automatic transmission fluid indicates that a failure has occurred in the internal radiator walls that separate the coolant from the transmission fluid (see Chapter 3).
   c) If the fluid is foaming, drain it and refill the transaxle, then check for coolant in the fluid, or a high fluid level.

3 Make sure the engine idle speed is correct (see Chapter 4). If the idle speed is incorrect, have it adjusted by a dealer service department before proceeding.
4 Inspect the shift cable (see Section 3). Make sure that it's properly adjusted and operates smoothly.

Fluid leak diagnosis

5 Most fluid leaks are easy to locate visually. Repair usually consists of replacing a seal or gasket. If a leak is difficult to find, the following procedure may help.
6 Identify the fluid. Make sure it's transmission fluid and not engine oil or brake fluid (automatic transmission fluid is a deep red color).
7 Try to pinpoint the source of the leak. Drive the vehicle several miles, then park it over a large sheet of cardboard. After a minute or two, you should be able to locate the leak by determining the source of the fluid dripping onto the cardboard.
8 Make a careful visual inspection of the suspected component and the area immediately around it. Pay particular attention to gasket mating surfaces. A mirror is often helpful for finding leaks in areas that are hard to see.
9 If the leak still cannot be found, clean the suspected area thoroughly with a degreaser or solvent, then dry it.
10 Drive the vehicle for several miles at normal operating temperature and varying speeds. After driving the vehicle, visually inspect the suspected component again.
11 Once the leak has been located, the cause must be determined before it can be properly repaired. If a gasket is replaced but the sealing flange is bent, the new gasket will not stop the leak. The bent flange must be straightened.
12 Before attempting to repair a leak, check to make sure that the following conditions are corrected or they may cause another leak. **Note:** *Some of the following conditions cannot be fixed without highly specialized tools and expertise. Such problems must be referred to a transmission shop or a dealer service department.*

Gasket leaks

13 Check the pan periodically. Make sure the bolts are tight, no bolts are missing, the gasket is in good condition and the pan is flat (dents in the pan may indicate damage to the valve body inside).
14 If the pan gasket is leaking, the fluid level or the fluid pressure may be too high, the vent may be plugged, the pan bolts may be too tight, the pan sealing flange may be warped, the sealing surface of the transaxle housing may be damaged, the gasket may be damaged or the transaxle casting may be cracked or porous. If sealant instead of gasket material has been used to form a seal between the pan and the transaxle housing, it may be the wrong sealant.

Seal leaks

15 If a transaxle seal is leaking, the fluid level or pressure may be too high, the vent may be plugged, the seal bore may be damaged, the seal itself may be damaged or improperly installed, the surface of the shaft protruding through the seal may be damaged or a loose bearing may be causing excessive shaft movement.
16 Make sure the dipstick tube seal is in good condition and the tube is properly seated. Periodically check the area around the speedometer gear or sensor for leakage. If transmission fluid is evident, check the O-ring for damage.

Case leaks

17 If the case itself appears to be leaking, the casting is porous and will have to be repaired or replaced.
18 Make sure the oil cooler hose fittings are tight and in good condition.

Fluid comes out vent pipe or fill tube

19 If this condition occurs, the transaxle is overfilled, there is coolant in the fluid, the case is porous, the dipstick is incorrect, the vent is plugged or the drain-back holes are plugged.

Diagnostic trouble codes

20 The computer for the automatic transaxle has a self-diagnostic capability; it continually monitors important information sensor and output actuator circuits for malfunctions. When a monitored circuit is damaged, shorted or disconnected, a diagnostic trouble code is stored in the computer's memory. At a dealer service department, stored trouble codes are extracted from computer memory with a proprietary diagnostic instrument known as CONSULT. Codes can also be extracted with some generic scanners. However, neither the CONSULT nor scanners are generally available to the home mechanic, so the following procedure is provided to enable you to extract any stored codes by using the OD OFF indicator light on the instrument cluster.

1993 and 1994 models

21 Start the engine and warm it up to its normal operating temperature.
22 Turn the ignition switch to the OFF position.
23 Place the A/T mode switch in the AUTO position.
24 Put the overdrive switch in the ON position.
25 Move the shift lever to the P position.
26 Turn the ignition switch to the ON position, but don't start the engine.
27 The OD OFF indicator light should come on for about two seconds.

   a) *If the OD OFF indicator light doesn't come on, there's something wrong with the transaxle computer or with the indicator light circuit. We don't recommend testing the resistance or voltage of the computer terminals; drawing too much current or putting too much voltage through the terminals can destroy the computer. Have the OD OFF indicator light circuit tested and repaired by a dealer service department or other qualified repair shop.*

b) If the OD OFF indicator light comes on, proceed to the next Step.
28 Turn the ignition key to OFF.
29 Move the shift lever to the D position.
30 Put the overdrive switch in the OFF position.
31 Turn the ignition switch to the ON position, but do not start the engine.
32 Wait for at least two seconds after the ignition switch is turned to ON, then move the shift lever to the 2 position.
33 Put the overdrive switch in the ON position.
34 Move the shift lever to the 1 position.
35 Put the overdrive switch in the OFF position.
36 Depress the accelerator pedal all the way to the floor, then release it.
37 The computer is now in its output mode. It will begin displaying any stored trouble code(s) by flashing the OD OFF indicator light in a sequence of long and short flashes specific to each stored code. Proceed to Step 71.

1995 through 1997 models

38 Start the engine and warm it up to its normal operating temperature.
39 Turn the ignition switch to the OFF position. Wait at least five seconds before proceeding.
40 Turn the ignition switch to the ACC position.
41 Put the overdrive switch in the ON position.
42 Move the shift lever to the P position.
43 Turn the ignition switch to the ON position, but don't start the engine.
44 The OD OFF indicator light should come on for about two seconds.

a) If the OD OFF indicator light doesn't come on, there's something wrong with the transaxle computer or with the indicator light circuit. We don't recommend testing the resistance or voltage of the computer terminals; drawing too much current or putting too much voltage through the terminals can destroy the computer. Have the OD OFF indicator light circuit tested and repaired by a dealer service department or other qualified repair shop.
b) If the OD OFF indicator light comes on, proceed to the next Step.

45 Turn the ignition key to OFF.
46 Turn the ignition switch to the ON position, but do not start the engine.
47 Move the shift lever to the D position.
48 Turn the ignition switch OFF.
49 Put the overdrive switch in the OFF position.
50 Turn the ignition switch ON, wait for at least two seconds, then move the shift lever to the 2 position.
51 Put the overdrive switch in the ON position.
52 Move the shift lever to the 1 position.
53 Put the overdrive switch in the OFF position.
54 Depress the accelerator pedal all the way to the floor, then release it.
55 The computer is now in its output mode. It will begin displaying any stored trouble code(s) by flashing the OD OFF indicator light in a sequence of long and short flashes specific to each stored code. Proceed to Step 71.

1998 and later models

56 Start the engine and warm it up to its normal operating temperature.
57 Turn the ignition switch to the OFF position. Wait at least five seconds before proceeding.
58 Move the shift lever to the P position.
59 Turn the ignition switch to the ON position, but don't start the engine.
60 The OD OFF indicator light should come on for about two seconds.

a) If the OD OFF indicator light doesn't come on, there's something wrong with the transaxle computer or with the indicator light circuit. We don't recommend testing the resistance or voltage of the computer terminals; drawing too much current or putting too much voltage through the terminals can destroy the computer. Have the OD OFF indicator light circuit tested and repaired by a dealer service department or other qualified repair shop.
b) If the OD OFF indicator light comes on, proceed to the next Step.

61 Turn the ignition key to OFF.
62 Turn the ignition switch to the ACC position.
63 Move the shift lever to the D position.
64 Turn the ignition switch to the ON position, but do not start the engine.
65 Push the OD switch in the OFF position and hold it there, then turn the ignition key OFF, then ON again. Release the OD switch (the light should go out).
66 Wait for at least two seconds after the ignition switch is turned to ON, then move the shift lever to the 2 position.
67 Push the overdrive switch in and hold it depressed, move the shift lever to the 1 position, then release the switch.
68 Put the overdrive switch in the OFF position - the light should go out.
69 Depress the accelerator pedal all the way to the floor, then release it.
70 The computer is now in its output mode. It will begin displaying any stored trouble code(s) by flashing the OD OFF indicator light in a sequence of long and short flashes specific to each stored code. Proceed to Step 71.

All models

71 If all monitored circuits are operating correctly, you will see one long flash, followed by one long pause, followed by ten short flashes of equal duration with short pauses of equal duration between them.
72 Each code is represented by the position of a long flash in the ten-consecutive-flash sequence:

a) If the first flash of the ten-flash sequence is long, the revolution sensor circuit is shorted or disconnected.
b) If the second of the ten flashes is long, the vehicle speed sensor circuit is shorted or disconnected.
c) If the third flash is long, the throttle position sensor circuit is shorted or disconnected.
d) If the fourth flash is long, the shift solenoid valve A circuit is shorted or disconnected.
e) If the fifth flash is long, the shift solenoid valve B circuit is shorted or disconnected.
f) If the sixth flash is long, the overrun clutch solenoid circuit is shorted or disconnected.
g) If the seventh flash is long, the torque converter clutch solenoid valve circuit is shorted or disconnected.
h) If the eighth flash is long, the fluid temperature sensor is disconnected or the computer power source circuit is damaged.
i) If the ninth flash is long, the engine speed signal circuit is shorted or disconnected.
j) If the tenth flash is long, the line pressure solenoid valve circuit is shorted or disconnected.
k) If the OD OFF indicator light flashes on and off, alternating back and forth between long flashes and long pauses, all of equal length, either the battery power is low, the battery has been disconnected for a long time, or the battery has been connected incorrectly.
l) 1998 and earlier models: If the OD OFF indicator light doesn't come on, either the neutral start/back-up light switch, or the overdrive switch or the throttle position switch circuit is disconnected, or the computer is damaged.
m) 1999 models: If the OD OFF indicator light comes on but doesn't flash, either the neutral start/back-up light switch, or the overdrive switch or the throttle position switch circuit is disconnected, or the computer is damaged.

73 Except for the vehicle speed sensor and throttle position sensor, both of which are covered in Chapter 6, repairing the rest of the malfunctions listed above is beyond the scope of the home mechanic. If one of more of these codes is displayed, have the transaxle repaired by a dealer service department or automatic transmission shop.

Erasing a diagnostic trouble code

74 If you have the transaxle repaired at a dealer or transmission shop, they will erase the trouble code when they're done making the repair. If you make a repair yourself, here's how to erase the code when you're done:
75 If the ignition switch remains on after a repair, turn it off once, wait at least five seconds, then turn it on again.
76 Perform the self-diagnostic procedure described above.
77 Change the diagnostic test mode from Mode II to Mode I by turning the mode selector on the ECM (see Chapter 6).

3 Shift cable - check, adjustment and replacement

Check

1 Move the shift lever from the "P" position to the "1" position. You should be able to feel the detents in each range. If you can't feel the detents, or if the pointer indicating the ranges is incorrectly aligned, adjust the shift cable.

Adjustment

Refer to illustrations 3.4 and 3.5

2 Place the shift lever in the "P" position.
3 Raise the vehicle and support it securely on jackstands.
4 On 1993 and 1994 models, loosen the locknuts that secure the cable to the shifter rod. On 1995 and later models, loosen the shift cable-to-manual lever locknut **(see illustration)**. place the manual shaft lever in the "P" position.
5 On 1993 and 1994 models, tighten the rear locknut until it contacts the shifter rod (just enough to take the slack out of the cable), then tighten the front locknut. On 1995 and later models, pull down on the cable **(see illustration)** and, holding the cable and manual shaft lever in this position, install and tighten the locknut.
6 Move the shift lever from "P" to "1" again. Make sure that it moves smoothly and quietly.
7 Remove the jackstands and lower the vehicle.

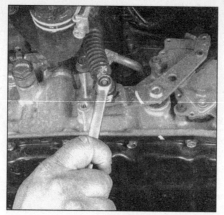

3.4 Loosen the shift cable locknut

3.5 Pull down on the shift cable and, holding the cable and manual shaft lever in this position, tighten the locknut (it's not really necessary to remove the locknut; it's removed in this photo so you can see the slotted adjustment hole in the end of the cable)

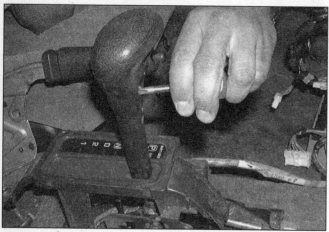

3.9a Remove the two Phillips screws in the front of the shift lever handle . . .

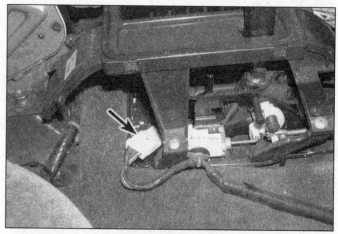

3.9b . . . unplug the electrical connector for the overdrive switch leads . . .

3.9c . . . disengage the overdrive switch leads from the connector with an awl . . .

3.9d . . . and carefully pull the shift lever handle straight up to remove it

Chapter 7 Part B Automatic transaxle

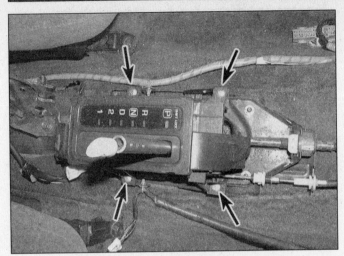

3.10 Remove the shift indicator panel screws

3.11 To disconnect the shift cable from the shift lever, remove this nut

Replacement

Refer to illustrations 3.9a, 3.9b, 3.9c, 3.9d, 3.10, 3.11, 3.12 and 3.15

Warning: *Some models covered by this manual are equipped with airbags. Always disable the airbag system when working in the vicinity of airbag system components (see Chapter 12).*

8 Remove the center console and the center dash trim bezel (see Chapter 11).
9 Remove the shift lever handle **(see illustrations)**.
10 Remove the shift lever indicator panel **(see illustration)**.
11 Disconnect the shift cable from the shift lever **(see illustration)**.
12 Disconnect the shift cable from the shift lever base **(see illustration)**.
13 Raise the vehicle and support it securely on jackstands.
14 Disconnect the shift cable from the manual shaft lever. On 1993 and 1994 models, it's retained by a clip. On 1995 and later models, remove the nut retaining the cable to

the lever **(see illustration 3.4)**.
15 Remove the C-clip from the shift cable bracket **(see illustration)** and disengage the cable from the bracket.
16 Installation is the reverse of removal. Be sure to adjust the cable when you're done.

4 Shift lever - removal and installation

Refer to illustration 4.3

Warning: *Some models covered by this manual are equipped with airbags. Always disable the airbag system when working in the vicinity of airbag system components (see Chapter 12).*

1 Refer to Section 3 and perform Steps 8 through 12.
2 Disconnect the key interlock cable from the shift lock solenoid and from its bracket at the front of the shifter base (see Section 5).
3 Remove the two nuts at the front of the shifter assembly **(see illustration)**.

3.12 To detach the shift cable from the shift lever base, pry off this retaining clip

4 Raise the vehicle and support it securely on jackstands.
5 Remove the two nuts that secure the rear of the shifter assembly to the floor.
6 From inside the vehicle, unplug any

3.15 To detach the shift cable from its bracket on the transaxle, remove this C-clip (arrow)

4.3 Remove these two nuts at the front of the shifter base, then, from under the vehicle, remove the two nuts securing the rear of the base to the floor

Chapter 7 Part B Automatic transaxle

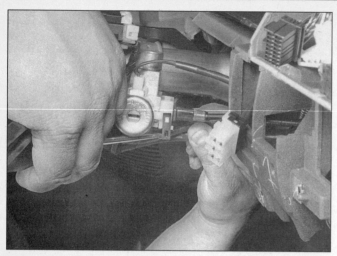

5.11a To disconnect the upper end of the key interlock cable from the key lock cylinder, remove the lock plate from the lock cylinder . . .

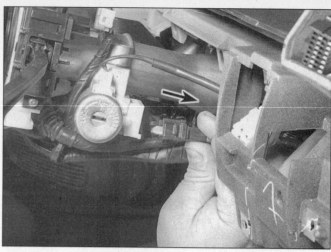

5.11b . . . and pull the cable out of the key lock cylinder

electrical connectors that may interfere with shifter removal, then detach the shifter assembly from the floor.

7   Installation is the reverse of removal. Be sure to adjust the shift cable (see Section 3).

5 Shift interlock system - description, check and component replacement

Warning: Some models covered by this manual are equipped with airbags. Always disable the airbag system when working in the vicinity of airbag system components (see Chapter 12).

Description

1   The shift lock system prevents the shift lever from being shifted out of Park or Neutral until the brake pedal is applied. Other than the following simple component checks, diagnosis of the shift lock system should be left to a dealer service department or other qualified repair shop.

Check

2   Remove the center console (see Chapter 11).
3   Follow the wiring harness from the shift lock solenoid back to the electrical connector, then unplug the connector **(see illustration 3.9b)**. Using a pair of jumper wires, momentarily apply battery voltage and ground to the solenoid terminals and verify that there's an audible "click." **Caution:** *Don't apply battery voltage any longer than necessary to perform this check.*
4   If the shift lock solenoid doesn't click when energized, replace it.

Component replacement

Shift lock solenoid

5   Remove the center console (see Chapter 11).
6   Remove the gear position indicator.
7   Unplug the electrical connector for the shift lock solenoid **(see illustration 5.3)**. Using a thin probe inserted into the front of the connector, disengage the terminal retaining tangs and pull the wires from the connector.
8   Remove the solenoid screws and remove the solenoid.
9   Installation is the reverse of removal.

Key interlock cable

Refer to illustrations 5.11a, 5.11b, 5.12 and 5.13

10 If the key interlock cable breaks, you'll have to remove the steering column cover and the center console to replace it (see Chapter 11).
11 Up at the key lock cylinder, remove the lock plate from the cylinder and disconnect the upper end of the shift lock cable **(see illustrations)**.

5.12 Pull the key interlock cable casing from the shift lever base, then detach the cable from the pin on the shift lever

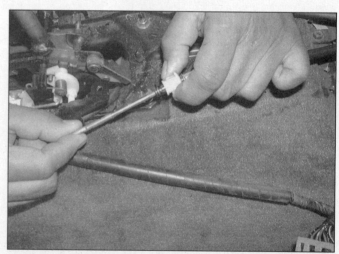

5.13 Unlock the slider from the adjuster holder by squeezing the two tabs on the slider and pull out the key interlock rod; to attach the key interlock rod to the new cable, insert it into the adjuster holder and push it in until it locks into place

Chapter 7 Part B Automatic transaxle

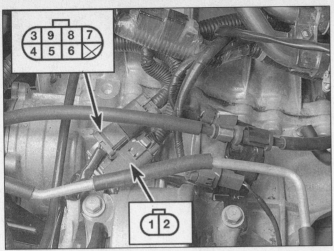

6.2 Unplug the electrical connectors for the neutral start/back-up light switch and check continuity at the indicated terminals

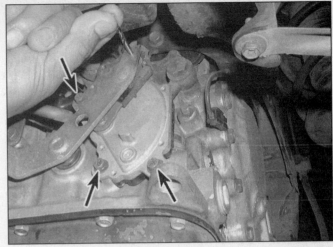

6.4 Disconnect the shift cable, loosen the switch retaining screws (arrows), and insert a 5/32-inch drill bit through the adjusting holes in the manual shaft lever and the switch

12 Down at the shift lever, detach the key interlock cable casing from the shift lever base **(see illustration)**, then remove the clip and disconnect it from the shift lever.
13 Unlock the slider from the adjuster holder and remove the key interlock rod; to insert the key interlock rod into the adjuster holder on the new cable, push it in until it locks into place **(see illustration)**.
14 Installation is otherwise the reverse of removal.

6 Park Neutral Position switch - check, adjustment and replacement

Check

Refer to illustration 6.2

1 If the engine will start with the shift lever in any position other than Park or Neutral, check and adjust the neutral start switch.
2 Unplug the switch electrical connectors and check continuity as follows **(see illustration)**:
   a) With the shift lever in Park, there should be continuity between terminals 1 and 2, and between terminals 3 and 4.
   b) With the shift lever in Reverse, there should be continuity between terminals 3 and 5.
   c) With the shift lever in Neutral, there should be continuity between terminals 1 and 2, and between terminals 3 and 6.
   d) With the shift lever in Drive, there should be continuity between terminals 3 and 7.
   e) With the shift lever in 2, there should be continuity between terminals 3 and 8.
   f) With the shift lever in 1, there should be continuity between terminals 3 and 9.
3 If the switch fails any of these continuity checks, disconnect the shift cable from the manual lever and retest the switch.
   a) If the switch passes all the continuity checks this time, reconnect the shift cable and adjust it, then retest the switch.
   b) If the switch still fails any of the continuity tests, remove it from the transaxle and try testing it again on the bench.
   c) If the switch passes all the continuity tests on the bench, install it and adjust it (see below).
   d) If the switch still fails any of the continuity tests, replace it (see below).

Adjustment

Refer to illustration 6.4

4 Disconnect the shift cable from the manual lever (see Section 3), loosen the switch retaining screws, set the manual lever at the Neutral position and insert a 0.16-inch pin (or a 5/32-inch drill bit) through the adjustment holes in both the manual shaft lever and the switch **(see illustration)**. Make sure the pin is perpendicular to the switch and the lever. Tighten the switch retaining screws securely.
5 Recheck the switch continuity as described above. If switch continuity is still not as specified, replace the switch.

7.5 To remove the left transaxle mount, remove the indicated fasteners (arrows)

Replacement

6 Disconnect the negative cable from the battery.
7 Shift the transaxle into Neutral.
8 Unplug the switch electrical connectors **(see illustration 6.2)**.
9 Remove the switch retaining screws **(see illustration 6.4)** and remove the switch.
10 Installation is the reverse of removal. Don't tighten the retaining screws until you have adjusted the switch as described in Step 4.

7 Automatic transaxle - removal and installation

Removal

Refer to illustrations 7.5, 7.10, 7.13 and 7.14

Note: *When removing the transaxle mounting bolts, record the position and length of each bolt so they can be returned to their original locations.*

1 Remove the battery and the battery tray (see Chapter 5).
2 Remove the air cleaner and resonator (see Chapter 4).
3 Unplug the electrical connectors for the automatic transaxle solenoid, the revolution sensor, the neutral start/back-up light switch and, on 1995 and later models, the crankshaft position sensor. Unplug the connector for the vehicle speed sensor (see Chapter 6). It's a good idea to remove the crankshaft position sensor so that it won't be damaged during transaxle removal.
4 Disconnect the vent hose.
5 Remove the left transaxle mount **(see illustration)**.
6 Remove the upper transaxle-to-engine bolts.
7 Remove the starter motor (see Chapter 5).
8 Loosen the wheel lug nuts. Raise the

Chapter 7 Part B Automatic transaxle

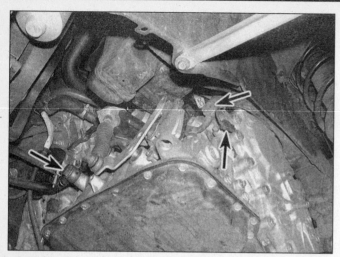

7.10 Disconnect the oil cooler lines (A) and the ground wire (B)

7.13 Remove the torque converter access cover bolts (arrows); the four big bolts (arrows) secure the engine oil pan casting to the transaxle

7.14 Mark the relationship of the torque converter to the driveplate, then remove the torque converter-to-driveplate bolts (arrow) by rotating the crankshaft to bring each bolt into the access hole

vehicle and support it securely on jackstands. Remove the wheels. Drain the transaxle/differential fluid (see Chapter 1).
9   Remove the dipstick tube.
10 Disconnect the oil cooler lines **(see illustration)**. Plug the fittings to keep out dirt and moisture. Disconnect the ground wire next to the left cooler line fitting.
11 Disconnect the shift cable from the transaxle (see Section 3).
12 Remove the driveaxles (see Chapter 8).
13 Remove the torque converter access cover **(see illustration)**.
14 Mark the relationship of the torque converter to the driveplate so they can be installed in the same position **(see illustration)**.
15 Remove the torque converter mounting bolts. Turn the crankshaft for access to each one in turn.
16 Support the engine using a hoist or an engine support fixture from above, or a jack and a wood block under the oil pan to spread the load.
17 Support the transaxle with a jack, preferably a special jack made for this purpose. Safety chains will help steady the transaxle on the jack.
18 Remove the center member (see Chapter 2).
19 Remove the oil pan-to-transaxle bolts **(see illustration 7.13)**. Remove the front transaxle-to-engine bolt and the rear engine-to-transaxle bolt.
20 Lower the transaxle slightly and make a final check for any wiring harnesses or lines that may still be connected.
21 Move the transaxle to the side to disengage it from the engine block dowel pins and make sure the torque converter is detached from the driveplate. Secure the torque converter to the transaxle so that it will not fall out during removal. Lower the transaxle from the vehicle.

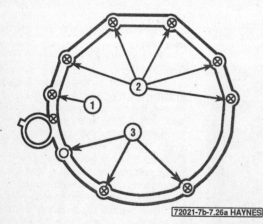

7.26a Transaxle mounting bolt location guide - 1993 models

1   2-3/8 inches (60mm)   3   1-inch (25 mm)
2   1-3/4 inch (45 mm)

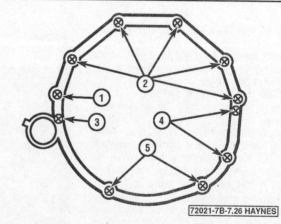

7.26b Transaxle mounting bolt location guide - 1994 models

1   2-3/8 inches (60mm)   4   13/16-inch (20 mm)
2   1-3/4 inch (45 mm)   5   1-1/8 inch (28 mm)
3   1-inch (25 mm)

Chapter 7 Part B Automatic transaxle

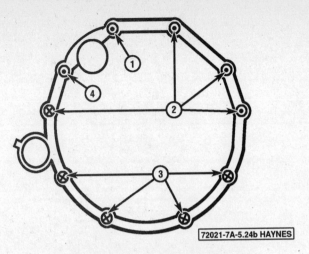

7.26c Transaxle mounting bolt location guide - 1995 through 1997 models

1 2-9/16 inches (65 mm)
2 2-1/16 inches (52 mm)
3 1-9/16 inch (40 mm)
4 4-7/8 inches (124 mm)

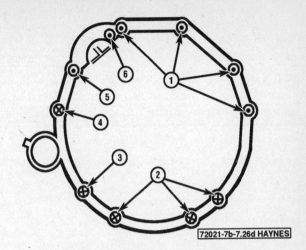

7.26d Transaxle mounting bolt location guide - 1998 and later models

1 1-3/4 inch (45 mm)
2 1-3/16 inch (30 mm)
3 1-9/16 inch (40 mm)
4 1-3/4 inch (45 mm)
5 3-5/32 inches (80 mm)
6 2-9/16 inches (65 mm)

Installation

Refer to illustrations 7.26a, 7.26b, 7.26c and 7.26d

22 Make sure that the torque converter is securely engaged in the transaxle prior to installation. Measure the distance from a converter bolt hole to the mating surface of the transaxle; it should be at least 3/4-inch (19 mm) on 1993 models, 23/32-inch on 1994 models, 9/16 inch (14 mm) on 1995 through 1997 models, and 3/4-inch (19 mm) on 1998 and later models. If it isn't, turn the torque converter while pushing it in until it "clunks" into place.

23 With the transaxle secured to the jack, raise it into position. Be sure to keep it level so the torque converter does not slide forward.

24 Move the transaxle carefully into place until the dowel pins are engaged and the torque converter is engaged.

25 Turn the torque converter to line up the bolt holes with the holes in the driveplate. The match marks on the torque converter and driveplate, made during step 14, must line up.

26 Install the transaxle mounting bolts **(see illustrations)**. Tighten them to the torque values listed in this Chapter's Specifications.

27 Install the torque converter-to-driveplate bolts. Tighten them to the torque listed in this Chapter's Specifications. Install the torque converter cover and tighten the bolts securely.

28 Install the center member and tighten the bolts securely.

29 Install the left transaxle mount. Tighten all bolts and nuts securely.

30 Remove the jacks supporting the transaxle and the engine.

31 Install the dipstick tube.

32 Install the starter motor (see Chapter 5).

33 Connect the shift cable (see Section 3).

34 Plug in the electrical connectors for the automatic transaxle solenoid, revolution sensor, vehicle speed sensor, neutral start/back-up light switch and, on 1995 and later models, the crankshaft position sensor.

35 Connect the driveaxles to the transaxle (see Chapter 8).

36 Remove the jackstands and lower the vehicle.

37 Fill the transaxle with the proper type and amount of fluid (see Chapter 1). Run the vehicle and check for fluid leaks.

Notes

Chapter 8
Clutch and driveaxles

Contents

	Section		Section
Clutch - description and check	2	Clutch start switch - check and replacement	9
Clutch components - removal, inspection and installation	6	Driveaxle boot - replacement	13
Clutch hydraulic system - bleeding	5	Driveaxle boot check	See Chapter 1
Clutch master cylinder - removal and installation	3	Driveaxles - general information and inspection	10
Clutch pedal - adjustment	8	Driveaxles - removal and installation	11
Clutch release bearing and lever - removal, inspection and installation	7	Flywheel - removal and installation	See Chapter 2
		General information	1
Clutch release cylinder - removal and installation	4	Support bearing assembly - check and replacement	12

Specifications

Clutch pedal
Pedal height
  1993 through 1999 models .... 168 to 175 mm (6 5/8 to 6 7/8 inches)
  2000 and later models .... 179-190 mm (7 to 7-1/2 inches)
Pedal freeplay
  1996 and earlier .... 1 to 3 mm (3/64 to 1/8 inch)
  1997 and later .... 9 to 16 mm (11/32 to 5/8 inch)

CV joint boot length
Inner joint .... 97 to 99 mm (3-13/16 to 3-29/32 inches)
Outer joint
  1993 through 1999 models .... 96 to 98 mm (3-25/32 to 3-7/8 inches)
  2000 and later models .... 103 mm (4-1/16 inches)

Torque specifications

	Nm	Ft-lbs (unless otherwise indicated)
Clutch master cylinder retaining nuts	8 to 11	69 to 96 in-lbs
Clutch release cylinder retaining bolts	30 to 40	22 to 30
Clutch hose-to-release cylinder banjo bolt	17 to 19	144 to 168 in-lbs
Clutch pressure plate bolts	34 to 44	25 to 33
Support bearing-to-bracket bolts	13 to 19	108 to 168 in-lbs
Driveaxle/hub nut	235 to 314	174 to 231

1 General information

The information in this Chapter deals with the components from the rear of the engine to the front wheels, except for the transaxle, which is dealt with in Chapter 7A and 7B. For the purposes of this Chapter, these components are grouped into two categories: Clutch and driveaxles. Separate Sections within this Chapter offer general descriptions and checking procedures for both groups.

Since nearly all the procedures covered in this Chapter involve working under the vehicle, make sure it's securely supported on sturdy jackstands or a hoist where the vehicle can be easily raised and lowered.

2 Clutch - description and check

1 All vehicles with a manual transaxle use a single dry plate, diaphragm spring type clutch. The clutch disc has a splined hub which allows it to slide along the splines of the transaxle input shaft. The clutch and pressure plate are held in contact by spring pressure exerted by the diaphragm in the pressure plate.

2 The clutch release system is hydraulically operated. The release system consists of the clutch pedal, the clutch master cylinder, the clutch release cylinder, the hydraulic line between the master cylinder and release cylinder, and the clutch release bearing.

3 When pressure is applied to the clutch pedal to release the clutch, the clutch master cylinder transmits this movement to the clutch release cylinder, which moves the clutch release lever. As the lever pivots, the shaft fingers push against the release bearing. The bearing pushes against the fingers of the diaphragm spring of the pressure plate assembly, which in turn releases the clutch plate.

4 Terminology can be a problem regarding the clutch components because common names have in some cases changed from that used by the manufacturer. For example, the clutch release cylinder is sometimes referred to as a slave cylinder, the driven plate is also called the clutch plate or disc, the pressure plate assembly is also known as the clutch cover, and the clutch release bearing is sometimes called a throw-out bearing.

5 Other than replacing components that have obvious damage, some preliminary checks should be performed to diagnose a clutch system failure.

a) Before proceeding, check and, if necessary, adjust clutch pedal freeplay and height (see Section 8).
b) To check clutch "spin down" time, run the engine at normal idle speed with the transaxle in Neutral (clutch pedal up, engaged). Disengage the clutch (pedal down), wait several seconds and shift the transaxle into Reverse. No grinding noise should be heard. A grinding noise would most likely indicate a problem in the pressure plate or the clutch disc.
c) To check for complete clutch release, run the engine (with the parking brake applied to prevent movement) and hold the clutch pedal approximately 1/2-inch from the floor. Shift the transaxle between 1st gear and Reverse several times. If the shift is not smooth, component failure is indicated.
d) Visually inspect the clutch pedal bushing at the top of the clutch pedal to make sure there is no sticking or excessive wear.
e) Under the vehicle, verify that the clutch release lever is solidly mounted on the ball stud.
f) Make sure that the hydraulic lines aren't leaking at either the master cylinder or the release cylinder (see Sections 3 and 4). Bleed the system if necessary (see Section 5).

3.2 To disconnect the clutch master cylinder pushrod from the clutch pedal, remove this retaining clip and clevis pin (arrow)

3 Clutch master cylinder - removal and installation

Removal

Refer to illustrations 3.2, 3.3a, 3.3b, 3.3c and 3.4

1 Disconnect the negative cable from the battery.

2 Under the dashboard, disconnect the pushrod from the clutch pedal. It's held in place with a clevis pin. To remove the clevis pin, remove the clip (see illustration).

3 Remove the ASCD (cruise control) pump (see illustrations). Disconnect the hydraulic line at the clutch master cylinder (see illustration). Use a flare-nut wrench on the fitting to protect it from being rounded off. Have rags handy, some fluid will be spilled when the line is disconnected. Caution: Don't allow brake fluid to spill onto the paint; it will damage the finish. If you do, rinse it off immediately with clean water.

3.3a To gain access to the hydraulic line and mounting nuts for the clutch master cylinder, remove this screw (arrow) that secures the Automatic Speed Control Device (ASCD) pump at the top . . .

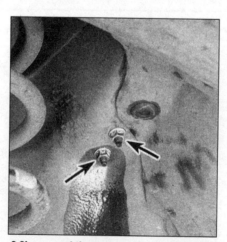

3.3b . . . and these two nuts (arrows) that secure the ASCD pump inside the wheelhousing

3.3c Unscrew the hydraulic line fitting from the clutch master cylinder with a flare-nut wrench to protect the fitting nut

Chapter 8 Clutch and driveaxles

3.4 To detach the clutch master cylinder from the firewall, remove these two nuts (arrows)

5.3 To bleed the clutch hydraulic release system, push one end of a clear hose over the bleeder screw on the release cylinder (arrow) and submerge the other end of the hose in a container of clean brake fluid

4   Remove the nuts (see illustration) that attach the master cylinder to the firewall. As you remove the master cylinder, make sure you don't spill any fluid.

Installation

5   Position the master cylinder on the firewall and install the mounting nuts, but don't tighten them yet.
6   Connect the hydraulic line to the master cylinder, moving the cylinder slightly as necessary to thread the fitting properly into the bore. Don't cross-thread the fitting as it's installed.
7   Tighten the master cylinder mounting nuts and the hydraulic line fitting securely.
8   Connect the pushrod to the clutch pedal.
9   Fill the clutch master cylinder reservoir with brake fluid conforming to DOT 3 specifications and bleed the clutch system (see Section 5).
10 Check the clutch pedal height and freeplay (see Section 8).

4 Clutch release cylinder - removal and installation

Removal

1   Disconnect the negative cable from the battery.
2   If necessary for access, remove the battery (see Chapter 5).
3   Loosen the clutch hydraulic line banjo bolt. Remove the release cylinder mounting bolts.
4   Disconnect the hydraulic line at the release cylinder. Have a small can and rags handy, some fluid will be spilled when the line is disconnected.
5   Remove the release cylinder.

Installation

6   Connect the hydraulic line to the release cylinder before bolting the release cylinder in place. Use new sealing washers. Tighten the banjo bolt finger tight, but don't torque it until the release cylinder is securely bolted in place.
7   Install the release cylinder on the transaxle. Make sure the pushrod is seated in the release lever pocket. Install the retaining bolts and tighten them to the torque listed in this Chapter's Specifications.
8   Tighten the hydraulic line banjo bolt to the torque listed in this Chapter's Specifications.
9   Fill the clutch master cylinder with brake fluid (conforming to DOT 3 specifications).
10 Bleed the system (see Section 5).
11 If removed, reinstall the battery (see Chapter 5).

5 Clutch hydraulic system - bleeding

Refer to illustration 5.3

1   The hydraulic system should be bled of all air whenever any part of the system has been removed or if the fluid level has been allowed to fall so low that air has been drawn into the master cylinder. The procedure is similar to bleeding a brake system.
2   Fill the master cylinder with new brake fluid conforming to DOT 3 specifications. **Caution:** *Do not re-use any of the fluid coming from the system during the bleeding operation or use fluid which has been inside an open container for an extended period of time.*
3   Locate the bleeder valve on the clutch release cylinder (see illustration). Remove the dust cap which fits over the bleeder valve and push a length of snug-fitting (preferably clear) hose over the valve. Place the other end of the hose into a clear container with about two inches of brake fluid in it. The hose end must be submerged in the fluid.
4   Have an assistant depress the clutch pedal and hold it. Open the bleeder valve on the release cylinder, allowing fluid to flow through the hose. Close the bleeder valve when fluid stops flowing from the hose. Once closed, have your assistant release the pedal.
5   Continue this process until all air is evacuated from the system, indicated by a full, solid stream of fluid being ejected from the bleeder valve each time and no air bubbles in the hose or container. Keep a close watch on the fluid level inside the clutch master cylinder reservoir; if the level drops too low, air will be sucked back into the system and the process will have to be started over again.
6   Install the dust cap on the bleeder valve. Check carefully for proper operation before placing the vehicle in normal service.

6 Clutch components - removal, inspection and installation

Warning: *Dust produced by clutch wear and deposited on clutch components may contain asbestos, which is hazardous to your health. DO NOT blow it out with compressed air and DO NOT inhale it. DO NOT use gasoline or petroleum-based solvents to remove the dust. Brake system cleaner should be used to flush the dust into a drain pan. After the clutch components are wiped clean with a rag, dispose of the contaminated rags and cleaner in a labeled, covered container.*

Removal

Refer to illustrations 6.5 and 6.7

1   Access to the clutch components is normally accomplished by removing the transaxle, leaving the engine in the vehicle. If, of course, the engine is being removed for major overhaul, then the opportunity should always be taken to check the clutch for wear and replace worn components as necessary. However, the relatively low cost of the clutch components compared to the time and labor involved in gaining access to them warrants their replacement any time the engine or transaxle is removed, unless they are new or in near-perfect condition. The following procedures assume that the engine will stay in place.
2   Remove the transaxle from the vehicle (see Chapter 7A). Support the engine while the transaxle is out. Preferably, an engine hoist or support fixture should be used to support it from above. However, if a jack is used underneath the engine, make sure a piece of wood is used between the jack and oil pan to spread the load. **Caution:** *The pickup for the oil pump is very close to the bottom of the oil pan. If the pan is bent or distorted in any way, engine oil starvation could occur.*
3   The release fork and release bearing can remain attached to the transaxle; however, you should inspect them (see Section 7) while the transaxle is removed.
4   To support the clutch disc during removal, install a clutch alignment tool through the clutch disc hub.

Chapter 8 Clutch and driveaxles

6.5 Mark the relationship of the pressure plate to the flywheel (in case you're going to re-use the same pressure plate)

6.7 Remove the pressure plate and the clutch disc

5 Carefully inspect the flywheel and pressure plate for indexing marks. The marks are usually an X, an O or a white letter. If they cannot be found, scribe marks yourself so the pressure plate and the flywheel will be in the same alignment during installation **(see illustration)**.

6 Slowly loosen the pressure plate-to-flywheel bolts. Work in a criss-cross pattern and loosen each bolt a little at a time until all spring pressure is relieved.

7 Hold the pressure plate securely and completely remove the bolts, followed by the pressure plate and clutch disc **(see illustration)**. Use a small slide hammer to remove the pilot bearing **(see illustration)**.

Inspection

Refer to illustrations 6.10, 6.12a and 6.12b

8 Ordinarily, when a problem occurs in the clutch, it can be attributed to wear of the clutch driven plate assembly (clutch disc). However, all components should be inspected at this time.

9 Inspect the flywheel for cracks, heat checking, score marks and other damage. If the imperfections are slight, a machine shop can resurface it to make it flat and smooth. Refer to Chapter 2 for the flywheel removal procedure.

10 Inspect the lining on the clutch disc. There should be at least 1/16-inch of lining above the rivet heads. Check for loose rivets, distortion, cracks, broken springs and other obvious damage **(see illustration)**. As mentioned above, ordinarily the clutch disc is replaced as a matter of course, so if in doubt about the condition, replace it with a new one.

11 The release bearing should be replaced along with the clutch disc (see Section 7).

12 Check the machined surface and the diaphragm spring fingers of the pressure plate **(see illustrations)**. If the surface is grooved or otherwise damaged, replace the pressure plate assembly. Also check for obvious damage, distortion, cracking, etc. Light glazing can be removed with emery cloth or sandpaper. If a new pressure plate is indicated, new or factory rebuilt units are available.

Installation

Refer to illustration 6.13

13 Position the clutch disc and pressure plate with the clutch held in place with an alignment tool **(see illustration)**. Make sure it's installed properly (most replacement clutch plates will be marked "flywheel side"

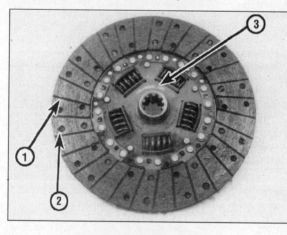

6.10 The clutch disc
1 **Lining** - this will wear down in use
2 **Rivets** - these secure the lining and will damage the flywheel or pressure plate if allowed to contact the surfaces
3 **Marks** - "Flywheel side" or something similar

NORMAL FINGER WEAR

EXCESSIVE FINGER WEAR

BROKEN OR BENT FINGERS

6.12a Replace the pressure plate if any of these conditions are noted

Chapter 8 Clutch and driveaxles

6.12b Examine the pressure plate friction surface for score marks, cracks and evidence of overheating (blue spots)

6.13 Center the clutch disc in the pressure plate with a clutch alignment tool, then tighten the pressure plate-to-flywheel bolts in a criss-cross pattern

7.3 Disengage the clutch release lever from the ball stud, then remove the release bearing and release lever

7.4 To check the bearing, hold it by the outer race and rotate the inner race while applying pressure; if the bearing doesn't turn smoothly or if it's noisy, replace it

7.5 Apply a light coat of high-temperature grease to the surface of the input shaft bearing retainer (before installing the transaxle, apply a light film of grease to the input shaft splines)

or something similar, if not marked, install the clutch disc with the damper springs or cushion toward the transaxle).

14 Tighten the pressure plate-to-flywheel bolts only finger tight, working around the pressure plate.

15 Center the clutch disc by ensuring the alignment tool is through the splined hub and into the recess in the crankshaft. Wiggle the tool up, down or side-to-side as needed to bottom the tool. Tighten the pressure plate-to-flywheel bolts a little at a time, working in a criss-cross pattern to prevent distortion of the cover. After all of the bolts are snug, tighten them to the torque listed in this Chapter's Specifications. Remove the alignment tool.

16 Using high-temperature grease, lubricate the inner groove of the release bearing (see Section 7). Also place grease on the release lever contact areas and the transaxle input shaft.

17 Install the clutch release bearing (see Section 7A).

18 Install the transaxle and all components removed previously, tightening all fasteners to the proper torque specifications.

7 Clutch release bearing and lever - removal, inspection and installation

Warning: *Dust produced by clutch wear and deposited on clutch components may contain asbestos, which is hazardous to your health. DO NOT blow it out with compressed air and DO NOT inhale it. DO NOT use gasoline or petroleum-based solvents to remove the dust. Brake system cleaner should be used to flush it into a drain pan. After the clutch components are wiped clean with a rag, dispose of the contaminated rags and cleaner in a labeled, covered container.*

Removal
Refer to illustration 7.3

1 Disconnect the negative cable from the battery.
2 Remove the transaxle (see Chapter 7A).
3 Disengage the clutch release lever from the ball stud, then remove the bearing and lever (see illustration).

Inspection
Refer to illustration 7.4

4 Hold the bearing by the outer race and rotate the inner race while applying pressure (see illustration). If the bearing doesn't turn smoothly or if it's noisy, replace the bearing/hub assembly with a new one. Wipe the bearing with a clean rag and inspect it for damage, wear and cracks. Don't immerse the bearing in solvent; it's sealed for life and to do so would ruin it. Also check the release lever for cracks and bends.

Installation
Refer to illustrations 7.5, 7.6a, 7.6b, 7.6c, 7.7a and 7.7b

5 Lubricate the surface of the input shaft bearing retainer with high-temperature grease (see illustration).
6 Lubricate the release lever ball socket,

Chapter 8 Clutch and driveaxles

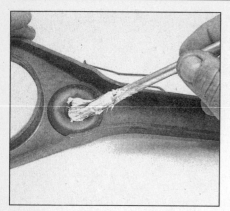

7.6a Using high-temperature grease, lubricate the ball stud socket in the back of the release lever . . .

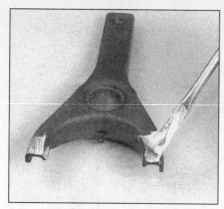

7.6b . . . the lever ends, the depression for the release cylinder pushrod . . .

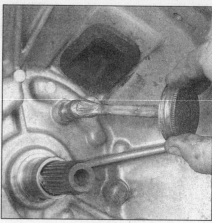

7.6c . . . and the ball stud

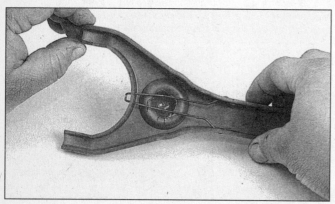

7.7a Install the retainer spring onto the release lever

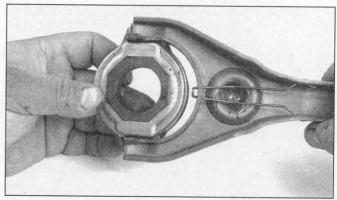

7.7b Slide the release bearing onto the release lever, engaging the clips on the ends

lever ends and release cylinder pushrod socket with high-temperature grease **(see illustrations)**.

7 Attach the retainer spring and the release bearing to the release lever **(see illustrations)**.

8 Slide the release bearing onto the transaxle input shaft bearing retainer while passing the end of the release lever through the opening in the clutch housing. Push the release lever onto the ball stud until it's firmly seated.

9 Apply a light coat of high-temperature grease to the face of the release bearing where it contacts the pressure plate diaphragm fingers.

10 The remainder of installation is the reverse of removal.

8 Clutch pedal - adjustment

Pedal height

Refer to illustrations 8.1 and 8.3

1 With the clutch pedal fully released, measure the distance from the top of the pad to the floor **(see illustration)**.

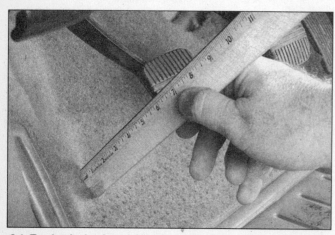

8.1 To check clutch pedal height, measure the distance between the pedal pad and the floor. To check pedal freeplay, measure the distance between the natural resting place of the pedal and the point at which you encounter resistance

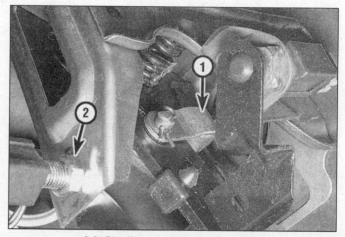

8.3 Clutch pedal adjustment details

1 Pushrod
2 Pedal stopper or cruise control switch

Chapter 8 Clutch and driveaxles

11.3 Loosen the driveaxle/hub nut while the wheel is still on the ground

11.5 Remove the driveaxle/hub nut and washer. To prevent the hub from turning, wedge a prybar between two of the wheel studs and allow the prybar to rest against the ground or the floorpan of the vehicle

2 If the height is not as listed in the Specifications at the beginning of this Chapter it must be adjusted.
3 Loosen the locknut on the pedal stopper or cruise control (ASCD) switch **(see illustration)**.
4 Turn the pedal stopper or cruise control switch until the pedal height is correct.
5 Tighten the locknut and recheck the clutch pedal height.

Pedal freeplay

6 Press down lightly on the clutch pedal and measure the distance that it moves freely before the clutch resistance is felt **(see illustration 8.1)**. The freeplay should be within the specified limits listed at the beginning of this Chapter. If it isn't, it must be adjusted.
7 Loosen the locknut on the master cylinder pushrod **(see illustration 8.3)** and turn the rod clockwise or counterclockwise to adjust the clutch pedal freeplay.
8 Tighten the locknut and recheck the clutch pedal freeplay.

9 Clutch start switch - check and replacement

1 The clutch start switch is located near the top of the clutch pedal, facing the opposite direction of the cruise control switch (or pedal stopper), shown in illustration 8.3.
2 Verify that the engine will not start when the clutch pedal is released.
3 Verify that the engine will start when the clutch pedal is depressed all the way.
4 If the clutch start switch doesn't perform as described above, adjust the clutch pedal (see Section 8). The switch should now operate properly. If it doesn't, check switch continuity.
5 Verify that there is continuity between the clutch start switch terminals when the pedal is depressed.
6 Verify that no continuity exists between the switch terminals when the pedal is released.
7 If the switch fails either of these continuity tests, replace it: Loosen the nut near the body of the switch, then unscrew the switch. Unplug the electrical connector. Installation is the reverse of removal.
8 Adjust the clutch pedal (see Section 8).
9 Verify that the engine doesn't start when the clutch pedal is released.

10 Driveaxles - general information and inspection

1 Power is transmitted from the transaxle to the wheels through a pair of driveaxles. The inner end of each driveaxle is splined into the differential side gears. The outer ends of the driveaxles are splined to the axle hubs and locked in place by a large nut.
2 The inner ends of the driveaxles are equipped with sliding constant velocity joints, which are capable of both angular and axial motion. The inner joints are the "double-offset" type, which consists of ball bearings running between an inner race and an outer cage. They can also be disassembled, cleaned and inspected, but they must be replaced as a single unit if defective.
3 The outer CV joints are the "Rzeppa" (pronounced "sheppa") or "Birfield" type, which also consists of ball bearings running between an inner race and an outer cage. However, Rzeppa/Birfield joints are capable of angular, but not axial, movement. These outer joints should be cleaned, inspected and repacked whenever replacing an outer CV joint boot, but they cannot be disassembled. If an outer joint is damaged, it must be replaced.
4 The boots should be inspected periodically for damage and leaking lubricant. Torn CV joint boots must be replaced immediately or the joints can be damaged. Boot replacement involves removal of the driveaxle (see Section 11). **Note:** *Some auto parts stores carry "split" type replacement boots, which can be installed without removing the driveaxle from the vehicle. This is a convenient alternative; however, the driveaxle should be removed and the CV joint disassembled and cleaned to ensure the joint is free from contaminants such as moisture and dirt which will accelerate CV joint wear. The most common symptom of worn or damaged CV joints, besides lubricant leaks, is a clicking noise in turns, a clunk when accelerating after coasting and vibration at highway speeds. To check for wear in the CV joints and driveaxle shafts, grasp each axle (one at a time) and rotate it in both directions while holding the CV joint housings, feeling for play indicating worn splines or sloppy CV joints. Also check the driveaxle shafts for cracks, dents and distortion.*

11 Driveaxles - removal and installation

Removal

Refer to illustrations 11.3, 11.5, 11.6, 11.9, 11.10a, 11.10b and 11.10c
1 Remove the wheel cover or hub cap.
2 Remove the cotter pin, the nut lock and the felt washer from the driveaxle/hub nut.
3 Break the hub nut loose with a socket and large breaker bar **(see illustration)**. **Note:** *If the socket won't fit through the opening in the center of the wheel, remove the wheel and install the spare.*
4 Loosen the wheel lug nuts, raise the vehicle and support it securely on jackstands. Remove the wheel.
5 Remove the driveaxle/hub nut and washer. If the hub turns, wedge a prybar between two of the wheel studs and allow the prybar to rest against the ground or the floorpan of the vehicle **(see illustration)**.
6 If the driveaxle splines are "frozen," free

11.6 If the driveaxle splines are "frozen," knock the driveaxle loose with a hammer and brass punch

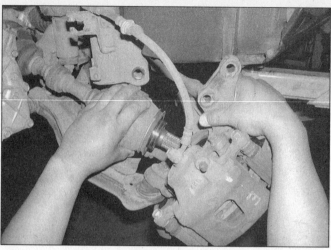

11.9 Pull out on the steering knuckle and detach the driveaxle from the hub

11.10a If you're removing the left driveaxle, carefully pry the inner CV joint out of the transaxle and remove the driveaxle assembly

11.10b If you're removing the right driveaxle, remove these three bolts (arrows) (third bolt not visible in this photo) . . .

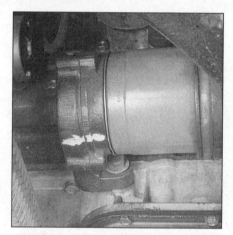

11.10c . . . mark the relationship of the bearing retainer to the support bracket and remove the driveaxle assembly

them by tapping the end of the driveaxle with a soft-faced hammer or a hammer and a brass punch **(see illustration)**.

7 Remove the engine splash shields. Place a drain pan underneath the transaxle to catch the lubricant that may spill out when the driveaxles are removed.

8 Remove the clip that secures the brake hose to the strut, then disconnect the strut from the steering knuckle (see Chapter 10).

9 Pull out on the steering knuckle and detach the driveaxle from the hub **(see illustration)**. Be careful not to pull on the brake hose. **Caution:** *If the brake hose prevents you from pulling the knuckle out far enough for driveaxle removal, remove the caliper (don't disconnect the hose) and hang it with a piece of wire from the coil spring.*

10 If you're removing a left driveaxle, carefully pry the inner CV joint out of the transaxle **(see illustration)** and remove the driveaxle assembly. The inner CV joint housing on the right (passenger side) driveaxle terminates at a support bracket. To detach the right driveaxle assembly from the bracket, remove the three bracket-to-bearing bolts, mark the relationship of the bearing to the support bracket and pull out the driveaxle assembly **(see illustrations)**. **Note:** *When removing the left (driver's) side driveaxle on models with an automatic transaxle, it may not be possible to pry the inner CV joint out of the transaxle. In this case, it will be necessary to remove the right side driveaxle and insert a screwdriver through the differential side gears and knock the left side shaft free.*

11 Install a new driveaxle oil seal (see Chapter 7).

Installation

Refer to illustration 11.12

12 Installation is the reverse of the removal procedure, but with the following additional points:

a) When installing the left driveaxle, be sure to fit a new retaining clip on the splines of the inner CV joint **(see illustration)**. Push the driveaxle in sharply to seat the retaining ring on the inner CV joint in its groove in the differential side gear (the right driveaxle assembly has no retaining ring, it's secured by the three bolts which attach it to the support bracket).

b) If you're installing the right driveaxle, be sure to tighten the support bearing retainer bolts to the torque listed in this Chapter's Specifications.

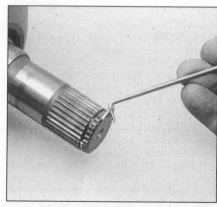

11.12 Always replace the retaining ring on the inner CV joint stub shaft

Chapter 8 Clutch and driveaxles

13.2a Cut off the old boot clamps with a pair of diagonal cutters

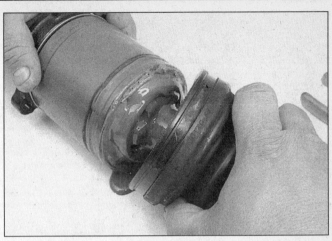

13.2b Slide the boot back and wipe off as much of the old grease as possible

13.3 Mark the shaft, inner race, cage and outer race (housing) so they can be reassembled in the original relationship to each other

    d) *Tighten the driveaxle/hub nut to the torque listed in this Chapter's Specifications, then install the felt washer, nut lock and a new cotter pin.*
    e) *Check the transaxle or differential lubricant and add, if necessary, to bring it to the proper level (see Chapter 1).*

12 Support bearing assembly - check and replacement

Note: *This procedure applies to the support bearing for the right driveaxle assembly.*

1   Remove the right driveaxle assembly (see Section 11).
2   Listen to the bearing while rotating it. It should operate smoothly and quietly.
3   If the bearing is rough or noisy, take the driveaxle assembly, a new bearing and three new dust shields (small one on the end of the extension shaft and two larger ones, on either side of the bearing) to an automotive machine shop. They will have the right tools to press off the old bearing and press the new one on.
4   Install the driveaxle (see Section 11).

    c) *Install the wheel and lug nuts, lower the vehicle and tighten the lug nuts to the torque listed in the Chapter 1 Specifications.*

13 Driveaxle boot - replacement

Note: *If the CV joints must be overhauled (usually due to torn boots), explore all options before beginning the job. Complete rebuilt driveaxles are available on an exchange basis, which eliminates much time and work. Whichever route you choose to take, check on the cost and availability of parts before disassembling the vehicle.*

All units

1   Remove the driveaxle (see Section 11).

Inner CV joint

Disassembly

Refer to illustrations 13.2a, 13.2b, 13.3, 13.4, 13.5, 13.6 and 13.9

2   Remove both boot clamps and discard them, then slide the boot out of the way **(see illustrations).**
3   Mark the shaft, the inner race, the cage and the outer race (housing) so they can be reassembled in the original position **(see illustration).**
4   Pry the wire ring bearing retainer from the housing **(see illustration).**
5   Pull the housing off the inner bearing

13.4 Pry the retainer from the housing with a small screwdriver

13.5 Slide the housing off the bearing assembly; some of the bearings may fall out when the race is removed, so be ready to catch them

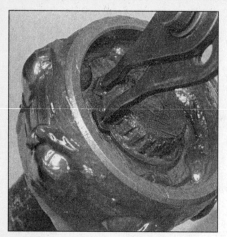

13.6 Remove the snap-ring from the groove in the axleshaft with a pair of snap-ring pliers

13.9 Remove the stop ring for the inner race

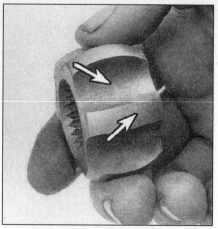

13.10a Inspect the inner race lands and grooves for pitting, score marks, cracks and other signs of wear and damage

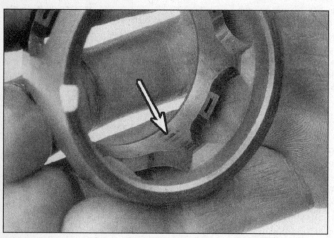

13.10b Inspect the cage for cracks, pitting and score marks (shiny, polished spots are normal and will not adversely affect CV joint performance

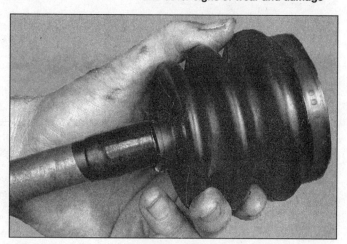

13.11 Wrap the axleshaft splines with tape to protect the boot, then slide the small boot clamp and boot onto the axleshaft and remove the tape

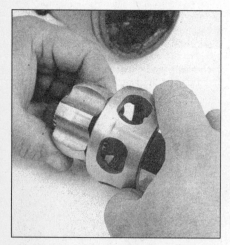

13.13 Install the stop ring, then slide the cage and the inner race onto the shaft splines until the inner race butts against the stop ring (make sure you have the cage on first - it can't be installed after the inner race is installed because it won't fit over the inner race)

assembly **(see illustration)**.

6 Remove the snap-ring from the groove in the axleshaft with a pair of snap-ring pliers **(see illustration)**.

7 Slide the inner race off the axleshaft. If the splines are stuck, apply some penetrant and give it a few careful taps with a brass hammer or a hammer and brass punch.

8 Using a screwdriver or piece of wood, pry the ball bearings from the cage. Be careful not to scratch the inner race, the ball bearings or the cage. Remove the cage.

9 Remove the stop ring for the inner race **(see illustration)**. Remove the boot from the shaft.

Inspection

Refer to illustrations 13.10a and 13.10b

10 Clean the components with solvent to remove all traces of grease. Inspect the cage and races for pitting, score marks, cracks and other signs of wear and damage. Shiny, polished spots are normal and will not adversely affect CV joint performance **(see illustrations)**.

Reassembly

Refer to illustrations 13.11, 13.13, 13.16, 13.17a, 13.17b, 13.18, 13.19a, 13.19b, 13.20a, 13.20b, 13.20c, 13.20d and 13.20e

11 Wrap the axleshaft splines with tape to avoid damaging the boot. Slide the small boot clamp and boot onto the axleshaft **(see illustration)**, then remove the tape. Slide the large boot clamp over the boot.

12 Install the cage on the axleshaft with the smaller diameter side of the cage facing toward the center of the shaft.

13 Install a new stop ring for the inner race **(see illustration 13.9)**, then install the inner race onto the axleshaft **(see illustration)**, with the matchmark on the race (or the larger diameter side) aligned with the mark on the end of the axleshaft.

14 Install the snap-ring that retains the inner race. Make sure it's completely seated in its groove by trying to push the inner race off the shaft.

15 Move the cage up over the inner race, aligning the match marks. Press the ball bearings into the cage windows with your

Chapter 8 Clutch and driveaxles

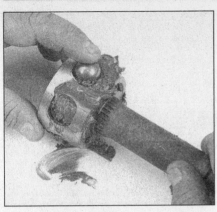

13.16 Pack the inner race and cage assembly with grease, by hand, until grease is worked completely into the assembly

13.17a Slide the housing over the assembled inner CV joint - make sure none of the balls fall out

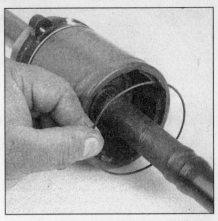

13.17b Install the retainer ring, making sure it seats in its groove

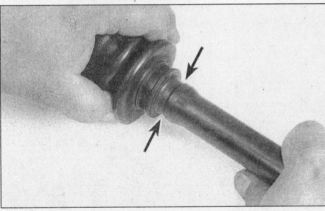

13.18 Make sure the small diameter end of the boot seats in this groove (arrows)

13.19a Adjust the length of the CV joint to the dimension listed in this Chapter's Specifications . . .

thumbs. If they won't stay in place, apply CV joint grease to hold them.

16   Fill the outer race and boot with CV joint grease (normally included with the new boot kit). Pack the inner race and cage assembly with grease, by hand, until grease is worked completely into the assembly **(see illustration)**.

17   Slide the inner race, balls and cage into the CV joint housing and install the wire ring bearing retainer **(see illustrations)**.

18   Wipe any excess grease from the axle boot groove on the outer race. Seat the small diameter of the boot in the recessed area on the axleshaft **(see illustration)**. Push the other end of the boot onto the CV joint housing and move the race in or out until there's no deformation (distortion or dents) in the boot.

19   Adjust the length of the CV joint (measured from one end of the boot to the other) **(see illustration)**. Equalize the pressure in the boot by inserting a dull screwdriver between the boot and the outer race **(see illustration)**. Don't damage the boot with the tool.

20   Install the boot clamps **(see illustrations)**.

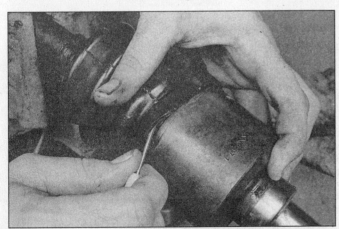

13.19b . . . then equalize the pressure inside the boot with atmospheric pressure by inserting a *dull* screwdriver between the boot and the outer race; make sure you don't damage the boot

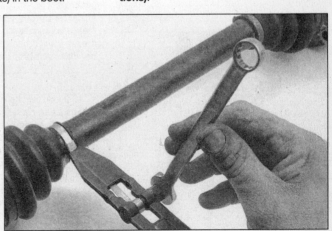

13.20a You'll need a special tightening tool to install "band" type boot clamps: Install the band with its end pointing in the direction of axle rotation and tighten it securely . . .

8-12   Chapter 8   Clutch and driveaxles

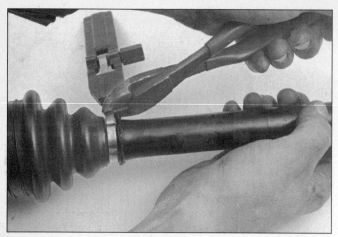

13.20b . . . then bend down the end of the clamp back and cut off the excess

13.20c If you're installing crimp-type boot clamps, you'll need a pair of special crimping pliers (available at most auto parts stores)

13.20d To install fold-over type boot clamps, bend the tang down . . .

13.20e . . . then tap the tabs over to hold it in place

13.24a To remove the outer CV joint, clamp the axleshaft in a bench vise and tap the joint off with a brass hammer - if the joint doesn't come off fairly easily, use a hammer and brass punch positioned on the inner race

21   Install a new retaining ring on the inner CV joint stub axle **(see illustration 11.12)**.
22   Install the driveaxle (see Section 11).

Outer CV joint

Disassembly

Refer to illustrations 13.24a and 13.24b

23   Remove the boot clamps and separate the boot from the outer CV joint **(see illustration 13.2a)**.
24   Clamp the axleshaft in a bench vise (equipped with protective jaws) and drive off the outer CV joint with a brass hammer **(see illustration)**, then remove the retainer ring for the inner race **(see illustration)**. Slide off the old boot.
25   Thoroughly wash the outer CV joint in clean solvent and blow it dry with compressed air, if available. The outer joint can't be disassembled, so it's difficult to wash away all the old grease, and to rid the bearing of solvent once it's clean. But it's imperative that the job be done thoroughly, so take your time and do it right.

Inspection

Refer to illustration 13.26

26   Rotate the outer CV joint housing at an

13.24b Remove the retainer ring for the outer CV joint assembly

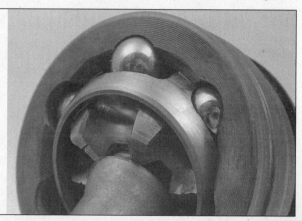

13.26 After the old grease has been rinsed away and the cleaning solvent has been blown out with compressed air, move the inner race through its full range of motion and inspect the bearing surfaces for wear or damage - if any of the balls, the race or the cage look damaged, replace the outer joint assembly

Chapter 8 Clutch and driveaxles   8-13

13.28 Install a new retainer ring for the outer CV joint assembly

13.29 Apply CV joint grease through the splined hole, then insert a wooden dowel (slightly smaller in diameter than the hole) into the hole and push down - the dowel will force the grease into the joint

13.30 To install the outer CV joint, put the axleshaft in a bench vise (equipped with protective jaws) and tap on the CV joint with a hammer and a block of wood; drive the joint onto the axleshaft splines until the retainer ring on the shaft seats in the groove in the inner race of the joint

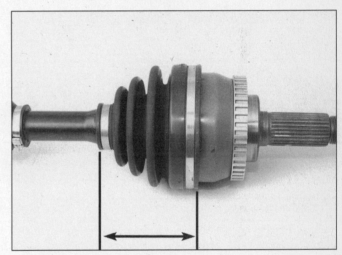

13.31 Adjust the boot to the length indicated in this Chapter's Specifications and tighten the clamps

angle to the driveaxle to expose the bearings, inner race and cage **(see illustration)**. Inspect the bearing surfaces for signs of wear. If the CV joint is worn, replace it.

Reassembly

Refer to illustrations 13.28, 13.29, 13.30 and 13.31

27 Slide the new outer boot and small clamp onto the driveaxle. It's a good idea to wrap vinyl tape around the shaft splines to prevent damage to the boot **(see illustration**

13.11). Remove the tape.
28 Install the inner race retainer ring **(see illustration)**.
29 Pack the CV joint with CV joint grease through the splined hole in the inner race. Force the grease into the joint by inserting a wooden dowel through the splined hole and pushing it to the bottom of the joint. Repeat this procedure until the bearing is completely packed **(see illustration)**.
30 Tap the outer CV joint into place with a hammer and a wood block **(see illustration)**.

31 Slide the boot into position. When the boot is in position add the remainder of the grease in the boot replacement kit to the CV joint boot. Slide the boot on and adjust the length **(see illustration)**, equalize the pressure inside the boot **(see illustration 13.19b)** and install the new clamps **(see illustrations 13.20a through 13.20e)**.

All units

32 Install the driveaxle (see Section 11).

Notes

Chapter 9 Brakes

Contents

Section		Section	
Anti-lock Brake System (ABS) - general information and trouble codes	2	Disc brake pads - replacement	3
		Drum brake shoes - replacement	6
Brake check	See Chapter 1	General information	1
Brake disc - inspection, removal and installation	5	Master cylinder - removal and installation	8
Brake hoses and lines - inspection and replacement	10	Parking brake - adjustment	13
Brake hydraulic system - bleeding	11	Parking brake cables - replacement	14
Brake light switch - check and replacement	16	Power brake booster - check, replacement and adjustment	12
Brake pedal - adjustment	15	Proportioning valve - replacement	9
Disc brake caliper - removal and installation	4	Wheel cylinder - removal and installation	7

Specifications

General
Brake fluid type .... See Chapter 1

Disc brakes
Minimum pad thickness	See Chapter 1
Brake disc minimum thickness	Cast into disc
Maximum disc runout	
Front	0.0028 inch (0.07 mm)
Rear	
1993 through 1999 models	0.0059 inch (0.01 mm)
2000 and later models	0.0028 inch (0.07 mm)
Maximum disc thickness variation	
Front	0.0004 inch (0.01 mm)
Rear	0.0008 inch (0.02 mm)

Rear drum brakes
Shoe friction material minimum thickness	See Chapter 1
Maximum inside diameter	Cast into drum
Maximum out-of-round	0.0012 inch (0.03 mm)

Power brake booster
Output rod length	0.4045 to 0.4144 inch (10.275 to 10.525 mm)
Booster-to-clevis dimension	4-59/64 inches (125 mm)

Brake pedal adjustments
Released height	
Manual transaxle	6-17/64 to 6-1/2 inches (159 to 165 mm)
Automatic transaxle	6-21/32 to 6-27/32 inches (169 to 174 mm)
Freeplay	3/64 to 1/8 inch (0.04 to 0.12 mm)
Depressed height	
1993 and 1994 models	3-35/64 inches (90 mm)
1995 and later models	
Manual transaxle	2-49/64 inch (70 mm)
Automatic transaxle	2-61/64 inch (75 mm)

Brake light switch
Plunger-to-pedal stopper clearance .... 0.012 to 0.039 inch (0.3 to 1.0 mm)

Torque specifications

	Nm	Ft-lbs (unless otherwise indicated)
Brake booster-to-body mounting nuts		
1993 and 1994 models	8 to 11	70 to 96 in-lbs
1995 and later models	13 to 16	108 to 144 in-lbs
Brake caliper		
Caliper mounting bolts (front and rear)	22 to 31	16 to 23
Caliper mounting bracket bolts		
Front	72 to 97	53 to 72
Rear	38 to 52	28 to 38
Brake hose-to-caliper banjo bolt	17 to 20	144 to 168 in-lbs
Caliper and wheel cylinder bleeder screws	7 to 9	61 to 78 in-lbs
Master cylinder-to-brake booster retaining nuts		
1993 models	8 to 11	72 to 96 in-lbs
1994 and later models	12 to 15	108 to 132 in-lbs
Wheel cylinder retaining bolts	6 to 11	52 to 96 in-lbs

1 General information

The vehicles covered by this manual are equipped with hydraulically operated front and rear brake systems. The front brakes are disc type and the rear brakes are disc or drum type (only 1993 and 1994 models were available with rear drum brakes). Both the front and rear brakes are self adjusting. The disc brakes automatically compensate for pad wear, while the drum brakes incorporate an adjustment mechanism which is activated as the parking brake is applied.

Hydraulic system

The hydraulic system consists of two separate circuits. The master cylinder has separate reservoirs for the two circuits, and, in the event of a leak or failure in one hydraulic circuit, the other circuit will remain operative. A pair of proportioning valves on non-ABS models, or a dual proportioning valve on ABS-equipped models, provides brake balance between the front and rear brakes.

Power brake booster

The power brake booster, utilizing engine manifold vacuum and atmospheric pressure to provide assistance to the hydraulically operated brakes, is mounted on the firewall in the engine compartment.

Parking brake

The parking brake operates the rear brakes only, through cable actuation. It's activated by a lever mounted in the center console.

Service

After completing any operation involving disassembly of any part of the brake system, always test drive the vehicle to check for proper braking performance before resuming normal driving. When testing the brakes, perform the tests on a clean, dry, flat surface. Conditions other than these can lead to inaccurate test results.

Test the brakes at various speeds with both light and heavy pedal pressure. The vehicle should stop evenly without pulling to one side or the other. Avoid locking the brakes, because this slides the tires and diminishes braking efficiency and control of the vehicle.

Tires, vehicle load and wheel alignment are factors which also affect braking performance.

2 Anti-lock Brake System (ABS) - general information and trouble codes

1 The Anti-lock Brake System (ABS) is designed to maintain vehicle steerability, directional stability and optimum deceleration

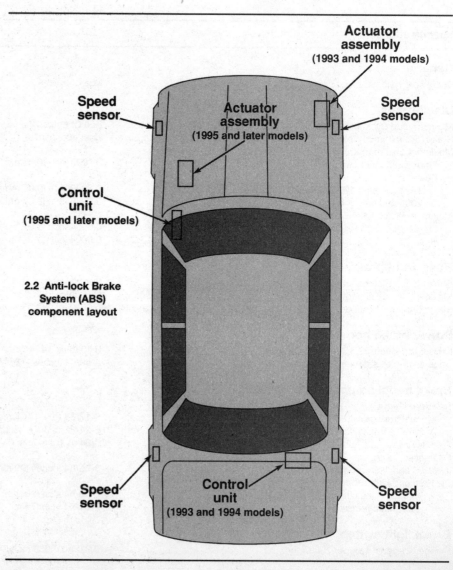

2.2 Anti-lock Brake System (ABS) component layout

Chapter 9 Brakes

under severe braking conditions and on most road surfaces. It does so by monitoring the rotational speed of each wheel and controlling the brake line pressure to each wheel during braking. This prevents the wheels from locking up.

Components

Actuator assembly

Refer to illustration 2.2

2 The actuator assembly **(see illustration)** consists of an electric hydraulic pump and three solenoid valves: front left, front right and rear. The electric pump provides hydraulic pressure to charge the reservoirs in the actuator, which supplies pressure to the braking system. The solenoid valves modulate brake line pressure during ABS operation. The body contains four valves - one for each wheel. The pump, the reservoirs and the solenoid valves are all housed in the actuator assembly.

Speed sensors

3 The speed sensors, which are located at each wheel, generate a small sine wave current when the toothed sensor rotors are turning. This analog voltage signal is monitored by the ABS control unit, which converts it to a digital signal from which it can determine wheel rotational speed.

4 The front speed sensors are mounted on the steering knuckles in close relationship to the toothed sensor rotors, which are pressed onto the outer constant velocity (CV) joints.

5 The rear wheel sensors are bolted to the rear knuckles. The toothed sensor rotors are integral with the rear wheel hub/bearing assemblies.

ABS computer

6 The ABS control unit, which is mounted in the trunk (1993 and 1994 models) or behind the left kick panel (1995 and later models), is the "brain" of the ABS system. The function of the control unit is to monitor and process information received from the wheel speed sensors to control the hydraulic line pressure, avoiding wheel lock up. The control unit also monitors the system for malfunctions, even when the ABS system is inactive during normal driving conditions.

7 Each time you start the engine, the system turns on the ABS warning light on the instrument panel for about a second. As soon as the engine is running, the light should go off. The system then performs a self-test the first time the vehicle speed exceeds four mph. You may hear a mechanical noise during the test; this is normal. If the system detects a problem, the ABS light will come on and remain on. A diagnostic code will also be stored in the control unit, which indicates the problem area or component.

Diagnosis and repair

8 If the ABS warning light on the dash comes on and stays on while the vehicle is in operation, the ABS system requires attention. Diagnosis is quite complex, involving a number of lengthy diagnostic procedures, so we don't recommend attempting to fix the ABS system at home. However, if you're willing to do a little work, you can obtain a diagnostic trouble code - which will indicate the general area of the problem - as follows:

1993 and 1994 models

a) *Drive the vehicle above 20 mph for at least one minute.*
b) *Stop the vehicle but keep the engine running.*
c) *Open the trunk and count the number of flashes of the ABS LED light on the control unit. If a trouble code is present, it will flash continuously, separated by a five to ten second pause between each series of flashes.*
d) *Refer to the accompanying table **for code diagnosis**.*
e) *Turn off the engine.*

ABS trouble codes - 1993 and 1994 models

No. of LED flashes	Defective part or circuit
1	Left front actuator solenoid circuit
2	Right front actuator solenoid circuit
3	Right rear actuator solenoid circuit
4	Left rear actuator solenoid circuit
5	Left front wheel sensor circuit
6	Right front wheel sensor circuit
7	Right rear wheel sensor circuit
8	Left rear wheel sensor circuit
9	Motor and motor relay
10	Solenoid valve relay
16 (or continuous)	Control unit
ABS warning light On, but LED Off	Ground circuit or power supply for control unit

1995 and later models

Refer to illustrations 2.8a and 2.8b

a) *Drive the vehicle above 20 mph for at least one minute.*
b) *Stop the vehicle and turn off the engine.*
c) *Ground terminal L on the Data Link Connector **(see illustration)**.*
d) *With terminal L grounded, turn the ignition switch to On.*
e) *After 3.6 seconds, the ABS warning light will begin flashing any stored trouble codes. The code is determined by counting the number of on-and-off flashes **(see illustration)**. The sequence always begins with a 3.6-second "off" period, followed by a flash, then a 1.6-second off period, then two flashes. This*

2.8a To put the control unit into output mode, ground the L terminal on the Data Link Connector

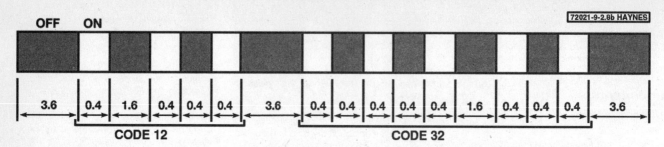

2.8b The ABS light on the dash indicates a problem in a particular circuit by the number of flashes. For example, three 0.4-second flashes, followed by a 1.6-second interval, followed by two more 0.4-second flashes indicates a code 32, which means there's a short circuit in the rear right sensor circuit

"Code 12" (the "start" code, not a trouble code) is followed by another 3.6-second off period, then the trouble codes are displayed, in the order in which they were stored, starting with the latest stored code. All codes are two-digit codes, so the first flash(es) indicate the tens place, followed by a longer delay, followed by the single-digit flash(es). For example, a sequence of four flashes, then a pause, followed by a sequence of five flashes, would indicate a Code 45 (actuator front left outlet solenoid valve).

f) Count the number of flashes of the ABS light, then refer to the accompanying table.

9 A trouble code can be set by a simple malfunction. Although you can't troubleshoot

ABS trouble codes - 1995 and later models

Code	Defective part or circuit	Code	Defective part or circuit
12	No malfunctions detected	51	Actuator rear right outlet solenoid valve
16	Brake light switch circuit	52	Actuator rear right inlet solenoid valve
18	Sensor rotor	55	Actuator rear left outlet solenoid valve
21	Front right sensor (open circuit)	56	Actuator rear left inlet solenoid valve
22	Front right sensor (short circuit)	57	Power supply (low or high voltage)
25	Front left sensor (open circuit)	61	Actuator motor or motor relay
26	Front left sensor (short circuit)	63	Solenoid valve relay or circuit
31	Rear right sensor (open circuit)	71	Control unit
32	Rear right sensor (short circuit)	81	Engine speed signal
35	Rear left sensor (open circuit)	85	ECM detects problem with ABS control unit
36	Rear left sensor (short circuit)	87	Engine operating in fail-safe mode
41	Actuator front right outlet solenoid valve	92	LAN communication start procedures not complete
42	Actuator front right inlet solenoid valve	94	Continued reception after LAN communication starts
45	Actuator front left outlet solenoid valve	96	LAN is monitoring
46	Actuator front left inlet solenoid valve	98	LAN communication system failure

Warning light stays on when ignition switch is turned on .... Control unit power supply circuit
(1995-1998) Warning light bulb circuit
Control unit or control unit connector
Solenoid valve relay stuck
Power supply for solenoid valve relay coil

SLIP indicator stays on when engine is running (1999)........................... Control unit power supply circuit
Warning light bulb circuit
Control unit or control unit connector
Solenoid valve relay stuck
Power supply for solenoid valve relay coil

Chapter 9 Brakes

Warning light stays on during self-diagnosis..	Control module
Warning light does not come on when ignition switch is turned on........ (1995 - 1998)	Fuse, warning light bulb or warning light circuit Control module
SLIP indicator does not come on when engine is running...................... (1999)	Fuse, warning light bulb or warning light circuit Control module
Warning light does not come on during self-diagnosis...........................	Control module

3.5a Before disassembling the brake, wash it thoroughly with brake system cleaner and allow it to dry - position a drain pan under the brake to catch the residue - DO NOT use compressed air to blow off brake dust!

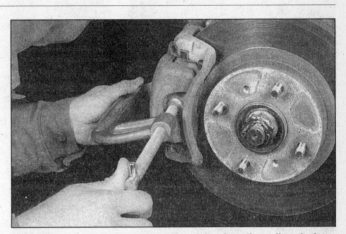

3.5b Use a C-clamp to depress the piston into the caliper before removing the caliper and pads

all of the malfunctions listed in the accompanying trouble code table, you *can* check the following things:

a) Check the brake fluid level in the reservoir.
b) Verify that all electrical connectors are securely connected.
c) Check the fuses.
d) Check the brake system (see Chapter 1).
e) Check the brake pads (see Section 3).
f) Check the brake pedal (see Section 15).

10 After verifying that all of the above are okay, try to erase the stored trouble code(s) as follows: Unground terminal L; the ABS warning light should remain on. Ground the L terminal three successive times within 12.5 seconds; each ground must last more than one second. The ABS light should now go out.

a) If the light stays on, take the vehicle to a dealer service department or other qualified repair shop and have the ABS system repaired.
b) If the light doesn't stay on, drive the vehicle above 20 mph for at least one minute and verify that the warning light on the dash doesn't come on again. If it does, take the vehicle to a dealer service department or other qualified repair shop and have the ABS system repaired.

3 Disc brake pads - replacement

Refer to illustrations 3.5a through 3.5p or 3.6a through 3.6m

Warning: *Disc brake pads must be replaced on both front or both rear wheels at the same time - never replace the pads on only one wheel. Also, the dust created by the brake system may contain asbestos, which is harmful to your health. Never blow it out with compressed air and don't inhale any of it. An approved filtering mask should be worn when working on the brakes. Do not, under any circumstances, use petroleum-based solvents to clean brake parts. Use brake system cleaner only!*

1 Remove the cap from the brake fluid reservoir. Remove about two-thirds of the fluid from the reservoir.
2 Loosen the front or rear wheel lug nuts, raise the front or rear of the vehicle and support it securely on jackstands. Block the wheels at the opposite end.
3 Remove the wheels. Work on one brake assembly at a time, using the assembled brake for reference if necessary.
4 Inspect the brake disc carefully as outlined in Section 5. If machining is necessary, follow the information in that Section to remove the disc, at which time the pads can be removed as well.
5 If you are replacing the front brake pads, follow the first photo sequence **(see illustrations 3.5a through 3.5p).** Be sure to stay in order and read the caption under each illustration.
6 If you're replacing the rear brake pads, follow the second photo sequence **(see illustrations 3.6a through 3.6m).** Be sure to stay in order and read the caption under each illustration.
7 When reinstalling the caliper, be sure to tighten the mounting bolt to the torque listed in this Chapter's Specifications.
8 After the job has been completed, firmly depress the brake pedal a few times to bring the pads into contact with the disc. Check the level of the brake fluid, adding some if necessary. Check the operation of the brakes carefully before placing the vehicle into normal service.

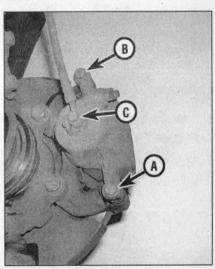

3.5c Remove the lower caliper mounting bolt (A); the upper bolt (B) doesn't need to be removed unless the caliper is going to be removed completely. The brake hose banjo fitting bolt (C) doesn't have to be removed unless the caliper or hose is to be replaced

3.5d Pivot the caliper up and support it in this position for access to the brake pads

3.5e On 1995 and later models, remove the anti-rattle springs

3.5f Remove the inner brake pad

3.5g Remove the outer brake pad

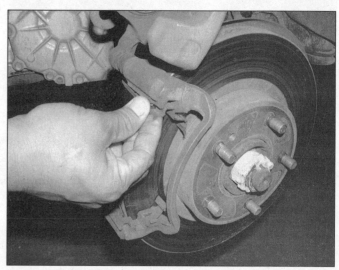

3.5h Remove the upper and lower pad retainers from the caliper mounting bracket

3.5i Remove the shims from the brake pads

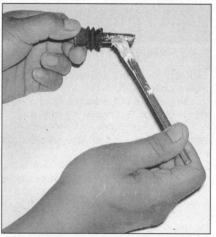

3.5j Clean the caliper pin, inspect it for scoring and corrosion; coat the pin with high-temperature grease and install it in the caliper mounting bracket

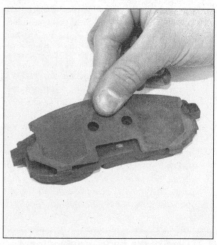

3.5k Install the shims on the pads . . .

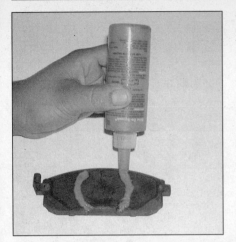

3.5l ... then apply anti-squeal compound to the back of both pads (let the compound "set up" a few minutes before installing them)

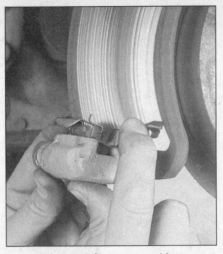

3.5m Install the upper and lower pad retainers

3.5n Install the inner brake pad (it's the one with the wear indicator on it)

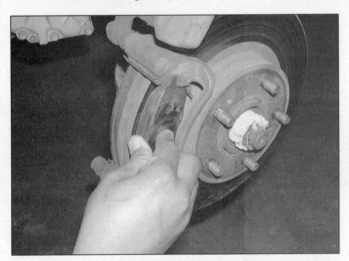

3.5o Install the outer brake pad ...

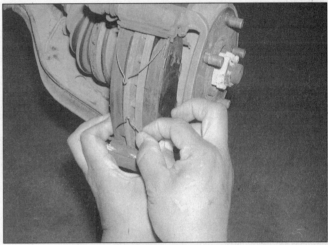

3.5p ... and, on 1995 and later models, the anti-rattle springs. Swing the caliper down over the pads and install the lower mounting bolt, tightening it to the torque listed in this Chapter's Specifications. **Note:** *If the caliper won't fit over the pads, use a C-clamp to push the piston into the caliper a little further*

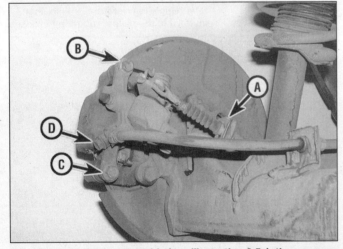

3.6a Clean the brake assembly (see illustration 3.5a), then remove the parking brake cable clip (A) and unhook the cable from the lever. On 1995 and later models unscrew the upper caliper bolt (B) (on earlier models unscrew bolt C); the brake hose banjo fitting bolt (D) doesn't have to be removed for pad replacement

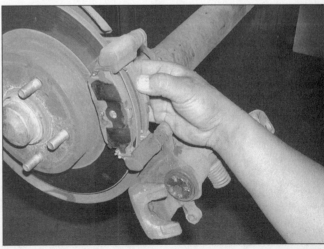

3.6b Pivot the caliper out of the mounting bracket and remove the inner pad ...

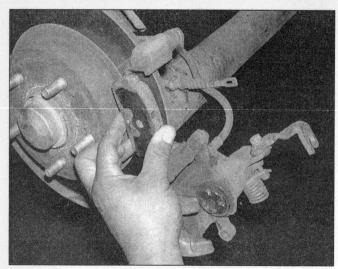
3.6c ... and the outer pad

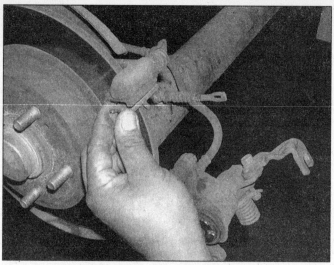
3.6d Remove the upper and lower pad retainers

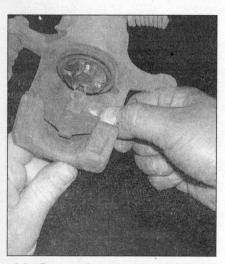
3.6e Remove the anti-rattle spring from the caliper and inspect it; if it's cracked or otherwise damaged, replace it

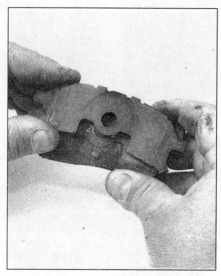
3.6f Remove the shims from the pads

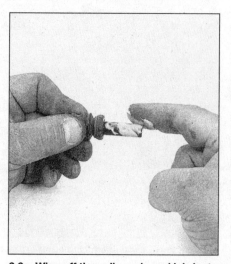
3.6g Wipe off the caliper pin and lubricate it with high-temperature grease, then reinstall it in the caliper mounting bracket

3.6h Install the shims on the pads

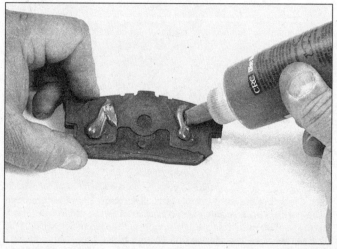
3.6i Apply anti-squeal compound to the backs of both pads (let the compound "set up" a few minutes before installing the pads)

Chapter 9 Brakes

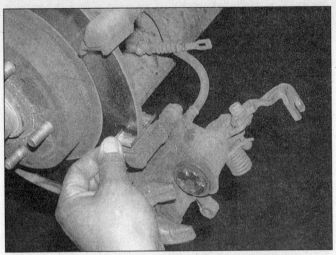
3.6j Install the upper and lower pad retainers

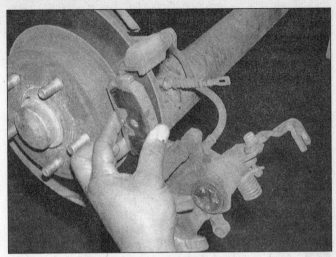
3.6k Install the outer pad

3.6l Install the inner pad

3.6m Turn the piston to retract it - make sure one of the notches in the piston engages with the projection on the pad. The tool shown here is available at most auto parts stores, but you can also turn it with a pair of needle-nose pliers. Swing the caliper back into position and install the mounting bolt, tightening it to the torque listed in this Chapter's Specifications, then reconnect the parking brake cable

4 Disc brake caliper - removal and installation

Refer to illustration 4.3

Warning: *The dust created by the brake system may contain asbestos, which is harmful to your health. Never blow it out with compressed air and don't inhale any of it. An approved filtering mask should be worn when working on the brakes. Do not, under any circumstances, use petroleum-based solvents to clean brake parts. Use brake system cleaner only!*

Removal

1 Loosen the front or rear wheel lug nuts, raise the front or rear of the vehicle and place it securely on jackstands. Block the wheels at the opposite end. Remove the front or rear wheel.

2 To disconnect the parking brake cable from the rear caliper, unbolt the cable bracket from the caliper and disengage the cable from the toggle lever.

3 Remove the banjo bolt and discard the old sealing washers **(see illustration 3.5c or 3.6a)**. Disconnect the brake hose from the caliper. Plug the brake hose to keep contaminants out of the brake system and to prevent losing any more brake fluid than is necessary **(see illustration)**.

4 Remove the caliper mounting bolts and detach the caliper from the mounting bracket.

Installation

5 Installation is the reverse of removal. Don't forget to use new sealing washers on each side of the brake hose banjo fitting and be sure to tighten the banjo fitting bolt and the caliper mounting bolts to the torque listed in this Chapter's Specifications.

6 Bleed the brake system (see Section 11). Make sure there are no leaks from the hose connections. Test the brakes carefully before returning the vehicle to normal service.

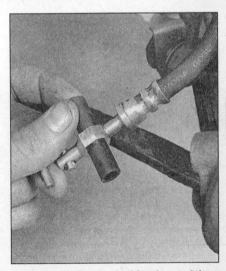
4.3 Using a piece of rubber hose of the appropriate size, plug the brake line

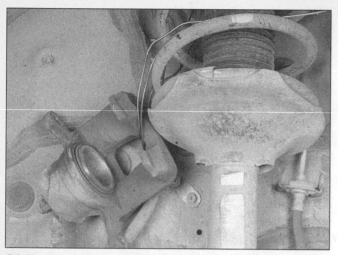

5.2 Hang the caliper out of the way with a piece of wire - don't let it hang by the brake hose!

5.3 The brake pads on this vehicle were obviously neglected, as they wore down completely and cut deep grooves into the disc - wear this severe means the disc must be replaced

5.4a To check disc runout, mount a dial indicator as shown and rotate the disc

5.4b Using a swirling motion, remove the glaze from the disc surface with sandpaper or emery cloth

5.5 Use a micrometer to measure disc thickness

5 Brake disc - inspection, removal and installation

Inspection

Refer to illustrations 5.2, 5.3, 5.4a, 5.4b and 5.5

1 Loosen the wheel lug nuts, raise the vehicle and support it securely on jackstands. Remove the wheel and install the lug nuts to hold the disc in place. **Note:** *If the lug nuts don't contact the disc when screwed on all the way, install washers under them.*

2 Remove the brake caliper (see Section 4). It isn't necessary to disconnect the brake hose. After removing the caliper bolts, suspend the caliper out of the way with a piece of wire **(see illustration)**.

3 Visually inspect the disc surface for score marks and other damage. Light scratches and shallow grooves are normal after use and may not always be detrimental to brake operation, but deep scoring requires disc removal and refinishing by an automotive machine shop. Be sure to check both sides of the disc **(see illustration)**. If pulsating has been noticed during application of the brakes, suspect disc runout.

4 To check disc runout, place a dial indicator at a point about 1/2-inch from the outer edge of the disc **(see illustration)**. Set the indicator to zero and turn the disc. The indicator reading should not exceed the specified allowable runout limit. If it does, the disc should be refinished by an automotive machine shop. **Note:** *When replacing the brake pads, it's a good idea to resurface the discs regardless of the dial indicator reading, as this will impart a smooth finish and ensure a perfectly flat surface, eliminating any brake pedal pulsation or other undesirable symptoms related to questionable discs. At the very least, if you elect not to have the discs resurfaced, remove the glaze from the surface with emery cloth or sandpaper, using a swirling motion* **(see illustration)**.

5 It's absolutely critical that the disc not be machined to a thickness under the specified minimum thickness. The minimum wear (or discard) thickness is cast into the disc. The disc thickness can be checked with a micrometer **(see illustration)**.

Removal

Refer to illustrations 5.6a, 5.6b and 5.7

6 Remove the two caliper mounting bracket bolts and detach the mounting bracket **(see illustrations)**.

7 Remove the lug nuts which you installed to hold the disc in place and remove the disc from the hub **(see illustration)**. If the disc is stuck to the hub and won't come off, install bolts of the proper diameter and thread pitch into the threaded holes between the wheel studs and tighten them, which will force the disc off the hub.

Chapter 9 Brakes 9-11

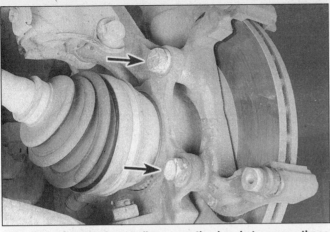

5.6a To remove the front caliper mounting bracket, remove these bolts (arrows)

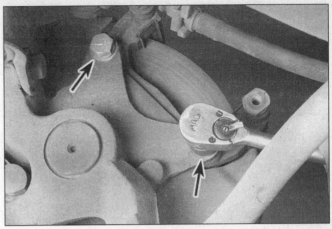

5.6b To remove the rear caliper bracket, remove these bolts (arrows)

Installation

8 Place the disc in position over the threaded studs.

9 Install the mounting bracket and tighten the bolts to the torque listed in this Chapter's Specifications. Install the brake pads.

10 Install the caliper onto the mounting bracket, tightening the bolts to the torque listed in this Chapter's Specifications.

11 Install the wheel and lug nuts. Lower the vehicle and tighten the lug nuts to the torque listed in the Chapter 1 Specifications. Depress the brake pedal a few times to bring the brake pads into contact with the disc. Bleeding won't be necessary unless the brake hose was disconnected from the caliper. Check the operation of the brakes carefully before driving the vehicle.

6 Drum brake shoes - replacement

Refer to illustrations 6.2a, 6.2b, 6.4, 6.6, 6.7, 6.11, 6.12, 6.14, 6.15a, 6.15b, 6.16, 6.17a, 6.17b, 6.18a, 6.18b, 6.20, 6.21a, 6.21b and 6.21c

Warning: *Drum brake shoes must be replaced on both wheels at the same time - never replace the shoes on only one wheel. Also, the dust created by the brake system may contain asbestos, which is harmful to your health. Never blow it out with compressed air and don't inhale any of it. An approved filtering mask should be worn when working on the brakes. Do not, under any circumstances, use petroleum-based solvents to clean brake parts. Use brake system cleaner only!*

Caution: *Whenever the brake shoes are replaced, the return and hold-down springs should also be replaced. Due to the continuous heating/cooling cycle the springs are subjected to, they lose tension over a period of time and may allow the shoes to drag on the drum and wear at a much faster rate than normal.*

1 Loosen the wheel lug nuts, raise the rear of the vehicle and support it securely on jackstands. Block the front wheels to keep the vehicle from rolling. Remove the wheels.

2 Release the parking brake and remove the brake drums. If the shoes have worn into the drum, preventing drum removal, remove the access plug from the backing plate, insert

5.7 Remove the brake disc; if it won't come off, thread bolts of the proper size and thread pitch into the two holes (arrows) and tighten them, which will force the disc off

a small screwdriver through the hole, lift the adjuster lever off the adjusting wheel and turn the wheel with another screwdriver to back off the brake shoes **(see illustration)**. If the

6.2a Lift the adjuster lever off the adjuster wheel with the screwdriver and rotate the adjuster wheel until the shoes are retracted sufficiently to allow drum removal (brake drum removed for clarity)

6.2b If a brake drum is stuck to the hub, insert two bolts in the holes provided and tighten them until they push against the hub flange, which will force the drum off - notice the maximum allowable diameter cast into the drum surface

6.4 Before disassembling the brake, wash it thoroughly with brake system cleaner and allow it to dry - position a drain pan under the brake to catch the residue - DO NOT use compressed air to blow the brake dust off!

6.6 Remove the hold-down springs from the brake shoes; to release a hold-down spring with a brake spring tool, push it in and rotate it 90-degrees, then release pressure and remove the retainer, spring and pin

6.7 Disengage the lower return spring from the shoes

6.11 After removing the front shoe, remove the upper return spring from the rear shoe

6.12 Detach the cable from the parking brake lever

drum is rusted to the hub, install a two bolts of the proper size and thread pitch into the threaded holes provided **(see illustration)** and turn them in. As the bolts are tightened, they will contact the surface of the hub flange and push the drum off.

3 **Note:** *All four rear brake shoes must be replaced at the same time, but to avoid mixing up parts, work on only one brake assembly at a time.*

4 Before disassembling anything, clean off the brake assembly with brake system cleaner **(see illustration)**.

5 Remove the hub and wheel bearing assembly (see Chapter 10).

6 Remove the hold-down retainers and springs from the brake shoes **(see illustration)**.

7 Pull the front shoe toward the front of the vehicle far enough to disengage it from the wheel cylinder and the anchor plate, then remove the lower return spring **(see illustration)**.

8 Pull the rear shoe toward the rear of the vehicle and disengage it from the wheel cylinder.

9 Remove the adjuster assembly, noting which end faces forward (mark it if necessary).

10 Unhook the upper return spring from the front shoe and remove the front shoe.

11 Unhook the upper return spring from the rear shoe **(see illustration)**.

12 Detach the parking brake cable from the lever on the rear shoe **(see illustration)**, then remove the rear shoe.

13 Peel back the dust boots on the wheel cylinder and check for leakage. Replace the wheel cylinder if it is leaking (see Section 7).

14 Clean the brake backing plate with

6.14 Clean, then lubricate the contact areas on the brake backing plate with high-temperature brake grease

Chapter 9 Brakes   9-13

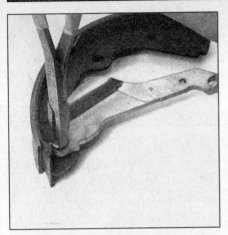

6.15a Pop off the retaining clip from the parking brake lever pivot pin

6.15b Secure the parking brake lever to the new rear shoe with a *new* retaining clip

6.16 Pull back the spring and insert the plug on the end of the parking brake cable into the slot in the parking brake lever

6.17a Insert the hold-down spring pin through the backing plate and install the rear brake shoe

6.17b Install the hold-down spring and retainer, push the spring in then rotate the retainer 90-degrees to lock it onto the pin

brake system cleaner, then lubricate the shoe contact points with high-temperature brake grease **(see illustration)**.

15 Remove the retaining clip from the parking brake lever pivot pin **(see illustration)**. Separate the lever from the brake shoe and transfer it to the new shoe. Be sure to install the washer between the lever and the shoe. Secure the pivot pin with a new retaining ring **(see illustration)**.

16 Connect the parking brake cable to the parking brake lever **(see illustration)**.

17 Place the rear shoe into position on the backing plate, inserting the hold-down pin through the hole in the shoe. Install the hold down spring and retainer **(see illustrations)**.

18 Clean the adjuster screw assembly, then lubricate the threads and the pivot with high-temperature brake grease **(see illustrations)**.

19 Install the adjuster screw assembly into the notch in the rear brake shoe, making sure

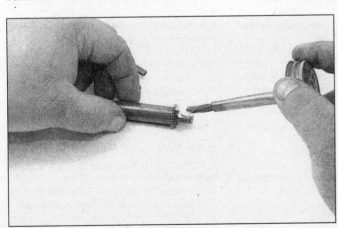

6.18a Lubricate the adjuster screw pivot end . . .

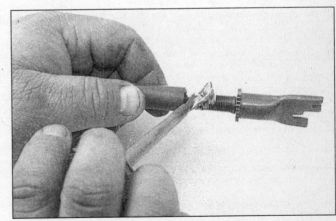

6.18b . . . and the threads with high-temperature brake grease

6.20 Install the front shoe hold-down pin, spring and retainer and lock them into place

6.21a Install the upper return spring into the holes in the shoes

6.21b Connect the lower return spring to the hole in the bottom of each shoe

it is facing the proper direction.
20 Install the front shoe and secure it with the hold-down spring **(see illustration)**. Make sure the adjuster screw assembly correctly engages the notches in both shoes.
21 Install the return springs **(see illustrations)**.
22 Before reinstalling the drum, it should be checked for cracks, score marks, deep scratches and hard spots, which will appear as small discolored areas. If the hard spots cannot be removed with fine emery cloth or if any of the other conditions listed above exist, the drum must be taken to an automotive machine shop to have it resurfaced. **Note:** *Professionals recommend resurfacing the drums each time a brake job is done. Resurfacing will eliminate the possibility of out-of-round drums.* If the drums are worn so much that they can't be resurfaced without exceeding the maximum allowable diameter, which is stamped into the drum **(see illustration 6.2b)**, then new ones will be required. At the very least, if you elect not to have the drums resurfaced, remove the glaze from the surface with emery cloth using a swirling motion.
23 Turn the adjuster screw wheel so the drum just slips over the shoes. Now, working through the backing plate, turn the adjuster screw wheel until the shoes drag on the drum when the drum is turned. Finally, back off the adjuster screw wheel so the shoes don't drag. Depress the brake pedal firmly several times, then rotate the drum to ensure that the brakes are not dragging. If they are, back off the star wheel a little more.
24 Mount the wheel, install the lug nuts, then lower the vehicle. Tighten the wheel lug nuts to the torque listed in the Chapter 1 Specifications.
25 Make a number of forward and reverse stops and operate the parking brake to adjust the brakes until satisfactory pedal action is obtained.
26 Check the operation of the brakes carefully before driving the vehicle.

6.21c Push the lower return spring into place behind the anchor plate

7 Wheel cylinder - removal and installation

Removal

Refer to illustration 7.4

1 Raise the rear of the vehicle and support it securely on jackstands. Block the front wheels to keep the vehicle from rolling.
2 Remove the brake shoe assembly (see Section 6).
3 Remove all dirt and foreign material from around the wheel cylinder.
4 Disconnect the brake line **(see illustration)** with a flare-nut wrench, if available. Don't pull the brake line away from the wheel cylinder.
5 Remove the wheel cylinder mounting bolts.
6 Detach the wheel cylinder from the brake backing plate and place it on a clean workbench. Immediately plug the brake line to prevent fluid loss and contamination.

Installation

7 Place the wheel cylinder in position and install the bolts finger tight. Connect the

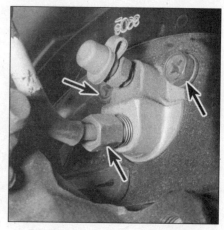

7.4 To remove the wheel cylinder, disconnect the brake line fitting (lower arrow), then remove the two wheel cylinder bolts (upper arrows)

brake line to the cylinder, being careful not to cross-thread the fitting. Tighten the wheel cylinder bolts to the torque listed in this Chapter's Specifications.
8 Tighten the brake line securely and install the brake shoes (see Section 6).
9 Bleed the brakes (see Section 11).
10 Check the operation of the brakes carefully before driving the vehicle.

8 Master cylinder - removal and installation

Refer to illustration 8.2

Removal

1 Disconnect the cable from the negative battery terminal.
2 Unplug the electrical connector for the fluid level warning switch **(see illustration)**.
3 Remove as much fluid as possible from the reservoir with a syringe.
4 Place rags under the fittings and pre-

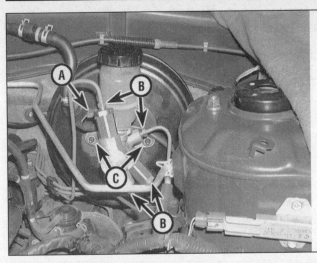

8.2 Master cylinder mounting details

A Electrical connector for fluid level switch
B Brake line fittings
C Mounting nuts

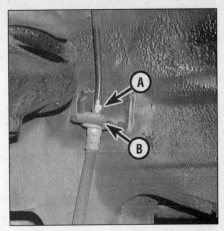

10.3 Loosen the threaded fitting on the brake line (A), then pull off the U-clip (B) with a pair of pliers

pare caps or plastic bags to cover the ends of the lines once they're disconnected. **Caution:** *Brake fluid will damage paint. Cover all body parts and be careful not to spill fluid during this procedure.* Loosen the fittings at the ends of the brake lines where they enter the master cylinder. To prevent rounding off the flats, use a flare-nut wrench, which wraps around the fitting hex.

5 Pull the brake lines away from the master cylinder and plug the ends to prevent contamination.

6 Remove the nuts attaching the master cylinder to the power booster (see illustration 8.2). Pull the master cylinder off the studs to remove it. Again, be careful not to spill the fluid as this is done. Remove and discard the old gasket between the master cylinder and the power brake booster.

Installation

7 Install the master cylinder over the studs on the power brake booster and tighten the nuts only finger-tight at this time. Don't forget to use a new gasket.

8 Thread the brake line fittings into the master cylinder. Since the master cylinder is still a bit loose, it can be moved slightly so the fittings thread in easily. Don't strip the threads as the fittings are tightened.

9 Tighten the mounting nuts to the torque listed in this Chapter's Specifications. Tighten the brake line fittings securely.

10 Fill the master cylinder reservoir with fluid, then bleed the lines at the master cylinder, followed by bleeding the remainder of the brake system (see Section 11). To bleed the lines at the master cylinder, have an assistant depress the brake pedal and hold it down. Loosen the fitting to allow air and fluid to escape. Tighten the fitting, then allow your assistant to return the pedal to its rest position. Repeat this procedure on both fittings until the fluid is free of air bubbles, then bleed the rest of the system. Check the operation of the brake system carefully before driving the vehicle. **Warning:** *If you do not have a firm brake pedal at the end of the bleeding procedure, or have any doubts as to the effectiveness of the brake system, DO NOT drive the vehicle. Have it towed to a dealer service department or other qualified repair shop for diagnosis.*

9 Proportioning valve - replacement

Note: *On 1993 and 1994 models, the proportioning valve is integral with the master cylinder.*

1 On 1995 and later models without an Anti-lock Brake System, two proportioning valves are threaded into the master cylinder. On models with ABS, a remote dual proportioning valve is used.

2 The valves are not serviceable; if you suspect it's malfunctioning (indicated by the rear wheels locking up prematurely), have it checked by a dealer service department or other repair shop equipped with the necessary pressure gauges.

3 If the valve is leaking or has been determined to be defective, replace it by unscrewing the brake lines with a flare-nut wrench and unbolting the valve from its mounting bracket or unscrewing it from the master cylinder. After the new valve is installed, bleed the brake system (see Section 11).

10 Brake hoses and lines - inspection and replacement

Inspection

About every six months, with the vehicle raised and supported securely on jackstands, the rubber hoses which connect the steel brake lines with the front and rear brake assemblies should be inspected for cracks, chafing of the outer cover, leaks, blisters and other damage. These are important and vulnerable parts of the brake system and inspection should be complete. A light and mirror will be helpful for a thorough check. If a hose exhibits any of the above conditions, replace it with a new one.

Replacement

Flexible brake hose

Refer to illustration 10.3

2 Loosen the wheel lug nuts, raise the vehicle and support it securely on jackstands. Remove the wheel.

3 At the bracket, unscrew the brake line fitting from the hose (see illustration). Use a flare-nut wrench to prevent rounding off the corners.

4 Remove the U-clip from the female fitting at the bracket with a pair of pliers, then pass the hose through the bracket.

5 At the caliper end of the hose, remove the banjo bolt, then separate the hose from the caliper. Note that there are two copper sealing washers on either side of the banjo fitting - they should be replaced with new ones during installation.

6 Remove the U-clip from the strut bracket, then detach the hose from the bracket.

7 To install the hose, pass the caliper fitting end through the strut bracket, then connect the fitting to the caliper with the banjo bolt and new copper washers.

8 Make sure the hose isn't twisted between the caliper and the strut bracket.

9 Route the hose into the frame bracket, again making sure it isn't twisted, then connect the brake line fitting, starting the threads by hand. Install the U-clip, then tighten the fitting securely.

10 Bleed the brakes (see Section 11).

11 Install the wheel and lug nuts, lower the vehicle and tighten the lug nuts to the torque listed in the Chapter 1 Specifications.

Metal brake lines

12 When replacing brake lines, be sure to use the correct parts. Don't use copper tubing for any brake system components. Purchase steel brake lines from a dealer or auto parts store.

13 Prefabricated brake line, with the tube ends already flared and fittings installed, is available at auto parts stores and dealer parts

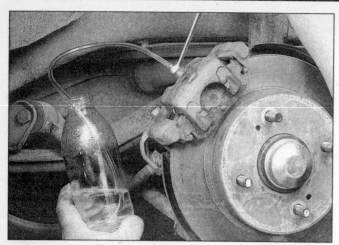

11.9 When bleeding the brakes, a hose is connected to the bleed screw at the caliper or wheel cylinder and then submerged in brake fluid - air will be seen as bubbles in the tube and container (all air must be expelled before moving to the next wheel)

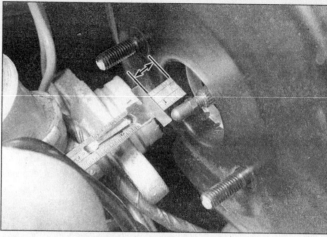

12.5 Measure the length of the booster output rod from the top of the rod to the master cylinder mounting surface and compare your measurement to the specified length listed in this Chapter's Specifications

departments. These lines must be bent to the proper shapes using a tubing bender.

14 When installing the new line, make sure it's securely supported in the brackets and has plenty of clearance between moving or hot components.

15 After installation, check the master cylinder fluid level and add fluid as necessary. Bleed the brake system (see Section 11) and test the brakes carefully before driving the vehicle in traffic.

11 Brake hydraulic system - bleeding

Refer to illustration 11.9

Warning: *Wear eye protection when bleeding the brake system. If the fluid comes in contact with your eyes, immediately rinse them with water and seek medical attention.*

Note: *Bleeding the hydraulic system is necessary to remove any air that manages to find its way into the system when it's been opened during removal and installation of a hose, line, caliper or master cylinder.*

1 You'll probably have to bleed the system at all four brakes if air has entered it due to low fluid level, or if the brake lines have been disconnected at the master cylinder.

2 If a brake line was disconnected only at a wheel, then only that caliper or wheel cylinder must be bled.

3 If a brake line is disconnected at a fitting located between the master cylinder and any of the brakes, that part of the system served by the disconnected line must be bled.

4 On models equipped with an Anti-lock Brake System (ABS), turn the ignition switch Off and disconnect the four electrical connectors from the ABS actuator.

5 Remove any residual vacuum from the brake power booster by applying the brake several times with the engine off.

6 Remove the master cylinder reservoir cover and fill the reservoir with brake fluid. Reinstall the cover. **Note:** *Check the fluid level often during the bleeding operation and add fluid as necessary to prevent the fluid level from falling low enough to allow air bubbles into the master cylinder.*

7 Have an assistant on hand, as well as a supply of new brake fluid, a clear plastic container partially filled with clean brake fluid, a length of clear tubing to fit over the bleeder valve and a wrench to open and close the bleeder valve.

8 Beginning at the right rear wheel, loosen the bleeder valve slightly, then tighten it to a point where it's snug but can still be loosened quickly and easily.

9 Place one end of the tubing over the bleeder valve and submerge the other end in brake fluid in the container **(see illustration)**.

10 Have the assistant depress the brake pedal slowly, then hold the pedal down firmly.

11 While the pedal is held down, open the bleeder valve just enough to allow a flow of fluid to leave the valve. Watch for air bubbles to exit the submerged end of the tube. When the fluid flow slows after a couple of seconds, close the valve and have your assistant release the pedal.

12 Repeat Steps 10 and 11 until no more air is seen leaving the tube, then tighten the bleeder valve and proceed to the left front wheel, the left rear wheel and the right front wheel, in that order, and perform the same procedure. Be sure to check the fluid in the master cylinder reservoir frequently.

13 Never use old brake fluid. It contains moisture which can cause the fluid to boil, rendering the brake system inoperative.

14 Refill the master cylinder with fluid at the end of the operation. If you're working on a model with ABS, be sure to reconnect the electrical connectors to the ABS actuator.

15 Check the operation of the brakes. The pedal should feel solid when depressed, with no sponginess. If necessary, repeat the entire process. **Warning:** *Do not operate the vehicle if you're in doubt about the effectiveness of the brake system.*

12 Power brake booster - check, replacement and adjustment

Check

Operating check

1 Depress the brake pedal several times with the engine off and make sure there's no change in the pedal reserve distance.

2 Depress the pedal and start the engine. If the pedal goes down slightly, operation is normal.

Airtightness check

3 Start the engine and turn it off after one or two minutes. Depress the brake pedal slowly several times. If the pedal depresses less each time, the booster is airtight.

4 Depress the brake pedal while the engine is running, then stop the engine with the pedal depressed. If there's no change in the pedal reserve travel after holding the pedal for 30 seconds, the booster is airtight.

Output rod length check

Refer to illustration 12.5

5 Remove the master cylinder (see Section 8). It isn't necessary to disconnect the lines from the master cylinder, as long as you can move it forward far enough to provide clearance for the following measurement. But make sure you don't kink the metal lines. Apply about 20 in-Hg of vacuum to the brake booster with a hand-operated vacuum pump, push in on the output rod with approximately 4-1/2 pounds of force, then measure the length of the output rod **(see illustration)** and compare your measurement to the dimensions listed in this Chapter's Specifications. If the rod length is outside specifications, replace the booster.

Chapter 9 Brakes

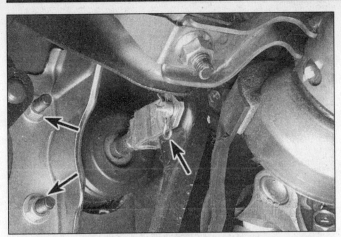

12.8 Remove this retaining clip (right arrow), pull out the clevis pin and detach the pushrod from the brake pedal; two of the power brake booster mounting nuts (left arrows) are visible in this photo

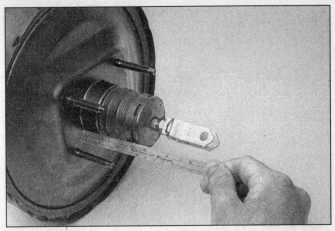

12.13 Measure the distance between the power brake booster and the hole in the clevis and compare your measurement to the dimension listed in this Chapter's Specifications; if they're not the same, adjust the clevis before installing the power brake booster

Replacement

Refer to illustration 12.8

Note: *Power brake booster units shouldn't be disassembled; if a problem with the booster develops, replace it with a new or rebuilt one.*

6 Remove the brake master cylinder, if you haven't already done so (see Section 8).
7 Disconnect the vacuum hose leading from the engine to the booster. Be careful not to damage the hose when removing it from the booster fitting.
8 Remove the under-dash panel. Locate the pushrod clevis connecting the booster to the brake pedal **(see illustration)**.
9 Remove the clevis pin retaining clip with pliers and pull out the clevis pin.
10 Remove the four nuts and washers holding the brake booster to the firewall **(see illustration 12.8)**.
11 Slide the booster straight out from the firewall until the studs clear the holes.
12 Installation is the reverse of removal. But be sure to measure the following dimension before installing the power brake booster assembly.

Adjustment

Refer to illustration 12.13

13 Measure the distance between the power brake booster and the hole in the clevis **(see illustration)** and compare it to the booster-to-clevis dimension listed in this Chapter's Specifications. If it isn't the same, loosen the adjusting nut and turn the clevis in or out to the specified length, then tighten the nut.

13 Parking brake - adjustment

Refer to illustration 13.4

Warning: *Some models covered by this manual are equipped with airbags. Always disable the airbag system when working in the vicinity of airbag system components (see Chapter 12).*

1 The parking brake lever, when properly adjusted, should travel eight to ten clicks (1993 and 1994 models) or 10 to 11 clicks (1995 and later models) when a moderate pulling force is applied.
2 If the parking brake lever travels less than the specified minimum number of clicks, it might not be releasing completely and the shoes or pads could even be dragging against the drum or disc. If the lever can be pulled up more than the specified maximum number of clicks, the parking brake may not hold adequately on an incline, allowing the car to roll.
3 To gain access to the parking brake cable adjuster, remove the center console (see Chapter 11).
4 Loosen or tighten the adjusting nut **(see illustration)** until the desired travel is attained. Turn the nut clockwise to tighten the cable and decrease the number of clicks at the parking brake lever, or turn it counterclockwise to loosen the cable and increase the number of clicks at the lever.
5 Install the console (see Chapter 11).

13.4 To adjust parking brake lever travel, turn this adjusting nut until the specified number of clicks is obtained

14 Parking brake cables - replacement

Warning: *Some models covered by this manual are equipped with airbags. Always disable the airbag system when working in the vicinity of airbag system components (see Chapter 12).*

Rear cables

Refer to illustrations 14.3a 14.3b, 14.5 and 14.7

1 Make sure the parking brake is completely released.
2 Loosen the rear wheel lug nuts, raise the rear of the vehicle and support it securely on jackstands. Block the front wheels. Remove the wheel.
3 On models with rear drum brakes, remove the brake drum and brake shoes, and disconnect the cable from the parking brake levers on the rear shoes (see Section 6). Remove the spring from the cable (see illus-

14.3a On models with rear drum brakes, remove the spring from the parking brake cable . . .

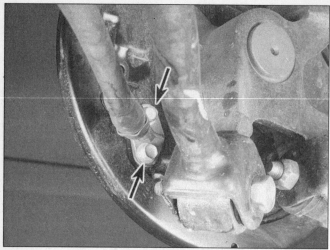

14.3b ... then remove these two bolts (arrows) to detach the cable housing from the backing plate

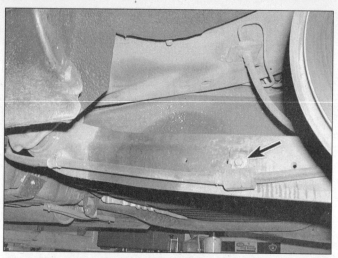

14.5 Follow the parking brake cable from the back to the front, unbolting any brackets along the way

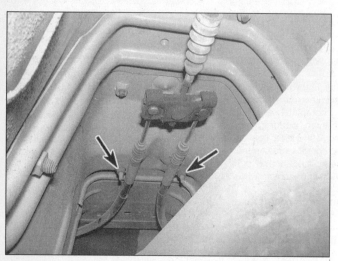

14.7 To disconnect the rear parking brake cables from the equalizer, remove these bracket nuts (arrows), slide the cables forward slightly and disengage the cable ends from the equalizer

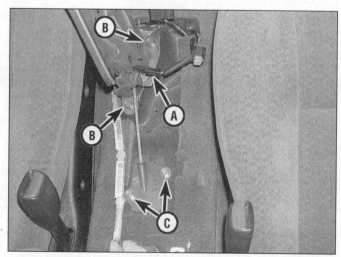

14.11 Unplug the electrical connector for the parking brake warning light switch (A), remove the lever mounting nuts (B) and detach the cable from the lever, then remove the bolts from the cable plate (C)

tration), unbolt the cable retainer from the backing plate (see illustration) and pull the cable through the backing plate.

4   On models with rear disc brakes, remove the locking plate that attaches the cable to the bracket bolted to the caliper (see illustration 3.6a) and disengage the cable from the toggle lever.

5   Unbolt any cable brackets from the frame or the suspension pieces (see illustration).

6   Remove the catalytic converter heat shield.

7   Remove the cable housing nuts just behind the equalizer and disengage the cables from the equalizer (see illustration).

8   Installation is the reverse of removal. Apply a light coat of grease to the portion of the cable end that engages with the equalizer.

9   Adjust the parking brake assembly when you're done (see Section 13).

Front cable

Refer to illustration 14.11

10   Remove the center console (see Chapter 11).

11   Unplug the electrical connector (see illustration) for the parking brake warning light switch.

12   Remove the parking brake cable adjusting nut (see illustration 13.4).

13   Remove the two bolts from the parking brake lever base (see illustration 14.11).

14   Raise the rear of the vehicle and support it securely on jackstands. Block the front wheels. Remove the catalytic converter heat shield.

15   Remove the two cable housing nuts just behind the equalizer and disengage the rear cables from the equalizer (see illustration 14.7).

16   Working inside the vehicle, remove the two cable retaining plate bolts and pull the cable and equalizer through the floorpan (see illustration 14.11).

17   Installation is the reverse of removal.

18   Adjust the parking brake assembly when you're done.

15 Brake pedal - adjustment

Brake pedal released height

Refer to illustrations 15.1 and 15.3

1   Peel back the carpet and insulator pad. With the brake pedal fully released, measure the distance from the top of the pad to the floor (see illustration).

2   If the height is not as listed in the Specifications Section at the beginning of this Chapter it must be adjusted.

3   Loosen the locknut just in front of the power brake booster clevis (see illustration).

4   Turn the booster input rod until the

Chapter 9 Brakes

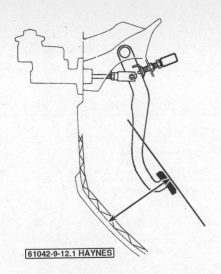

15.1 With the brake pedal fully released, measure the distance from the top of the pad to the floor

Brake pedal depressed height

11 After checking and, if necessary, adjusting the pedal released height and freeplay, the pedal depressed height must be checked.
12 With the engine running, press the brake pedal fully and measure the pedal pad-to-floor distance.
13 If the minimum depressed height is below that listed in the Specifications Section listed at the beginning of this Chapter, check the brake system for leaks or other damage.

16 Brake light switch - check and replacement

Refer to illustration 16.1

Check

1 The brake light switch **(see illustration)** is located on a bracket near the top of the brake pedal (it's the switch on the left side; the one on the right is for the cruise control system). The switch activates the brake lights at the rear of the vehicle when the pedal is depressed.
2 To check the brake light switch, simply note whether the brake lights come on when the pedal is depressed and go off when the pedal is released.
3 If the brake lights don't come on when the brake pedal is depressed, make sure the brake pedal is correctly adjusted (see Section 15). Then try adjusting the switch as follows.
4 A locknut secures the switch to the bracket. Loosen the locknut, then screw the switch in or out to provide a 0.012 to 0.039-inch (0.3 to 1.0mm) clearance between the switch plunger and the pedal stopper, with no pressure on the plunger, and tighten the locknut. Recheck the clearance to verify that it didn't change when you tightened the lock-

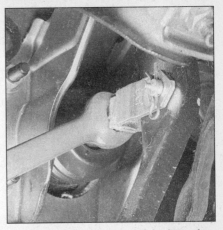

15.3 To adjust brake pedal released height, loosen the locknut in front of the brake booster clevis and turn the input rod until free height is correct (this procedure is also used to adjust brake pedal freeplay)

nut. The switch should now function properly.
5 If the switch still doesn't work properly, either it isn't getting voltage, or the switch itself is defective. Use a voltmeter or test light to verify that there's voltage at the switch connector. With the pedal at rest, voltage should be present at one of the terminals of the switch. With the pedal depressed, voltage should be present at both terminals. If voltage isn't present at both terminals when the pedal is depressed, replace the switch.

Replacement

6 Unplug the electrical connector from the brake light switch.
7 Loosen the locknut and unscrew the switch from the bracket.
8 Installation is the reverse of removal.
9 Adjust the brake pedal (see Section 15), then adjust the switch (see above).

pedal height is correct.
5 Tighten the locknut.
6 After adjusting the pedal height, check the freeplay.

Brake pedal freeplay

Refer to illustration 15.7

7 Press down lightly on the brake pedal and measure the distance that it moves freely before resistance is felt **(see illustration)**. The freeplay should be within the specified limits. If it isn't, it must be adjusted.
8 Loosen the locknut for the brake booster clevis **(see illustration 15.3)**.
9 Turn the booster input rod until the pedal freeplay is correct.
10 Tighten the locknut.

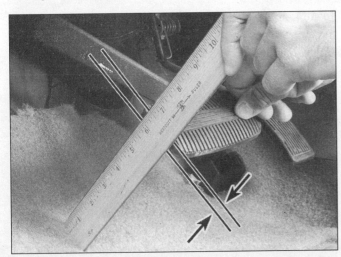

15.7 To measure brake pedal freeplay, press down lightly on the brake pedal and measure the distance that it moves freely before resistance is felt

16.1 The brake light switch (arrow) is located on a bracket near the top of the brake pedal (instrument panel removed for clarity)

Notes

Chapter 10
Suspension and steering systems

Contents

	Section
Balljoints - replacement	8
Control arm (front) - removal, inspection and installation	7
General information	1
Hub and wheel bearing assembly (front) - removal and installation	5
Hub and wheel bearing assembly (rear) - removal and installation	10
Power steering fluid level check	See Chapter 1
Power steering pump - removal and installation	18
Power steering system - bleeding	19
Rear axle beam, lateral link and control rod (1995 and later models) - removal and installation	13
Rear suspension arms (1993 and 1994 models) - removal and installation	12
Stabilizer bar (front) - removal and installation	6
Stabilizer bar (rear) (1993 and 1994 models) - removal and installation	11

	Section
Steering gear - removal and installation	17
Steering gear boots - replacement	16
Steering knuckle and hub - removal and installation	4
Steering wheel - removal and installation	14
Strut or coil spring - replacement	3
Strut or shock absorber/coil spring assembly (rear) - removal, inspection and installation	9
Strut/coil spring assembly (front) - removal, inspection and installation	2
Tie-rod ends - removal and installation	15
Tire and tire pressure checks	See Chapter 1
Tire rotation	See Chapter 1
Wheel alignment - general information	22
Wheel studs - replacement	20
Wheels and tires - general information	21

Specifications

Power steering fluid type .... See Chapter 1

Torque specifications

	Nm	Ft-lbs (unless otherwise indicated)
Front suspension		
Balljoint-to-control arm nuts (1993 and 1994 models)	76 to 109	56 to 80
Balljoint-to-steering knuckle nut		
1993 and 1994	71 to 86	52 to 64
1995 and later	62 to 76	46 to 56
Control arm		
Pivot bolt nut		
1993 and 1994	88 to 118	65 to 87
1995 and later	118 to 147	87 to 108
Bushing pin/gusset/rear bushing clamp bolts (1993 and 1994)	118 to 147	87 to 108
Bushing pin bolts (1995 and later)	118 to 147	87 to 108
Rear bushing clamp bolts (1995 and later)	118 to 147	87 to 108
Stabilizer bar		
Link-to-stabilizer bar nut		
1993 and 1994	41 to 51	30 to 38
1995 and later	16 to 22	144 to 192 in-lbs
Link-to-control arm nut		
1993 and 1994	16 to 22	144 to 192 in-lbs
1995 and later	41 to 47	30 to 35
Stabilizer bar clamp bolts		
1993 and 1994	31 to 42	23 to 31
1995 and later	49 to 59	36 to 43

Torque specifications

Front suspension (continued)

	Nm	Ft-lbs (unless otherwise indicated)
Strut/coil spring assembly		
Strut-to-steering knuckle bolts/nuts		
1993 and 1994	157 to 167	116 to 123
1995 and later	140 to 159	103 to 117
Strut upper mounting nuts	39 to 54	29 to 40
Piston rod nut		
1993 through 1988	59 to 78	43 to 58
1999 and later	59 to 88	43 to 65

Rear suspension

1993 and 1994 models

	Nm	Ft-lbs
Brake backing plate-to-rear knuckle bolts (drum brake models)	38 to 52	28 to 38
Disc brake splash shield bolts	38 to 52	28 to 38
Hub and bearing assembly retaining nut	186 to 255	137 to 188
Rear suspension arms		
Lateral arms		
Lateral arm-to-crossmember nuts	88 to 118	65 to 87
Lateral arm-to-rear knuckle nuts	88 to 118	65 to 87
Trailing arm bolts/nuts (both ends)	88 to 108	65 to 80
Trailing arm bracket-to-strut bolts	59 to 78	43 to 58
Stabilizer bar		
Stabilizer bar bushing clamp nuts	31 to 42	23 to 31
Stabilizer bar link-to-chassis nuts	41 to 47	30 to 35
Stabilizer bar-to-trailing arm bracket bolts	59 to 78	43 to 58
Strut/coil spring assembly		
Strut upper mounting nuts	42 to 54	31 to 40
Piston rod nut	59 to 78	43 to 58

1995 and later models

	Nm	Ft-lbs
Shock absorber assembly		
Piston rod nut	18 to 24	156 to 204 in-lbs
Upper mounting nuts		
1995 through 1998	16 to 19	144 to 168 in-lbs
1999 and later	16 to 22	144 to 192 in-lbs
Lower mounting bolt/nut		
1995 through 1998	98 to 118	72 to 87
1999 and later	108 to 127	80 to 94
Lateral link and control rod		
Lateral link-to-axle nut	98 to 118	72 to 87
Lateral link-to-chassis bolt/nut	98 to 118	72 to 87
Control rod-to-lateral link bolt/nut		
1995 through 1998	59 to 78	43 to 58
1999 and later	79 to 98	58 to 72
Control rod-to-axle nut	98 to 118	72 to 87
Trailing arm-to-chassis bolt/nut		
1995 through 1998	98 to 118	72 to 87
1999 and later	108 to 127	80 to 94
Disc brake splash shield bolts	22 to 29	16 to 22
Hub and bearing assembly retaining nut	186 to 255	137 to 188

Steering

	Nm	Ft-lbs
Airbag module Torx bolts	15 to 25	132 to 216 in-lbs
Steering gear mounting bracket bolts and nuts	73 to 97	54 to 72
Steering shaft universal joint pinch bolt	24 to 29	17 to 22
Steering wheel nut		
1993 and 1994	29 to 39	22 to 29
1995 and later	33 to 39	25 to 29
Tie-rod end-to-steering knuckle nut		
1993 and 1994	29 to 39	22 to 29
1995 and 1996	63 to 73	46 to 54
1997 through 1999	29 to 39	22 to 29
Power steering pump pressure line banjo bolt	69	51

Chapter 10 Suspension and steering systems

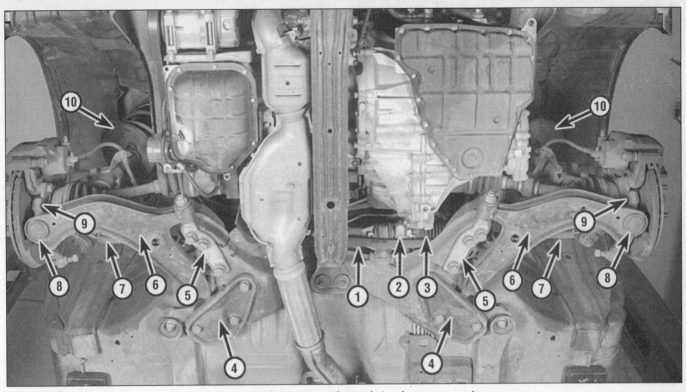

1.1 Typical front suspension and steering components

1. Stabilizer bar
2. Steering gear
3. Steering gear boot
4. Control arm bushing clamp
5. Control arm bushing pin
6. Control arm
7. Tie-rod end
8. Balljoint
9. Steering knuckle
10. Strut/coil spring assembly

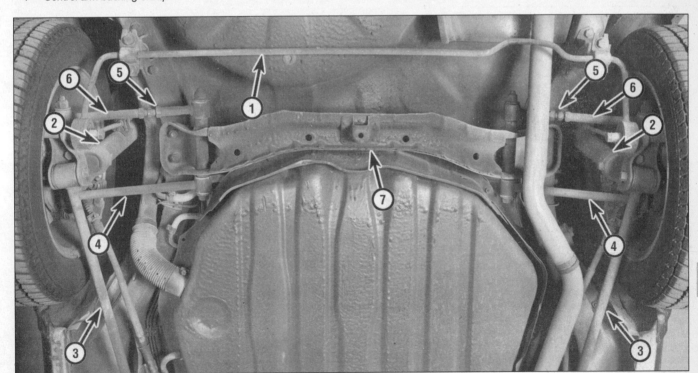

1.2 Typical rear suspension components (1993 and 1994 models)

1. Stabilizer bar
2. Strut/coil spring assembly
3. Trailing arm (radius rod)
4. Front lateral arm
5. Toe adjuster
6. Rear lateral arm
7. Suspension crossmember

Chapter 10 Suspension and steering systems

1.3 Rear suspension components (1995 and later models)

| 1 Axle beam | 3 Lateral link | 5 Shock absorber |
| 2 Control rod | 4 Trailing arm | 6 Coil spring |

1 General information

Refer to illustrations 1.1, 1.2 and 1.3

The front suspension system **(see illustration)** is a strut/coil spring design. The upper end of each strut is attached to the vehicle body. The lower end of the strut is connected to the upper end of the steering knuckle. The bottom of the steering knuckle is attached to a balljoint on the outer end of the control arm. The balljoint is an integral part of the control arm; if the balljoint is worn, the control arm must be replaced. A stabilizer bar is used on all models. The bar is attached to the frame with a pair of clamps and to the control arms with link rods.

The rear suspension system on 1993 and 1994 models **(see illustration)** also uses strut/coil spring assemblies, a pair of lateral suspension arms and a trailing arm, or radius rod, at each corner. The upper ends of the struts are attached to the vehicle body and their lower ends are attached to the upper ends of the rear knuckles. The lower ends of the knuckles are attached to the outer ends of the lateral arms; the strut/knuckle/hub assemblies are positioned laterally by the trailing arms. A stabilizer bar is attached to the vehicle by a pair of brackets and to the struts by link rods.

The rear suspension on 1995 and later models consists of a beam axle located by trailing arms and a lateral link, and is suspended by shock absorber/coil spring assemblies **(see illustration)**. The lateral link is fastened to the chassis and the axle and pivots on an internal control link.

The rack-and-pinion steering gear is located behind the engine/transaxle assembly at the bottom of the firewall (it's bolted to the lower rear crossmember). The steering gear actuates the tie-rods, which are attached to the steering knuckles. The inner ends of the tie-rods are protected by rubber boots which should be inspected periodically for secure attachment, tears and leaking lubricant.

The power assist system consists of a belt-driven pump and associated lines and hoses. The fluid level in the power steering pump reservoir should be checked periodically (see Chapter 1).

The steering wheel operates the steering shaft, which actuates the steering gear through universal joints. Looseness in the steering can be caused by wear in the steering shaft universal joints, the steering gear, the tie-rod ends and loose retaining bolts.

Frequently, when working on the suspension or steering system components, you may come across fasteners which seem impossible to loosen. These fasteners on the underside of the vehicle are continually subjected to water, road grime, mud, etc., and can become rusted or "frozen," making them extremely difficult to remove. In order to unscrew these stubborn fasteners without damaging them (or other components), be sure to use lots of penetrating oil and allow it to soak in for a while. Using a wire brush to clean exposed threads will also ease removal of the nut or bolt and prevent damage to the threads. Sometimes a sharp blow with a hammer and punch will break the bond between a nut and bolt threads, but care must be taken to prevent the punch from slipping off the fastener and ruining the threads. Heating the stuck fastener and surrounding area with a torch sometimes helps too, but isn't recommended because of the obvious dangers associated with fire. Long breaker bars and extension, or "cheater," pipes will increase leverage, but never use an extension pipe on a ratchet - the ratcheting mechanism could be damaged. Sometimes tightening the nut or bolt first will help to break it loose. Fasteners that require drastic measures to remove should always be replaced with new ones.

Since most of the procedures dealt with in this Chapter involve jacking up the vehicle and working underneath it, a good pair of jackstands will be needed. A hydraulic floor jack is the preferred type of jack to lift the vehicle, and it can also be used to support certain components during various operations. **Warning:** *Never, under any circumstances, rely on a jack to support the vehicle while working on it. Whenever any of the suspension or steering fasteners are loosened or removed they must be inspected and, if necessary, replaced with new ones of the same part number or of original equipment quality and design. Torque specifications must be followed for proper reassembly and component retention. Never attempt to heat or straighten any suspension or steering components. Instead, replace any bent or damaged part with a new one.*

Chapter 10 Suspension and steering systems

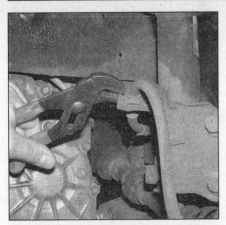

2.2 Remove the retaining clip with a pair of pliers and detach the brake hose from the strut

2.3 To detach the strut assembly from the steering knuckle, remove the two nuts (arrows), then drive out the strut-to-knuckle bolts with a hammer and punch

2.5 To detach the upper end of the strut from the body, remove the upper mounting nuts (arrows). Warning: *Don't unscrew the center nut!*

2 Strut/coil spring assembly (front) - removal, inspection and installation

Removal

Refer to illustrations 2.2, 2.3 and 2.5

1   Loosen the front wheel lug nuts, raise the front of the vehicle and support it securely on jackstands. Remove the wheels.
2   Unclip the brake hose from the strut bracket and detach it from the bracket **(see illustration)**.
3   Remove the strut-to-knuckle nuts **(see illustration)** and knock the bolts out with a hammer and punch.
4   Separate the strut from the steering knuckle. Be careful not to overextend the inner CV joint and don't let the knuckle fall outward, as this could damage the brake hose.
5   Remove the three strut upper mounting nuts **(see illustration)**. **Note:** *Support the strut with one hand before removing the last nut.* Remove the assembly out from the fenderwell.

Inspection

6   Check the strut body for leaking fluid, dents, cracks and other obvious damage which would warrant repair or replacement.
7   Check the coil spring for chips or cracks in the spring coating (this will cause premature spring failure due to corrosion). Inspect the spring seat for cuts, hardness and general deterioration.
8   If any undesirable conditions exist, proceed to the strut disassembly procedure (see Section 3).

Installation

9   Guide the strut assembly up into the fenderwell and insert the upper mounting studs through the holes in the shock tower. Once the studs protrude from the shock tower, install the nuts so the strut won't fall back through. This is most easily accomplished with the help of an assistant, as the strut is quite heavy and awkward.
10   Slide the steering knuckle into the strut flange and insert the bolts. Install *new* nuts and tighten them to the torque listed in this Chapter's Specifications.
11   Guide the brake hose through its bracket in the strut and install the retaining clip.
12   Install the wheel and lug nuts, then lower the vehicle and tighten the lug nuts to the torque listed in the Chapter 1 Specifications.
13   Tighten the upper mounting nuts to the torque listed in this Chapter's Specifications.

3 Strut or coil spring - replacement

1   If the struts or coil springs exhibit the telltale signs of wear (leaking fluid, loss of damping capability, chipped, sagging or cracked coil springs) explore all options before beginning any work. The strut/shock absorber assemblies are not serviceable and must be replaced if a problem develops. However, strut assemblies complete with springs may be available on an exchange basis, which eliminates much time and work. Whichever route you choose to take, check on the cost and availability of parts before disassembling your vehicle. **Warning:** *Disassembling a strut is potentially dangerous and utmost attention must be directed to the job, or serious injury may result. Use only a high-quality spring compressor and carefully follow the manufacturer's instructions furnished with the tool. After removing the coil spring from the strut assembly, set it aside in a safe, isolated area.*

Disassembly

Refer to illustrations 3.3, 3.4, 3.5, 3.6 and 3.7

2   Remove the strut and spring assembly following the procedure described in the previous Section. Mount the strut assembly in a vise. Line the vise jaws with wood or rags to

3.3 Install the spring compressor according to the tool manufacturer's instructions and compress the spring until all pressure is relieved from the upper spring seat

prevent damage to the unit and don't tighten the vise excessively.
3   Following the tool manufacturer's instructions, install the spring compressor (which can be obtained at most auto parts stores or equipment yards on a daily rental basis) on the spring and compress it suffi-

3.4 Remove the piston rod nut

10-6   Chapter 10   Suspension and steering systems

3.5 Lift the upper mount off the piston rod

3.6 Remove the spring seat from the piston rod

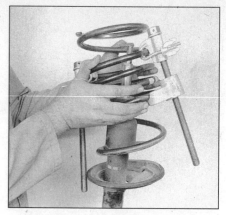

3.7 Remove the compressed spring assembly - keep the ends of the spring pointed away from your body

ciently to relieve all pressure from the upper spring seat **(see illustration)**. This can be verified by wiggling the spring.

4   Remove the piston rod nut **(see illustration)**.

5   Remove the upper mount **(see illustration)**. Inspect the bearing in the mount for smooth operation. If it does not turn smoothly, replace it. Check the rubber portion of the mount for cracking and general deterioration. If there is any separation of the rubber, replace it.

6   Lift the spring seat and upper insulator from the piston rod **(see illustration)**. Check the rubber spring seat for cracking and hardness, replacing it if necessary.

7   Carefully lift the compressed spring from the assembly **(see illustration)** and set it in a safe place. **Warning:** *Carry the spring carefully and never place any part of your body near the end of the spring!*

8   Slide the dust boot off the piston rod.

9   Check the lower insulator (if equipped) for wear, cracking and hardness and replace it if necessary.

Reassembly

Refer to illustration 3.11, 3.12a and 3.12b

10 If the lower insulator is being replaced, set it into position with the dropped portion seated in the lowest part of the seat. Extend the damper rod to its full length and install the dust boot.

11 Carefully place the coil spring onto the lower insulator, with the end of the spring resting in the lowest part of the insulator **(see illustration)**.

12 Install the upper insulator and the spring seat. Make sure the cutout on the spring seat is facing out (away from the vehicle), in line with the strut-to-knuckle attachment points **(see illustrations)**.

13 Install the dust seal and upper mount to the piston rod.

14 Install the nut and tighten it to the torque listed in this Chapter's Specifications.

15 Install the strut/shock absorber and coil spring assembly (see Section 2).

4 Steering knuckle and hub - removal and installation

Warning: *Dust created by the brake system may contain asbestos, which is harmful to your health. Never blow it out with compressed air and don't inhale any of it. Do not, under any circumstances, use petroleum-based solvents to clean brake parts. Use brake system cleaner only.*

Removal

1   Remove the hub cap or wheel cover, remove the cotter pin and nut lock, then loosen the driveaxle/hub nut (see Chapter 8). **Note:** *If the socket won't fit through the opening in the center of the wheel, remove the wheel and install the spare.*

2   Loosen the wheel lug nuts, raise the vehicle and support it securely on jackstands. Remove the wheel.

3   Remove the brake caliper and the brake disc (see Chapter 9), and disconnect the brake hose from the strut **(see illustration 2.2)**. If the vehicle is equipped with ABS, disconnect and remove the wheel speed sensor.

4   Remove the strut-to-steering knuckle nuts, but don't remove the bolts yet (see Section 2).

5   Separate the tie-rod end from the steering knuckle arm (see Section 15).

6   Separate the balljoint from the steering knuckle (see Section 7).

7   Remove the strut-to-knuckle bolts and separate the knuckle from the strut while pushing the driveaxle from the hub (if the driveaxle sticks in the hub, tap of the end of the driveaxle with a hammer and brass punch). Support the end of the driveaxle with a length of wire.

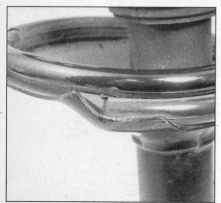

3.11 When installing the spring, make sure the end fits into the recessed portion of the lower seat (arrow)

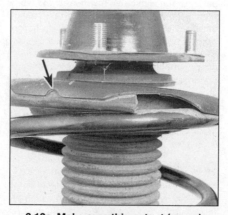

3.12a Make sure this cutout (arrow) in the upper seat . . .

3.12b . . . is facing out (toward the strut-to-knuckle flanges)

Chapter 10 Suspension and steering systems

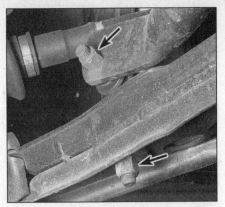

6.2a To disconnect the stabilizer bar link from the control arm on 1993 and 1994 models, remove the nut indicated by the lower arrow; to disconnect the link from the stabilizer bar, remove the upper nut (upper arrow)

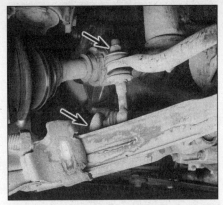

6.2b To disconnect the stabilizer bar link from the control arm on 1995 and later models, remove the nut indicated by the lower arrow; to detach the link from the stabilizer bar, remove the nut indicated by the upper arrow

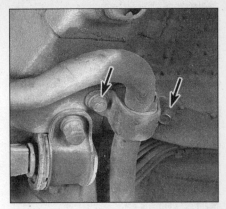

6.3a To disconnect the stabilizer bar from the body on a 1993 or 1994 model, remove the bushing clamp nuts and bolts (arrows) from both sides

Installation

8 Lubricate the splines of the driveaxle with multi-purpose grease. Guide the knuckle and hub assembly into position, inserting the driveaxle into the hub.
9 Push the knuckle into the strut flange and install the bolts and two *new* nuts, but don't tighten them yet.

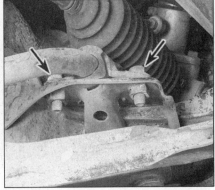

6.3b To remove the stabilizer bar clamps on 1995 and later models, remove these bolts (don't try to unscrew the nuts - they're welded to the subframe)

7.3a On 1993 and 1994 models, the balljoint is secured to the lower arm by three bolts/nuts (arrows)

10 Attach the control arm to the steering knuckle (see Section 7).
11 Attach the tie-rod end to the steering knuckle arm (see Section 15). Tighten the strut-to-knuckle nuts to the torque listed in this Chapter's Specifications.
12 Place the brake disc on the hub and install the caliper (see Chapter 9).
13 Install the driveaxle/hub nut and tighten it securely (but not completely yet).
14 Install the wheel and lug nuts, then lower the vehicle.
15 Tighten the lug nuts to the torque listed in the Chapter 1 Specifications. Tighten the driveaxle/hub nut to the torque listed in the Chapter 8 Specifications, then install the felt washer, nut lock, and a new cotter pin.

5 Hub and wheel bearing assembly (front) - removal and installation

Due to the special tools and expertise required to press the hub and bearing from the steering knuckle, this job should be left to a professional mechanic. However, the steering knuckle and hub may be removed and the assembly taken to an automotive machine

7.3b Remove the cotter pin . . .

shop or other qualified repair facility. See Section 4 for the steering knuckle and hub removal procedure.

6 Stabilizer bar (front) - removal and installation

Refer to illustrations 6.2a, 6.2b, 6.3a and 6.3b
1 Loosen the wheel lug nuts, raise the front of the vehicle, support it securely on jackstands and remove the wheels.
2 Disconnect the stabilizer bar links from both control arms **(see illustrations)**.
3 Remove both stabilizer bar clamps **(see illustrations)**.
4 Remove the stabilizer assembly.
5 Inspect the clamp bushings and the link bushings. If they're cracked or torn, replace them.
6 Installation is the reverse of removal. When installing the clamp bushings, make sure the slits in the bushings face forward. Be sure to tighten all fasteners to the torque listed in this Chapter's Specifications.

7 Control arm (front) - removal, inspection and installation

Removal
Refer to illustrations 7.3a, 7.3b, 7.3c, 7.3d, 7.3e, 7.4 and 7.5
1 Loosen the wheel lug nuts on the side to be dismantled, raise the front of the vehicle, support it securely on jackstands and remove the wheel.
2 Disconnect the stabilizer bar link from the control arm (see Section 6). On 1993 and 1994 models, remove the stabilizer bar (see Section 6).
3 If you're working on a 1993 or 1994 model, unbolt the balljoint from the control arm **(see illustration)**. If you're working on a 1995 and later model, remove the cotter pin and loosen the balljoint stud nut **(see illustrations)**. Separate the balljoint stud from the

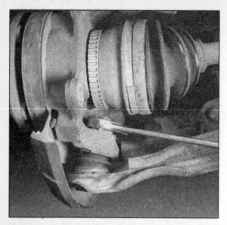

7.3c ... then loosen the balljoint stud nut and back it off as far as it will go (without actually removing it)

7.3d Separating the balljoint from the steering knuckle with a "picklefork"-type balljoint separator

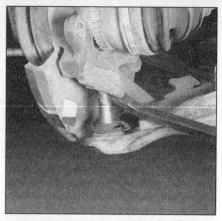

7.3e Separate the control arm from the steering knuckle by prying it down with a prybar or large screwdriver

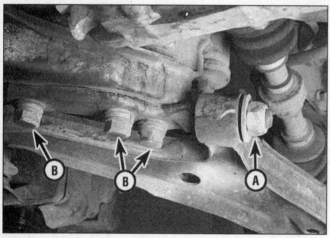

7.4 Loosen the nut from the bushing pin (A), then unscrew the bolts from the control arm bushing pin (B) (shown here is a 1995 model; 1993 and 1994 models only have two bolts securing the bushing pin)

7.5 Remove the rear bushing clamp bolts (arrows) - 1995 model shown

steering knuckle with a balljoint separator **(see illustration)**. **Note:** *The use of a picklefork-type balljoint separator will almost certainly damage the balljoint boot.* Separate the arm from the steering knuckle **(see illustration)**.

4 If the control arm is going to be replaced, loosen the nut from the control arm bushing pin. Remove the bolts from the control arm bushing pin **(see illustration)**.

5 Remove the rear bushing clamp bolts **(see illustration)**, then remove the control arm. If necessary, on 1993 and 1994 models, remove the control arm gusset.

Inspection

6 Inspect the front and rear bushings for cracks and tears. If either bushing is damaged or worn, check on the availability of replacement bushings. If bushings are available, take the arm to an automotive machine shop or other qualified repair facility to have the bushings replaced. If replacement bushings are not available, replace the control arm.

7 Inspect the control arm for straightness. If it's bent, replace it. Do not attempt to straighten a bent control arm.

Installation

Refer to illustrations 7.8a and 7.8b

8 Installation is the reverse of removal. Tighten all of the fasteners to the torque listed in this Chapter's Specifications. **Note:** *The bushing pin bolts and the rear bushing clamp bolts (and the gusset bolts on 1993 and 1994 models) must be tightened in a specific order* **(see illustrations)**. *Also, before fully tightening the bushing clamp bolts or the nut on the end of the bushing pin, raise the outer end of the control arm with a floor jack to simulate normal ride height.* Be sure to install a new cotter pin through the balljoint stud.

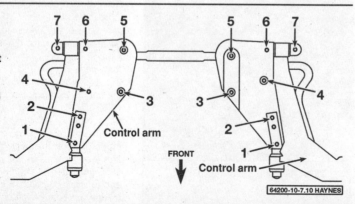

7.8a Control arm bushing pin, clamp and gusset bolt tightening sequence (1993 and 1994 models)

Chapter 10 Suspension and steering systems

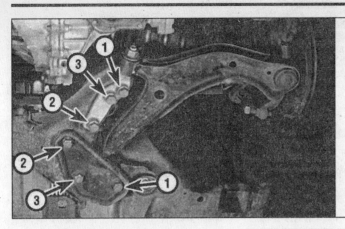

7.8b Control arm pin and bushing bolt tightening sequence (1995 and later models)

9 Install the wheel and lug nuts, lower the vehicle and tighten the lug nuts to the torque listed in the Chapter 1 Specifications.

8 Balljoints - replacement

1993 and 1994 models

1 Loosen the wheel lug nuts, raise the front of the vehicle and support it securely on jackstands. Remove the wheel.
2 Separate the balljoint from the steering knuckle (see Section 7).
3 Remove the three nuts that secure the balljoint to the steering knuckle (see illustration 7.3a). Detach the balljoint from the control arm.
4 Installation is the reverse of removal. Tighten the balljoint-to-control arm nuts, and the balljoint stud nut, to the torque values listed in this Chapter's Specifications. Also be sure to install a new cotter pin. If the cotter pin hole doesn't line up with the slots on the nut, tighten the nut until the next slot in the nut lines up with the hole in the balljoint stud.
5 Install the wheel and lug nuts. Lower the vehicle and tighten the lug nuts to the torque listed in the Chapter 1 Specifications.

1995 and later models

6 On these models the balljoint is an integral part of the control arm. If it's worn or damaged, replace the control arm (see Section 7).

9 Strut or shock absorber/coil spring assembly (rear) - removal, inspection and installation

1993 and 1994 models

Refer to illustrations 9.7a and 9.7b

Removal

1 Loosen the rear wheel lug nuts, raise the rear of the vehicle and support it securely on jackstands. Remove the wheels.
2 Unbolt the brake hose bracket from the strut. If the vehicle is equipped with ABS, detach the speed sensor wiring harness from the strut by removing the clamp bracket bolt.
3 Remove the brake caliper (without disconnecting the hose) and the brake disc (see Chapter 9). Support the caliper with a length of wire - don't let it hang by the hose.
4 If the strut is going to be replaced, remove the hub and wheel bearing assembly.
5 Detach the lateral arms and the trailing arm from the strut (see Section 12).
6 Open the trunk and remove the trim panels to get at the upper mounting nuts.
7 Remove the rubber cover (see illustration). Have an assistant support the strut and spring assembly while you remove the three strut upper mounting nuts (see illustration). Remove the assembly through the fenderwell. If the strut is going to be replaced, remove the trailing arm bracket from the bottom of the strut and transfer it to the new one, tightening the bolts to the torque listed in this Chapter's Specifications.

Inspection

8 Inspect the strut as outlined in Steps 6 and 7 in Section 2. If any undesirable conditions exist, replace the strut (see Section 3).

Installation

9 Guide the strut assembly up into the fenderwell and insert the upper mounting studs through the holes in the shock tower. Once the studs protrude from the shock tower, install the nuts so the strut won't fall back through. This is most easily accomplished with the help of an assistant, as the strut is quite heavy and awkward.
10 Connect the lateral arms and trailing arm to the strut (see Section 12). Raise the rear suspension with a floor jack placed under the trailing arm bracket, to simulate normal ride height, then tighten the lateral arm bolts and the trailing arm bolts to the torque listed in this Chapter's Specifications.
11 Connect the brake hose bracket to the strut and tighten the bolt securely. If the vehicle is equipped with ABS, install the speed sensor wiring harness bracket.
12 Install the wheel and lug nuts, then lower the vehicle and tighten the lug nuts to the torque listed in the Chapter 1 Specifications.
13 Tighten the upper mounting nuts to the torque listed in this Chapter's Specifications. Install the trim panels in the luggage compartment.

1995 and later models

Removal

Refer to illustrations 9.15 and 9.16

14 Loosen the rear wheel lug nuts, raise the rear of the vehicle and support it securely on jackstands. Remove the wheels.
15 Remove the lower shock absorber-to-axle nut, remove the bolt and separate the shock from the axle (see illustration).
16 The upper mounting nuts are located inside the trunk. Remove the trim panels from inside the trunk, then have an assistant support the shock and spring assembly while you remove the two mounting nuts (see illustration). Remove the assembly out from the fenderwell.

9.7a Pry off this rubber cover . . .

9.7b . . . and remove the strut upper mounting nuts (1993 and 1994 models)

9.15 To disconnect the lower end of the shock absorber from the axle, remove this nut and bolt

10-10 Chapter 10 Suspension and steering systems

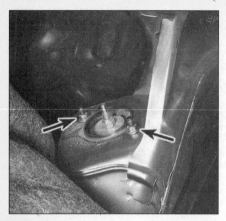

9.16 The shock absorber upper mounting nuts are located in the trunk

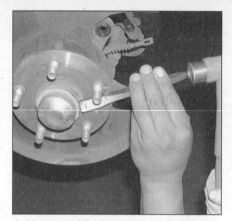

10.3 Remove the hub/bearing grease cap with a hammer and chisel

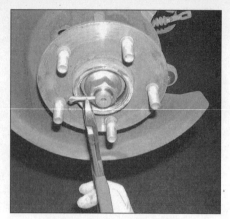

10.4 Remove the cotter pin

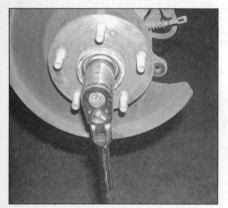

10.5 Remove the bearing/hub retaining nut

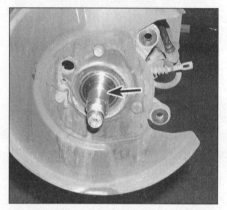

10.7 Apply a film of wheel bearing grease to the spindle (arrow) before installing the hub and bearing assembly

Inspection

17 Refer to Steps 6 and 7 in Section 2.
18 If any undesirable conditions exist, proceed to the disassembly procedure to replace the shock absorber or coil spring (see Section 3).

Installation

19 Guide the assembly up into the fenderwell and insert the upper mounting studs through the holes in the shock tower. Make sure the upper spring seat is aligned. Once the studs protrude from the shock tower,

install the nuts so the shock absorber won't fall back through. This is most easily accomplished with the help of an assistant, as the assembly is quite heavy and awkward.
20 Slide the lower end of the shock into the axle mount and insert the bolt. Raise the suspension with a floor jack to simulate normal ride height, install the nut and tighten it to the torque listed in this Chapter's Specifications. **Caution:** *Place the jack head under the trailing arm bracket - not under the trailing arm or lateral arms.*

21 Tighten the upper mounting nuts to the torque listed in this Chapter's Specifications.

10 Hub and wheel bearing assembly (rear) - removal and installation

Refer to illustrations 10.3, 10.4, 10.5 and 10.7

1 Loosen the rear wheel lug nuts, raise the rear of the vehicle, support it securely on jackstands and remove the wheels.
2 Remove the brake drum or disc (see Chapter 9).
3 Remove the grease cap **(see illustration)**.
4 On models so equipped, remove the cotter pin **(see illustration)**.
5 Remove the wheel bearing locknut and washer **(see illustration)**.
6 Remove the hub and bearing assembly from the spindle.
7 Before installing the hub and bearing assembly, clean the spindle and apply a coat of wheel bearing grease to the area on the spindle where the bearing rides **(see illustration)**.
8 Installation is the reverse of removal. Tighten the hub nut to the torque listed in this Chapter's Specifications. On models so equipped, install a new cotter pin. On all other models, stake the new hub nut against the spindle on two sides. Tighten the caliper mounting bolts to the torque listed in the Chapter 9 Specifications.
9 Install the wheel and lug nuts. Lower the vehicle and tighten the lug nuts to the torque listed in the Chapter 1 Specifications.

11 Stabilizer bar (rear) (1993 and 1994 models) - removal and installation

Refer to illustrations 11.2 and 11.3

1 Loosen the rear wheel lug nuts, raise the rear of the vehicle, support it securely on jackstands and remove the wheels.
2 Remove the bolts that attach the stabilizer bar to the trailing arm brackets **(see illustration)**.

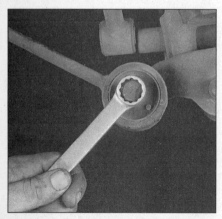

11.2 Remove the stabilizer bar-to-trailing arm bracket bolt

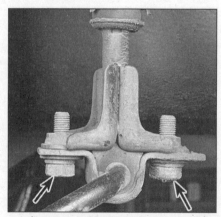

11.3 Remove the stabilizer bar bracket bolts

Chapter 10 Suspension and steering systems

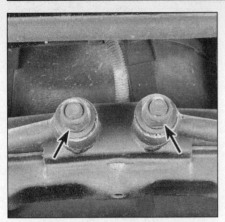

12.2 To disconnect the inner ends of the lateral arms, remove these nuts and bolts (arrows)

3   Remove the bushing clamp nuts and remove the stabilizer bar **(see illustration)**.
4   Inspect all clamp bushings. If they're cracked or torn, replace them.
5   Installation is the reverse of removal. Be sure to tighten all fasteners to the torque listed in this Chapter's Specifications.

12 Rear suspension arms (1993 and 1994 models) - removal and installation

1   Loosen the rear wheel lug nuts. Raise the rear of the vehicle, support it securely on jackstands and remove the wheels.

Lateral arms

Refer to illustration 12.2
2   Remove the inner pivot nuts **(see illustration)**.
3   Remove the bolts and nuts that attach the lateral arms to the knuckle.
4   Remove the arms.
5   Installation is the reverse of removal. Be sure to tighten all fasteners to the torque listed in this Chapter's Specifications. **Note:** *Raise the rear suspension with a floor jack to simulate normal ride height before tightening the fasteners.* **Caution:** *Place the jack head under the trailing arm bracket - not under the trailing arm or lateral arms.*
6   When you're done, drive the vehicle to an alignment shop and have the rear-wheel toe-in checked and, if necessary, adjusted.

Trailing arms

Refer to illustration 12.8
7   Remove the pivot bolt and nut from the forward end of the trailing arm.
8   Remove the bolt and nut that attach the rear end of the arm to its bracket **(see illustration)**.
9   Remove the trailing arm.
10 Installation is the reverse of removal. Place a floor jack under the rear knuckle and raise the suspension to simulate normal ride height, then tighten the fasteners to the torque listed in this Chapter's Specifications. **Caution:** *Place the jack head under the trailing arm bracket - not under the trailing arm or lateral arms.*

All arms

11 Install the wheels and lug nuts, lower the vehicle and tighten the lug nuts to the torque listed in the Chapter 1 Specifications.

13 Rear axle beam, lateral link and control rod (1995 and later models) - removal and installation

Refer to illustrations 13.7 and 13.8
1   Loosen the wheel lug nuts, raise the vehicle and support it securely on jackstands. Remove the wheels.
2   On vehicles with an Anti-lock Brake System (ABS), remove the rear wheel speed sensors. Also detach the speed sensor wiring harness by removing the clamp bracket bolt.
3   Disconnect the parking brake cables and remove the brake calipers and brake discs (see Chapter 9). Hang the calipers with lengths of wire from the coil springs.
4   If the axle assembly is being replaced, remove the rear hub and wheel bearing assemblies (see Section 10).
5   Place a floor jack underneath the center of the rear axle assembly. Make sure the jack doesn't contact the lateral link. Raise the jack just enough to support the axle assembly. **Note:** *If two floor jacks are available, place one at each end of the axle beam.*
6   Unbolt the lower ends of the shock absorbers from the axle (see Section 9).
7   Disconnect the lateral link from the chassis **(see illustration)**.
8   Remove the pivot bolts for the trailing arms **(see illustration)** and, with an assistant helping to balance the axle assembly, carefully lower the axle/lateral link assembly on the jack.
9   Check the condition of the lateral link, control rod and trailing arm bushings. If any of the bushings are deteriorated, replace them.
10 The lateral link and control rod can be removed from the axle beam by removing the fasteners and sliding the assembly off the pin on the axle. When reassembling the components, make sure the arrow on the lateral link points up, and the control rod bushing with the smaller diameter hole is connected to the lateral link.
11 Installation is the reverse of removal. Make sure you tighten all of the suspension fasteners to the torque listed in this Chapter's Specifications. **Note:** *The lateral link and shock absorber fasteners should be tightened with the suspension at normal ride height (this can be simulated by raising the rear suspension with a floor jack). The trailing arm pivot bolts should be tightened with the rear suspension in the unloaded (fully extended) position.*
12 Tighten the brake fasteners to the torque values listed in the Chapter 9 Specifications.
13 Install the wheels and lug nuts. Lower the vehicle and tighten the lug nuts to the torque listed in the Chapter 1 Specifications.

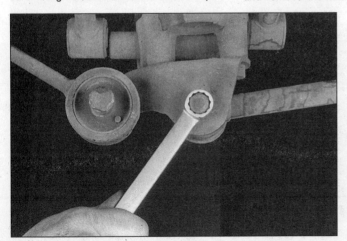

12.8 Removing the trailing arm-to-bracket bolt

13.7 Remove the nut and bolt (arrow) to detach the lateral link from the chassis

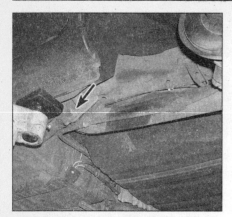

13.8 Remove the pivot bolt and nut (arrow) from the front of each trailing arm

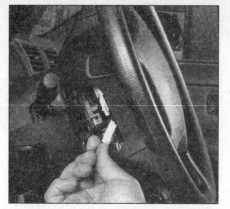

14.3 Pry off the access cover from the underside of the steering wheel and unplug the airbag module connector

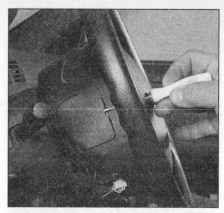

14.4a Pry off the two small side covers ...

14 Steering wheel - removal and installation

Airbag models

Warning: *To prevent accidental deployment (and possible injury) when working near airbag components, disconnect the negative battery cable, then the positive battery cable and wait at least 10 minutes before beginning work (the system has a back-up capacitor that must fully discharge). For more information on the airbag system, see Chapter 12.*

Removal

Refer to illustrations 14.3, 14.4a, 14.4b, 14.5a, 14.5b, 14.7 and 14.8

1   Disconnect the cable from the negative battery terminal, then the positive battery terminal and wait at least ten minutes before removing the steering wheel.
2   Turn the steering wheel so that the front wheels are pointing straight ahead.
3   Remove the small cover from the underside of the steering wheel and unplug the airbag module connector **(see illustration)**.
4   Remove the side covers and remove the Torx bolts behind them **(see illustrations)**. **Note:** *Some models are equipped with special tamper-resistant Torx screws, which require a special driver. The Torx bolts are coated with a special bonding agent, so they must be discarded; be sure to replace them with new ones during reassembly.* Lift the airbag module off the steering wheel. **Warning:** *Handle the airbag module with care, carry the module with the trim cover side facing away from your body and store it in a safe location with the trim side facing up. See the precautions in Chapter 12.*
5   Remove the steering wheel retaining nut, then mark the relationship of the steering wheel to the steering shaft **(see illustrations)**.
6   Disconnect the electrical connector for the cruise control wiring harness.
7   Use a steering wheel puller to separate the steering wheel from the steering shaft **(see illustration)**. When removing the wheel, make sure the electrical leads for the airbag module and the cruise control system don't snag on the wheel. **Warning:** *Do not turn the steering shaft while the steering wheel is removed.*
8   Remove the clockspring only if the steering column switches must be checked or replaced **(see illustration)**. **Caution:** *Regardless if the clockspring will be removed or not, tape the center hub of the clockspring to the outer ring so the center hub cannot rotate. This will retain the clockspring in the centered position. Also, don't allow the steer-*

14.4b ... and remove the Torx bolt (arrow); the Torx bolts are coated with a special bonding agent, so they must be discarded - be sure to replace them during reassembly

ing shaft to rotate with the steering wheel removed.
9   If necessary, remove the screws, unlock and disconnect the electrical connector, then separate the clockspring from the steering column.

Installation

Refer to illustrations 14.10a and 14.10b
10 Verify that the front wheels are pointing

14.5a Remove the steering wheel retaining nut

14.5b Mark the relationship of the steering wheel to the shaft before removing the wheel

14.7 Use a steering wheel puller to separate the steering wheel from the steering shaft

Chapter 10 Suspension and steering systems

14.8 If it is necessary to remove the clockspring, remove these screws, disconnect the electrical connector and detach it from the steering column

14.10a To align the clockspring in the neutral position, turn it clockwise until it stops (don't apply much force) . . .

14.10b . . . then turn it counterclockwise 2-1/2 turns and align the arrows

straight ahead. Turn the clockspring for the airbag clockwise by hand until it becomes hard to turn, then rotate the clockspring counterclockwise about two and one-half turns until the two arrows are aligned **(see illustrations)**.

11 Pull the electrical leads for the airbag module and the cruise control system through the steering wheel and install the wheel. Make sure the clockspring pin guides are properly engaged with their corresponding holes in the back of the steering wheel.

12 Install the steering wheel retaining nut and tighten it to the torque listed in this Chapter's Specifications.

13 Install the airbag module and secure it with new Torx bolts. Do not reuse the old bolts. Install cruise control switch and the side covers.

14 Plug in the airbag module connector. Install the lower cover.

15 Verify that the airbag circuit is operational by turning the ignition key to the On or Start position. The "AIR BAG" warning light should illuminate for about seven seconds, then turn off.

Non-airbag models
Removal

16 Disconnect the cable from the negative terminal of the battery.

17 To remove the horn pad, grasp it along the bottom edge and pull it off the steering wheel.

18 The remainder of removal is similar to that for an airbag-equipped model (see Steps 5, 6 and 7).

Installation

19 Before installing the wheel, lubricate the horn contact slip ring and the sliding portion of the turn signal cancel pin with multi-purpose grease.

20 Install the steering wheel nut and tighten it to the torque listed in this Chapter's Specifications.

21 Connect the horn wire and install the horn pad.

22 Connect the negative battery cable.

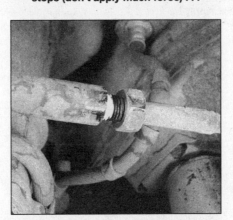

15.2 Loosen the jam nut, then mark the position of the tie-rod end in relation to the threads

15 Tie-rod ends - removal and installation

Refer to illustrations 15.2, 15.3a, 15.3b and 15.4

Removal

1 Loosen the wheel lug nuts. Raise the front of the vehicle, support it securely on jackstands, block the rear wheels and set the

15.3b . . . then loosen - but don't remove - the tie-rod end ballstud nut

15.3a Remove the cotter pin . . .

parking brake. Remove the front wheel.

2 Loosen the jam nut enough to mark the position of the tie-rod end in relation to the threads **(see illustration)**.

3 Remove the cotter pin and loosen, but don't remove, the nut on the tie-rod end stud **(see illustrations)**.

4 Disconnect the tie-rod end from the steering knuckle arm with a puller **(see illustration)**. Remove the nut and separate the tie-rod.

5 Unscrew the tie-rod end from the steering rod.

15.4 Disconnect the tie-rod end from the steering knuckle arm with a puller

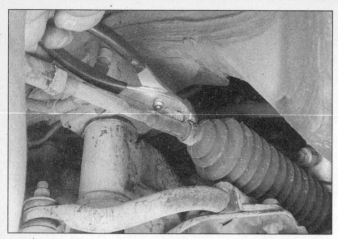

16.3a The outer end of each steering gear boot is secured by a spring-type clamp which can be slid off simply by pinching the ends together

16.3b The inner end of each steering gear boot is secured by a retaining wire (arrow) which must be cut off and discarded

Installation

6 Thread the tie-rod end on to the marked position and insert the tie-rod stud into the steering knuckle arm. Tighten the jam nut securely.

7 Install the castle nut on the stud and tighten it to the torque listed in this Chapter's Specifications. Install a new cotter pin.

8 Install the wheel and lug nuts. Lower the vehicle and tighten the lug nuts to the torque listed in the Chapter 1 Specifications.

9 Have the front end alignment checked and, if necessary, adjusted.

16 Steering gear boots - replacement

Refer to illustrations 16.3a and 16.3b

1 Loosen the lug nuts, raise the vehicle and support it securely on jackstands. Remove the wheel.

2 Remove the tie-rod end and jam nut (see Section 15).

3 Remove the outer steering gear boot clamp **(see illustration)** with a pair of pliers. Cut off the inner boot retaining wire **(see illustration)** with a pair of diagonal cutters. Slide the boot off.

4 Before installing the new boot, wrap the threads on the end of the steering rod with a layer of tape so the small end of the new boot isn't damaged.

5 Slide the new boot into position on the steering gear until it seats in the groove in the steering rod and install a new clamp. Wrap a new retaining wire around the inner end of the boot twice, then twist the ends together four turns while pulling with a force of approximately 22 pounds.

6 Remove the tape and install the tie-rod end (see Section 15).

7 Install the wheel and lug nuts. Lower the vehicle and tighten the lug nuts to the torque listed in the Chapter 1 Specifications.

17 Steering gear - removal and installation

Warning: *Some models covered by this manual have airbags. Always disconnect the negative battery cable, then the positive cable and wait 10 minutes before working in the vicinity of the impact sensors, steering column or instrument panel to avoid the possibility of accidental deployment of the airbag, which could cause personal injury (see Chapter 12).*

Removal

Refer to illustrations 17.3a, 17.3b, 17.4, 17.9a, 17.9b and 17.9c

1 Park the vehicle with the wheels pointing straight ahead. Disconnect the cable from the negative terminal of the battery. On airbag-equipped models, also disconnect the positive cable and wait 10 minutes before proceeding.

2 Remove the air filter housing and intake duct (see Chapter 4). Remove the charcoal canister and its bracket (see Chapter 6).

3 Remove the nuts and separate the steering shaft universal joint cover from the firewall **(see illustration)**. Mark the relationship of the universal joint to the steering gear input shaft, then remove the lower intermediate shaft pinch bolt **(see illustration)**. **Warning:** *Do not turn the steering wheel while the steering gear is removed on a model*

17.3a Unscrew the nuts (arrows) and detach the cover from the firewall

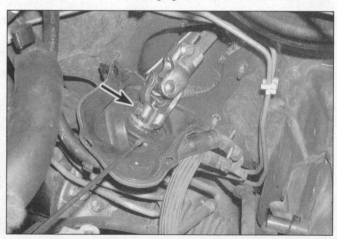

17.3b Mark the relationship of the universal joint to the steering gear input shaft, then remove the U-joint pinch bolt (arrow)

Chapter 10 Suspension and steering systems

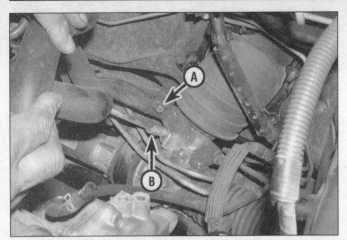

17.4 Loosen the clamp from the return hose (A) and disconnect the hose; use a flare-nut wrench to unscrew the pressure line fitting (B)

17.9a Remove this shield for access to the right side steering gear mount

17.9b The right side mounting clamp is retained by two bolts

17.9c The left side mounting clamp is retained by three bolts

equipped with an airbag. If the steering wheel is inadvertently turned, remove the steering wheel and center the clockspring (see Section 14). To prevent the steering wheel from turning, loop the seat belt through the steering wheel and fasten it into its latch.

4 Place a drain pan under the steering gear. Detach the power steering pressure and return lines and cap the ends to prevent excessive fluid loss and contamination **(see illustration)**.

5 Loosen the front wheel lug nuts, raise the front of the vehicle and support it securely on jackstands. Apply the parking brake and remove the wheels. Remove the engine splash shields.

6 Separate the tie-rod ends from the steering knuckle arms (see Section 15).

7 Remove the stabilizer bar (see Section 6).

8 Remove the center crossmember from underneath the engine (see Chapter 2).

9 Remove the steering gear mounting bolts **(see illustrations)**. Separate the intermediate shaft from the steering gear input shaft and maneuver the steering gear assembly out from under the vehicle.

10 Check the steering gear rubber mounts for wear or deterioration, replacing them if necessary.

Installation

Refer to illustration 17.12

Note: *Make sure the steering gear is centered from side-to-side before installing it.*

11 Guide the steering gear into position and connect the U-joint, aligning the marks.

12 Install the mounting clamps and bolts and tighten them in the recommended sequence to the torque listed in this Chapter's Specifications **(see illustration)**. Install the shield over the right side mount.

13 Install the center crossmember (see Chapter 2).

14 Install the stabilizer bar (see Section 6).

15 Connect the tie-rod ends to the steering knuckle arms (see Section 15).

16 Install the engine splash shields.

17 Install the wheels and lug nuts. Lower the vehicle and tighten the lug nuts to the torque listed in the Chapter 1 Specifications.

18 Install the U-joint pinch bolt and tighten it to the torque listed in this Chapter's Specifications. Install the universal joint cover.

19 Connect the power steering pressure and return lines to the steering gear and fill the power steering pump reservoir with the recommended fluid (see Chapter 1).

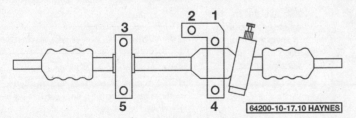

17.12 Steering gear mounting bolt tightening sequence

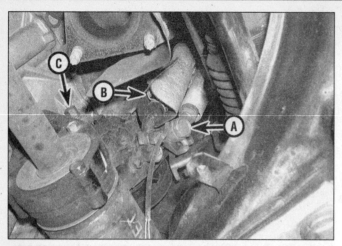

18.5 Unscrew the banjo fitting bolt from the power steering fluid pressure line (A), loosen the clamp on the return hose fitting (B) and disconnect the hose, then remove the pivot bolt (C)

18.8a Remove this bolt (arrow) that attaches the adjuster bracket to the power steering pump

20 Install the charcoal canister and the air filter housing and duct.
21 Bleed the steering system (see Section 19).

18 Power steering pump - removal and installation

Removal

Refer to illustrations 18.5, 18.8a and 18.8b

1 Disconnect the cable from the negative battery terminal.
2 Loosen the tensioner and remove the drivebelt (see Chapter 1).
3 Using a large syringe or suction gun, suck as much fluid out of the power steering fluid reservoir as possible. Place a drain pan under the vehicle to catch any fluid that spills out when the hoses are disconnected.
4 Loosen the right front wheel lug nuts, raise the front of the vehicle and support it securely on jackstands. Remove the wheel.
5 Working under the vehicle, remove the pressure line-to-pump banjo bolt **(see illustration)**, then detach the line from the pump. Remove and discard the copper sealing washers. They must be replaced when installing the pump. Wrap a plastic bag around the end of the hose to prevent fluid spillage.
6 Loosen the clamp and disconnect the fluid return hose from the pump **(see illustration 18.5)**. Plug the hose.
7 Detach the right tie-rod end from the steering knuckle arm (see Section 15) and swing the tie-rod to the rear, out of the way.
8 Remove the pump mounting bolts **(see accompanying illustration and illustration 18.5)**, then guide the pump out through the fenderwell **(see illustration)**.

Installation

9 Installation is the reverse of removal. Be sure to tighten the pressure line banjo bolt to the torque listed in this Chapter's Specifications (be sure to use new sealing washers). Adjust the drivebelt tension following the procedure described in Chapter 1.
10 Connect the tie-rod end to the steering knuckle (see Section 15).
11 Install the wheel and lug nuts. Lower the vehicle and tighten the lug nuts to the torque listed in the Chapter 1 Specifications.
12 Top up the fluid level in the reservoir (see Chapter 1) and bleed the system (see Section 19).

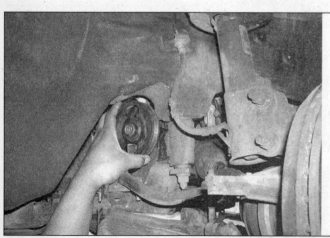

18.8b Guide the power steering pump out through this opening in the inner fender panel

19 Power steering system - bleeding

1 Following any operation in which the power steering fluid lines have been disconnected, the power steering system must be bled to remove all air and obtain proper steering performance.
2 With the front wheels in the straight ahead position, check the power steering fluid level and, if low, add fluid until it is between the Cold marks on the reservoir.
3 Raise the front of the vehicle and support it securely on jackstands.
4 Turn the steering wheel back-and-forth repeatedly, lightly hitting the stops.
5 Start the engine and allow it to run at fast idle. Recheck the fluid level and add more if necessary to reach the Cold marks.
6 Bleed the system by turning the wheels from side to side, just barely contacting the stops. This will work the air out of the system. Keep the reservoir full of fluid as this is done.
7 When the air is worked out of the system and the fluid level stabilizes, return the wheels to the straight ahead position and leave the vehicle running for several more minutes before shutting it off. Lower the vehicle.
8 Road test the vehicle to be sure the steering system is functioning normally and noise-free.
9 Recheck the fluid level to be sure it is between the Hot marks on the reservoir while the engine is at normal operating temperature. Add fluid if necessary (see Chapter 1).

20 Wheel studs - replacement

Refer to illustrations 20.3 and 20.4

1 Loosen the wheel lug nuts, raise the vehicle and support it securely on jackstands. Remove the wheel.
2 Remove the brake disc or drum (see Chapter 9).
3 Install a lug nut part way onto the stud

Chapter 10 Suspension and steering systems

20.3 Use a press tool to push the stud out of the hub flange

1 Hub flange
2 Lug nut on stud
3 Press tool

20.4 Install a spacer and a lug nut on the stud, then tighten the nut to draw the stud into place

1 Hub flange
2 Spacer

being replaced. Push the stud out of the hub flange with a press tool **(see illustration)**.

4 Insert the new stud into the hub flange from the back side and install a spacer or several flat washers and a lug nut on the stud **(see illustration)**.

5 Tighten the lug nut until the stud is seated in the flange.

6 Reinstall the brake drum or disc. Install the wheel and lug nuts. Lower the vehicle and tighten the lug nuts to the torque listed in the Chapter 1 Specifications.

21 Wheels and tires - general information

Refer to illustration 21.1

1 All vehicles covered by this manual are equipped with metric-sized steel belted radial tires **(see illustration)**. Use of other size or type of tires may affect the ride and handling of the vehicle. Don't mix different types of tires, such as radials and bias belted, on the same vehicle as handling may be seriously affected. It's recommended that tires be replaced in pairs on the same axle, but if only one tire is being replaced, be sure it's the same size, structure and tread design as the other.

2 Because tire pressure has a substantial effect on handling and wear, the pressure on all tires should be checked at least once a month or before any extended trips (see Chapter 1).

3 Wheels must be replaced if they are bent, dented, leak air, have elongated bolt holes, are heavily rusted, out of vertical symmetry or if the lug nuts won't stay tight. Wheel repairs that use welding or peening are not recommended.

4 Tire and wheel balance is important in the overall handling, braking and performance of the vehicle. Unbalanced wheels can adversely affect handling and ride characteristics as well as tire life. Whenever a tire is installed on a wheel, the tire and wheel should be balanced by a shop with the proper equipment.

22 Wheel alignment - general information

Refer to illustration 22.1

A wheel alignment refers to the adjustments made to the wheels so they are in proper angular relationship to the suspension and the ground. Wheels that are out of proper alignment not only affect vehicle control, but also increase tire wear. The front end angles normally measured are camber, caster and toe-in **(see illustration)**. Camber and caster are preset at the factory on the vehicle covered by this manual; toe-in is the only adjustable angle on these vehicles. The rear toe-in can also be adjusted, but the camber and caster cannot (however, camber and caster are usually measured to check for bent or worn suspension parts).

Getting the proper wheel alignment is a

21.1 Metric tire size code

very exacting process, one in which complicated and expensive machines are necessary to perform the job properly. Because of this, you should have a technician with the proper equipment perform these tasks. We will, however, use this space to give you a basic idea of what is involved with a wheel alignment so you can better understand the process and deal intelligently with the shop that does the work.

Toe-in is the turning in of the wheels. The purpose of a toe specification is to ensure parallel rolling of the wheels. In a vehicle with zero toe-in, the distance between the front edges of the wheels will be the same as the distance between the rear edges of the wheels. The actual amount of toe-in is normally only a fraction of an inch. On the front end, toe-in is controlled by the tie-rod end position on the tie-rod. On the rear end, it's controlled by a threaded adjuster on the rear lateral arm. Incorrect toe-in will cause the tires to wear improperly by making them scrub against the road surface.

Camber is the tilting of the wheels from vertical when viewed from one end of the vehicle. When the wheels tilt out at the top, the camber is said to be positive (+). When the wheels tilt in at the top the camber is negative (-). The amount of tilt is measured in degrees from vertical and this measurement is called the camber angle. This angle affects the amount of tire tread which contacts the road and compensates for changes in the suspension geometry when the vehicle is cornering or traveling over an undulating surface.

Caster is the tilting of the front steering axis from the vertical. A tilt toward the rear is positive caster and a tilt toward the front is negative caster.

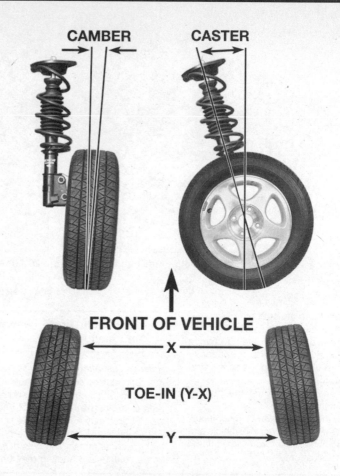

22.1 Camber, caster and toe-in angles

Chapter 11 Body

Contents

	Section		Section
Body - maintenance	2	Hood and hood support struts - removal, installation	
Body repair - major damage	6	and adjustment	10
Body repair - minor damage	5	Hood latch and release cable - removal and installation	11
Bumpers - removal and installation	12	Instrument panel - removal and installation	27
Center console - removal and installation	24	Mirrors - removal and installation	20
Cowl cover - removal and installation	14	Radiator grille - removal and installation	9
Dashboard trim panels - removal and installation	25	Rear package shelf - removal and installation	29
Door - removal, installation and adjustment	16	Seats - removal and installation	28
Door latch, lock cylinder and handle - removal and installation	17	Steering column covers - removal and installation	26
Door trim panel - removal and installation	15	Sunroof - adjustment	30
Door window glass - removal and installation	18	Trunk lid - removal, installation and adjustment	21
Door window glass regulator - removal and installation	19	Trunk lid latch and lock cylinder - removal and installation	22
Front fender - removal and installation	13	Trunk release and fuel door cables - removal and installation	23
General information	1	Upholstery and carpets - maintenance	4
Hinges and locks - maintenance	7	Vinyl trim - maintenance	3
		Windshield and fixed glass - replacement	8

1 General information

Warning: *The models covered by this manual are equipped with Supplemental Restraint Systems (SRS), more commonly known as airbags. Always disable the airbag system before working in the vicinity of any airbag system components to avoid the possibility of accidental deployment of the airbags, which could cause personal injury (see Chapter 12).*

These models feature a "unibody" construction, using a floor pan with front and rear frame side rails which support the body components, front and rear suspension systems and other mechanical components. Certain components are particularly vulnerable to accident damage and can be unbolted and repaired or replaced. Among these parts are the front fenders, bumpers, hood, doors, trunk lid and all glass.

Only general body maintenance practices and body panel repair procedures within the scope of the do-it-yourselfer are included in this Chapter.

2 Body - maintenance

1   The condition of your vehicle's body is very important, because the resale value depends a great deal on it. It's much more difficult to repair a neglected or damaged body than it is to repair mechanical components. The hidden areas of the body, such as the wheel wells, the frame and the engine compartment, are equally important, although they don't require as frequent attention as the rest of the body.
2   Once a year, or every 12,000 miles, it's a good idea to have the underside of the body steam cleaned. All traces of dirt and oil will be removed and the area can then be inspected carefully for rust, damaged brake lines, frayed electrical wires, damaged cables and other problems. The front suspension components should be greased after completion of this job.
3   At the same time, clean the engine and the engine compartment with a steam cleaner or water-soluble degreaser.
4   The wheel wells should be given close attention, since undercoating can peel away and stones and dirt thrown up by the tires can cause the paint to chip and flake, allowing rust to set in. If rust is found, clean down to the bare metal and apply an anti-rust paint.
5   The body should be washed about once a week. Wet the vehicle thoroughly to soften the dirt, then wash it down with a soft sponge and plenty of clean, soapy water. If the surplus dirt is not washed off very carefully, it can wear down the paint.
6   Spots of tar or asphalt thrown up from the road should be removed with a cloth soaked in tar remover or kerosene lamp oil.
7   Once every six months, wax the body and chrome trim. If a chrome cleaner is used to remove rust from any of the vehicle's plated parts, remember that the cleaner also removes part of the chrome, so use it sparingly.

3 Vinyl trim - maintenance

Don't clean vinyl trim with detergents, caustic soap or petroleum-based cleaners. Plain soap and water works just fine, with a soft brush to clean dirt that may be ingrained. Wash the vinyl as frequently as the rest of the vehicle.

After cleaning, application of a high quality rubber and vinyl protectant will help prevent oxidation and cracks. The protectant can also be applied to weatherstripping, vacuum lines and rubber hoses (which often fail as a result of chemical degradation) and to the tires.

4 Upholstery and carpets - maintenance

1   Every three months remove the floormats and clean the interior of the vehicle (more frequently if necessary). Use a stiff whiskbroom to brush the carpeting and loosen dirt and dust, then vacuum the upholstery and carpets thoroughly, especially along seams and crevices.
2   Dirt and stains can be removed from carpeting with basic household or automotive carpet shampoos available in spray cans. Follow the directions and vacuum again, then use a stiff brush to bring back the "nap" of the carpet.
3   Most interiors have cloth or vinyl upholstery, either of which can be cleaned and

These photos illustrate a method of repairing simple dents. They are intended to supplement *Body repair - minor damage* in this Chapter and should not be used as the sole instructions for body repair on these vehicles.

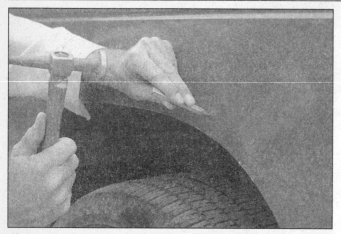

1 If you can't access the backside of the body panel to hammer out the dent, pull it out with a slide-hammer-type dent puller. In the deepest portion of the dent or along the crease line, drill or punch hole(s) at least one inch apart . . .

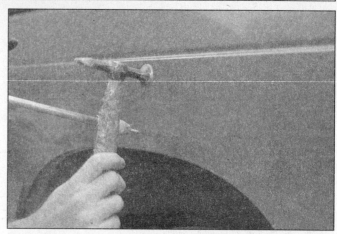

2 . . . then screw the slide-hammer into the hole and operate it. Tap with a hammer near the edge of the dent to help 'pop' the metal back to its original shape. When you're finished, the dent area should be close to its original contour and about 1/8-inch below the surface of the surrounding metal

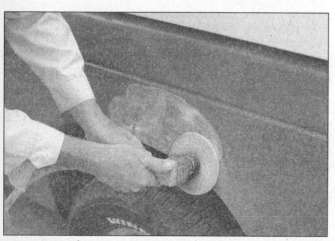

3 Using coarse-grit sandpaper, remove the paint down to the bare metal. Hand sanding works fine, but the disc sander shown here makes the job faster. Use finer (about 320-grit) sandpaper to feather-edge the paint at least one inch around the dent area

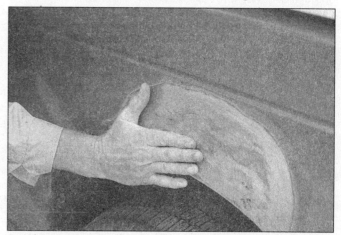

4 When the paint is removed, touch will probably be more helpful than sight for telling if the metal is straight. Hammer down the high spots or raise the low spots as necessary. Clean the repair area with wax/silicone remover

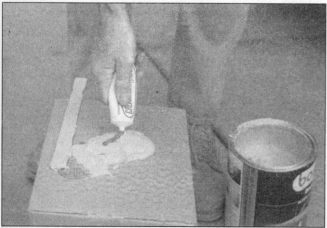

5 Following label instructions, mix up a batch of plastic filler and hardener. The ratio of filler to hardener is critical, and, if you mix it incorrectly, it will either not cure properly or cure too quickly (you won't have time to file and sand it into shape)

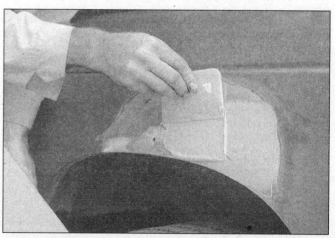

6 Working quickly so the filler doesn't harden, use a plastic applicator to press the body filler firmly into the metal, assuring it bonds completely. Work the filler until it matches the original contour and is slightly above the surrounding metal

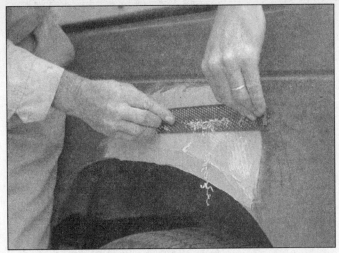

7   Let the filler harden until you can just dent it with your fingernail. Use a body file or Surform tool (shown here) to rough-shape the filler

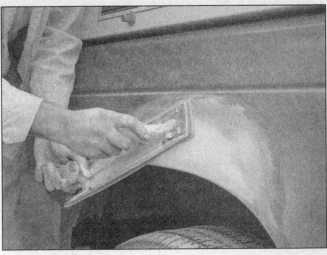

8   Use coarse-grit sandpaper and a sanding board or block to work the filler down until it's smooth and even. Work down to finer grits of sandpaper - always using a board or block - ending up with 360 or 400 grit

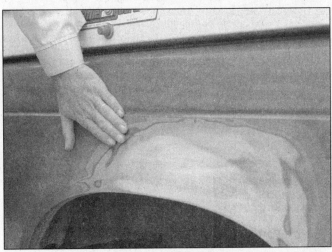

9   You shouldn't be able to feel any ridge at the transition from the filler to the bare metal or from the bare metal to the old paint. As soon as the repair is flat and uniform, remove the dust and mask off the adjacent panels or trim pieces

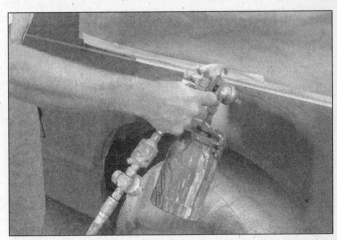

10   Apply several layers of primer to the area. Don't spray the primer on too heavy, so it sags or runs, and make sure each coat is dry before you spray on the next one. A professional-type spray gun is being used here, but aerosol spray primer is available inexpensively from auto parts stores

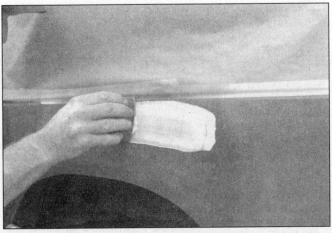

11   The primer will help reveal imperfections or scratches. Fill these with glazing compound. Follow the label instructions and sand it with 360 or 400-grit sandpaper until it's smooth. Repeat the glazing, sanding and respraying until the primer reveals a perfectly smooth surface

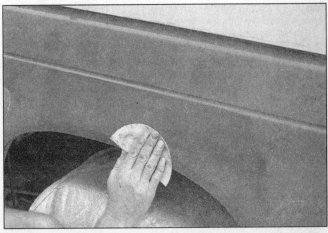

12   Finish sand the primer with very fine sandpaper (400 or 600-grit) to remove the primer overspray. Clean the area with water and allow it to dry. Use a tack rag to remove any dust, then apply the finish coat. Don't attempt to rub out or wax the repair area until the paint has dried completely (at least two weeks)

maintained with a number of material-specific cleaners or shampoos available in auto supply stores. Follow the directions on the product for usage, and always spot-test any upholstery cleaner on an inconspicuous area (bottom edge of a backseat cushion) to ensure that it doesn't cause a color shift in the material.

4 After cleaning, vinyl upholstery should be treated with a protectant. **Note:** *Make sure the protectant container indicates the product can be used on seats - some products may make a seat too slippery.* **Caution:** *Do not use protectant on vinyl-covered steering wheels.*

5 Leather upholstery requires special care. It should be cleaned regularly with saddlesoap or leather cleaner. Never use alcohol, gasoline, nail polish remover or thinner to clean leather upholstery.

6 After cleaning, regularly treat leather upholstery with a leather conditioner, rubbed in with a soft cotton cloth. Never use car wax on leather upholstery.

7 In areas where the interior of the vehicle is subject to bright sunlight, cover leather seating areas of the seats with a sheet if the vehicle is to be left out for any length of time.

5 Body repair - minor damage

See photo sequence

Repair of minor scratches

1 If the scratch is superficial and does not penetrate to the metal of the body, repair is very simple. Lightly rub the scratched area with a fine rubbing compound to remove loose paint and built-up wax. Rinse the area with clean water.

2 Apply touch-up paint to the scratch, using a small brush. Continue to apply thin layers of paint until the surface of the paint in the scratch is level with the surrounding paint. Allow the new paint at least two weeks to harden, then blend it into the surrounding paint by rubbing with a very fine rubbing compound. Finally, apply a coat of wax to the scratch area.

3 If the scratch has penetrated the paint and exposed the metal of the body, causing the metal to rust, a different repair technique is required. Remove all loose rust from the bottom of the scratch with a pocketknife, then apply rust inhibiting paint to prevent the formation of rust in the future. Using a rubber or nylon applicator, coat the scratched area with glaze-type filler. If required, the filler can be mixed with thinner to provide a very thin paste, which is ideal for filling narrow scratches. Before the glaze filler in the scratch hardens, wrap a piece of smooth cotton cloth around the tip of a finger. Dip the cloth in thinner and then quickly wipe it along the surface of the scratch. This will ensure that the surface of the filler is slightly hollow. The scratch can now be painted over as described earlier in this section.

Repair of dents

4 When repairing dents, the first job is to pull the dent out until the affected area is as close as possible to its original shape. There is no point in trying to restore the original shape completely as the metal in the damaged area will have stretched on impact and cannot be restored to its original contours. It is better to bring the level of the dent up to a point that is about 1/8-inch below the level of the surrounding metal. In cases where the dent is very shallow, it is not worth trying to pull it out at all.

5 If the backside of the dent is accessible, it can be hammered out gently from behind using a soft-face hammer. While doing this, hold a block of wood firmly against the opposite side of the metal to absorb the hammer blows and prevent the metal from being stretched.

6 If the dent is in a section of the body which has double layers, or some other factor makes it inaccessible from behind, a different technique is required. Drill several small holes through the metal inside the damaged area, particularly in the deeper sections. Screw long, self-tapping screws into the holes just enough for them to get a good grip in the metal. Now the dent can be pulled out by pulling on the protruding heads of the screws with locking pliers.

7 The next stage of repair is the removal of paint from the damaged area and from an inch or so of the surrounding metal. This is done with a wire brush or sanding disk in a drill motor, although it can be done just as effectively by hand with sandpaper. To complete the preparation for filling, score the surface of the bare metal with a screwdriver or the tang of a file, or drill small holes in the affected area. This will provide a good grip for the filler material. To complete the repair, see the subsection on filling and painting later in this Section.

Repair of rust holes or gashes

8 Remove all paint from the affected area and from an inch or so of the surrounding metal using a sanding disk or wire brush mounted in a drill motor. If these are not available, a few sheets of sandpaper will do the job just as effectively.

9 With the paint removed, you will be able to determine the severity of the corrosion and decide whether to replace the whole panel, if possible, or repair the affected area. New body panels are not as expensive as most people think and it is often quicker to install a new panel than to repair large areas of rust.

10 Remove all trim pieces from the affected area except those which will act as a guide to the original shape of the damaged body, such as headlight shells, etc. Using metal snips or a hacksaw blade, remove all loose metal and any other metal that is badly affected by rust. Hammer the edges of the hole in to create a slight depression for the filler material.

11 Wire brush the affected area to remove the powdery rust from the surface of the metal. If the back of the rusted area is accessible, treat it with rust inhibiting paint.

12 Before filling is done, block the hole in some way. This can be done with sheet metal riveted or screwed into place, or by stuffing the hole with wire mesh.

13 Once the hole is blocked off, the affected area can be filled and painted. See the following subsection on filling and painting.

Filling and painting

14 Many types of body fillers are available, but generally speaking, body repair kits which contain filler paste and a tube of resin hardener are best for this type of repair work. A wide, flexible plastic or nylon applicator will be necessary for imparting a smooth and contoured finish to the surface of the filler material. Mix up a small amount of filler on a clean piece of wood or cardboard (use the hardener sparingly). Follow the manufacturer's instructions on the package, otherwise the filler will set incorrectly.

15 Using the applicator, apply the filler paste to the prepared area. Draw the applicator across the surface of the filler to achieve the desired contour and to level the filler surface. As soon as a contour that approximates the original one is achieved, stop working the paste. If you continue, the paste will begin to stick to the applicator. Continue to add thin layers of paste at 20-minute intervals until the level of the filler is just above the surrounding metal.

16 Once the filler has hardened, the excess can be removed with a body file. From then on, progressively finer grades of sandpaper should be used, starting with a 180-grit paper and finishing with 600-grit wet-or-dry paper. Always wrap the sandpaper around a flat rubber or wooden block, otherwise the surface of the filler will not be completely flat. During the sanding of the filler surface, the wet-or-dry paper should be periodically rinsed in water. This will ensure that a very smooth finish is produced in the final stage.

17 At this point, the repair area should be surrounded by a ring of bare metal, which in turn should be encircled by the finely feathered edge of good paint. Rinse the repair area with clean water until all of the dust produced by the sanding operation is gone.

18 Spray the entire area with a light coat of primer. This will reveal any imperfections in the surface of the filler. Repair the imperfections with fresh filler paste or glaze filler and once more smooth the surface with sandpaper. Repeat this spray-and-repair procedure until you are satisfied that the surface of the filler and the feathered edge of the paint are perfect. Rinse the area with clean water and allow it to dry completely.

19 The repair area is now ready for painting. Spray painting must be carried out in a warm, dry, windless and dust free atmosphere. These conditions can be created if you have access to a large indoor work area, but if you are forced to work in the open, you will have to pick the day very carefully. If you

Chapter 11 Body

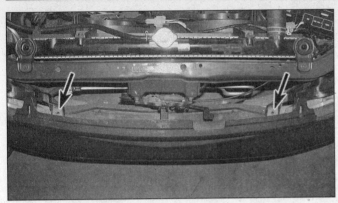

9.2 With the hood open, remove the two screws or plastic clips (arrows)

9.3 Insert a screwdriver through the small hole (A) in the center of the grille to turn the securing fastener 45 degrees - the two upper fasteners are loosened by inserting the screwdriver through slots (B)

are working indoors, dousing the floor in the work area with water will help settle the dust that would otherwise be in the air. If the repair area is confined to one body panel, mask off the surrounding panels. This will help minimize the effects of a slight mismatch in paint color. Trim pieces such as chrome strips, door handles, etc., will also need to be masked off or removed. Use masking tape and several thickness of newspaper for the masking operations.

20 Before spraying, shake the paint can thoroughly, then spray a test area until the spray painting technique is mastered. Cover the repair area with a thick coat of primer. The thickness should be built up using several thin layers of primer rather than one thick one. Using 600-grit wet-or-dry sandpaper, rub down the surface of the primer until it is very smooth. While doing this, the work area should be thoroughly rinsed with water and the wet-or-dry sandpaper periodically rinsed as well. Allow the primer to dry before spraying additional coats.

21 Spray on the top coat, again building up the thickness by using several thin layers of paint. Begin spraying in the center of the repair area and then, using a circular motion, work out until the whole repair area and about two inches of the surrounding original paint is covered. Remove all masking material 10 to 15 minutes after spraying on the final coat of paint. Allow the new paint at least two weeks to harden, then use a very fine rubbing compound to blend the edges of the new paint into the existing paint. Finally, apply a coat of wax.

6 Body repair - major damage

1 Major damage must be repaired by an auto body shop specifically equipped to perform unibody repairs. These shops have the specialized equipment required to do the job properly.

2 If the damage is extensive, the body must be checked for proper alignment or the vehicle's handling characteristics may be adversely affected and other components may wear at an accelerated rate.

3 Due to the fact that many of the major body components (hood, fenders, etc.) are separate and replaceable units, any seriously damaged components should be replaced rather than repaired. Sometimes the components can be found in a wrecking yard that specializes in used vehicle components, often at considerable savings over the cost of new parts.

7 Hinges and locks - maintenance

Once every 3000 miles, or every three months, the hinges and latch assemblies on the doors, hood and trunk should be given a few drops of light oil or lock lubricant. The door latch strikers should also be lubricated with a thin coat of grease to reduce wear and ensure free movement. Lubricate the door and trunk locks with spray-on graphite lubricant.

8 Windshield and fixed glass - replacement

Replacement of the windshield and rear-window glass requires the use of special fast-setting adhesive/caulk materials and some specialized tools. It is recommended that these operations be left to a dealer or a shop specializing in glass work.

9 Radiator grille - removal and installation

Refer to illustrations 9.2 and 9.3
Note: On 2000 and later models, the grille is an integral part of the front bumper and cannot be removed separately.
1 Open the hood.
2 Remove the plastic clips (bolts and nuts on some models) holding the sides of the grille to the headlight assemblies (see illustration).

3 Three square plastic fasteners secure the center and the top of the grille at each side (see illustration). Each of these fasteners can be turned 45 degrees with a screwdriver to loosen them for removal. The two upper fasteners are reached by inserting the screwdriver in a slot at the top of the grille.
4 Installation is the reverse of removal.

10 Hood and hood support struts - removal, installation and adjustment

Refer to illustrations 10.1, 10.4, 10.10 and 10.11
Note: The hood is heavy and somewhat awkward to remove and install - at least two people should perform this procedure.

Removal and installation

1 Make marks around the bolt heads or around the entire hinge to ensure proper alignment during installation (see illustration).
2 Use blankets or pads to cover the cowl area of the body and fenders. This will protect the body and paint as the hood is lifted off.
3 Disconnect any cables or wires that will interfere with removal.

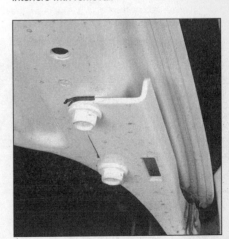

10.1 Before removing the hood, make marks around the hinge plate

10.4 The hood support struts have a captive stud-mount - unscrew it from the hood

10.10 Loosen the hood latch bolts, move the latch and retighten bolts, then close the hood to check the fit - repeat the procedure until the hood is flush with the fenders

10.11 Adjust the hood closing height by turning the hood bumpers in or out

4   With an assistant supporting the hood, use a wrench to unscrew the support strut studs from the hood **(see illustration)**.
5   Remove the hinge-to-hood screws or bolts and lift off the hood.
6   Installation is the reverse of removal.

Adjustment

7   Fore-and-aft and side-to-side adjustment of the hood is done by moving the hinge plate slot after loosening the bolts or nuts.
8   Scribe a line around the entire hinge plate so you can judge the amount of movement **(see illustration 10.1)**.
9   Loosen the bolts or nuts and move the hood into correct alignment. Move it only a little at a time. Tighten the hinge bolts and carefully lower the hood to check the position.
10   If necessary after installation, the entire hood latch assembly can be adjusted up-and-down as well as from side-to-side on the radiator support so the hood closes securely, flush with the fenders. To make the adjustment, scribe a line around the hood latch mounting bolts to provide a reference point, then loosen them and reposition the latch assembly, as necessary **(see illustration)**. Following adjustment, retighten the mounting bolts.
11   Finally, adjust the hood bumpers on the radiator support so the hood, when closed, is flush with the fenders **(see illustration)**.
12   The hood latch assembly, as well as the hinges, should be periodically lubricated with white, lithium-base grease to prevent binding and wear.

11   Hood latch and release cable - removal and installation

Refer to illustrations 11.1 and 11.4

Latch

1   Remove the bolts and detach the latch assembly **(see illustration 10.10)**. Unhook the spring and use a screwdriver to detach the cable end from the latch **(see illustration)**.
2   Installation is the reverse of removal.

Cable

3   Disconnect the cable from the latch **(see illustration 11.1)**. Detach the cable from the fasteners along the left inner fender.
4   Working in the passenger compartment, remove the screws and detach the hood-release cable and handle assembly from the instrument panel **(see illustration)**.
5   Pull the cable through the firewall grommet into the passenger compartment.
6   Insert the end of new cable through the firewall grommet into the engine compartment.
7   Pull the cable through from the engine compartment side.
8   The remainder of installation is the reverse of removal.

12   Bumpers - removal and installation

1   Apply the parking brake, raise the vehicle and support it securely on jackstands.
2   Disconnect the cable from the negative battery terminal and disconnect any wiring that would interfere with bumper removal.

Front (1993 through 1999 models)

Refer to illustrations 12.5a and 12.5b, 12.6

3   From below, remove the splash panel below the radiator and bumper.
4   Refer to Section 9 and remove the radiator grille. Refer to Chapter 12 and remove the headlight housings, turn signal lamps and cornering lamps.
5   Refer to Chapter 12 and remove the headlights and cornering lamps. Refer to Section 13 and remove the fascia-to-fender fasteners, then remove the remaining fasten-

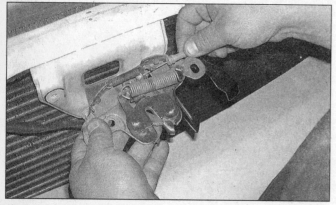

11.1 Use needle-nose pliers or a screwdriver to release the hood latch cable from the back of the latch assembly

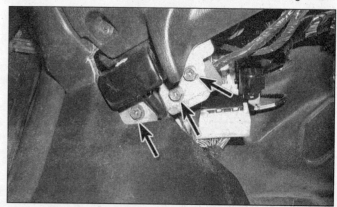

11.4 Remove the three bolts (arrows) securing the interior hood latch handle to the bottom of the instrument panel

Chapter 11 Body   11-7

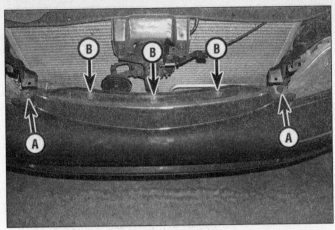

12.5a In the grille opening, remove the two plastic pins (A), then the three bolts (B)

12.5b With the lights removed, there is access to the remaining front fascia fasteners (arrows)

ers and the bumper cover, taking care to avoid damaging the cover and the fenders **(see illustrations)**.

6 Working under the vehicle, remove the bumper beam mounting bolts **(see illustration)**. **Note:** *The bumper beam brackets can be removed from the chassis by removing the nuts from the studs at the back of the brackets, or removing the foam inserts to access the front mounting bolts once the bumper cover is removed.*

7 Detach the bumper assembly from the vehicle.

8 Installation is the reverse of the removal procedure.

Front bumper (2000 and later models)

9 Apply the parking brake, raise the vehicle and support it securely on jackstands.

10 Disconnect the negative battery cable. If equipped with airbag(s), disconnect the positive battery cable and wait 10 minutes before proceeding any further.

11 Working under the vehicle, detach the screws securing the sides of the right and left front fender protectors.

12 Detach the clips securing the front

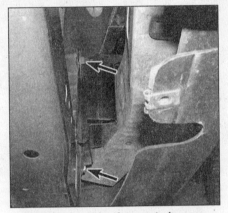

12.6 Remove the three nuts (arrows indicate two, there is one on the other side) to separate the front bumper beam brackets from the chassis

bumper protectors to the wheel well.

13 Remove the clips and protector securing the front bumper to the hood cross brace.

14 Pull the bumper assembly part way out from the vehicle and disconnect the fog lamp and side marker electrical connectors. Remove the front bumper.

15 Installation is the reverse of removal.

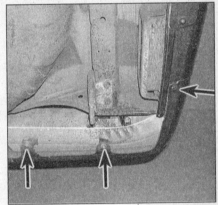

12.16a On each side, remove these three screws (arrows) . . .

Rear (1993 through 1999 models)

Refer to illustrations 12.16a, 12.16b, 12.17a, 12.17b, 12.17c, 12.17d and 12.18

16 Working underneath the vehicle, remove the fasteners along the bottom edge of the rear bumper fascia, and in the license plate opening **(see illustrations)**.

17 Open the trunk lid and remove the trim for access to the upper mounting bolts for the rear bumper fascia **(see illustrations)**.

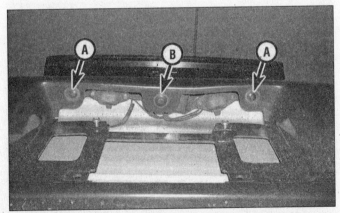

12.16b . . . then remove the two screws (A) and the plastic pushpin (B) above the license plate

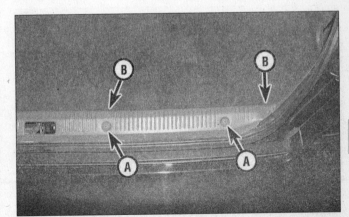

12.17a To remove the carpet shield at the trunk opening, remove the four screws (A indicates two of the four), and the inner plastic "twist" fasteners (B indicates two of the four)

Chapter 11 Body

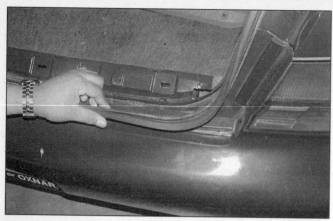

12.17b Pull off the lower part of the trunk weatherstrip . . .

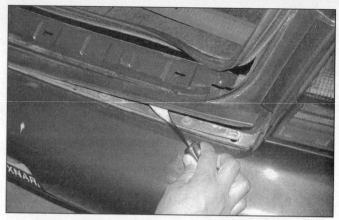

12.17c . . . to have room to pry up this plastic strip . . .

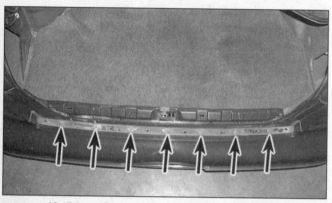

12.17d . . . then remove the bolts (arrows) at the top of the bumper fascia

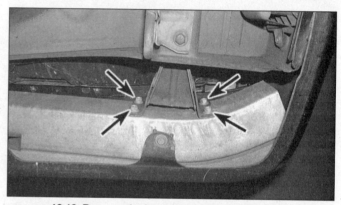

12.18 Remove the bumper beam mounting bolts (arrows, one side shown)

18 Remove the mounting bolts and remove the rear bumper beam, if required (see illustration).

Rear bumper (2000 and later models)

19 Apply the parking brake, raise the vehicle and support it securely on jackstands.
20 Disconnect the negative battery cable. If equipped with airbag(s), disconnect the positive battery cable and wait 10 minutes before proceeding any further.
21 Remove the rear trunk floor trim.
22 Working inside the trunk, remove the jack and the jack bracket. Remove the bolts securing the right and left rear fender to the rear bumper. Remove the nuts securing the rear bumper to the rear panel.
23 Remove the perimeter clamps securing the rear bumper to the body panel below the trunk opening and pull the bumper assembly out and away from the vehicle.
24 Installation is the reverse of removal.

13 Front fender - removal and installation

Refer to illustrations 13.3, 13.4a, 13.4b, 13.4c, 13.4d and 13.4e

1 Open the hood, raise the vehicle, support it securely on jackstands and remove the front wheel. Disconnect the negative cable from the battery (see Chapter 1).
2 Refer to Chapter 12 and remove the cornering lamp from the fender.

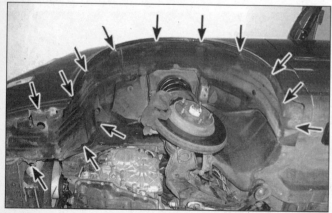

13.3 Remove the screws and plastic fasteners (arrows) securing the fenderwell liner

13.4a With the liner removed, remove the nuts (arrows) where the fender meets the front bumper fascia - the other fasteners are plastic alignment studs

Chapter 11 Body

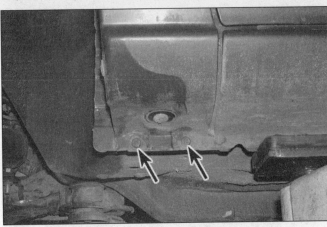

13.4b Remove the two bolts (arrows) where the lower rear of the fender is mounted to the rocker panel

13.4c Open the door to access the two fender bolts (arrow indicates the upper one, there is also a lower bolt)

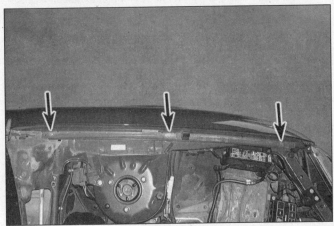

13.4d Remove the three upper fender-mounting bolts (arrows) on the body

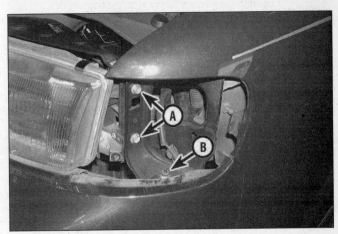

13.4e With the cornering lamp removed, remove the two headlight mounting bolts (A), and the fender brace nut (B) (1993 through 1999 model shown)

3 Pry out the plastic rivets, remove the screws and remove the inner fenderwell liner **(see illustration)**.

4 Remove the fender mounting bolts **(see illustrations)**.

5 Detach the fender. It's a good idea to have an assistant support the fender while it's being moved away from the vehicle to prevent damage to the surrounding body panels. Once the fender is pulled out slightly at the rear, slide it back to clear the plastic studs at the fender-to-front bumper fascia flange and allow the forward brace of the fender to clear the bumper cover support.

6 Installation is the reverse of removal. If a new fender is being installed, transfer the plastic fenderwell lip, if applicable, to the new fender.

7 Tighten all nuts, bolts and screws securely.

14 Cowl cover - removal and installation

Refer to illustration 14.3

1 Mark the position of the windshield wiper blades on the windshield with a wax marking pencil.

2 Remove the wiper arms (see Chapter 12).

3 Carefully pry up under the rubber hood seal, the plastic clips that hold it onto the cowl also secure the cowl cover **(see illustration)**. Once the hood seal and clips are removed, push down at the rear of the cowl cover to release its two upper clips from plastic pins on the cowl. Remove the cowl cover. **Note:** *The windshield washer hose on the right side of the cowl snaps into a slot on the cowl cover.*

4 Installation is the reverse of removal. Make sure to align the wiper blades with the marks made during removal.

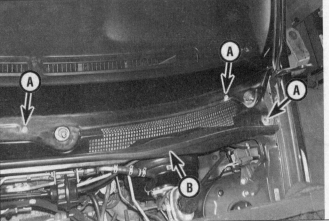

14.3 Remove the plastic pins (A indicates three on the left side) from the cowl cover, then pry up the plastic pins securing the hood seal strip (B)

Chapter 11 Body

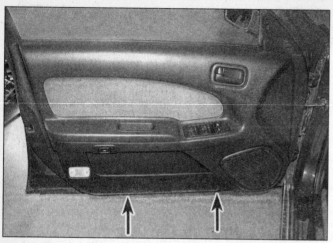

15.2 There are several clips securing the door panel to the door, and these screws (arrows) at the bottom

15.3 Carefully pry up at the front of the power window switch (driver's side shown) and lift it out far enough to disconnect the electrical connector

15 Door trim panel - removal and installation

Refer to illustrations 15.2, 15.3, 15.4, 15.5, 15.6 and 15.7

1 Disconnect the negative cable from the battery.
2 At the bottom of the door panel, remove the two screws (see illustration).
3 Pry up at the front edge of the power window switch pod and lift it out, then unplug the electrical connector (see illustration).
4 Remove the inside door handle escutcheon (see illustration).
5 Remove the screw in the door pull pocket (see illustration).
6 Pull the door panel away from the door far enough to disconnect the electrical connectors (see illustration).
7 For access to the inner door, remove the plastic watershield. Peel back the watershield, taking care not to tear it (see illustration). To install the trim panel, first press the watershield back into place. If necessary, add more sealant to hold it in place.
8 Prior to installation of the door panel, be sure to reinstall any clips in the panel which may have come out during the removal procedure and stayed in the door.
9 Plug in any electrical connectors and place the panel in position. Press it into place until the clips are seated and install any retaining screws and armrest/door pulls. Install the power window switch assembly.

16 Door - removal, installation and adjustment

Removal and installation

Refer to illustrations 16.3 and 16.4

1 Remove the door trim panel (see Section 15), disconnect any electrical connectors and push them through the door opening so they won't interfere with removal.
2 Position a jack under the door or have an assistant on hand to support the door when the hinge bolts are removed. **Note:** *If a jack or stand is used, place a rag between it and the door to protect the door's paint.*
3 Remove the bolt securing the door stop strut at the body (see illustration).
4 Scribe or draw a line around the door hinge bolts (see illustration).
5 Remove the hinge-to-door bolts and carefully detach the door. Installation is the reverse of removal.

Adjustment

Refer to illustration 16.6

6 Following installation, make sure the door is aligned properly. Adjust it if necessary as follows:

  a) Up-and-down and forward-and-backward adjustments are made by loosening the hinge-to-body bolts and moving the door, as necessary. A special offset tool may be required to reach some of the bolts.
  b) In-and-out and up-and-down adjustments are made by loosening the door side hinge bolts and moving the door, as necessary. A special offset tool may be required to reach some of the bolts.

15.4 Carefully pry out the door handle escutcheon

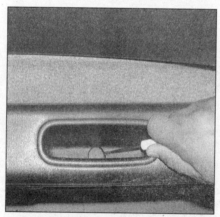

15.5 In the door pull pocket, pry up the plastic cap and remove the screw underneath

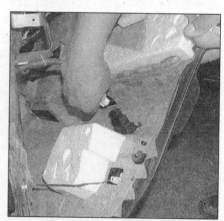

15.6 Pull the door panel away to disengage the clips, then disconnect any electrical connectors on the door side of the panel

Chapter 11 Body

15.7 If the plastic watershield is peeled off carefully, it can be reused

16.3 Remove the bolt (arrow) to release the body-end of the door stop strut

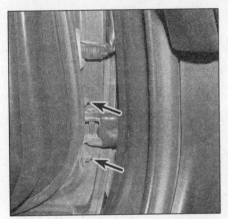

16.4 Before loosening or removing the door hinge bolts (arrows), draw alignment marks around the hinge location with a marking pen (lower door hinge bolts shown, upper are similar)

16.6 Adjust the door lock striker by loosening the mounting screws and gently tapping the striker in the desired direction

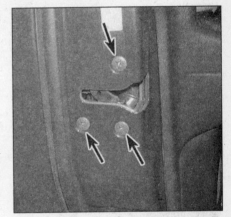

17.2 Remove the latch screws (arrows) from the end of the door

c) The door lock striker can also be adjusted both up-and-down and sideways to provide a positive engagement with the locking mechanism. This is done by loosening the screws and moving the striker, as necessary **(see illustration)**.

17 Door latch, lock cylinder and handle - removal and installation

1 Remove the door trim panel and the plastic watershield (see Section 15).

Door latch

Refer to illustration 17.2

2 Remove the latch retaining screws from the end of the door, then reach inside the door to release the latch from the control rods **(see illustration)**.
3 On models with power door locks, disconnect the electrical connector.
4 Detach the door latch and (if equipped) the door lock solenoid.
5 Installation is the reverse of removal.

Lock cylinder and outside handle

Refer to illustration 17.6

6 Disconnect the control link rod from the outside handle, remove the outside handle retention nuts and pull the handle from the door **(see illustration)**.
7 Detach the electrical connector from the lock cylinder, use a screwdriver or pliers to pry the retaining clip off and remove the lock cylinder from the door **(see illustration 17.6)**.
8 Installation is the reverse of removal.

Inside handle

Refer to illustration 17.10

9 Remove the door trim panel.
10 Remove the retaining screws, pull the

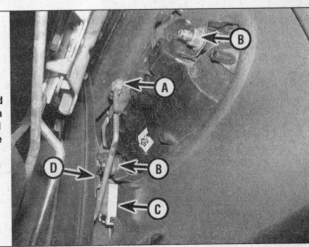

17.6 The link rod (A) and outside handle retention nuts (B) can be reached through the access hole in the doorframe - the electrical connector (C) is for the door lock cylinder (D)

17.10 Remove the handle retaining screws (arrows) and pull the handle out, then detach the control link(s) (1995 and later model shown with two links)

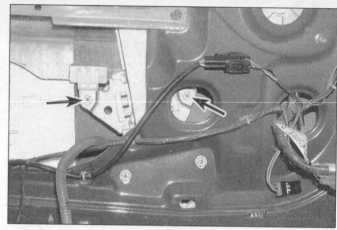

18.3 Remove the two mounting bolts (arrows), detach the glass and lift it out of the door - place a rag over the glass to help prevent scratching it

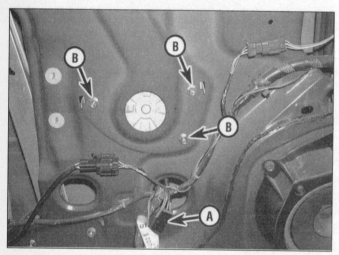

19.3 Detach the electrical connector (A) and remove the regulator motor mounting bolts (B)

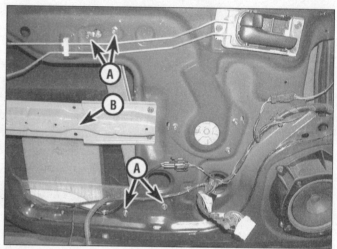

19.4 Remove the four regulator frame mounting bolts (A) - the door panel support brace (B) may need to be removed in order to extract the regulator and motor through the door opening

handle free, disconnect the link(s) from the inside handle control and remove the handle from the door **(see illustration)**.
11 Installation is the reverse of removal.

20.3 With the door panel removed, remove the screw (arrow) at the bottom of the mirror cover and pull to release the cover the from the two clips on the door

18 Door window glass - removal and installation

Refer to illustration 18.3

1 Remove the door trim panel and the plastic watershield (see Section 15).
2 Lower the window glass.
3 Place a rag inside the door panel to help prevent scratching the glass and remove the two glass mounting bolts **(see illustration)**.
4 Remove the glass by pulling it straight up and to the outside of the door.
5 Installation is the reverse of the removal procedure.

19 Door window glass regulator - removal and installation

Refer to illustrations 19.3 and 19.4

1 Remove the door trim panel and the plastic watershield (see Section 15).
2 Remove the window glass (see Section 18).

3 Unclip the power window electrical connector and remove the regulator motor mounting bolts **(see illustration)**.
4 Remove the bolts at the top and bottom of the regulator frame **(see illustration)**.
5 Pull the regulator through the service hole to remove it.
6 Installation is the reverse of removal.

20 Mirrors - removal and installation

Refer to illustrations 20.3, 20.4 and 20.6

Outside mirrors

1 Remove the door trim panel (see Section 15).
2 On power mirrors, unplug the electrical connector.
3 Remove the triangular cover over the mirror mounting **(see illustration)**.
4 Remove the bolts and detach the mirror from the door **(see illustration)**.
5 Installation is the reverse of removal.

Chapter 11 Body

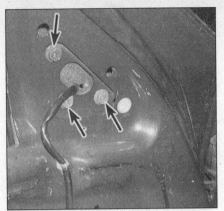

20.4 Remove the three screws (arrows) securing the outside mirror to the door - trace the wiring down to the connector and disconnect it

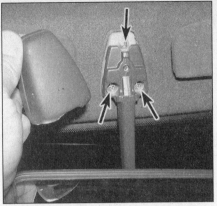

20.6 Pry off the plastic cover, then remove the three inside mirror mounting screws (arrows)

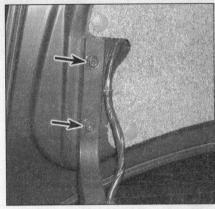

21.3 Make alignment marks around the trunk hinge, then remove the mounting bolts (arrows)

Inside mirror

6 Remove the plastic cover at the top of the mirror mount, then remove the mirror mounting screws **(see illustration)**.

7 Installation is the reverse of removal.

21 Trunk lid - removal, installation and adjustment

Refer to illustrations 21.3 and 21.9

1 Open the trunk lid and cover the edges of the trunk compartment with pads or cloths to protect the painted surfaces when the lid is removed. **Note:** *The trunk lid is heavy and awkward to remove, have an assistant to help you with this procedure.*

2 Disconnect any cables or wire harness connectors attached to the trunk lid that would interfere with removal.

3 Make alignment marks around the hinge mounting bolts with a marking pen **(see illustration)**.

4 While an assistant supports the lid, remove the hinge-to-trunk bolts on both sides and lift it off **(see illustration 21.3)**.

5 Installation is the reverse of removal. **Note:** *When reinstalling the trunk lid, align the hinge with the marks made during removal before final tightening of the bolts.*

6 After installation, close the lid and make sure it's in proper alignment with the surrounding panels.

7 Forward-and-backward and side-to-side adjustments are made by loosening the hinge-to-lid bolts and gently moving the lid into correct alignment.

8 Vertical adjustments to the lid are made by adding or removing the shims used on the hinge.

9 To adjust the lid so it is flush with the body when closed, loosen the mounting bolts and move the striker **(see illustration)**.

22 Trunk lid latch and lock cylinder - removal and installation

Trunk lid latch

Refer to illustration 22.4

1 Open the trunk and use a trim tool to pry up the plastic buttons securing the lower portion of the trunk lid inner panel to expose the latch assembly, then slide off the plastic cover over the latch.

2 Scribe a line around the trunk lid latch assembly for a reference point to aid the installation procedure.

3 Disconnect the electrical connector and pop the link from the rod connecting the latch and the lock cylinder.

4 Detach the two retaining bolts and remove the latch **(see illustration)**.

5 Installation is the reverse of removal.

Trunk lock cylinder

Refer to illustration 22.7

6 Open the trunk and remove the trunk lid inner panel, using a trim tool to pry up the plastic buttons.

7 Look upward through the trunk lid access hole near the right taillight. Remove the lock cylinder rod from its clip and remove the lock's mounting plate **(see illustration)**. On models so equipped, disconnect the electrical connector from the lock.

8 Twist the lock about 45 degrees and remove it from the trunk.

9 Installation is the reverse of removal.

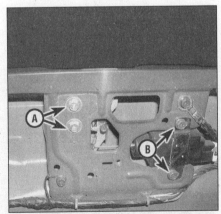

21.9 The trunk striker bolts (A) can be loosened to allow minor adjustment of the trunk closing - the trunk release actuator is secured with two bolts (B)

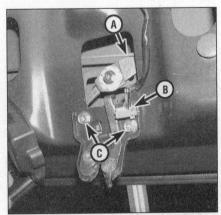

22.4 Pull off the plastic cover, disconnect the linkage rod (A) and the electrical connector (B), then remove the two latch mounting bolts (C)

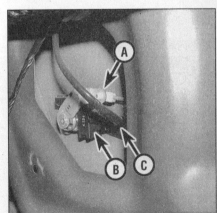

22.7 To remove the lock cylinder, disconnect the linkage rod (A) and the electrical connector (B), then use pliers or a screwdriver to push off the mounting plate (C)

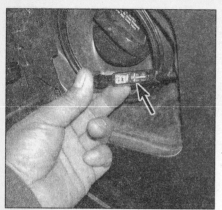

23.2 The end-housing of the fuel door cable has been pushed out of its body socket, showing where the cable eye (arrow) can easily be removed from the end-housing

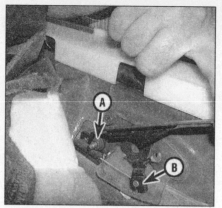

23.4 Remove the nut (A) securing the "flag" tab on the cable(s), then swing the cable down to release the cable eye (B) from the lever

24.3 Carefully pry up the trim panel around the shift lever - it's retained by clips

23 Trunk release and fuel door cables - removal and installation

Refer to illustrations 23.2 and 23.4

1 All models have a cable for operation of the fuel door opener, while remote trunk operation is accomplished by a cable on some models or an electric solenoid on other models.

2 Inside the trunk, remove the left-side panel for access to the fuel door cable end. With the fuel door open, push the cable from inside until its plastic end-housing comes through into the fuel door compartment, then pry the cable end out and slips the end-housing off the old cable (see illustration).

3 Peel back the carpeting to access the release cable. Open all of the clips holding the cable to the body. Refer to Section 22 for the location of the trunk release cable at the trunk end. Some models will have an electric solenoid in place of the cable, but they both attach to the trunk release at the same point.

4 Pull back the carpeting next to the driver's seat to reveal the release lever mechanism (see illustration). Remove the nut securing the cable end(s) to the stud, then remove the cable eye(s) from the bottom of the lever.

5 Attach a piece of thin wire to the end of the cable.

6 Working in the trunk compartment, pull the cable assembly towards the rear of the vehicle until you can see the wire.

7 Attach the wire to the front of the new cable and fish it back through the body until it can be attached to the lever. The remainder of the installation is the reverse of removal.

24 Center console - removal and installation

Refer to illustrations 24.3, 24.4a and 24.4b

Warning: *The models covered by this manual are equipped with Supplemental Restraint Systems (SRS), more commonly known as airbags. Always disable the airbag system before working in the vicinity of any airbag system components to avoid the possibility of accidental deployment of the airbags, which could cause personal injury (see Chapter 12).*

1 Disconnect the cable from the negative terminal of the battery.

2 Pull the parking brake on.

3 Pry up the plastic trim panel surrounding the shift lever (see illustration).

4 Remove the console mounting screws at the front and rear of the console (see illustrations).

5 Unplug any electrical connections and remove the console assembly from the vehicle.

6 Installation is the reverse of the removal procedure.

25 Dashboard trim panels - removal and installation

Warning: *The models covered by this manual are equipped with Supplemental Restraint Systems (SRS), more commonly known as airbags. Always disable the airbag system before working in the vicinity of any airbag system components to avoid the possibility of accidental deployment of the airbags, which could cause personal injury (see Chapter 12).*

Note: *The following procedures are shown on a 1993 through 1999 model. The shape of the components and some fastener locations vary slightly on the 2000 and later models but the removal and installation procedures are identical.*

1 Disconnect the cable from the negative terminal of the battery.

24.4a Where the instrument panel meets the console, remove the two console mounting screws (arrows)

24.4b Remove the two mounting screws at the rear of the console (arrow indicates screw on right side, left side similar)

25.3 Remove the two screws (arrows) at the top of the instrument cluster bezel, then pull out the bottom of the bezel where it is retained by clips

25.4 Remove the two screws (arrows) at the bottom of the knee bolster

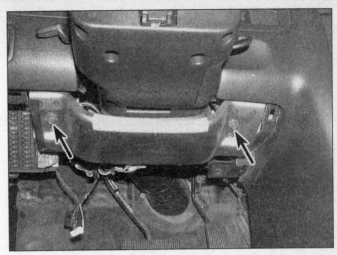

25.5 Remove the two screws (arrows) and the knee bolster reinforcement panel

Instrument cluster bezel

Refer to illustration 25.3

2 Remove the steering column covers (see Section 26).
3 Remove the two screws at the top of the instrument cluster bezel, then pull out the bezel at the bottom to release the clips (see illustration). **Note:** *On models with cruise control or theft system switches in the bezel, disconnect the electrical connectors at the back before final removal of the bezel.*

Knee bolster

Refer to illustrations 25.4 and 25.5

4 Pull off the cover over the interior fuse panel, then remove the two screws at the bottom edge of the knee bolster and pull the bolster off (see illustration).
5 If the knee bolster reinforcement panel must be removed for another procedure, remove the mounting bolts and lower the panel (see illustration).
6 Installation is the reverse of the removal procedure.

Center bezel panel

Refer to illustration 25.7

7 Use a trim tool or a screwdriver with tape over the tip to pry up and remove the panel (see illustration).
8 With the panel pulled out partway, disconnect the electrical connectors at the back of the switches mounted on the panel (see Chapter 12).

Glove box

Refer to illustrations 25.9a and 25.9b

9 With the glove box open, remove the four screws at the top, two screws at the bottom, and pull the glove box housing from the instrument panel (see illustrations).
10 Installation is the reverse of the removal procedure.

26 Steering column covers - removal and installation

Refer to illustrations 26.2 and 26.3
Warning: *The models covered by this manual are equipped with Supplemental Restraint Systems (SRS), more commonly known as airbags. Always disable the airbag system before working in the vicinity of any airbag system components to avoid the possibility of accidental deployment of the airbags, which could cause personal injury (see Chapter 12).*
1 Refer to Section 25 and remove the knee bolster and knee bolster reinforcement panel.
2 On the right side of the column, pull out the small cover around the ignition switch (see illustration).

25.7 Pull the center bezel out far enough to disconnect the electrical connectors at the back

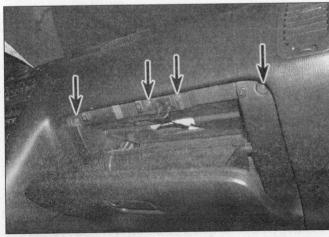

25.9a Remove the four glove box screws (arrows) at the top . . .

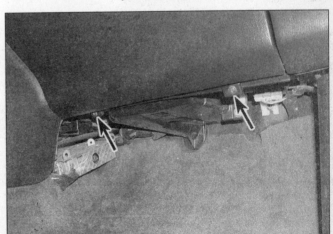

25.9b . . . and the two screws at the bottom (arrows)

11-16   Chapter 11 Body

26.2 Pull the cover from around the ignition switch

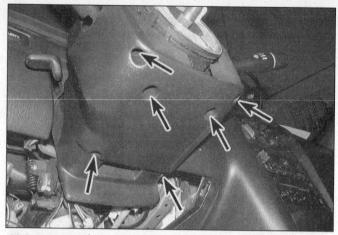

26.3 Remove the two lower and four upper screws (arrows) from the lower column cover and separate the upper and lower covers

3   Remove the steering column cover screws from the lower cover (see illustration).
4   Separate the cover halves and detach them from the steering column.
5   Installation is the reverse of the removal procedure.

27 Instrument panel - removal and installation

Refer to illustrations 27.8, 27.9, 27.10, 27.11, 27.12, 27.13, 27.14a, 27.14b, 27.16a and 27.16b

Warning: *The models covered by this manual are equipped with Supplemental Restraint Systems (SRS), more commonly known as airbags. Always disable the airbag system before working in the vicinity of any airbag system components to avoid the possibility of accidental deployment of the airbags, which could cause personal injury (see Chapter 12).*

Note: *This is a difficult procedure for the home mechanic, involving tedious disassembly and the disconnection/reconnection of numerous electrical connectors. If you do attempt this procedure, make sure you take good notes and mark all matching connectors (and their mounting points) to aid reassembly.*

27.8 Pull out the kick panels at each side

1   Disconnect the cable from the negative terminal of the battery.
2   Remove the center console (see Section 24).
3   Remove the steering column covers (see Section 26).
4   Remove the instrument cluster bezel (see Section 25).
5   Remove the instrument cluster (see Chapter 12).
6   Remove the dashboard trim panels (see Section 25).

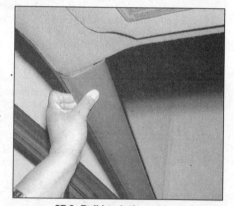

27.9 Pull back the rubber weatherstripping, then remove the windshield post trim strips (retained by clips)

7   Remove the radio and heater/air conditioner controls (see Chapter 3 and Chapter 12).
8   Remove the kick panels at each side (see illustration). The panels simply pull out.
9   Remove the left and right trim strips along the interior of each windshield post (see illustration).
10 The passenger airbag is attached to the

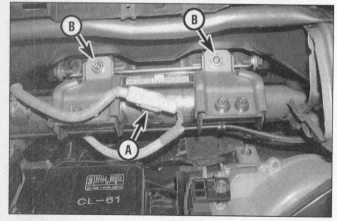

27.10 Disconnect the passenger airbag connector (A), then remove the two bolts (B)

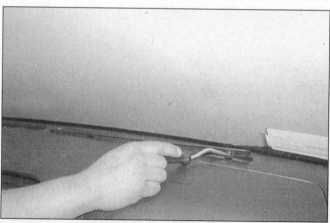

27.11 Access the upper instrument panel bolts by prying up the defroster vent grilles

Chapter 11 Body

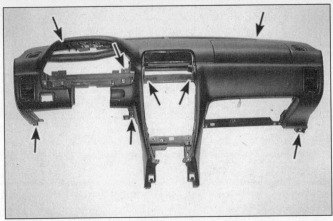

27.12 Instrument panel bolt locations (arrows)

27.13 To remove the BCM from the front of the console area, disconnect the electrical connectors (A indicates the main connector), then remove the mounting nuts (B) - the module in front of the BCM here is the ECM, which must also be removed (see Chapter 6)

27.14a Remove both supports (right-side shown) by removing the mounting nuts (A) and removing any electrical harness clips (B) attached to the supports

27.14b Lower the steering column (Chapter 10), drop the fuse panel (at left here), and remove the bolt (arrow) securing the beam to the firewall brace

back of the instrument panel. Disconnect the electrical connector and remove the bolts securing the bottom of the airbag to the instrument panel reinforcement beam **(see illustration)**. Note: *The bolts are "tamper-proof" Torx type and require a special socket to remove or install*.

11 At the top of the instrument panel where it meets the windshield, pry up the defroster grilles to access the upper bolts of the instrument panel **(see illustration)**.

12 Remove all of the fasteners securing the instrument panel, and any electrical connectors still attached to the panel **(see illustration)**. Pull the panel back and out of the vehicle.

13 At the center of the dash/console support structure, remove the Body Control Module (BCM), marking all the connectors **(see illustration)**. Refer to Chapter 6 and remove the ECM.

14 Remove the two upright supports at the center of the reinforcement beam, removing the bolts at the floorpan and the beam **(see illustration)**. Remove the bolts and drop the interior fuse panel and hood release handle from the reinforcement beam **(see illustration)**.

15 Refer to *Heater core - replacement* in Chapter 3 and lower the steering column from the reinforcement beam.

16 Mark and disconnect all wiring harness clips and connectors attached to the instrument panel reinforcement beam, then remove the bolts and angle the reinforcement beam out of the vehicle **(see illustrations)**.

17 Installation is the reverse of removal.

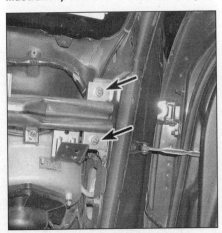

27.16a Remove the two nuts (arrows) at the right side of the reinforcement beam . . .

27.16b . . . and the two bolts (A) and two nuts (B) at the left side of the reinforcement beam - angle the beam out of the vehicle by pulling the right side out first

Chapter 11 Body

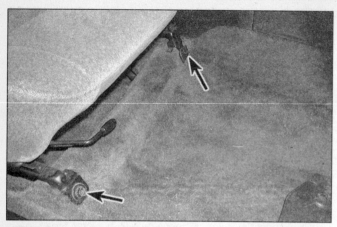

28.1a Remove the two front bolts (arrows) for the front seat . . .

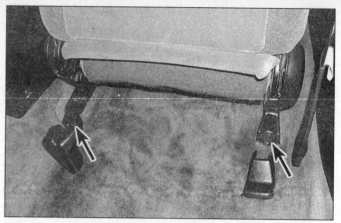

28.1b . . . and the plastic covers and bolts (arrows) at the rear of the front seat

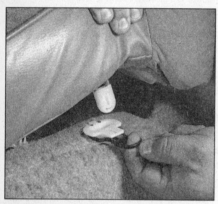

28.3a Pull out on the clip retainer ring while pulling sharply upward to detach the back seat lower cushion

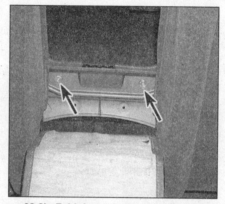

28.3b Fold down the center armrest in the back seat and remove these bolts (arrows) . . .

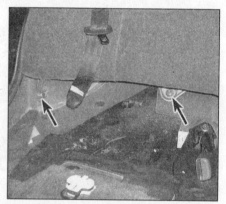

28.3c . . . then remove the four lower mounting bolts (arrows indicate the two right-side bolts)

28 Seats - removal and installation

Front

Refer to illustrations 28.1a and 28.1b

1 Remove the retaining bolts, detach any clips, unplug any electrical connectors and lift the seat from the vehicle **(see illustrations)**.
2 Installation is the reverse of removal.

Rear

Refer to illustrations 28.3a, 28.3b and 28.3c

3 Remove the seat cushion by pulling out on the retainer, then pulling up sharply on the cushion **(see illustration)**. After removing the cushion, remove the bolts and lift the seat back out **(see illustrations)**.
4 Installation is the reverse of removal.

29 Rear package shelf - removal and installation

Refer to illustrations 29.2 and 29.3

1 Refer to Section 28 and remove the back seat.
2 Using a trim tool or a taped screwdriver to pry the rear roof pillar trim panels out **(see illustration)**. The rear shoulder belts are routed through the panels. If you are just removing the package shelf to replace the rear speakers, just slide the panels along the belts, out of the way.
3 The shelf can be pried up carefully from the body with a trim tool **(see illustration)**.
4 Installation is the reverse of the removal.

30 Sunroof - adjustment

1 Access to the sunroof motor is by prying out the cover for the overhead light and sunroof switches.
2 Close the sunroof to its flush-with roof position.
3 Unbolt the sunroof motor from the roof and disconnect the drive cables.
4 With the motor's electrical connector still in place, run the motor until it reaches the end of its travel, then stop.
5 Reattach the two cables and bolt the motor back in.
6 Installation is the reverse of removal.

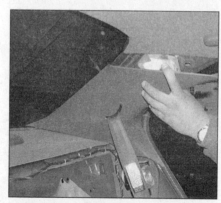

29.2 Pry out the rear roof pillar trim panels, and slide them along the seat belts and out of the way

29.3 Pry up the package shelf with a trim tool - slide the tool along between the shelf and the body until you locate the plastic pins, then pry up

Chapter 12
Chassis electrical system

Contents

Section		Section	
Airbag system - general information	29	Ignition switch and key lock cylinder - check and replacement	8
Antenna - removal and installation	13	Instrument cluster - removal and installation	11
Bulb replacement	17	Instrument panel gauges - check	10
Circuit breakers - general information	4	Instrument panel switches - check and replacement	9
Cruise control system - description and check	22	In-Vehicle Multiplexing System - description	28
Daytime Running Lights (DRL) - general information	20	Power door lock system - description and check	24
Electric side view mirrors - description and check	25	Power seats - description and check	27
Electric sunroof - description and check	26	Power window system - description and check	23
Electrical troubleshooting - general information	2	Radio and speakers - removal and installation	12
Fuses and fusible links - general information	3	Rear window defogger - check and repair	21
General information	1	Relays - general information and testing	5
Headlight bulb - replacement	14	Steering column switches - check and replacement	7
Headlight housing - replacement	16	Turn signal/hazard flashers - check and replacement	6
Headlights - adjustment	15	Wiper motor - removal and installation	18
Horn - check and replacement	19	Wiring diagrams - general information	30

1 General information

The electrical system is a 12-volt, negative ground type. Power for the lights and all electrical accessories is supplied by a lead/acid-type battery that is charged by the alternator.

This Chapter covers the various electrical components not associated with the engine. Information on the battery, alternator, distributor and starter motor can be found in Chapter 5.

It should be noted that when portions of the electrical system are serviced, the cable should be disconnected from the negative battery terminal to prevent electrical shorts and/or fires.

Warning: *The models covered by this manual are equipped with Supplemental Restraint Systems (SRS), more commonly known as airbags. Always disable the airbag system before working in the vicinity of any airbag system components to avoid the possibility of accidental deployment of the airbags, which could cause personal injury (see Section 29).*

2 Electrical troubleshooting - general information

Refer to illustrations 2.5a, 2.5b, 2.6, 2.9 and 2.15

A typical electrical circuit consists of an electrical component, any switches, relays, motors, fuses, fusible links or circuit breakers related to that component and the wiring and connectors that link the component to both the battery and the chassis. To help you pinpoint an electrical circuit problem, wiring diagrams are included at the end of this Chapter.

Before tackling any troublesome electrical circuit, first study the appropriate wiring diagrams to get a complete understanding of what makes up that individual circuit. Trouble spots, for instance, can often be narrowed down by noting if other components related to the circuit are operating properly. If several components or circuits fail at one time, chances are the problem is in a fuse or ground connection, because several circuits are often routed through the same fuse and ground connections.

Electrical problems usually stem from simple causes, such as loose or corroded connections, a blown fuse, a melted fusible link or a failed relay. Visually inspect the condition of all fuses, wires and connections in a problem circuit before troubleshooting the circuit.

If test equipment and instruments are going to be utilized, use the diagrams to plan ahead of time where you will make the necessary connections in order to accurately pinpoint the trouble spot.

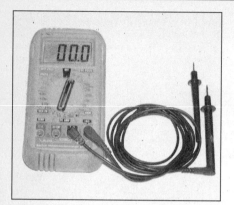

2.5a The most useful tool for electrical troubleshooting is a digital multimeter that can check volts, amps, and test continuity

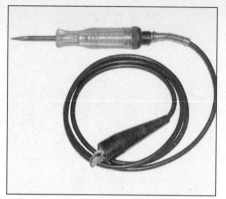

2.5b A simple test light is very handy, especially when testing for voltage

2.6 In use, a basic test light's lead is clipped to a known good ground, then the pointed probe can test connectors, wires or electrical sockets - if the bulb lights, the part being tested has battery voltage

The basic tools needed for electrical troubleshooting include a circuit tester or voltmeter (a 12-volt bulb with a set of test leads can also be used), a continuity tester, which includes a bulb, battery and set of test leads, and a jumper wire, preferably with a circuit breaker incorporated, which can be used to bypass electrical components **(see illustrations)**. Before attempting to locate a problem with test instruments, use the wiring diagram(s) to decide where to make the connections.

Voltage checks

Voltage checks should be performed if a circuit is not functioning properly. Connect one lead of a circuit tester to either the negative battery terminal or a known good ground. Connect the other lead to a connector in the circuit being tested, preferably nearest to the battery or fuse **(see illustration)**. If the bulb of the tester lights, voltage is present, which means that the part of the circuit between the connector and the battery is problem free. Continue checking the rest of the circuit in the same fashion. When you reach a point at which no voltage is present, the problem lies between that point and the last test point with voltage. Most of the time the problem can be traced to a loose connection. **Note:** *Keep in mind that some circuits receive voltage only when the ignition key is in the Accessory or Run position.*

Finding a short

One method of finding shorts in a circuit is to remove the fuse and connect a test light or voltmeter in place of the fuse terminals. There should be no voltage present in the circuit. Move the wiring harness from side-to-side while watching the test light. If the bulb goes on, there is a short to ground somewhere in that area, probably where the insulation has rubbed through. The same test can be performed on each component in the circuit, even a switch.

Ground check

Perform a ground test to check whether a component is properly grounded. Disconnect the battery and connect one lead of a continuity tester or multimeter (set to the ohms scale), to a known good ground. Connect the other lead to the wire or ground connection being tested. If the resistance is low (less than 5 ohms), the ground is good. If the bulb on a self-powered test light does not go on, the ground is not good.

Continuity check

A continuity check is done to determine if there are any breaks in a circuit - if it is passing electricity properly. With the circuit off (no power in the circuit), a self-powered continuity tester or multimeter can be used to check the circuit. Connect the test leads to both ends of the circuit (or to the "power" end and a good ground), and if the test light comes on the circuit is passing current properly **(see illustration)**. If the resistance is low (less than 5 ohms), there is continuity; if the reading is 10,000 ohms or higher, there is a break somewhere in the circuit. The same procedure can be used to test a switch, by connecting the continuity tester to the switch terminals. With the switch turned On, the test light should come on (or low resistance should be indicated on a meter).

Finding an open circuit

When diagnosing for possible open circuits, it is often difficult to locate them by sight because the connectors hide oxidation or terminal misalignment. Merely wiggling a connector on a sensor or in the wiring harness may correct the open circuit condition. Remember this when an open circuit is indicated when troubleshooting a circuit. Intermittent problems may also be caused by oxidized or loose connections.

Electrical troubleshooting is simple if you keep in mind that all electrical circuits are basically electricity running from the battery, through the wires, switches, relays, fuses and fusible links to each electrical component (light bulb, motor, etc.) and to ground, from which it is passed back to the battery. Any electrical problem is an interruption in the flow of electricity to and from the battery.

Connectors

Most electrical connections on these vehicles are made with multiwire plastic connectors. The mating halves of many connectors are secured with locking clips molded into the plastic connector shells. The mating halves of large connectors, such as some of those under the instrument panel, are held together by a bolt through the center of the connector.

To separate a connector with locking clips, use a small screwdriver to pry the clips apart carefully, then separate the connector halves. Pull only on the shell, never pull on the wiring harness as you may damage the individual wires and terminals inside the connectors. Look at the connector closely before trying to separate the halves. Often the locking clips are engaged in a way that is not immediately clear. Additionally, many connectors have more than one set of clips.

Each pair of connector terminals has a male half and a female half. When you look at the end view of a connector in a diagram, be

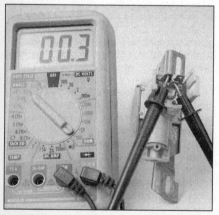

2.9 With a multimeter set to the ohms scale, resistance can be checked across two terminals - when checking for continuity, a low reading indicates continuity, a high reading indicates lack of continuity

Chapter 12 Chassis electrical system

2.15 To backprobe a connector, insert a small, sharp probe (such as a straight-pin) into the back of the connector alongside the desired wire until it contacts the metal terminal inside; connect your meter leads to the probes - this allows you to test a functioning circuit

sure to understand whether the view shows the harness side or the component side of the connector. Connector halves are mirror images of each other, and a terminal shown on the right side end-view of one half will be on the left side end view of the other half.

It is often necessary to take circuit voltage measurements with a connector connected. Whenever possible, carefully insert a small straight pin (not your meter probe) into the rear of the connector shell to contact the terminal inside, then clip your meter lead to the pin. This kind of connection is called "backprobing" **(see illustration)**. When inserting a test probe into a male terminal, be careful not to distort the terminal opening. Doing so can lead to a poor connection and corrosion at that terminal later. Using the small straight pin instead of a meter probe results in less chance of deforming the terminal connector.

3 Fuses and fusible links - general information

Fuses

Refer to illustrations 3.1a, 3.1b and 3.3

The electrical circuits of the vehicle are protected by a combination of fuses, circuit breakers and fusible links. The fuse blocks are located under the instrument panel on the left side of the dashboard and in the engine compartment **(see illustrations)**.

Each of the fuses is designed to protect a specific circuit, and the various circuits are identified on the fuse panel itself.

Miniaturized fuses are employed in the fuse blocks. These compact fuses, with blade terminal design, allow fingertip removal and replacement. If an electrical component fails, always check the fuse first. The best way to check the fuses is with a test light. Check for power at the exposed terminal tips of each fuse. If power is present at one side of the fuse but not the other, the fuse is blown. A blown fuse can also be identified by visually inspecting it **(see illustration)**.

Be sure to replace blown fuses with the correct type. Fuses of different ratings are physically interchangeable, but only fuses of the proper rating should be used. Replacing a fuse with one of a higher or lower value than specified is not recommended. Each electrical circuit needs a specific amount of protection. The amperage value of each fuse is molded into the fuse body.

If the replacement fuse immediately fails, don't replace it again until the cause of the problem is isolated and corrected. In most cases, this will be a short circuit in the wiring caused by a broken or deteriorated wire.

Fusible links

Some circuits are protected by fusible links. The links are used in circuits that are not ordinarily fused, such as the ignition circuit.

The fusible links on these models are located in the engine compartment fuse block next to the battery **(see illustration 3.1b)** and are similar to fuses, but larger.

To replace a fusible link, first disconnect the negative cable from the battery. Unplug the burned-out link and replace it with a new one (available from your dealer or auto parts store). Always determine the cause for the overload that melted the fusible link before installing a new one.

4 Circuit breakers - general information

Circuit breakers protect the power windows, power door locks, power seat and power sunroof. There are two, each located under the left end of the instrument panel.

Because the circuit breakers reset automatically, an electrical overload in a circuit breaker protected system will cause the circuit to fail momentarily, then come back on. If the circuit does not come back on, check it immediately.

5 Relays - general information and testing

General information

Refer to illustrations 5.1a, 5.1b and 5.1c

1 Several electrical accessories in the vehicle, such as the fuel injection system, horns, starter, and fog lamps use relays to transmit the electrical signal to the component. Relays use a low-current circuit (the

3.1a The interior fuse box is located under the left (driver's) side of the instrument panel, under a cover

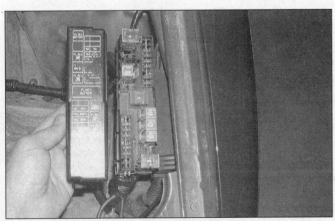

3.1b The engine compartment fuse and fusible link box is located next to the battery

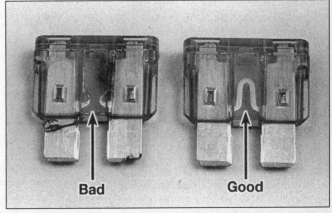

3.3 When a fuse blows, the element between the terminals melts - the fuse on the left is blown, the fuse on the right is good

Chapter 12 Chassis electrical system

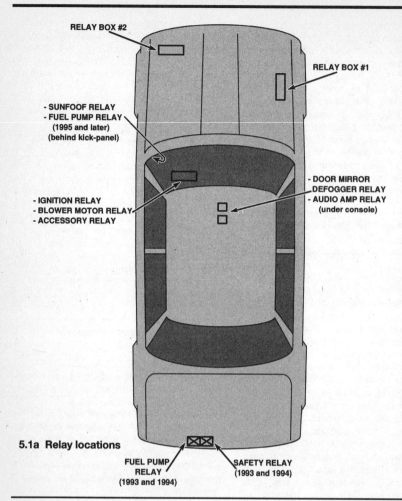

5.1a Relay locations

5.1b Underhood relay box #1 is located near the power steering reservoir

5.1c Underhood relay box #2 is located in front of the battery - the covers of the relay boxes have a printed key to identifying the relays

control circuit) to open and close a high-current circuit (the power circuit). If the relay is defective, that component will not operate properly. Most relays are mounted in the engine compartment and interior fuse/relay boxes, with some specialized relays located in other locations around the vehicle **(see illustrations)**. If a faulty relay is suspected, it can be removed and tested using the procedure below or by a dealer service department or a repair shop. Defective relays must be replaced as a unit.

Testing

Refer to illustrations 5.3a, 5.3b and 5.6

2 Refer to the wiring diagrams for the circuit to determine the proper connections for the relay you're testing. If you can't determine the correct connection from the wiring diagrams, however, you may be able to determine the test connections from the information that follows.

3 There are four basic types of relays used on the Maxima **(see illustrations)**. Some are normally open type and some normally closed, while others include a circuit of each type.

4 On most relays, two of the terminals are the relay control circuit (they connect to the relay coil which, when energized, closes the large contacts to complete the circuit). The other terminals are the power circuit (they are connected together within the relay when the control-circuit coil is energized).

5 Some relays may be marked as an aid to help you determine which terminals are the control circuit and which are the power circuit. If the relay is not marked, refer to the wiring diagrams at the end of this Chapter to

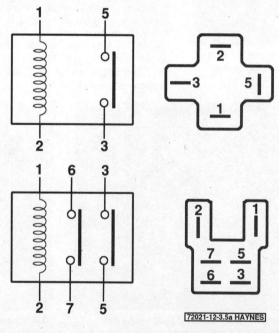

5.3a These two relays are typical normally open types, the one above completes a single circuit (terminal 5 to terminal 3) when energized, the lower relay type completes two circuits (6 and 7, and 3 and 5)

Chapter 12 Chassis electrical system

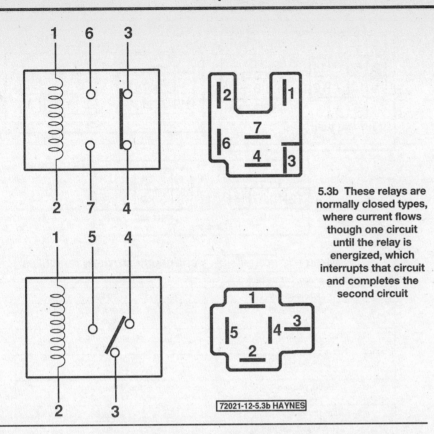

5.3b These relays are normally closed types, where current flows though one circuit until the relay is energized, which interrupts that circuit and completes the second circuit

determine the proper hook-ups for the relay you're testing.

6 To test a relay connect an ohmmeter across the two terminals of the power circuit, continuity should not be indicated (see illustration). Now connect a fused jumper wire between one of the two control circuit terminals and the positive battery terminal. Connect another jumper wire between the other control circuit terminal and ground. When the connections are made, the relay should click and continuity should be indicated on the meter. On some relays, polarity may be critical, so, if the relay doesn't click, try swapping the jumper wires on the control circuit terminals.

7 If the relay fails the above test, replace it.

6 Turn signal/hazard flashers - check and replacement

Refer to illustration 6.4

Warning: *The models covered by this manual are equipped with Supplemental Restraint Systems (SRS), more commonly known as airbags. Always disable the airbag system before working in the vicinity of any airbag system components to avoid the possibility of accidental deployment of the airbags, which could cause personal injury (see Section 29).*

1 The turn signal and hazard flasher is a single combination unit.

2 When the flasher unit is functioning properly, an audible click can be heard during its operation. If the turn signals fail on one side or the other and the flasher unit does not make its characteristic clicking sound, or if a bulb on one side of the vehicle flashes much faster than normal but the bulb at the other end of the vehicle (on the same side) doesn't light at all, a faulty turn signal bulb may be indicated.

3 If both turn signals fail to blink, the problem may be due to a blown fuse, a faulty flasher unit, a broken switch or a loose or open connection. If a quick check of the fuse box indicates that the turn signal fuse has blown, check the wiring for a short before installing a new fuse.

4 To replace the flasher, disconnect the electrical connector and remove the flasher unit from its mounting bracket located under the instrument panel to the right of the steering column (see illustration).

5 Make sure that the replacement unit is identical to the original. Compare the old one to the new one before installing it.

6 Installation is the reverse of removal.

7 Steering column switches - check and replacement

Warning: *The models covered by this manual are equipped with Supplemental Restraint Systems (SRS), more commonly known as airbags. Always disable the airbag system before working in the vicinity of any airbag system components to avoid the possibility of accidental deployment of the airbags, which could cause personal injury (see Section 29).*

Combination switch
Check

Refer to illustrations 7.4a and 7.4b

1 Disconnect the cable from the negative terminal of the battery.

2 Refer to the replacement procedure below to remove the combination switch for testing.

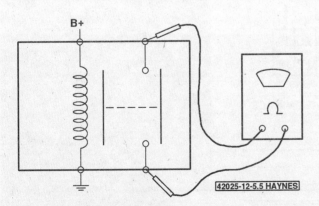

5.6 To test a typical four-terminal normally open relay, connect an ohmmeter to the power side and test with the relay energized and not energized

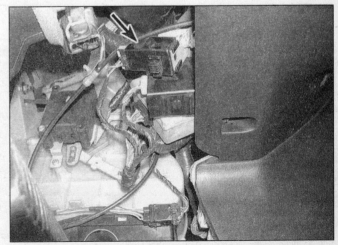

6.4 Turn signals and hazard flasher unit location (arrow)

3 Using an ohmmeter, check for continuity between the indicated terminals with the various switches in each of the indicated positions.
4 If the continuity is not as specified, replace the defective switch (see illustrations).

Replacement

Refer to illustrations 7.8, 7.9 and 7.10

5 Disconnect the cable from the negative terminal of the battery.
6 Remove the steering wheel (see Chapter 10).
7 Remove the driver's knee bolster and steering column covers (see Chapter 11).
8 Remove the four screws retaining the airbag clockspring and remove the clockspring from the combination switch (see Chapter 10). Remove the combination switch retaining screws (see illustration).
9 Remove the combination switch. Slide the switch up off the column and unplug the connectors (see illustration).
10 Remove the retaining screws from the switch being replaced and remove the defective switch from the switch body (see illustration).
11 Insert the terminals from the new switch into the connector, pushing in until they are securely locked in place.
12 The remainder or installation is the reverse of removal. Refer to Chapter 10 and center the clockspring before installing the steering wheel.

Cruise control switches

Check

Refer to illustration 7.14

13 The cruise control switch pod (all cruise control switches except the On/Off switch, which is on the instrument panel) is located on the right side of the steering wheel. Refer to the replacement procedure below to remove the cruise control switch for testing.
14 Using an ohmmeter, check for continuity between the indicated terminals with the various switches in each of the indicated positions (see illustration).
15 If the continuity is not as specified, replace the defective switch.

Replacement

Refer to illustration 7.18

16 Disconnect the cable from the negative terminal of the battery.
17 Remove the cruise control switch pod cover by prying it off.
18 Remove the switch mounting screws, remove the switch and unplug the electrical connector (see illustration).
19 Installation is the reverse of removal.

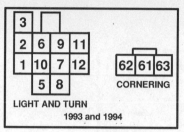
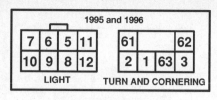
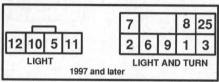

SWITCH POSITION	CONTINUITY BETWEEN TERMINALS
Headlight switch, Off, position C	5 and 6, 8 and 9
Park, positions A or B	11 and 12
Park, position C	5 and 6, 8 and 9 11 and 12
Head, position A or C	5 and 6, 8 and 9 11 and 12
Head, position B	5 and 7, 8 and 10 11 and 12
Head, position C	5 and 6, 8 and 9 11 and 12
Turn signal, right turn	1 and 2
Turn signal, left turn	1 and 3
Cornering lamp, right turn	61 and 62
Cornering lamp, left turn	61 and 63

7.4a Terminal identification and continuity chart for lighting portion of combination switch - all years

SWITCH POSITION	CONTINUITY BETWEEN TERMINALS
Wiper switch, Off	13 and 14
Intermittent	13 and 14, 15 and 17
Low	14 and 17
High	16 and 17
Wash	17 and 18

7.4b Terminal identification and continuity chart for wiper portion of combination switch - all years

7.8 The combination switch is held in place by three screws (arrows)

Chapter 12 Chassis electrical system

Check

Refer to illustrations 8.3 and 8.4

1 Disable the airbag system (see Section 29).
2 Refer to Chapter 11 and remove the driver's knee bolster and steering column covers.
3 Disconnect the ignition switch electrical connector that is about six inches down the column from the ignition switch **(see illustration)**.
4 Check the connector for continuity between the indicated terminals with the key in each position **(see illustration)**.
5 If the continuity is not as specified, replace the switch.
6 Check the lock cylinder in each position to make sure it isn't worn or loose and that the key position corresponds to the markings on the housing. If the lock cylinder is faulty, the entire steering column lock assembly will have to be replaced.

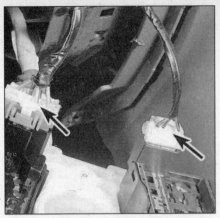

7.9 Unplug the electrical connectors (arrows) from the combination switch

7.10 Remove individual switches by removing the mounting screws (arrows)

8 Ignition switch and key lock cylinder - check and replacement

Warning: *The models covered by this manual are equipped with Supplemental Restraint Systems (SRS), more commonly known as airbags. Always disable the airbag system before working in the vicinity of any airbag system components to avoid the possibility of accidental deployment of the airbags, which could cause personal injury (see Section 29).*

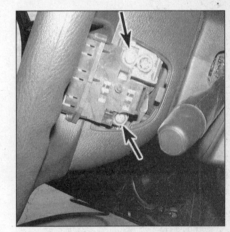

7.18 Pry the cover off with a screwdriver in the bottom slot, then remove these screws (arrows) and the cruise-control switch pod

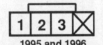

1993 and 1994 (w/airbag)   1995 and 1996 (1993 and 1994 w/o airbag)   1997 and later

SWITCH POSITION	CONTINUITY BETWEEN TERMINALS
Cruise control, Resume/Accel	1 and 3
Cruise control, Set/Coast	1 and 2
Cruise control, Cancel	1 and 3*, 1 and 2* (with diodes)

*Diodes are directional, test both directions

7.14 Terminal identification and continuity chart for steering column-mounted cruise control switches

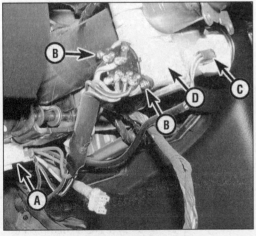

8.3 Ignition switch components

- A Connector to harness
- B Switch mounting screws
- C Key buzzer connector
- D Lock cylinder housing

1993 and 1994   1995 and later

SWITCH POSITION	CONTINUITY BETWEEN TERMINALS
Ignition switch, Off	none
Ignition switch, Acc	1 and 2
Ignition switch, On 1993 and 1994 models	1, 2 and 3
Ignition switch, On 1995 and later models	1, 2, 3 and 4
Ignition switch, Start 1993 and 1994 models	1, 3, 4 and 5
Ignition switch, Start 1995 and later models	1, 3, 5 and 6

8.4 Ignition switch terminal identification and continuity chart

Chapter 12 Chassis electrical system

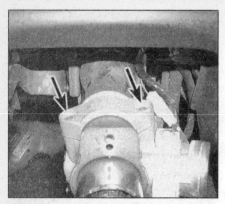

8.10 To remove the ignition/lock cylinder assembly, drill out the two retaining bolts (arrows)

Replacement

Refer to illustration 8.10

7 Follow Steps 1 through 3 above to access the ignition switch.
8 Disconnect the electrical connector and remove the switch mounting screws to remove the switch and its short harness **(see illustration 8.3)**.
9 Install the new switch and harness.
10 The lock cylinder can't be replaced by itself, the whole housing must be replaced. Remove the shear-head bolts retaining the ignition switch/lock cylinder assembly and separate the bracket halves from the steering column. This can be accomplished by drilling out the bolts and unscrewing them with a screw extractor **(see illustration)**.
11 Place the new switch assembly position, install the new shear-head bolts and tighten them until the heads snap off. Be sure to center the airbag clockspring (if equipped) before installing the steering wheel (refer to Chapter 10).

9 Instrument panel switches - check and replacement

Warning: *The models covered by this manual are equipped with Supplemental Restraint Systems (SRS), more commonly known as airbags. Always disable the airbag system before working in the vicinity of any airbag system components to avoid the possibility of accidental deployment of the airbags, which could cause personal injury (see Section 29).*

Instrument cluster bezel switches

Refer to illustrations 9.2, 9.4a, 9.4b and 9.4c

1 Disable the airbag system (see Section 29). The cluster bezel holds several switches. To the left of the steering wheel, they include: the outside power mirror switch, cruise control ON/Off switch, and if equipped, the theft control warning light and the fog light switch. To the right of the steering wheel is the instrument panel brightness switch, and on 1993 and 1994 models, the hazard warning switch (on the center panel on later models).
2 All of the switches can be removed from the front, by prying them out of the bezel with a screwdriver with the tip taped to prevent scratches **(see illustration)**. Simple switches such as the fog light switch should exhibit

9.2 Remove the switches in the instrument cluster bezel by prying them out from the front, without removing the bezel

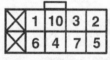

SWITCH POSITION 1993 and 1994 models	CONTINUITY BETWEEN TERMINALS
Mirror switch, Off	none
Mirror switch, Right	1 and 2
Mirror switch, Left	2 and 3
Mirror switch, Up	2 and 3
Mirror switch, Down	1 and 2
Mirror defogger switch,	1, 3 and 10

9.4a Power mirror switch terminal identification and continuity chart - 1993 and 1994 models

SWITCH POSITION 1995 and later models	CONTINUITY BETWEEN TERMINALS
Mirror switch, Off	none
Left mirror, move Right	1 and 2, 6 and 3
Left mirror, move Left	2 and 3, 1 and 6
Left mirror, move Up	2 and 3, 1 and 7
Left mirror, move Down	1 and 2, 3 and 6
Right mirror, move Right	1 and 2, 3 and 4
Right mirror, move Left	2 and 3, 1 and 4
Right mirror, move Up	2 and 3, 1 and 5
Right mirror, move Down	1 and 2, 3 and 5

9.4b Power mirror switch terminal identification and continuity chart - 1995 and later models

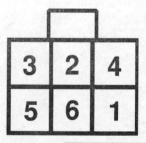

9.4c Cruise control On/Off switch terminal identification - with the switch On, there should be continuity between 1 and 3 and between 2 and 4

Chapter 12 Chassis electrical system

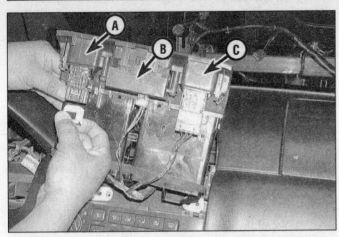

9.8 The components attached to the back of the center instrument panel bezel are the Hazard warning switch (A), the clock (B) and the rear window defogger switch (C) - all are screwed to the back of the bezel

11.3 Remove the mounting screws (arrows) and pull the instrument cluster away from the dash

continuity between the two terminals only when the switch is On. For the three-terminal dash illumination control switch, there should be continuity between all the terminals when the switch is in any position other than Off.

3 With the switch pulled out, disconnect the electrical connector.

4 Test the continuity of the switches **(see illustrations)**. If any switch fails the tests, replace it.

5 Installation is the reverse of the removal procedure.

Center instrument panel bezel switches

Refer to illustrations 9.8

6 Disable the airbag system (see Section 29). The center bezel holds the hazard flasher switch, clock and rear window defogger switch (1995 and later models).

7 Refer to Chapter 11 for removal of the bezel.

8 With the bezel pulled away from the instrument panel, disconnect the electrical connectors at each device on the bezel **(see illustration)**. The individual switches are retained to the bezel by screws, but they can be left in place for testing.

9 Refer to the wiring diagrams at the end of this Chapter to determine the proper continuity between the wires/terminals of the hazard switch and defogger switch.

10 Instrument panel gauges - check

Note: *This procedure applies to conventional analog type gauges (NON-digital) only.*

Fuel and temperature gauges

1 All tests below require the ignition switch to be turned to Off position before testing.

2 If the gauge pointer does not move from the empty or cold positions, check the fuse. If the fuse is OK, locate the particular sending unit for the circuit you're working on (see Chapter 4 for fuel sending unit location or Chapter 3 for the temperature gauge sending unit location). Connect the sending unit connector to ground with a jumper wire.

3 Turn the ignition key to On momentarily. If the pointer goes to the full or hot position replace the sending unit. **Note:** *Turn the key Off right away; grounding the sending unit for too long could damage the gauge.* If the pointer stays in same position, use a jumper wire to ground the sending unit terminal on the back of the gauge. If necessary, refer to the wiring diagrams at the end of this Chapter. If the pointer moves, the problem lies in the wiring between the gauge and the sending unit. If the pointer does not move with the sending unit terminal on the back of the gauge grounded, check for voltage at the other terminal of the gauge. There should not be voltage.

11 Instrument cluster - removal and installation

Refer to illustration 11.3

Warning: *The models covered by this manual are equipped with Supplemental Restraint Systems (SRS), more commonly known as airbags. Always disable the airbag system before working in the vicinity of any airbag system components to avoid the possibility of accidental deployment of the airbags, which could cause personal injury (see Section 29).*

1 Disable the airbag system (see Section 29).

2 Remove the instrument cluster bezel (see Chapter 11).

3 Remove the retaining screws and pull the cluster forward **(see illustration)**.

4 Unplug the electrical connectors and remove the cluster from the vehicle.

5 Installation is the reverse of the removal procedure.

12 Radio and speakers - removal and installation

Warning: *The models covered by this manual are equipped with Supplemental Restraint Systems (SRS), more commonly known as airbags. Always disable the airbag system before working in the vicinity of any airbag system components to avoid the possibility of accidental deployment of the airbags, which could cause personal injury (see Section 29).*

Radio/CD player

Refer to illustrations 12.3 and 12.4

1 Disable the airbag system (see Section 29).

2 Remove the center bezel panel from the dash and the plastic panel just below the ashtray (see Chapter 11).

3 Remove the screws and pull the air conditioning controls/radio/CD player assembly away from the dash **(see illustration)**.

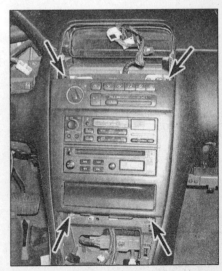

12.3 Remove the mounting screws (arrows)

Chapter 12 Chassis electrical system

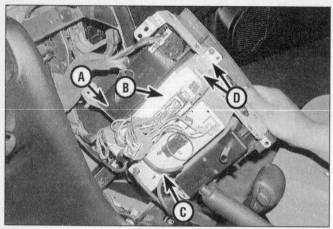

12.4 The center of the instrument panel holds three components as a unit: the air conditioning controls (A), the radio (B) and the CD player (C) if equipped - to remove an individual unit, remove the screws at each side (D indicates the left side screws for the radio)

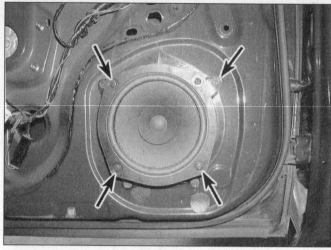

12.7 Remove the front speaker mounting screws (arrows)

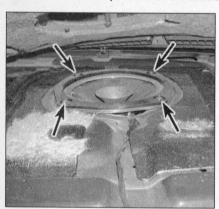

12.10 Rear speaker mounting screws (arrows)

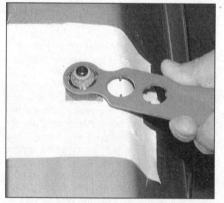

13.1 Remove the base bezel nut with snap-ring pliers or an antenna tool (shown, available in auto parts stores)

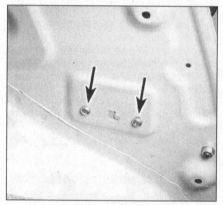

13.3 Remove the retaining screws (arrows), pull the antenna motor out and disconnect it

4 Disconnect the antenna lead and the electrical connectors, then remove the screws at each side bracket to remove the radio or CD player from the air conditioning controls/radio/CD player assembly (see illustration).
5 Installation is the reverse of removal.

Speakers

Front

Refer to illustration 12.7

6 Remove the front door trim panel (see Chapter 11).
7 Remove the speaker retaining screws. Unplug the electrical connector and remove the speaker (see illustration).
8 Installation is the reverse of removal.

Rear

Refer to illustration 12.10

9 Refer to Chapter 11 for removal of the rear package shelf.
10 Lift the package shelf up for access and remove the rear speaker mounting screws (see illustration).
11 Installation is the reverse of removal.

13 Antenna - removal and installation

Antenna motor

Refer to illustrations 13.1 and 13.3

1 Remove the antenna mast-retaining nut (see illustration). Apply masking tape around the antenna mount to avoid scratching the paint.
2 Working in the trunk, pry out the plastic clips securing the passenger side trunk finishing panels to allow access to the antenna motor.
3 Detach the motor retaining screws (see illustration). Disconnect the electrical connector and remove the antenna motor from the vehicle.
4 Installation is the reverse of removal.

Antenna mast

Refer to illustrations 13.6 and 13.7

Note: *At least two people should perform this task.*

5 Remove the antenna mast-retaining nut (see illustration 13.1).
6 With one person controlling the ignition switch and the second person holding the antenna mast, turn the ignition key and the radio to the ON position. This will enable the antenna mast to unwind itself from the motor assembly (see illustration).
7 When installing the antenna mast insert the antenna cable with the teeth facing the rear of the vehicle. Then have your assistant turn the ignition key and the radio to the ON

13.6 With the ignition key and the radio in the On position, guide the antenna mast and toothed drive cable out of the motor assembly

Chapter 12 Chassis electrical system

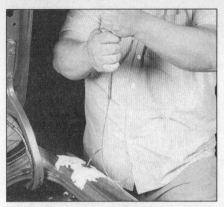

13.7 Insert the new antenna drive cable so that the teeth face the rear of the vehicle during installation

14.1 To access the bulb on the right headlight, remove the two screws (arrows) and the plastic shield

14.2a Disconnect the headlight electrical connector . . .

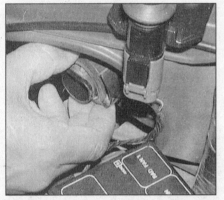

14.2b . . . then remove the retaining ring

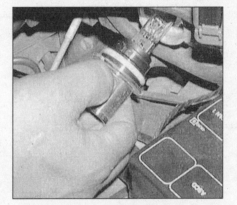

14.3 Pull out the bulb holder and replace the bulb

15.1 Use a Phillips screwdriver or small wrench to adjust the headlights - the adjuster closest to the fender (A) controls the horizontal movement and the one closest to the radiator (B), vertical movement

position. This will enable the antenna mast to wind itself back into the motor assembly **(see illustration)**.

8   The remainder of the installation is the reverse of removal.

Antenna grid

9   Most models have a wire-grid antenna in the rear window, in addition to the mast antenna.

10   The antenna grid occupies the upper portion of the rear window, with the separate rear window defogger grid below it. The antenna grid can be tested for continuity in the same manner as outlined in Section 21, and if there is a break in the grid, it can be repaired in the same manner as the rear defogger grid (see Section 21).

14 Headlight bulb - replacement

Refer to illustrations 14.1, 14.2a, 14.2b and 14.3

Warning: *These models are equipped with halogen gas-filled bulbs, which are under pressure and may shatter if the surface is scratched or the bulb is dropped. Wear eye protection and handle the bulbs carefully, grasping only the base whenever possible. Do not touch the surface of the bulb with your fingers because the oil from your skin could cause it to overheat and fail prematurely. If*

you do touch the bulb surface, clean it with rubbing alcohol.

1   Open the hood and locate the bulb assembly on the back of the headlight housing. On the right side headlight, remove the two screws and the plastic shield for access to the bulb **(see illustration)**.

1993 through 1999 models

2   Unplug the electrical connector and detach the retaining ring **(see illustrations)**.
3   Withdraw the bulb assembly from the headlight housing **(see illustration)**.
4   Without touching the glass with your bare fingers, insert the new bulb assembly into the headlight housing and secure it with the retaining ring.
5   Plug in the electrical connector. Test headlight operation, then close the hood.

2000 and later models

6   Unplug the electrical connector from the backside of the bulb.
7   Pull the rubber cap and bulb assembly from the back of the headlight lens.
8   Detach the retaining ring, then push straight on the back of the rubber cap and dislodge the bulb from the rubber cap. Remove the bulb.
9   Without touching the glass with your fingers, insert the new bulb into the rubber cap and secure with the retaining ring.

10   Install the rubber cap and bulb assembly onto the headlight lens and push it on until it bottoms.
11   Plug in the electrical connector. Test the headlight operation, then close the hood.

15 Headlights - adjustment

Refer to illustrations 15.1 and 15.2
Note: *It is important that the headlights are aimed correctly. If adjusted incorrectly they could blind the driver of an oncoming vehicle and cause a serious accident or seriously reduce your ability to see the road. The headlights should be checked for proper aim every 12 months and any time a new headlight is installed or front end body work is performed. It should be emphasized that the following procedure is only an interim step that will provide temporary adjustment until the headlights can be adjusted by a properly equipped shop.*

1   These models are equipped with composite headlights with two adjustment screws, one controlling left-and-right movement and one for up-and-down movement **(see illustration)**. Early models have one adjustment rod to the rear and one on top,

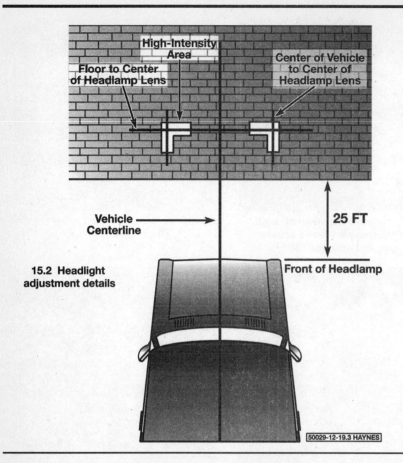

15.2 Headlight adjustment details

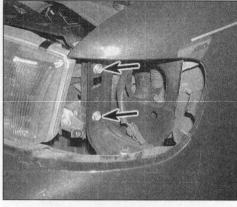

16.5a There are four headlight housing retaining bolts and nuts: two bolts (arrows) are on the fender brace . . .

16.5b . . . and the two nuts (arrows) are accessible from behind the radiator brace

while 1995 and later models have both rods pointing to the rear.

2 There are several methods of adjusting the headlights. The simplest method requires a blank wall 25 feet in front of the vehicle and a level floor **(see illustration)**.

3 Position masking tape vertically on the wall in reference to the vehicle centerline and the centerlines of both headlights.

4 Position a horizontal tape line in reference to the centerline of all the headlights.

Note: *It may be easier to position the tape on the wall with the vehicle parked only a few inches away.*

5 Adjustment should be made with the vehicle sitting level, the gas tank half-full and no unusually heavy load in the vehicle.

6 Starting with the low beam adjustment, position the high intensity zone so it is two inches below the horizontal line and two inches to the side of the vertical headlight line away from oncoming traffic. Twist the adjustment screws until the desired level has been achieved.

7 With the high beams on, the high intensity zone should be vertically centered with the exact center just below the horizontal line.

Note: *It may not be possible to position the headlight aim exactly for both high and low beams. If a compromise must be made, keep in mind that the low beams are the most used and have the greatest effect on driver safety.*

8 Have the headlights adjusted by a dealer service department at the earliest opportunity.

16 Headlight housing - replacement

Refer to illustrations 16.5a and 16.5b

1 Disconnect the cable from the negative battery terminal.
2 Remove the headlight bulb (Section 14).
3 Remove the radiator grille (Chapter 11).
4 Remove the parking light housing (Section 17).
5 Remove the retaining nuts and bolts, detach the housing and withdraw it from the vehicle **(see illustrations)**.
6 Installation is the reverse of removal.

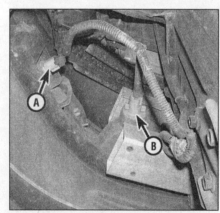

17.2 The side marker bulb holder (A) can be pulled straight out, but the turn signal bulb holder (B) must be rotated to remove it

17 Bulb replacement

Front side marker/turn signal lights

Refer to illustration 17.2

1 Raise the vehicle and suitably support it on jackstands.
2 From below, you can access both the

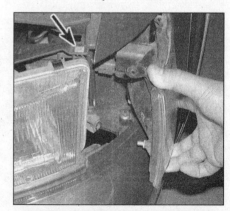

17.4 Remove the one top screw (arrow indicates location, screw is already removed here) and twist the clearance light until it is free of the clip, then pull straight out

Chapter 12 Chassis electrical system

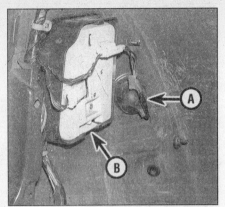

17.6a The outer bulb (A) is removed by pulling off the rubber cover, then twisting the bulb holder out - push up on the clip (B) . . .

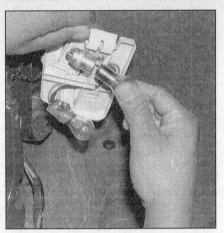

17.6b . . . and pull out the bulb holder to access the taillight bulbs

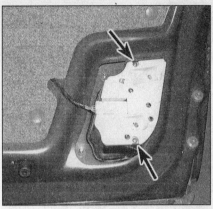

17.8 Remove the screws (arrows) in the trunk-lid-mounted taillight housing to access the bulbs under this cover

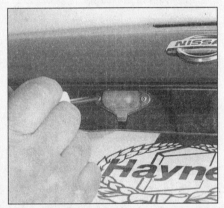

17.10 Remove the license plate light lens with a small screwdriver

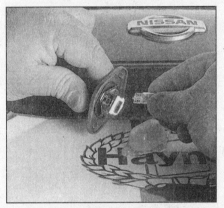

17.11 Pull the license plate bulb straight out

17.12 Turn the high-mounted brake light housing to the right and then pull toward the front of the car - arrow indicates the twist-to-remove bulb holder

turn signal and side marker lights (see illustration).

3   The side marker bulb holder pulls straight out of the socket, but the turn signal bulb holder must be rotated to remove it. Push in and rotate the bulb counterclockwise to remove it from the holder.

Clearance light

Refer to illustration 17.4

4   With the hood open, remove the screw near the outside corner of the headlight housing, then rock the clearance light (cornering lamp on early models) and pull it straight forward (see illustration).
5   Remove the bulb holder from the back of the housing and exchange the bulb.

Tail and back-up lights

Refer to illustrations 17.6a and 17.6b

Body-side lights

6   Open the trunk and pull back the trim cover for access to the body-side of the taillight housing. The outside bulb has a rubber cover over the bulb holder, while the taillight housing has a cover that lifts up for access to the bulbs (see illustrations).
7   Push in on the bulbs and rotate them counterclockwise to remove them from the holder.

Trunk-side lights

Refer to illustration 17.8

8   Open the trunk lid and remove the two screws securing the bulb holder plate at the back of the light assembly (see illustration).
9   Pull out the bulb holder plate and remove the bulbs by pushing in and rotating them counterclockwise.

License plate light

Refer to illustrations 17.10 and 17.11

10   Remove the screws, detach the lens and pull the bulb holder down for access to the bulb (see illustration).
11   Pull the bulb straight out to replace it (see illustration).

High-mounted brake light

Refer to illustration 17.12

12   Push the housing cover to the right, then pull to remove it. The bulb pulls straight out (see illustration).

Instrument cluster illumination

Refer to illustration 17.13

13   To gain access to the instrument cluster illumination lights, the instrument cluster will have to be removed (see Section 11). The bulbs can then be removed and replaced from the rear of the cluster (see illustration).

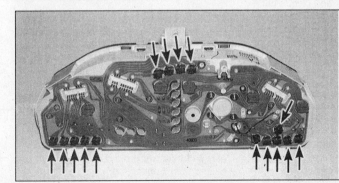

17.13 To remove an instrument cluster light bulb (arrows), depress it and turn counterclockwise to release it

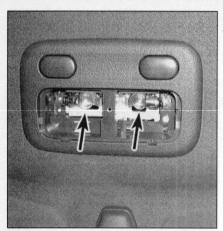

17.15a The map light bulbs (arrows) twist out...

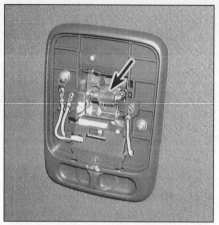

17.15b ... while the double-ended dome light bulb pulls straight out

18.2 Backprobe the connector (arrow) at the wiper motor to check for power, referring to the wiring diagrams

Interior lights

Refer to illustrations 17.15a and 17.15b

14 Remove the lenses for the map lights or dome light by prying the cover off with a small screwdriver.

15 Replace the bulb in the map lights by twisting the bulbs out of their holders, and remove the dome light bulb by then pulling it straight out **(see illustrations)**. The bulb in the dome light should be handled with a cloth to avoid leaving oils from you hands on the surface that could shorten the bulb life.

18 Wiper motor - removal and installation

Check

Refer to illustration 18.2

Note: *Refer to the wiring diagrams for wire colors and locations in the following checks. When checking for voltage, probe a grounded 12-volt test light to each terminal at a connector until it lights; this verifies voltage (power) at the terminal. If the following checks fail to locate the problem, have the system diagnosed by a dealer service department or other properly equipped repair facility.*

1 If the wipers work slowly, make sure the battery is in good condition and has a strong charge (see Chapter 1). If the battery is in good condition, remove the wiper motor (see below) and operate the wiper arms by hand. Check for binding linkage and pivots. Lubricate or repair the linkage or pivots as necessary. Reinstall the wiper motor. If the wipers still operate slowly, check for loose or corroded connections, especially the ground connection. If all connections look OK, replace the motor.

2 If the wipers fail to operate when activated, check the fuse. If the fuse is OK, connect a jumper wire between the wiper motor and ground, then retest. If the motor works now, repair the ground connection. If the motor still doesn't work, turn the wiper switch to the HI position and check for voltage at the

18.7 Use a small screwdriver to pry off the wiper arm nut cover, then remove the nut and pull the arm straight off its splined shaft

motor **(see illustration)**. Note: *The cowl cover will have to be removed (see Chapter 11).* If there's voltage at the connector, remove the motor and check it off the vehicle with fused jumper wires from the battery. If the motor now works, check for binding linkage (see Step 1 above). If the motor still doesn't work, replace it. If there's no voltage to the motor, check for voltage at the wiper control relays. If there's voltage at the wiper control relays and no voltage at the wiper motor, check the switch for continuity (see Section 7).

3 If the interval (delay) function is inoperative, check the continuity of all the wiring between the switch and wiper control module. If the wiring is OK, check the resistance of the delay control knob of the multi-function switch (see Section 7).

4 If the wipers stop at the position they're in when the switch is turned off (fail to park), check for a good ground on the connector side at the motor. With an ohmmeter connected between any of the black wire terminals and a known ground, resistance should be less than 5 ohms.

18.10 Pry off the linkage rod socket (arrow) at the rear of the motor

5 If the wipers won't shut off unless the ignition is OFF, disconnect the wiring from the wiper control switch. If the wipers stop, replace the switch. If the wipers keep running, there's a defective limit switch in the motor; replace the motor.

6 If the wipers won't retract below the hood line, check for mechanical obstructions in the wiper linkage or on the vehicle's body that would prevent the wipers from parking. If there are no obstructions, check the wiring between the switch and motor for continuity. If the wiring is OK, replace the wiper motor.

Wiper motor replacement

Refer to illustrations 18.7, 18.10 and 18.11

7 Remove the windshield wiper arms **(see illustration)**.

8 Remove the cowl cover (see Chapter 11).

9 Disconnect the electrical connector from the wiper motor **(see illustration 18.2)**.

10 Detach the wiper motor/linkage assembly from the cowl **(see illustration)**.

11 Remove the wiper motor retaining bolts and remove the motor **(see illustration)**.

12 Installation is the reverse of removal.

Chapter 12 Chassis electrical system

18.11 Unplug the electrical connector, remove the mounting bolts (arrows) and remove the wiper motor

19.3 Unplug the electrical connector (A) and remove the bolt (B), then detach the horn

19 Horn - check and replacement

Refer to illustration 19.3

Note: *Check the fuses before beginning electrical diagnosis.*

1 Unplug the electrical connector from the horn.
2 To test the horn, refer to the wiring diagrams and connect battery voltage and ground to the two terminals with a pair of jumper wires. If the horn doesn't sound, replace it. If it does sound, the problem lies in the switch, relay or the wiring between the components.
3 To replace the horn, unplug the electrical connector and remove the bracket bolt **(see illustration)**.
4 Installation is the reverse of removal.

20 Daytime Running Lights (DRL) - general information

The Daytime Running Lights (DRL) system used on Canadian models turns the headlights on whenever the engine is started. The only exception is when the engine is turned on when the parking brake is engaged. Once the parking brake is released, the lights will remain on as long as the ignition switch is on, even if the parking brake is later applied.

The DRL system supplies reduced power to the headlights so they won't be too bright for daytime use while prolonging headlight life.

21 Rear window defogger - check and repair

1 The rear window defogger consists of a number of horizontal heating elements baked onto the inside surface of the glass. Power is supplied through a large fuse from the power distribution box in the engine compartment. The heater is controlled by the instrument panel switch. Test the switch for continuity (see Section 9).
2 Small breaks in the element can be repaired without removing the rear window.

Check

Refer to illustrations 21.5 and 21.6

3 Turn the ignition switch and defogger switches to the ON position.
4 Using a voltmeter, place the positive probe against the defogger grid positive terminal and the negative probe against the ground terminal. If battery voltage is not indicated, check the fuse, defogger switch and related wiring. If voltage is indicated, but all or part of the defogger doesn't heat, proceed with the following tests.
5 When measuring voltage during the next two tests, wrap a piece of aluminum foil around the tip of the voltmeter positive probe and press the foil against the heating element with your finger **(see illustration)**. Place the negative probe on the defogger grid ground terminal.
6 Check the voltage at the center of each heating element **(see illustration)**. If the voltage is 5 to 6 volts, the element is okay (there is no break). If the voltage is 0 volts, the element is broken between the center of the element and the positive end. If the voltage is 10 to 12 volts the element is broken between the center of the element and the ground side. Check each heating element.
7 If none of the elements are broken, connect the negative probe to a good chassis ground. The voltage reading should stay the same, if it doesn't the ground connection is bad.
8 To find the break, place the voltmeter negative probe against the defogger ground terminal. Place the voltmeter positive probe with the foil strip against the heating element at the positive side and slide it toward the negative side. The point at which the voltmeter deflects from several volts to zero is the point where the heating element is broken.

Repair

Refer to illustration 21.14

9 Repair the break in the element using a repair kit specifically for this purpose, such as Dupont paste No. 4817 (or equivalent). The kit includes conductive plastic epoxy.

21.5 When measuring the voltage at the rear window defogger grid, wrap a piece of aluminum foil around the negative probe of the voltmeter and press the foil against the wire with your finger

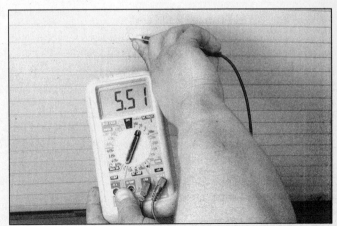

21.6 To determine if a heating element has broken, check the voltage at the center of each element - if the voltage is 6-volts, the element is unbroken

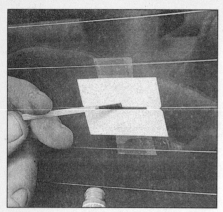

21.14 To use a defogger repair kit, apply masking tape to the inside of the window at the damaged area, then brush on the special conductive coating

22.5 The cruise control servo (A) is located on the firewall - make sure the cruise control cable (B) mounted on the throttle body is not damaged and that it operates smoothly when the throttle is opened

10 Before repairing a break, turn off the system and allow it to cool for a few minutes.
11 Lightly buff the element area with fine steel wool; then clean it thoroughly with rubbing alcohol.
12 Use masking tape to mask off the area being repaired.
13 Thoroughly mix the epoxy, following the kit instructions.
14 Apply the epoxy material to the slit in the masking tape, overlapping the undamaged area about 3/4-inch on either end **(see illustration)**.
15 Allow the repair to cure for 24 hours before removing the tape and using the system.

22 Cruise control system - description and check

Refer to illustration 22.5

1 The cruise control system maintains vehicle speed with a vacuum-actuated servo motor located on the firewall in the engine compartment, which is connected to the throttle linkage by a cable. The system consists of the servo motor, brake switch, vacuum pump, control switches, a relay and associated vacuum hoses. Some features of the system require special testers and diagnostic procedures that are beyond the scope of the home mechanic. Listed below are some general procedures that may be used to locate common problems.
2 Check the fuse (see Section 3).
3 The brake pedal position (BPP) switch (or brake light switch) deactivates the cruise control system. Have an assistant press the brake pedal while you check the brake light operation.
4 If the brake lights do not operate properly, correct the problem and retest the cruise control.
5 Check the control cable between the cruise control servo/amplifier and the throttle linkage and adjust/replace as necessary **(see illustration)**. See Chapter 4 for the cable adjustment procedure, which is the same for accelerator cable and cruise control cable.
6 The cruise control system uses a speed sensing device. The speed sensor is located in the transmission. To test the speed sensor, see Chapter 6.
7 The testing of the cruise control On/Off switch is covered in Section 9.
8 The cruise control actuator switch is mounted in the steering wheel.
9 To test the actuator switch, refer to Section 7. If the switch fails the tests, replace it.
Note: *Make sure the horn works before testing the cruise control system, as the cruise control switch gets power from the horn relay.*
10 Test-drive the vehicle to determine if the cruise control is now working. If it isn't, take it to a dealer service department or an automotive electrical specialist for further diagnosis.

23 Power window system - description and check

Refer to illustrations 23.12a and 23.12b

1 The power window system operates the electric motors mounted in the doors which lower and raise the windows. The system consists of the control switches, the motors (regulators), glass mechanisms and associated wiring.
2 Power windows are wired so they can be lowered and raised from the master control switch by the driver or by remote switches located at the individual windows. Each window has a separate motor that is reversible. The position of the control switch determines the polarity and therefore the direction of operation. Some systems are equipped with relays that control current flow to the motors.
3 Some vehicles are equipped with a separate circuit breaker for each motor in addition to the fuse or circuit breaker protecting the whole circuit. This prevents one stuck window from disabling the whole system.
4 The power window system will only operate when the ignition switch is ON. In addition, many models have a window lockout switch at the master control switch which, when activated, disables the switches at the rear windows and, sometimes, the switch at the passenger's window also. Always check these items before troubleshooting a window problem.
5 These procedures are general in nature, so if you can't find the problem using them, take the vehicle to a dealer service department.
6 If the power windows don't work at all, check the fuse or circuit breaker.
7 If only the rear windows are inoperative, or if the windows only operate from the master control switch, check the rear window lockout switch for continuity in the unlocked position. Replace it if it doesn't have continuity.
8 Check the wiring between the switches and fuse panel for continuity. Repair the wiring, if necessary.
9 If only one window is inoperative from the master control switch, try the other control switch at the window. **Note:** *This doesn't apply to the drivers door window*.
10 If the same window works from one switch, but not the other, check the switch for continuity.
11 If the switch tests OK, check for a short or open in the wiring between the affected switch and the window motor.
12 If one window is inoperative from both switches, remove the trim panel from the affected door and check for voltage at the

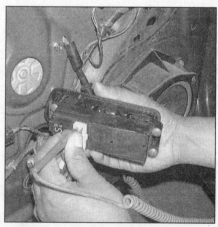

23.12a Using a voltmeter or a test light, check for power at the window switch by backprobing while operating the switch

Chapter 12 Chassis electrical system

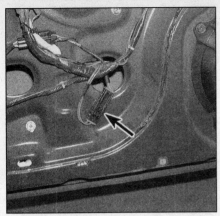

23.12b If there is power at the switch, check for power getting to the window motor by probing at the motor connector (arrow)

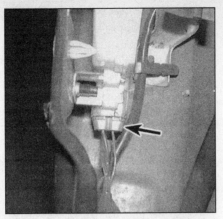

24.10 Check for power at the door lock actuator connector (arrow) with the switch depressed - check the door lock actuator itself by disconnecting the connector and using jumper wires to temporarily apply battery voltage and ground directly

switch and at the motor while the switch is operated **(see illustrations)**.
13 If voltage is reaching the motor, disconnect the glass from the regulator (see Chapter 11). Move the window up and down by hand while checking for binding and damage. Also check for binding and damage to the regulator. If the regulator is not damaged and the window moves up and down smoothly, replace the motor. If there's binding or damage, lubricate, repair or replace parts, as necessary.
14 If voltage isn't reaching the motor, check the wiring in the circuit for continuity between the switches and motors. You'll need to consult the wiring diagram for the vehicle. Some power window circuits are equipped with relays. If equipped, check that the relays are grounded properly and receiving voltage from the switches. Also check that each relay sends voltage to the motor when the switch is turned on. If it doesn't, replace the relay.
15 Test the windows after you are done to confirm proper repairs.

24 Power door lock system - description and check

Refer to illustration 24.10

1 The power door lock system operates the door lock actuators mounted in each door. The system consists of the switches, actuators and associated wiring. Diagnosis can usually be limited to simple checks of the wiring connections and actuators for minor faults that can be easily repaired.
2 Power door lock systems are operated by bi-directional solenoids located in the doors. The lock switches have two operating positions: Lock and Unlock. These switches activate a relay, which in turn connects voltage to the door lock solenoids. Depending on which way the relay is activated, it reverses polarity, allowing the two sides of the circuit to be used alternately as the feed (positive)

and ground side.
3 Some vehicles may have keyless entry, electronic control modules and anti-theft systems incorporated into the power locks. If you are unable to locate the trouble using the following general steps, consult your dealer service department. **Note:** *Some vehicles also have control switches connected to the key locks in the doors, which unlock all the doors when one is unlocked.*
4 Always check the circuit protection first. Some vehicles use a combination of circuit breakers and fuses.
5 Operate the door lock switches in both directions (Lock and Unlock) with the engine off. Listen for the faint click of the relay operating.
6 If there's no click, check for voltage at the switches. If no voltage is present, check the wiring between the fuse panel and the switches for shorts and opens.
7 If voltage is present but no click is heard, test the switch for continuity. Replace it if there's not continuity in both switch positions.
8 If the switch has continuity but the relay doesn't click, check the wiring between the switch and relay for continuity. Repair the wiring if there's no continuity.
9 If the relay is receiving voltage from the switch but is not sending voltage to the solenoids, check for a bad ground at the relay case. If the relay case is grounding properly, replace the relay.
10 If all but one lock solenoids operate, remove the trim panel from the affected door (see Chapter 11) and check for voltage at the solenoid while the lock switch is operated **(see illustration)**. One of the wires should have voltage in the Lock position; the other should have voltage in the Unlock position.
11 If the inoperative solenoid is receiving voltage, replace the solenoid.
12 If the inoperative solenoid isn't receiving voltage, check for an open or short in the wire between the lock solenoid and the relay.

Note: *It's common for wires to break in the portion of the harness between the body and door (opening and closing the door fatigues and eventually breaks the wires).*

25 Electric side view mirrors - description and check

1 Most electric side view mirrors use two motors to move the glass; one for up and down adjustments and one for left-right adjustments.
2 The control switch has a selector portion that sends voltage to the left or right side mirror. With the ignition ON but the engine OFF, roll down the windows and operate the mirror control switch through all functions (left-right and up-down) for both the left and right side mirrors.
3 Listen carefully for the sound of the electric motors running in the mirrors.
4 If the motors can be heard but the mirror glass doesn't move, there's probably a problem with the drive mechanism inside the mirror. Remove and disassemble the mirror to locate the problem.
5 If the mirrors don't operate and no sound comes from the mirrors, check the fuse (see Chapter 1).
6 If the fuse is OK, remove the mirror control switch from its mounting without disconnecting the wires attached to it. Turn the ignition ON and check for voltage at the switch. There should be voltage at one terminal. If there's no voltage at the switch, check for an open or short in the wiring between the fuse panel and the switch.
7 If there's voltage at the switch, disconnect it. Check the switch for continuity in all its operating positions (see Section 9). If the switch does not have continuity, replace it.
8 Re-connect the switch. Locate the wire going from the switch to ground. Leaving the switch connected, connect a jumper wire between this wire and ground. If the mirror works normally with this wire in place, repair the faulty ground connection.
9 If the mirror still doesn't work, remove the mirror and check the wires at the mirror for voltage. Check with ignition ON and the mirror selector switch on the appropriate side. Operate the mirror switch in all its positions. There should be voltage at one of the switch-to-mirror wires in each switch position (except the neutral "off" position).
10 If voltage isn't present in each switch position, check the wiring between the mirror and control switch for opens and shorts.
11 If there's voltage, remove the mirror and test it off the vehicle with jumper wires. Replace the mirror if it fails this test.

26 Electric sunroof - description and check

1 The electric sunroof is powered by a single motor located in the roof behind the over-

head console. The power circuit is protected by a circuit breaker. When sunlight isn't desired, an interior sliding panel can be closed.

2 The control switches (tilt and slide) send a ground signal to the sunroof motor when the switches are pressed. Power is supplied to the motor from the sunroof relay. With the ignition On but the engine Off, operate the sunroof control switch through the tilt and slide functions.

3 Listen carefully for the sound of the sunroof motor running in the roof.

4 If the motors can be heard but the sunroof glass doesn't move, there's probably a problem with the drive mechanism or drive cables.

5 If the sunroof does not operate and no sound comes from the motor, check the fuse (7.5-amp fuse number 12 in the interior fuse panel, and 30-amp fuse F in the engine compartment fuse/fusible link box).

6 If the fuses are OK, pull down the overhead interior light/switch panel. Turn the ignition On and check for voltage at the yellow/red wire at the motor. If there's voltage at the motor, check for power and ground at the switch. If power and ground exist at the motor and there's still no voltage at the switch replace the switch. If there's no voltage at the motor, check the sunroof relay or look for an open or short in the wiring between the ignition relay and the motor. **Note:** *The sunroof relay is located behind the left kick panel, next to the fuel pump relay.*

7 If there's voltage at the switch, disconnect it. Check the switch for continuity in all its operating positions. If the switch does not have continuity, replace it.

8 If the switch has continuity re-connect the switch. Locate the wire going from the switch to ground. Leaving the switch connected, connect a jumper wire between this wire and ground. If the motor works normally with this wire in place, repair the faulty ground connection.

9 The sunroof can be closed manually by inserting a wrench into the motor shaft and rotating it clockwise. If your vehicle is equipped with a factory sunroof, the wrench comes in the factory toolbag in the trunk.

27 Power seats - description and check

1 The optional power seats on these models adjust forward and backward, up and down and tilt forward and backward.

2 The power seat system consists of a motor, a switch on the seat, the #1 circuit breaker (under the left end of the instrument panel) and the 30-amp F fuse in the engine compartment fuse/fusible link block.

3 Look under the seat for any objects which may be preventing the seat from moving.

4 If the seat won't work at all, check the fuse (see Section 3).

5 With the engine off to reduce the noise level, operate the seat controls in all directions and listen for sound coming from the seat motor(s).

6 If the motor runs or clicks but the seat doesn't move, the integral the seat drive mechanism is damaged and the motor assembly must be replaced.

7 If the motor doesn't work or make noise, check for voltage at the motor while an assistant operates the switch.

8 If the motor is getting voltage but doesn't run, test it off the vehicle with jumper wires. If it still doesn't work, replace it.

9 If the motor isn't getting voltage, check for voltage at the switch. If there's no voltage at the switch, check the wiring between the fuse panel and the switch. If there's voltage at the switch, check the switch for continuity in all its operating positions. Replace the switch if there's no continuity.

10 If the switch is OK, check for a short or open in the wiring between the switch and motor.

11 Test the completed repairs.

28 In-Vehicle Multiplexing System - description

1 The vehicles covered by this manual have a complex electrical system, encompassing many power accessories. To reduce the amount of hard wiring going back and forth throughout the vehicle, the designers have incorporated the In-Vehicle Multiplexing System (IVMS).

2 The main components of the system are the Body Control Module (BCM, located under the front of the floor console) and five Local Control Units (LCUs) located around the vehicle. Two multiplex lines (A and B) connect the IVMS components, and are able to carry large amounts of data back and forth.

3 Some accessory systems in the vehicle are connected directly to the BCM, while others connect to the various LCUs, which control the On/Off function of those devices when directed by the computer in the BCM. Among the systems controlled by the IVMS are the power windows, power door locks, interior lighting, warning lights, theft warning systems (if applicable), and various other components.

4 The BCM also has diagnostic capability for the IVMS. If you suspect any electrical problem you have is traceable to the IVMS, bring your vehicle to a dealer or other qualified repair shop with the diagnostic tools to extract the trouble codes.

29 Airbag system - general information

1 These models are equipped with a Supplemental Restraint System (SRS), more commonly known as airbags, designed to protect the driver and front seat passenger from serious injury in the event of a head-on or frontal collision. Early models have only a driver's airbag, located in the center of the steering wheel. The 1995 and later models also have a passenger airbag module in the right side of the dash. All models have a diagnostic/control unit located inside the passenger compartment.

2 The 1998 and 1999 models have additional occupant protection with supplemental side airbags in the outside rear corners of the driver and passenger front seats. These airbags are designed to operate primarily in side-impact collisions, though they may also activate in some other types of collisions.

Airbag modules

3 The airbag modules consist of a housing incorporating the cushion (airbag) and inflator unit. The inflator assembly is mounted on the back of the housing over a hole through which gas is expelled, inflating the bag almost instantaneously when an electrical signal is sent from the system. The specially wound wire on the driver's side that carries this signal to the module is called a clockspring. The clockspring is a flat, ribbon-like electrically conductive tape that is wound many times so that it can transmit an electrical signal regardless of steering wheel position.

Diagnostic/control unit and tunnel/safing sensors

Refer to illustration 29.4

4 The diagnostic/control unit contains an

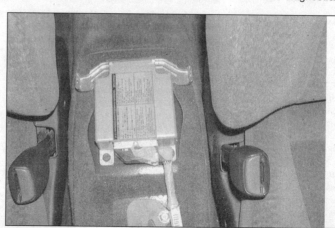

29.4 The diagnostic control unit is located under the rear of the floor console (1995 and later shown) - if it is removed for any procedure, new factory bolts must be used to install it

Chapter 12 Chassis electrical system

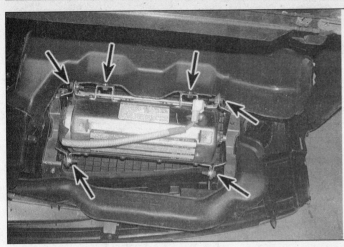

29.20 The passenger airbag is hard to access (this view is of the back of the instrument panel, removed from the vehicle for clarity) - working from below, disconnect the connector and remove the Torx fasteners (arrows)

on-board microprocessor which monitors the operation of the system, and also contains a crash sensor. It checks this system every time the vehicle is started, causing the "AIRBAG" light to go on then off, if the system is operating properly. If there is a fault in the system, the light will go on and stay on and the unit will store fault codes indicating the nature of the fault. If the AIRBAG light goes on and stays on, the vehicle should be taken to your dealer immediately for service. Models with side airbags also have a "satellite" crash sensor in each door "B" pillar. The diagnostic control unit is located under the rear of the floor console on most models, but is under the center of the instrument panel on some early models (see illustration).

Operation

5 For the airbag(s) to deploy, one or both impact sensors and the safing sensor must be activated. When this condition occurs, the circuit to the airbag inflator is closed and the airbag inflates. If the battery is destroyed by the impact, or is too low to power the inflator, a back-up power unit inside the control unit provides power.

Self-diagnosis system

6 A self-diagnosis circuit in the control unit displays a light on the instrument panel when the ignition switch is turned to the On position. If the system is operating normally, the light should go out after about seven seconds. If the light doesn't come on, or doesn't go out after seven seconds, or if it comes on while you're driving the vehicle, or if it blinks at any time, there's a malfunction in the SRS system. Have it inspected and repaired as soon as possible. Do not attempt to troubleshoot or service the SRS system yourself. Even a small mistake could cause the SRS system to malfunction when you need it.

Servicing components near the SRS system

7 Nevertheless, there are times when you need to remove the steering wheel, radio or service other components on or near the dashboard. At these times, you'll be working around components and wire harnesses for the SRS system. The SRS wiring harnesses are easy to identify: They're all bright yellow. Do not unplug the connectors for these wires. And do not use electrical test equipment on yellow wires; it could cause the airbag(s) to deploy. **ALWAYS DISABLE THE SRS SYSTEM BEFORE WORKING NEAR THE SRS SYSTEM COMPONENTS OR RELATED WIRING.**

Disabling the SRS system

Warning: *Any time you are working in the vicinity of airbag wiring or components, DISABLE THE SRS SYSTEM.*
8 Disconnect the battery negative and positive cables, then wait ten minutes before proceeding with any work.

Driver's side airbag

9 Remove the access panel in the steering wheel below the airbag and unplug the two-pin connector between the airbag and the clockspring (see Chapter 10).

Passenger's side airbag

10 Open the glove box.
11 At the top of the glove box opening, near the latch, disconnect the two-pin electrical connector (yellow harness) between the passenger side airbag and the SRS main wiring harness.

Side-impact airbags

12 Using a trim tool, carefully pry around the perimeter of the finish panel on the back of each front seat. **Warning:** *Work only from the side of the seat opposite the airbag, and be careful when prying near the airbag so you don't contact the airbag harness.*
13 With the seat back cover removed, disconnect the airbag connector in the yellow harness.

Enabling the system

14 After you've disabled the airbag and performed the necessary service, reconnect the two-pin airbag connector into the two-pin clockspring connector (driver's side), the SRS main harness (passenger's side) or the side-impact airbag (1998 and 1999 models). Reinstall the lid to the underside of the steering wheel or reinstall the glove box or seat back panel.
15 Turn the ignition switch to the Off position.
16 Reattach the battery cables (see Chapter 1).

Removal and installation

Driver's side airbag

17 Refer to Chapter 10 for removal and installation of the driver's side airbag.

Passenger side airbag

Refer to illustration 29.20
18 Disable the airbag system (beginning with Step 8).
19 Refer to Chapter 11 and remove the passenger-side lower dash panel and the glove box.
20 Disconnect the two-pin connector. Remove the special Torx tamper-proof bolts and gently pry the airbag unit from the top of the dashboard with a screwdriver **(see illustration)**. **Caution:** *The airbag assembly is heavier than it looks, use both hands when removing it from the dash.*
21 Installation is the reverse of the removal procedure. **Note:** *The bolts used throughout the airbag system to mount the modules and diagnostic unit are special tamper-proof Torx bolts, which have a special coating. These bolts are designed to be used once. Replace them with new factory bolts, and never use a substitute fastener.*

Side-impact airbags

22 Disable the airbag system (beginning with Step 8).
23 Refer to Steps 12 and 13 and disconnect the airbag connector. Remove the two special Torx nuts and remove the airbag.
24 Installation is the reverse of the removal procedure.

30 Wiring diagrams - general information

Since it isn't possible to include all wiring diagrams for every year covered by this manual, the following diagrams are those that are typical and most commonly needed.

Prior to troubleshooting any circuits, check the fuse and circuit breakers (if equipped) to make sure they are in good condition. Make sure the battery is properly charged and has clean, tight cable connections (see Chapter 1).

When checking the wiring system, make sure that all electrical connectors are clean, with no broken or loose pins. When unplugging an electrical connector, do not pull on the wires, only on the connector housings themselves.

12-20   Chapter 12   Chassis electrical system

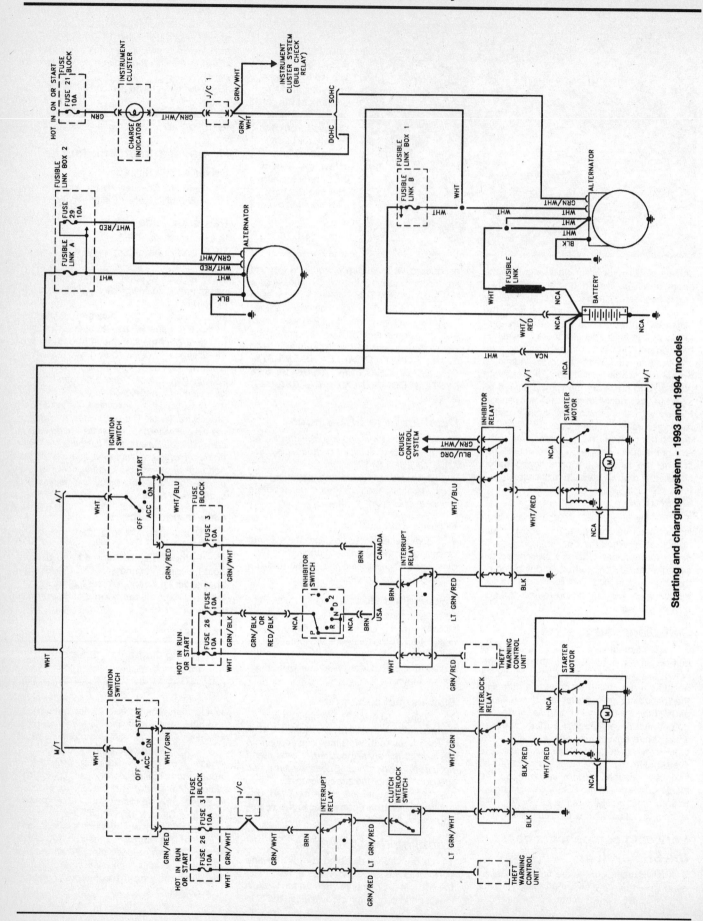

Starting and charging system - 1993 and 1994 models

Chapter 12 Chassis electrical system

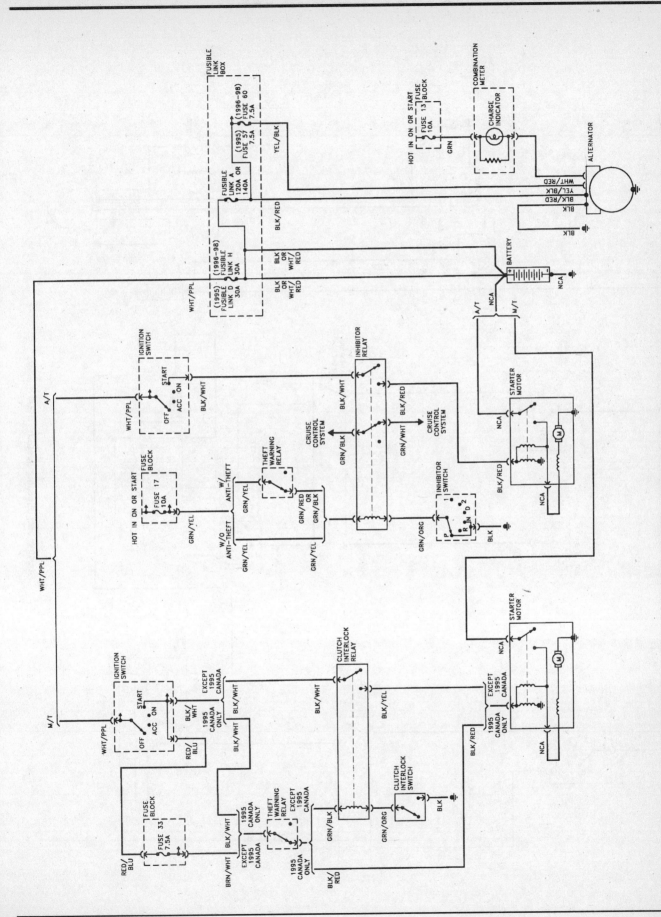

Starting and charging system - 1995 through 1998 models

12-22 Chapter 12 Chassis electrical system

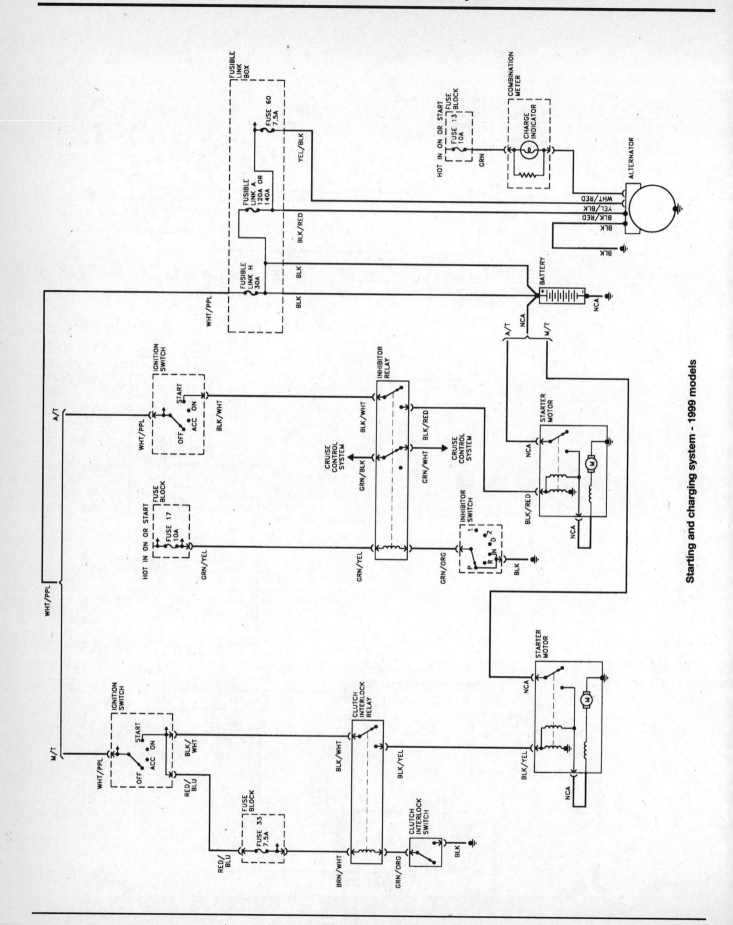

Starting and charging system - 1999 models

Chapter 12 Chassis electrical system

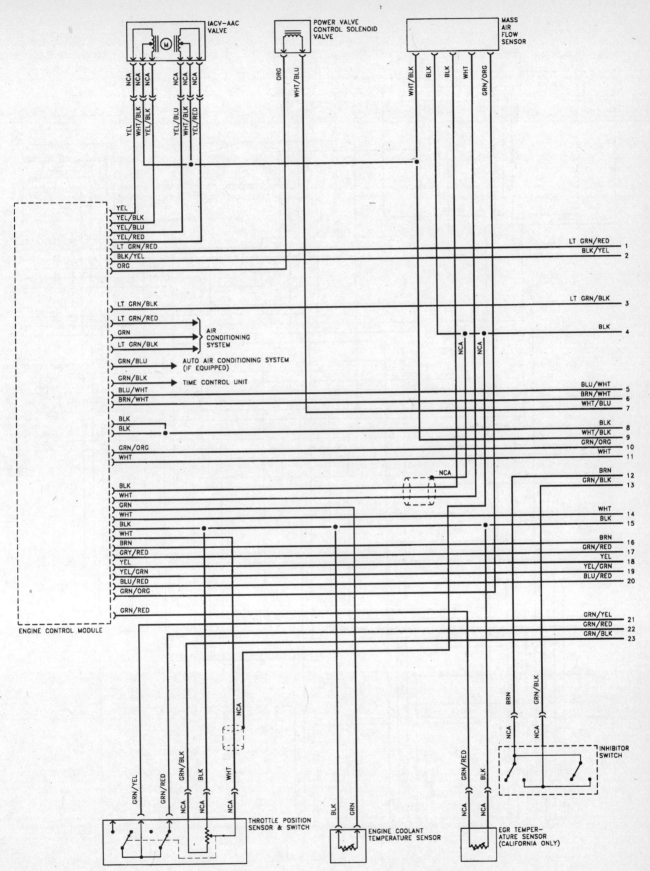

Engine control system - 1993 and 1994 models (Part 1 of 3)

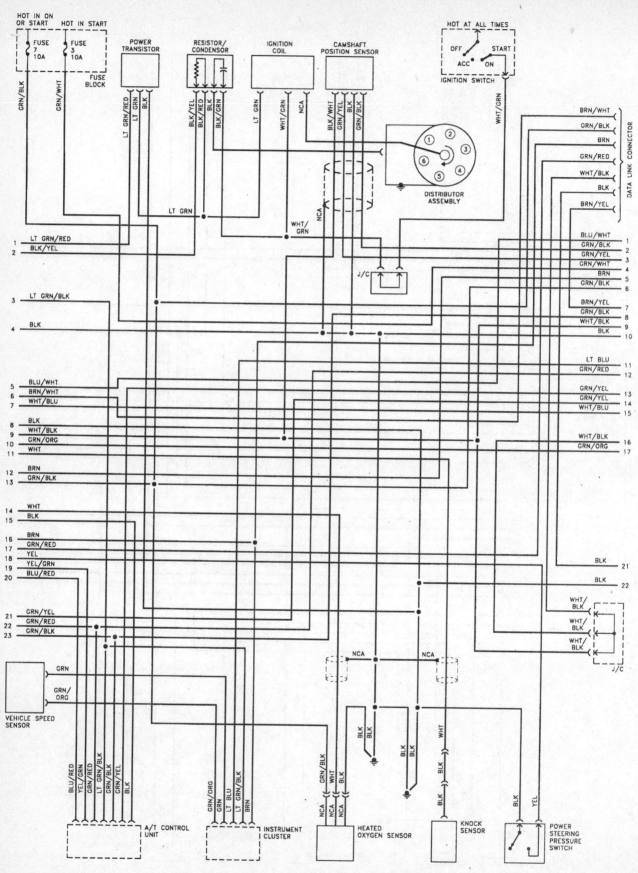

Chapter 12 Chassis electrical system

12-25

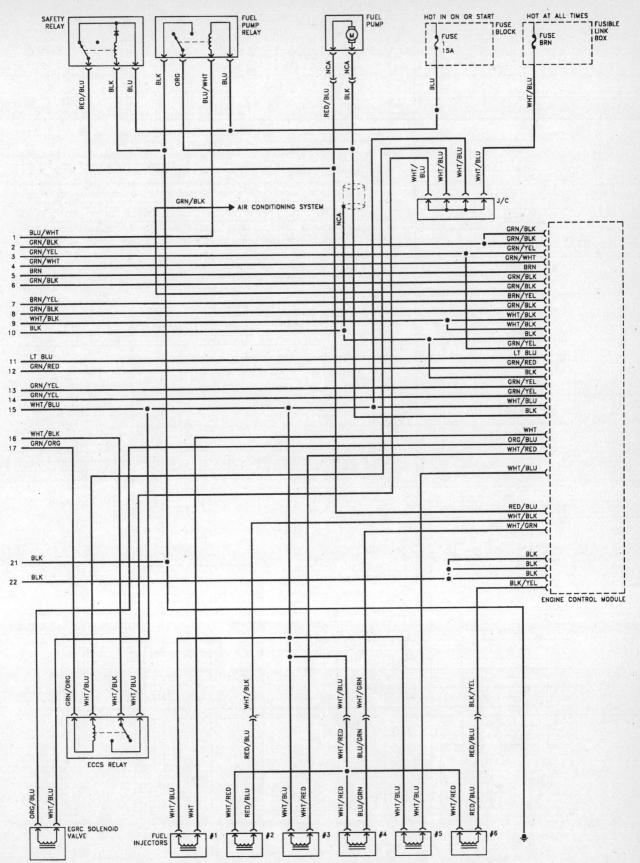

Engine control system - 1993 and 1994 models (Part 3 of 3)

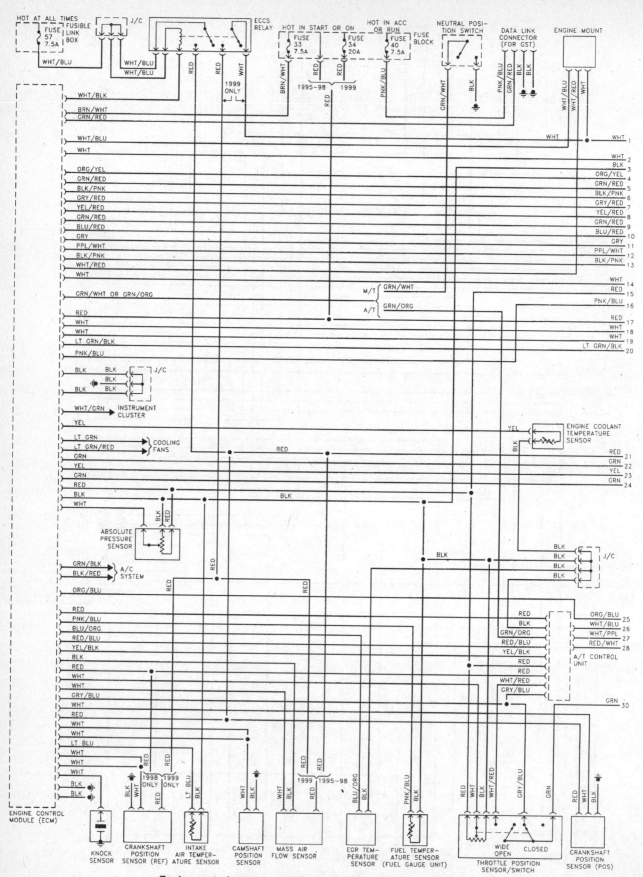

Engine control system - 1995 through 1999 models (Part 1 of 2)

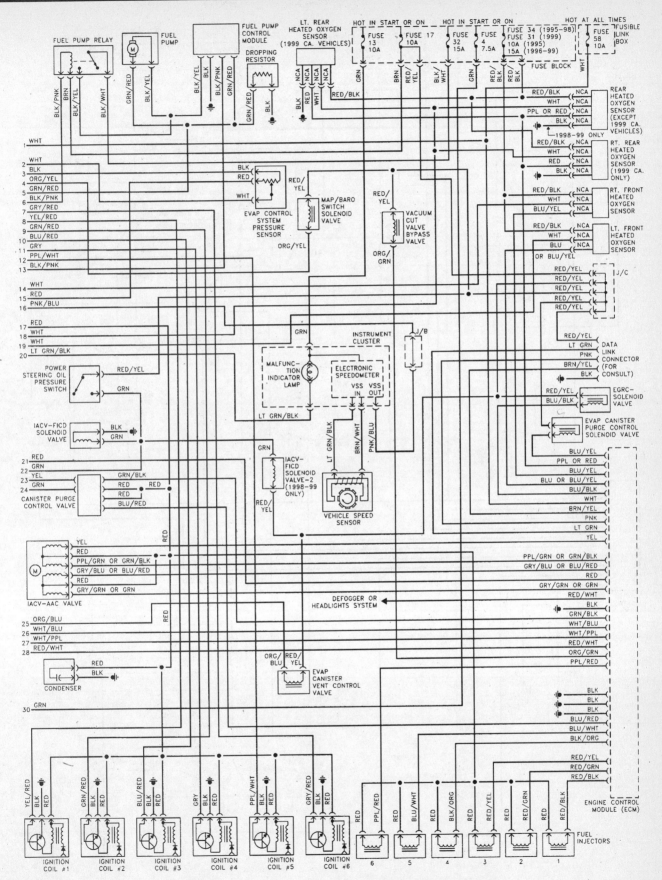

Engine control system - 1995 through 1999 models (Part 2 of 2)

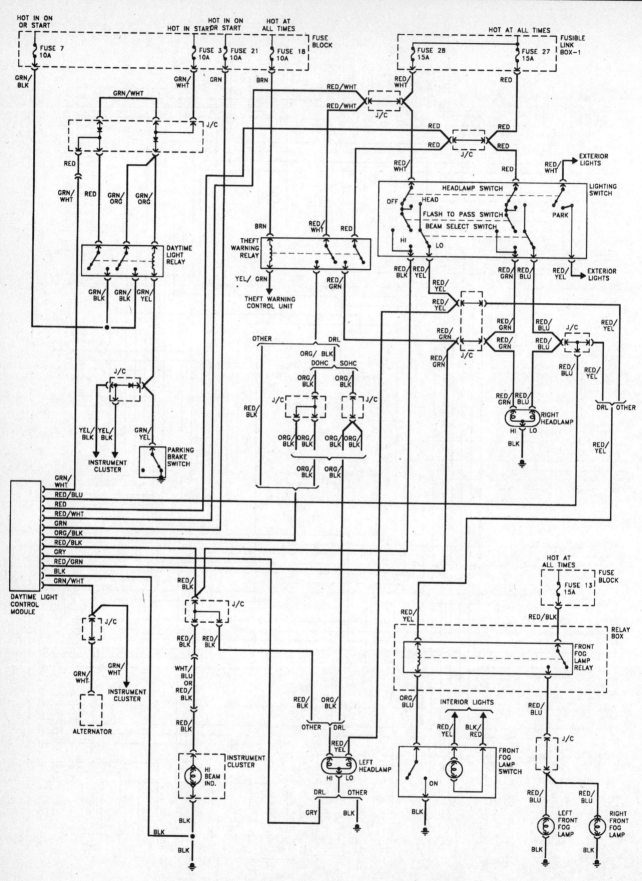

Headlight system (includes fog lights) - 1993 and 1994 models

Chapter 12 Chassis electrical system

12-29

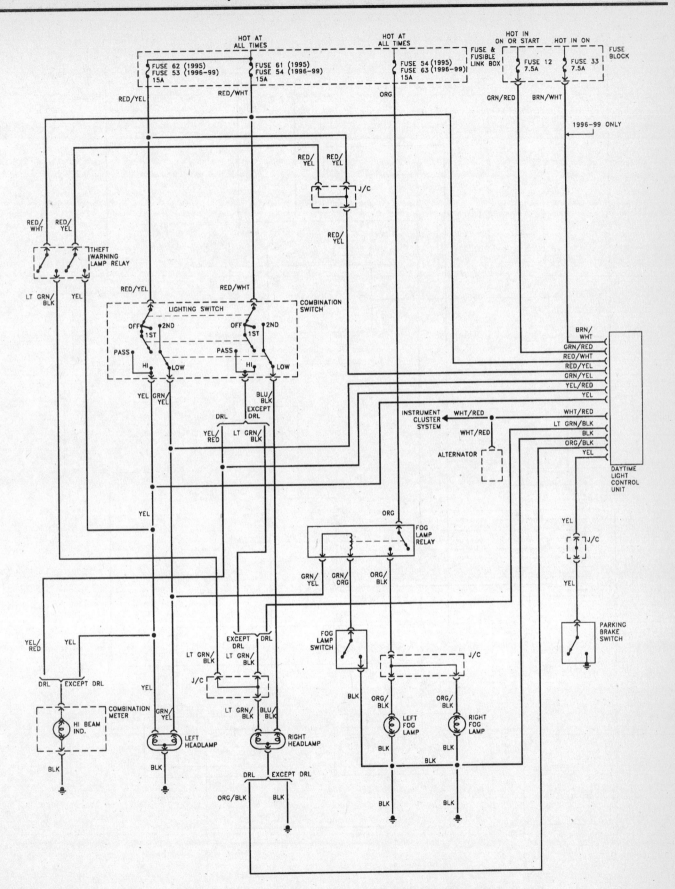

Headlight system (includes fog lights) - 1995 through 1999 models

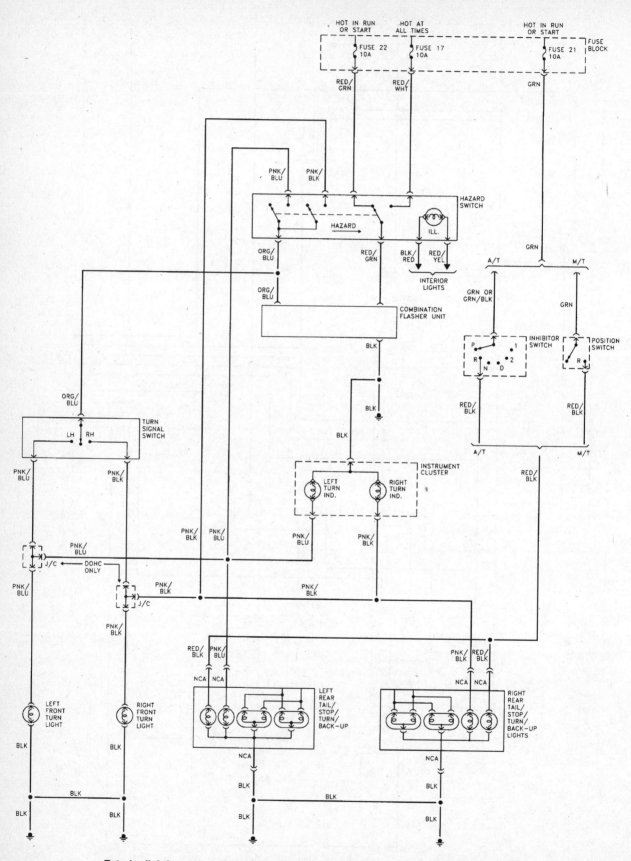

Exterior lighting system (backup and turn signals) - 1993 and 1994 models (Part 1 of 2)

Chapter 12 Chassis electrical system

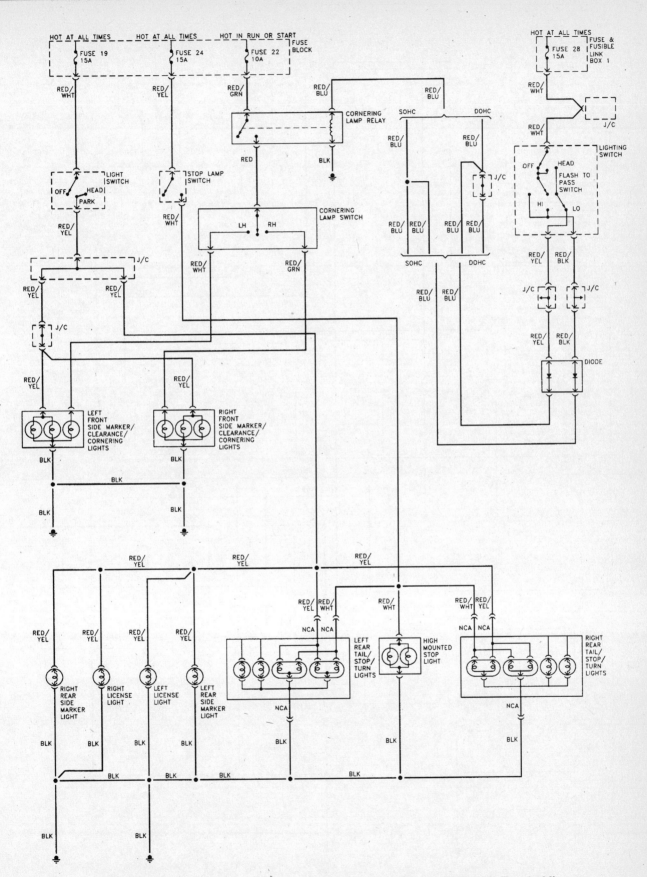

Exterior lighting system (stop, park and cornering lamps) - 1993 and 1994 models (Part 2 of 2)

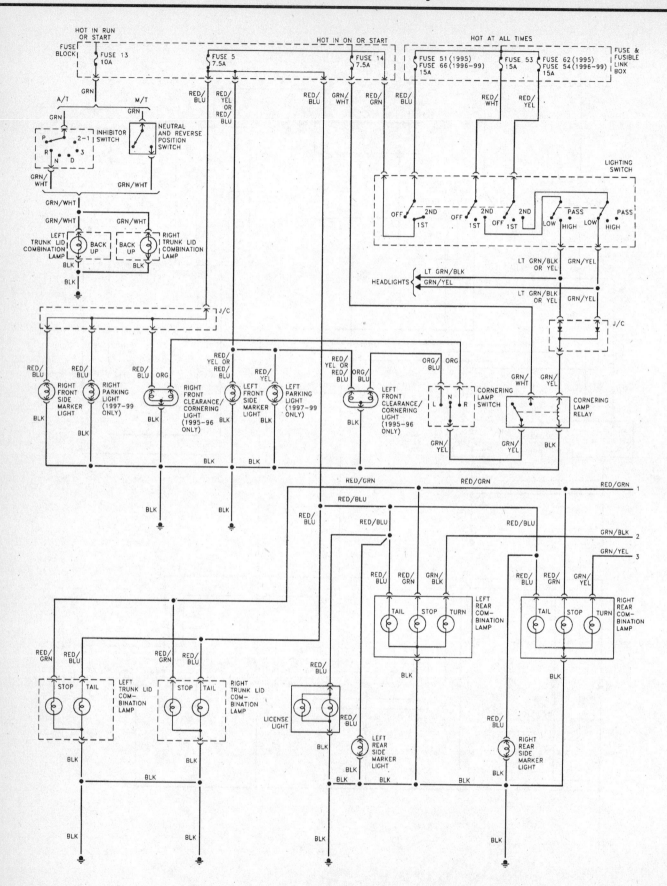

Exterior lighting system - 1995 though 1999 models (Part 1 of 2)

Chapter 12 Chassis electrical system

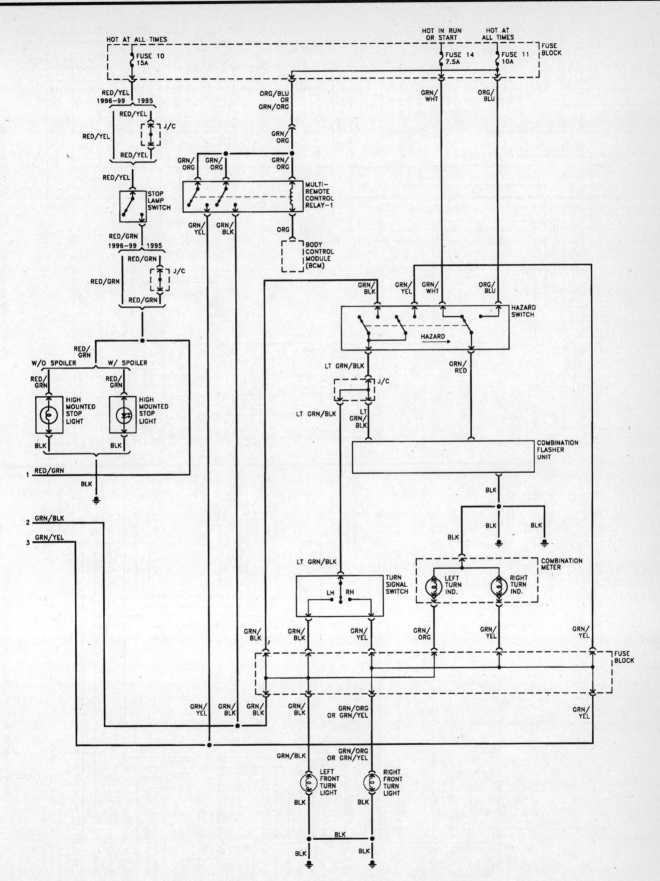

Exterior lighting system - 1995 though 1999 models (Part 2 of 2)

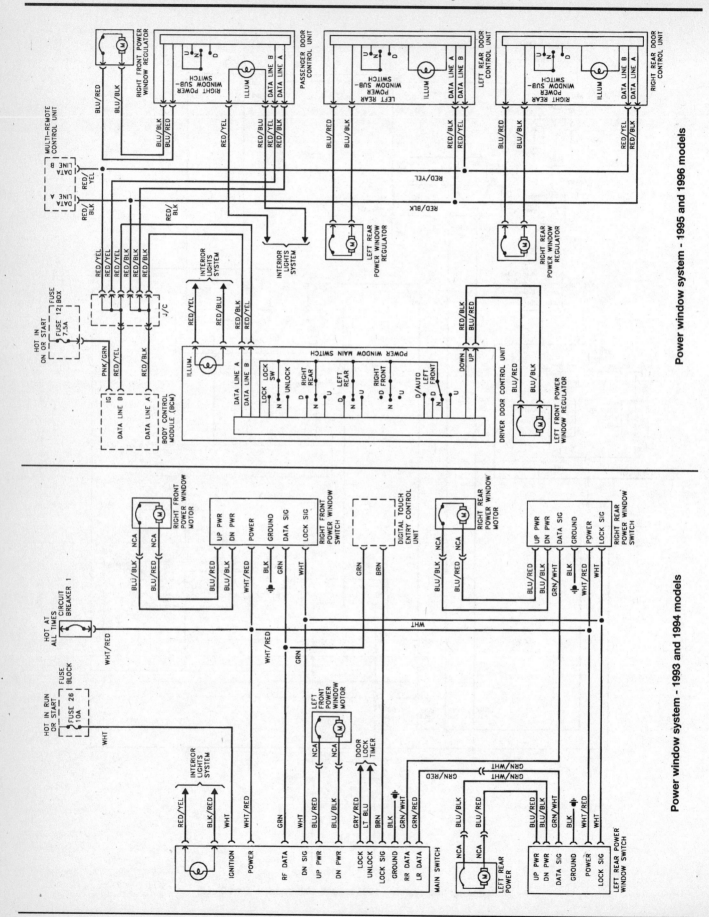

Chapter 12 Chassis electrical system

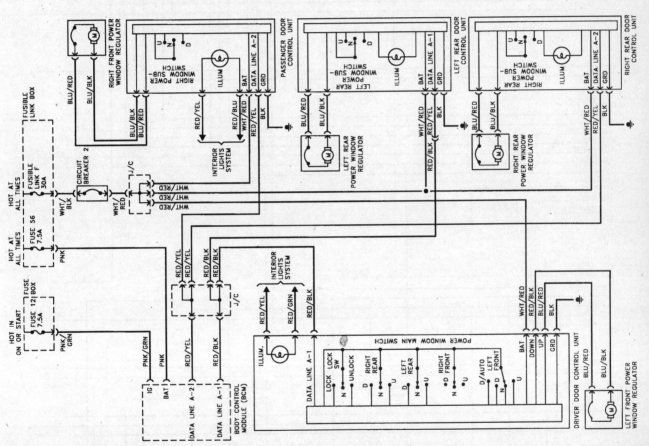

Power window system - 1997 through 1999 models

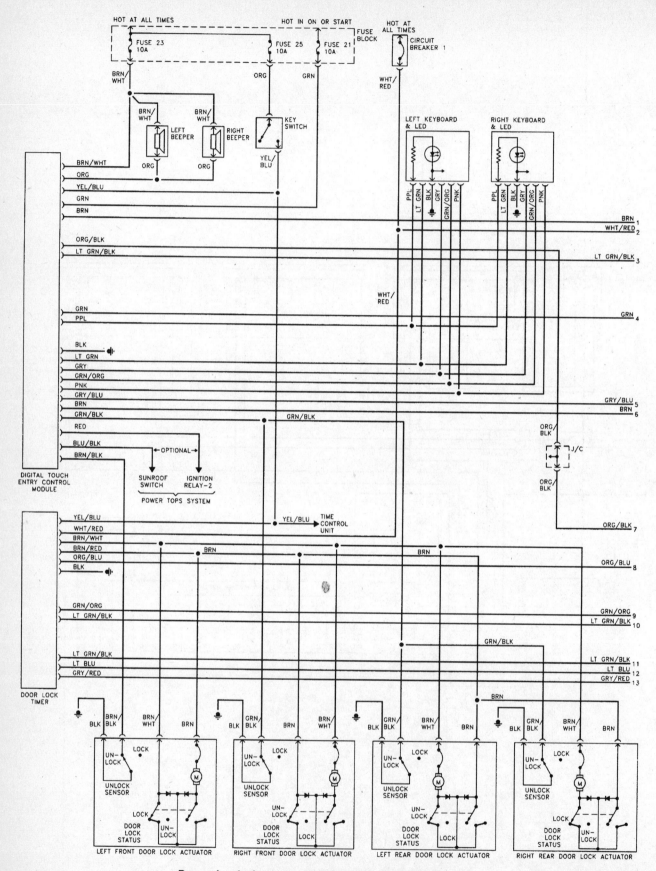

Power door lock system - 1993 and 1994 models (Part 1 of 2)

Chapter 12 Chassis electrical system

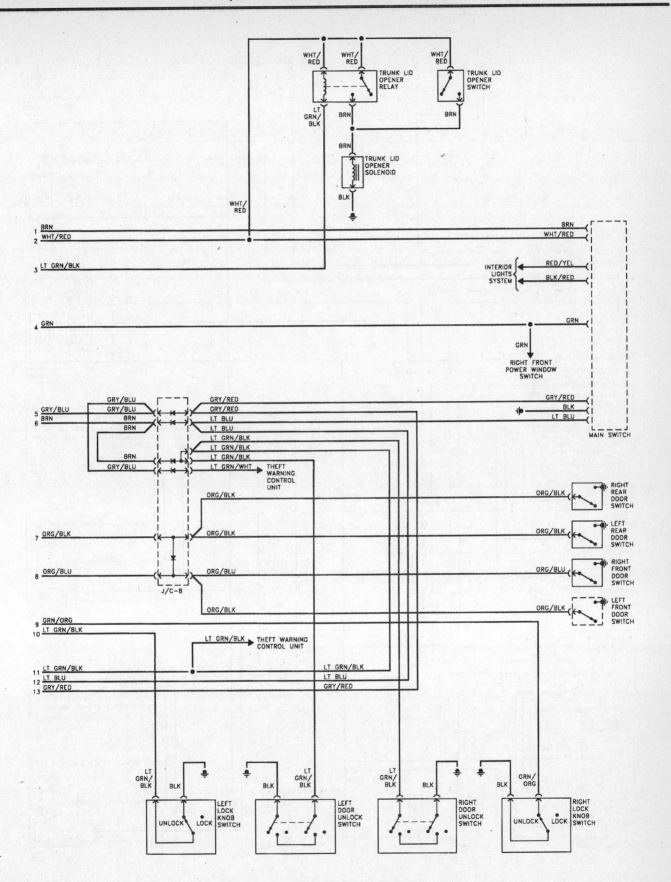

Power door lock system - 1993 and 1994 models (Part 2 of 2)

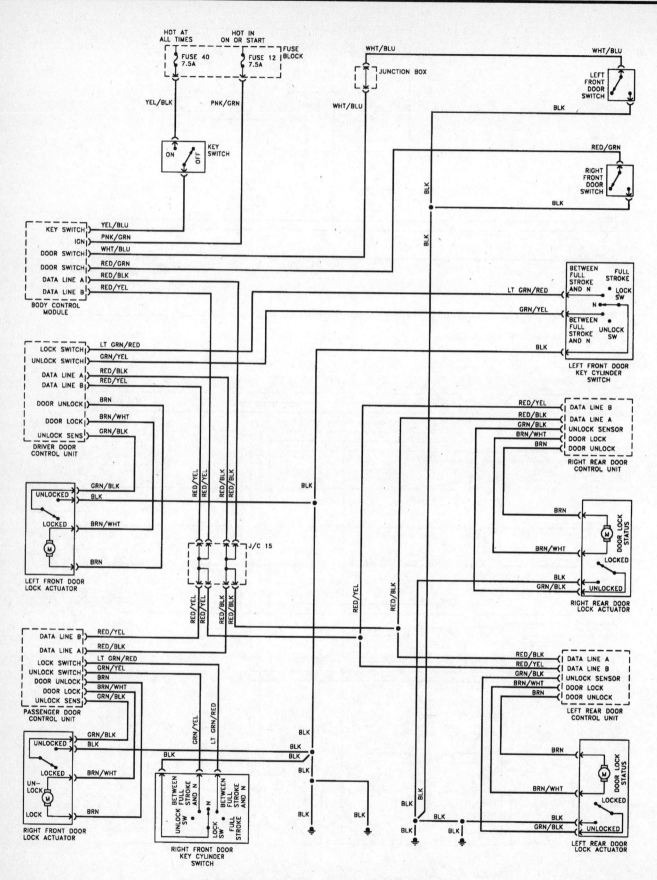

Power door lock system - 1995 and 1996 models

Chapter 12 Chassis electrical system

12-39

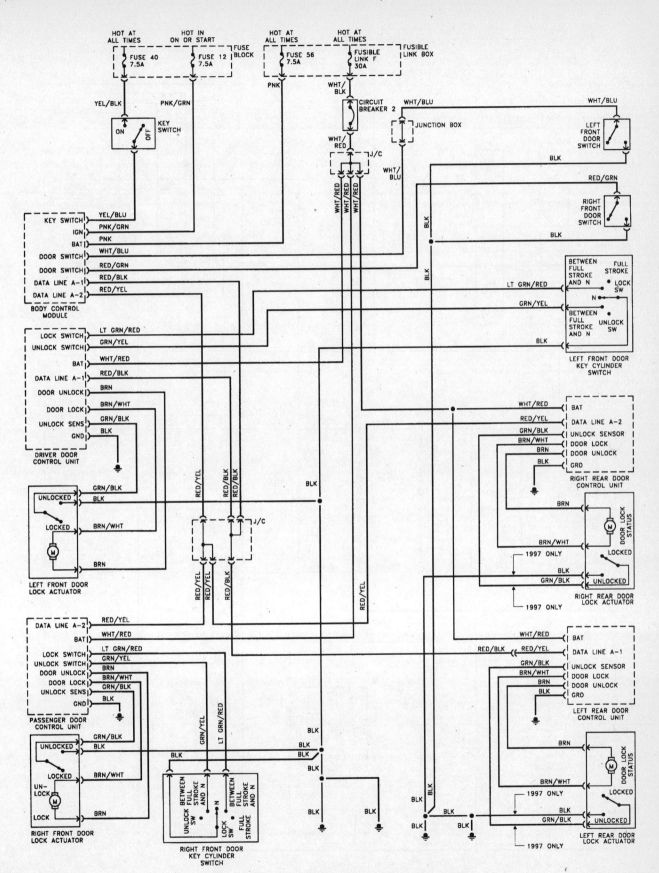

Power door lock system - 1997 through 1999 models

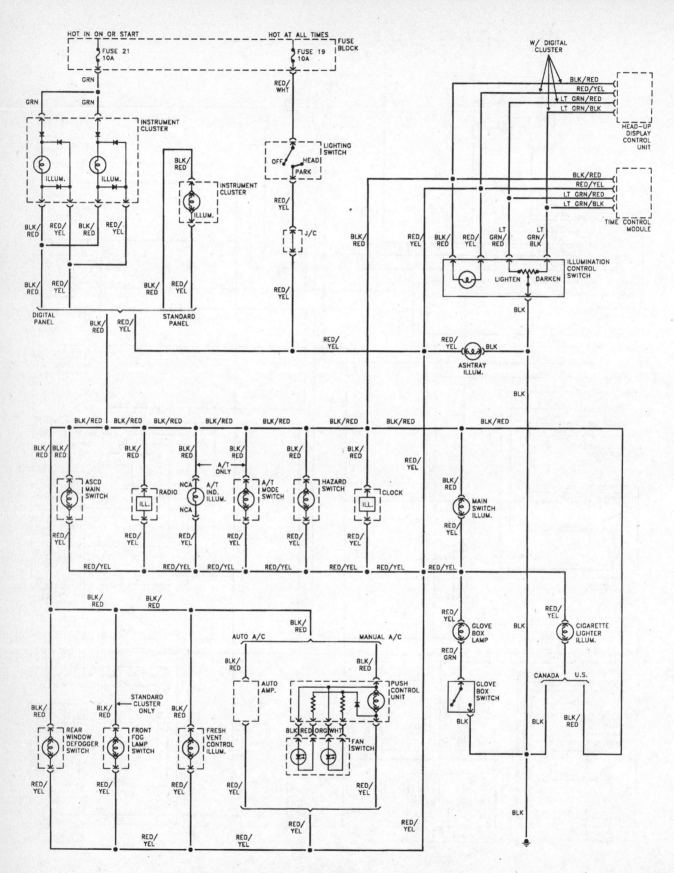

Interior lighting system - 1993 and 1994 models (Part 1 of 2)

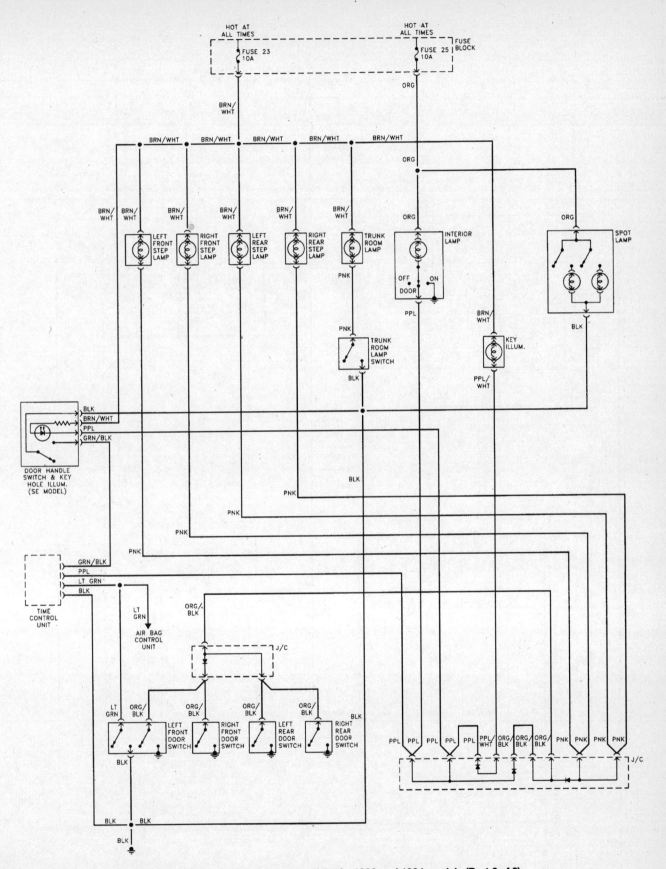

Interior lighting system (courtesy lights) - 1993 and 1994 models (Part 2 of 2)

Chapter 12 Chassis electrical system

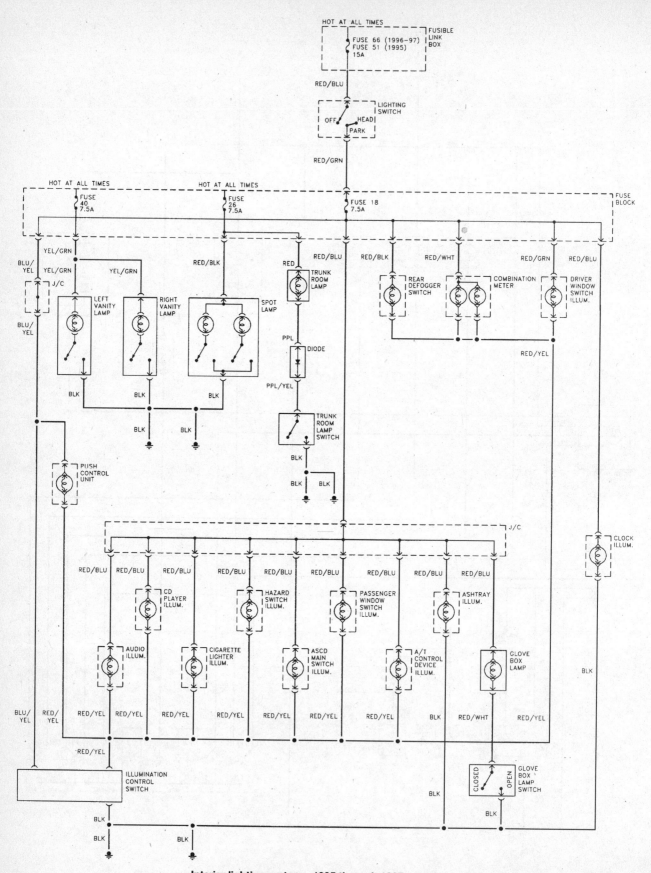

Interior lighting system - 1995 through 1997 models

Chapter 12 Chassis electrical system

12-43

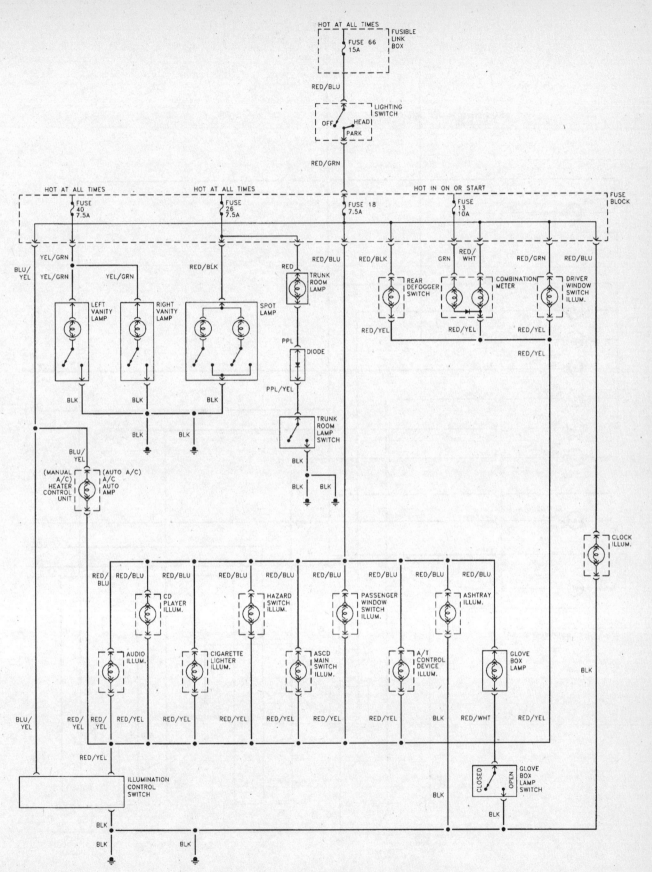

Interior lighting system - 1998 and 1999 models

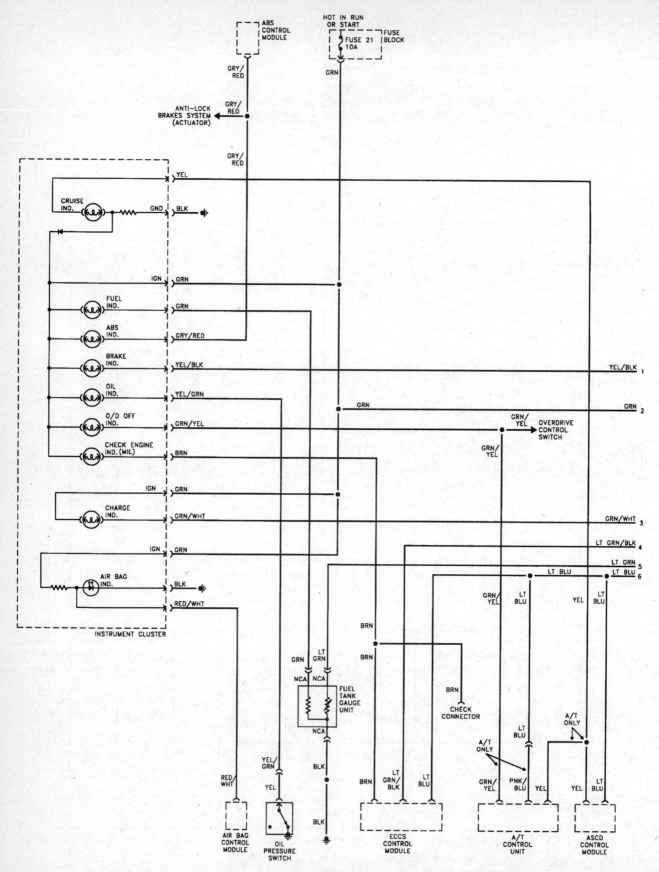

Warning lights and gauge system - 1993 and 1994 models with heads-up display (Part 1 of 2)

Chapter 12 Chassis electrical system

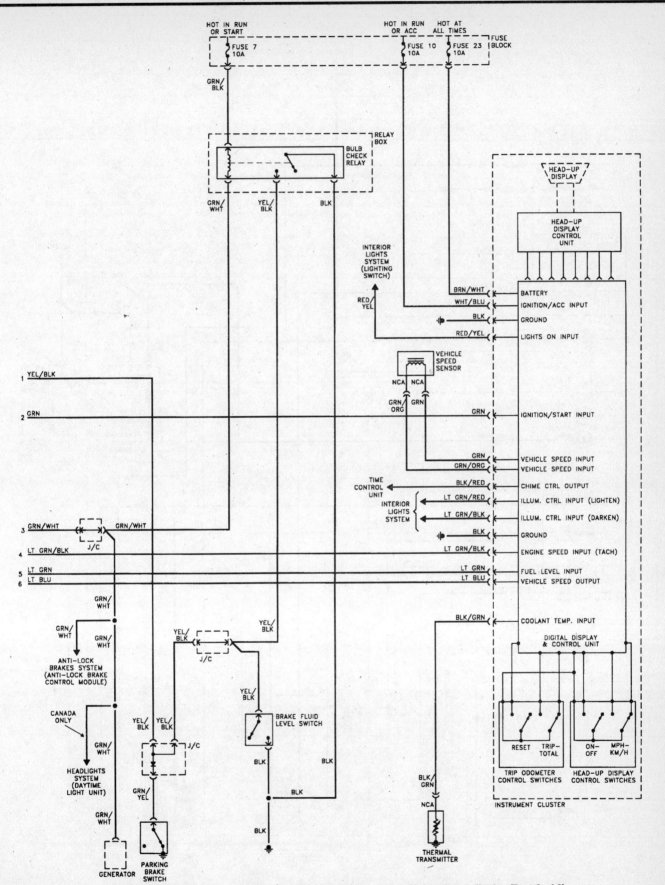

Warning lights and gauge system - 1993 and 1994 models with heads-up display (Part 2 of 2)

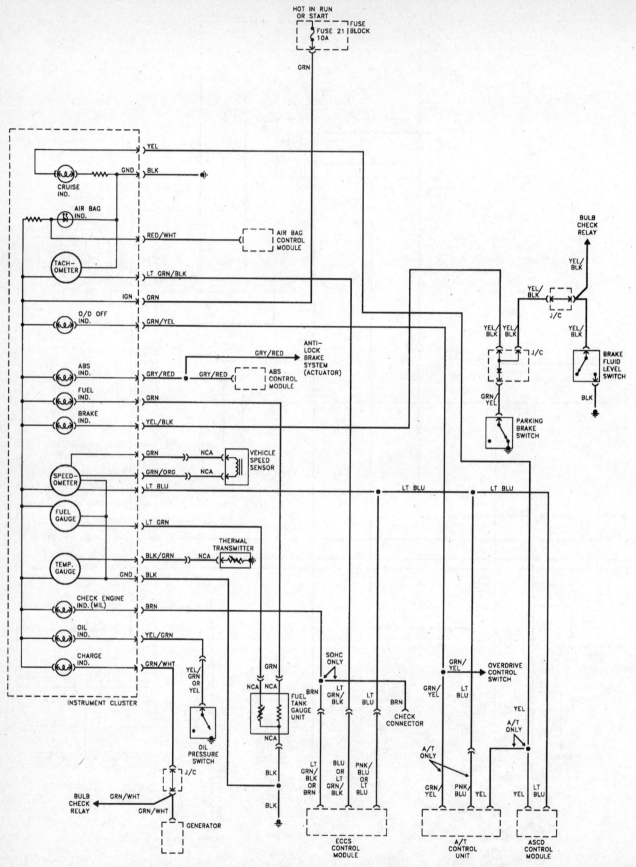

Warning lights and gauge system - 1993 and 1994 models without heads-up display

Chapter 12 Chassis electrical system

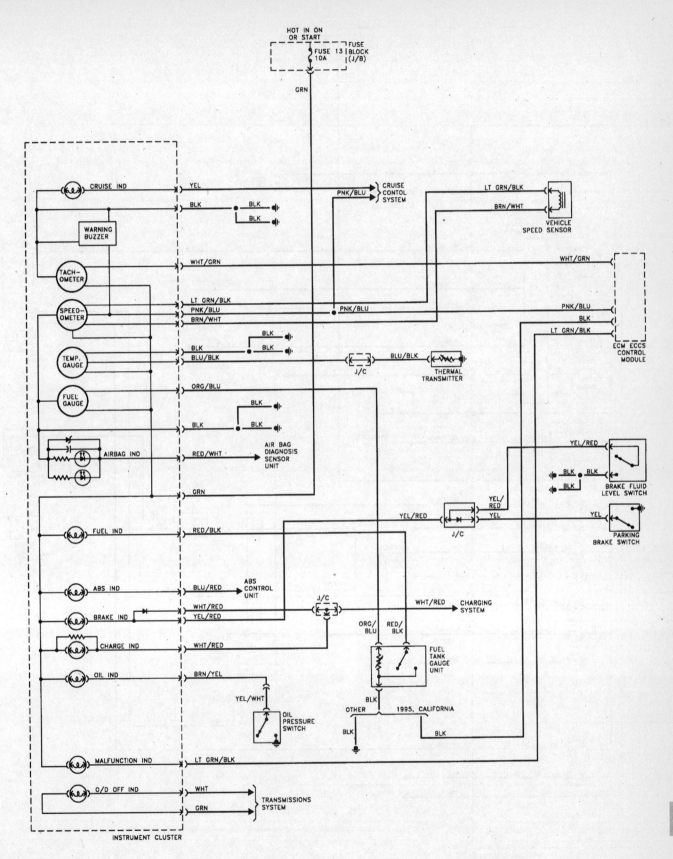

Warning lights and gauge system - 1995 through 1997 models

12-48   Chapter 12   Chassis electrical system

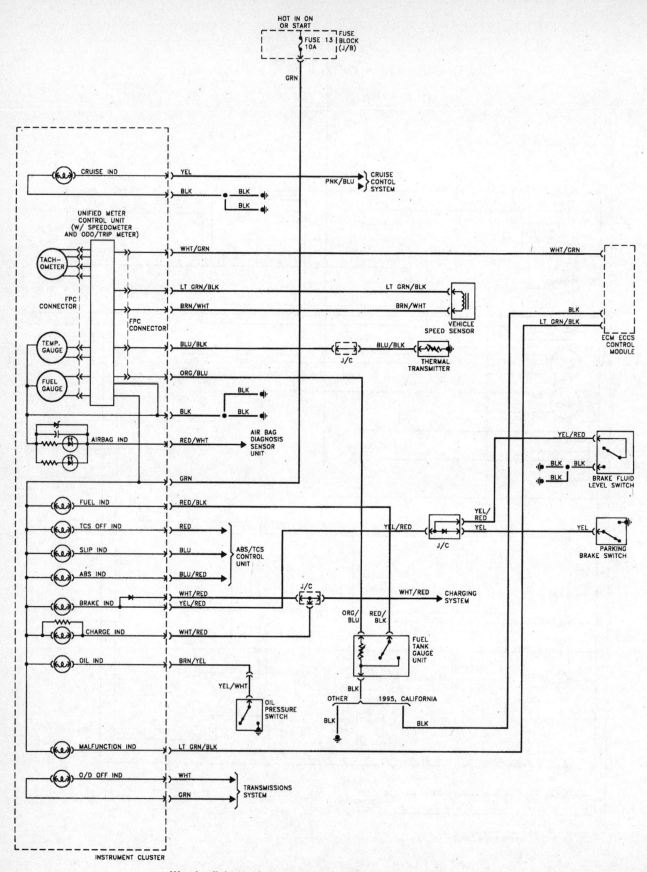

Warning lights and gauge system - 1998 and 1999 models

Chapter 12 Chassis electrical system

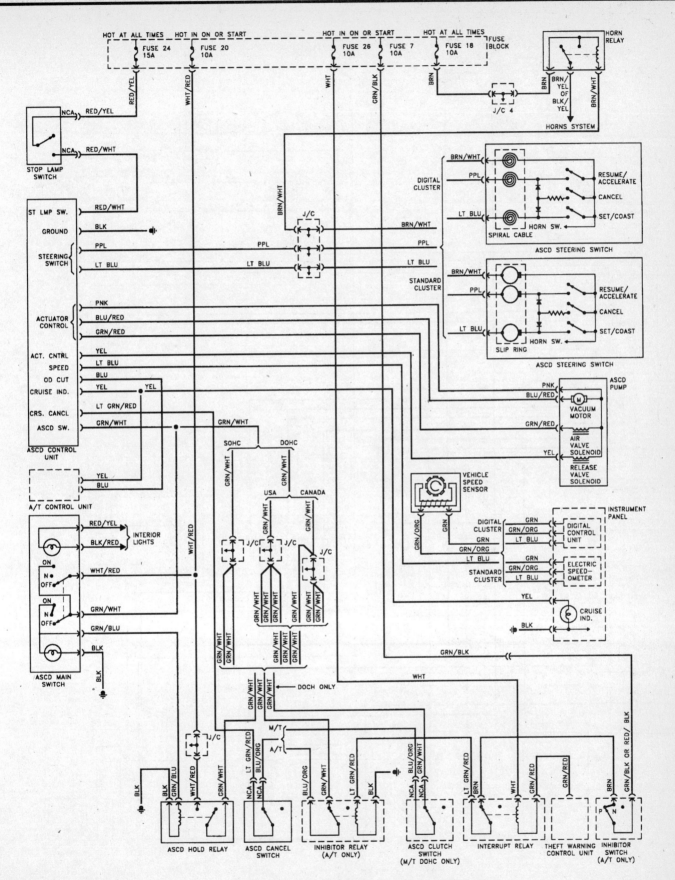

Cruise control system - 1993 and 1994 models

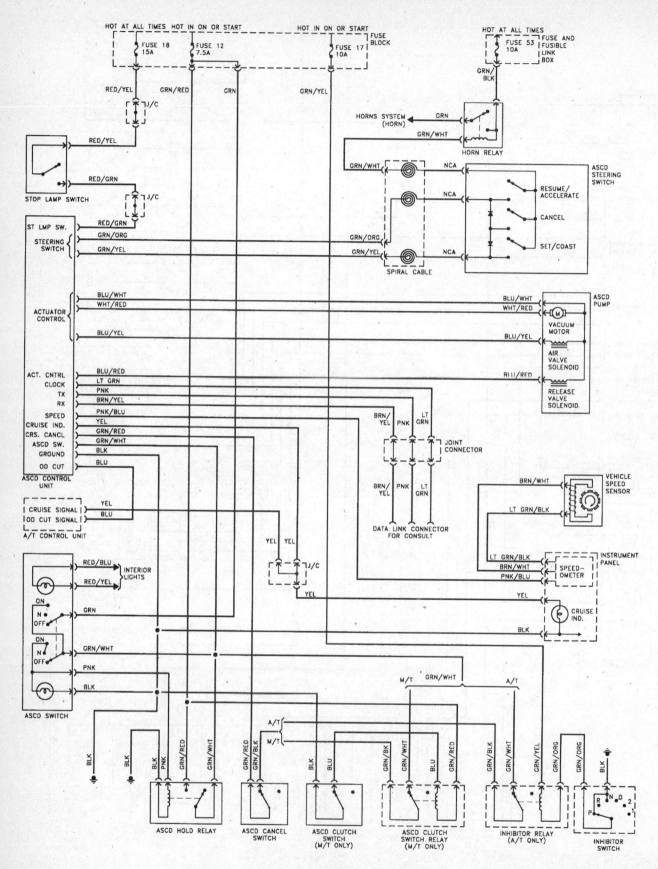

Cruise control system - 1995 models

Chapter 12 Chassis electrical system 12-51

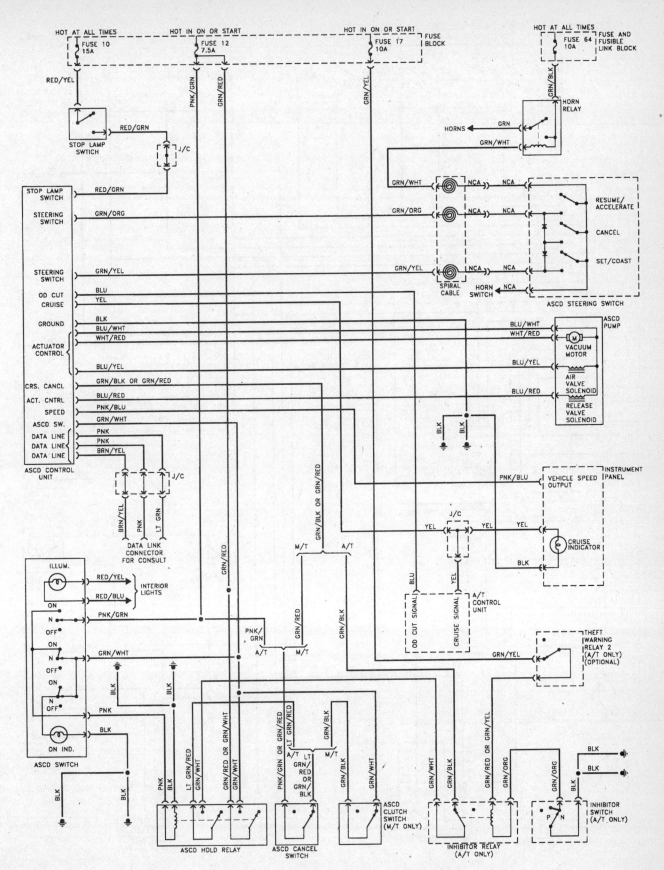

Cruise control system - 1996 through 1999 models

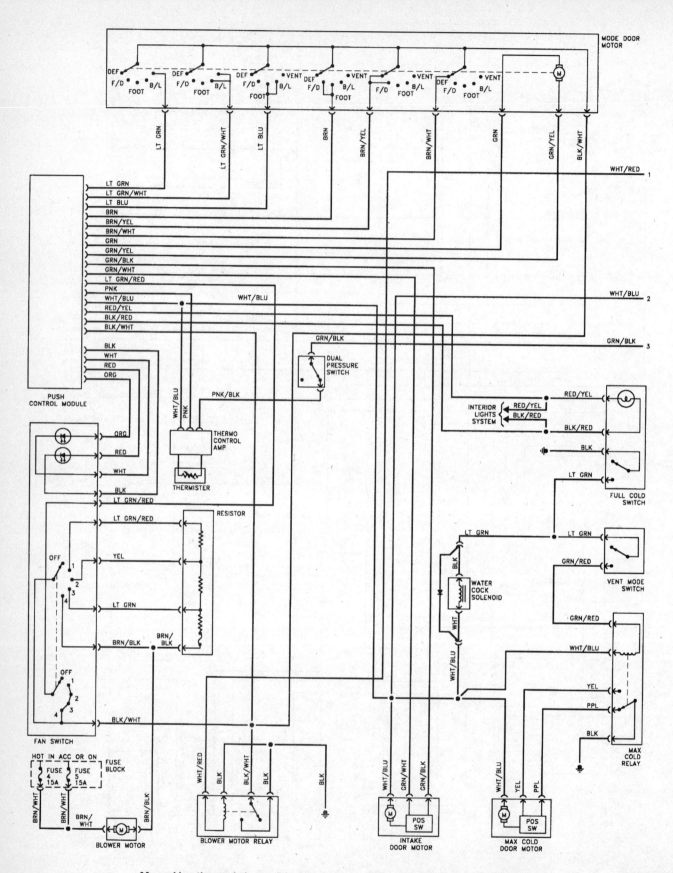

Manual heating and air conditioning system - 1993 and 1994 models (Part 1 of 2)

Chapter 12 Chassis electrical system

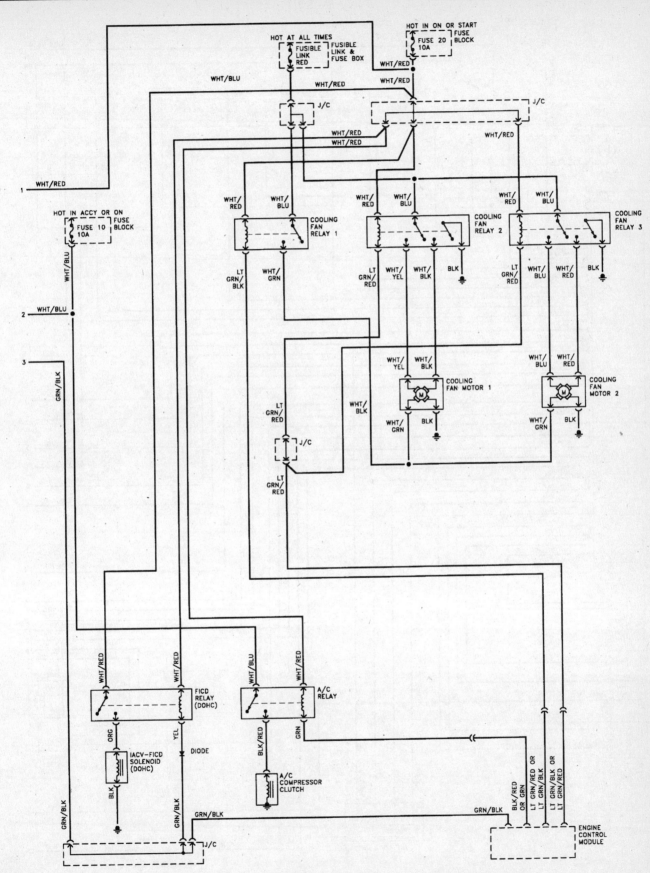

Manual heating and air conditioning system - 1993 and 1994 models (Part 2 of 2)

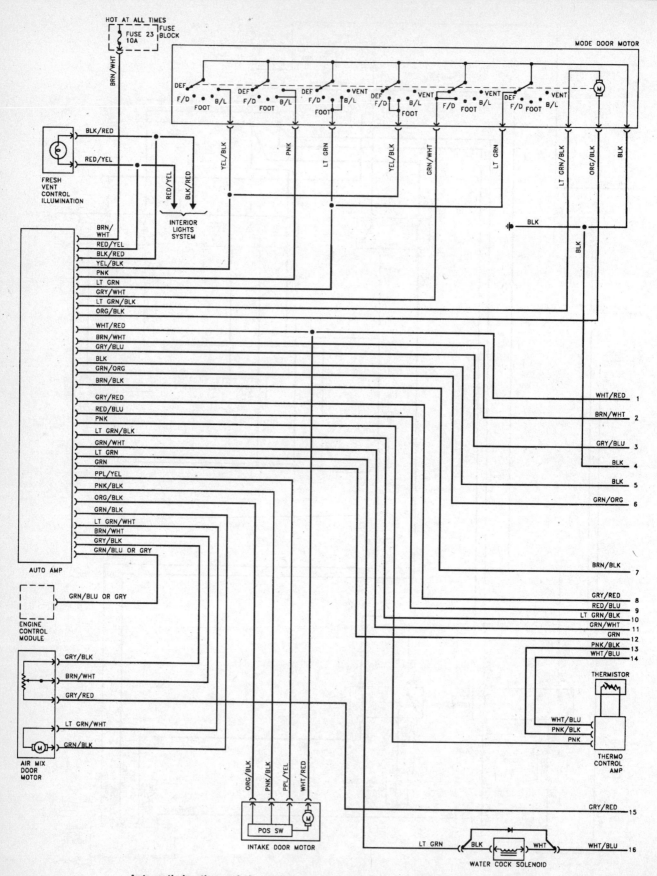

Automatic heating and air conditioning system - 1993 and 1994 models (Part 1 of 2)

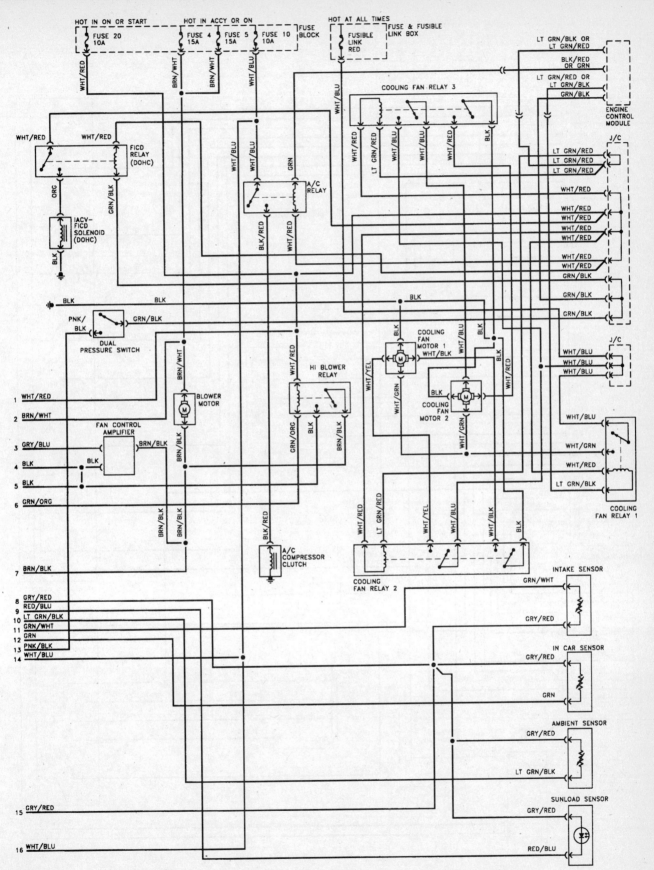

Automatic heating and air conditioning system - 1993 and 1994 models (Part 2 of 2)

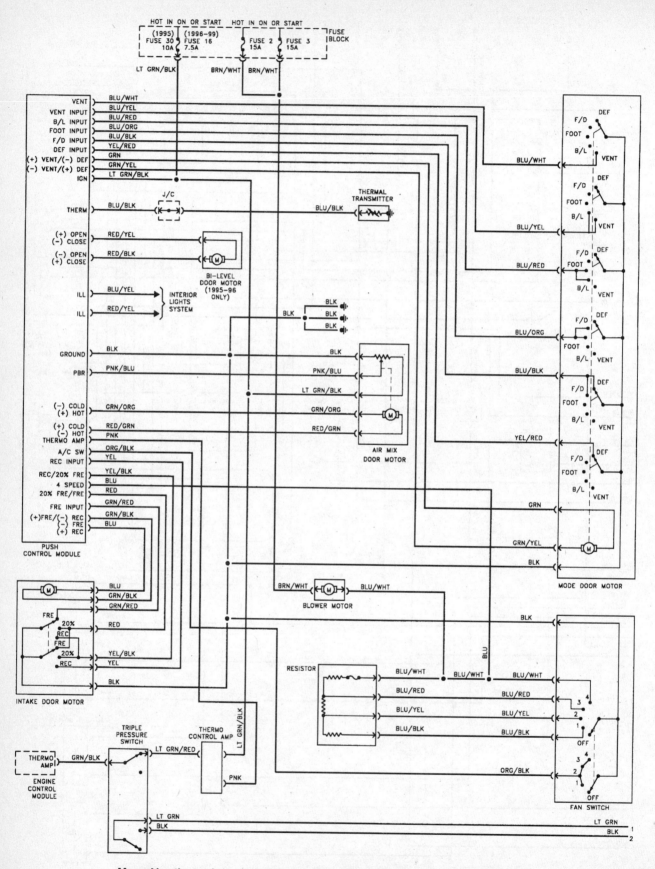

Manual heating and air conditioning system - 1995 through 1999 models (Part 1 of 2)

Chapter 12 Chassis electrical system

12-57

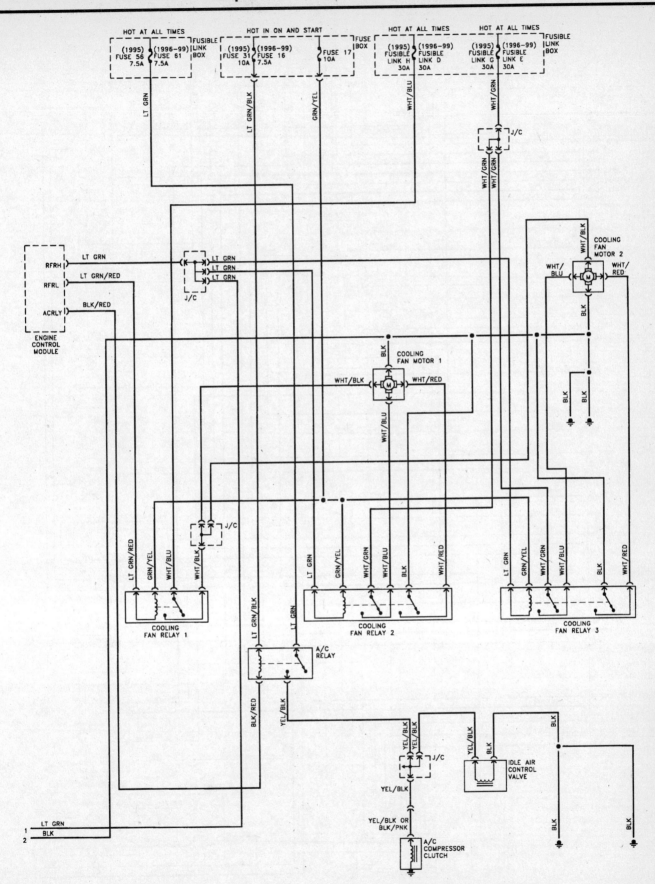

Manual heating and air conditioning system - 1995 through 1999 models (Part 2 of 2)

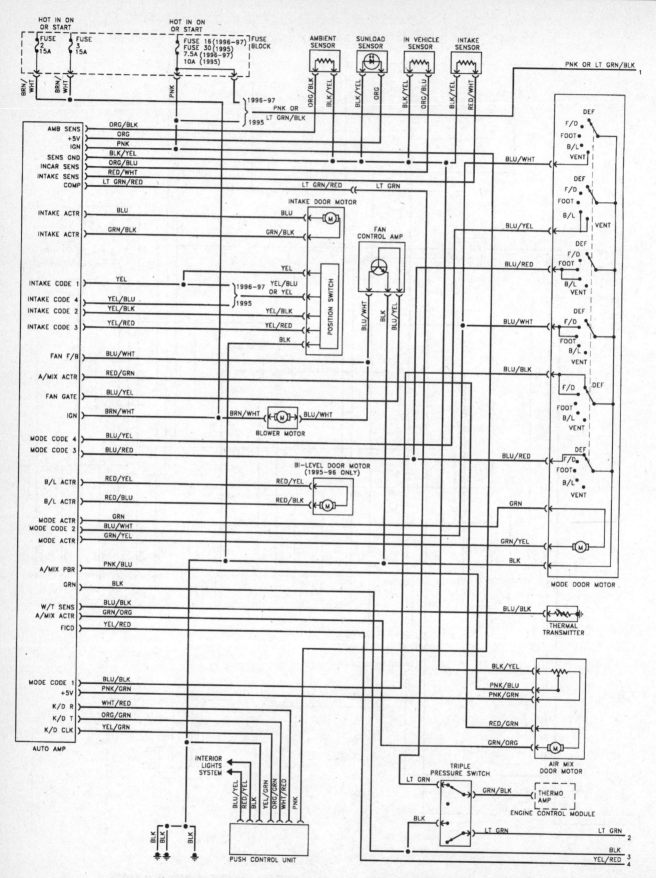

Automatic heating and air conditioning system - 1995 through 1997 models (Part 1 of 2)

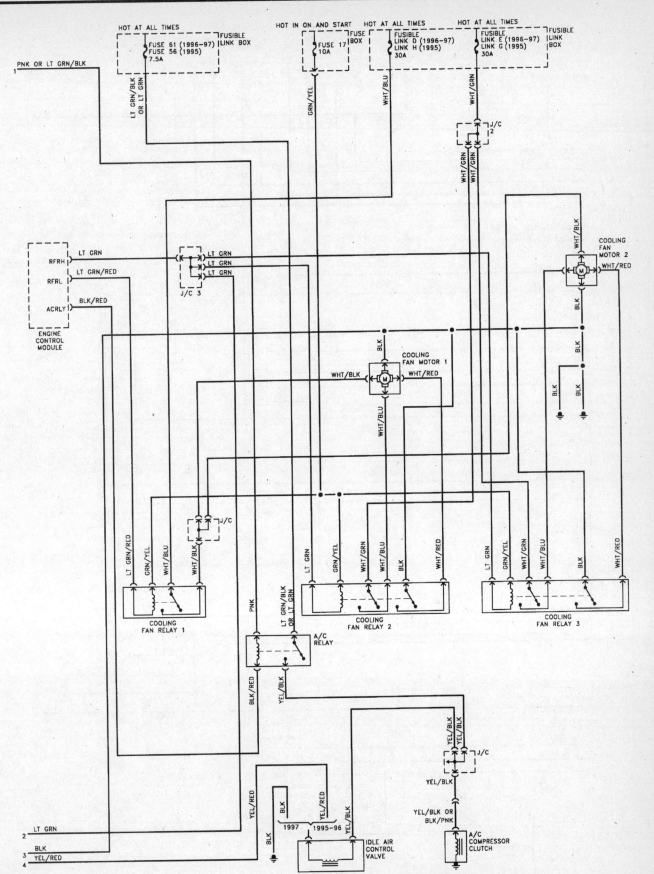

Automatic heating and air conditioning system - 1995 through 1997 models (Part 2 of 2)

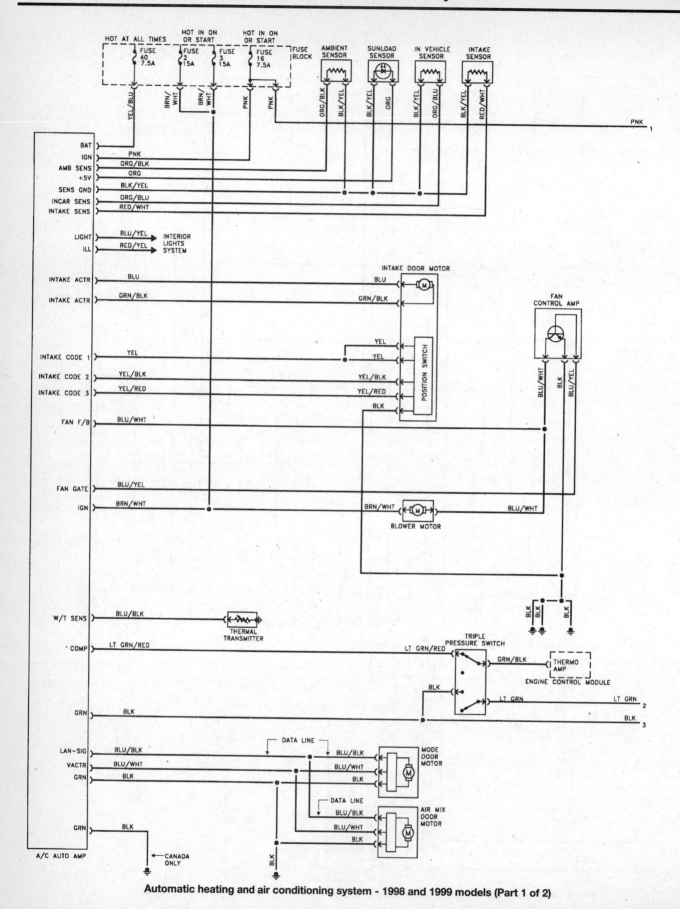

Automatic heating and air conditioning system - 1998 and 1999 models (Part 1 of 2)

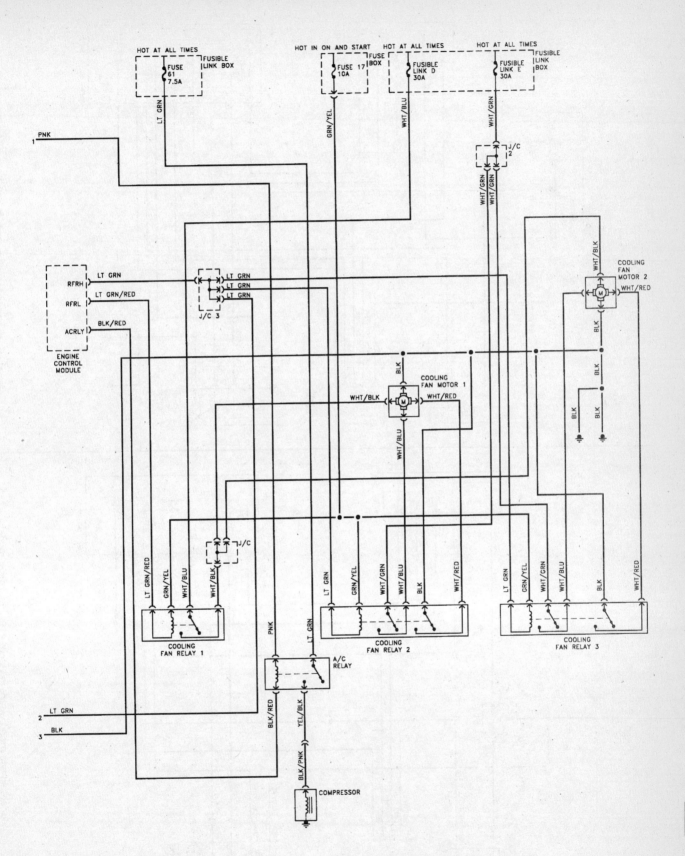

Automatic heating and air conditioning system - 1998 and 1999 models (Part 2 of 2)

12-62　Chapter 12　Chassis electrical system

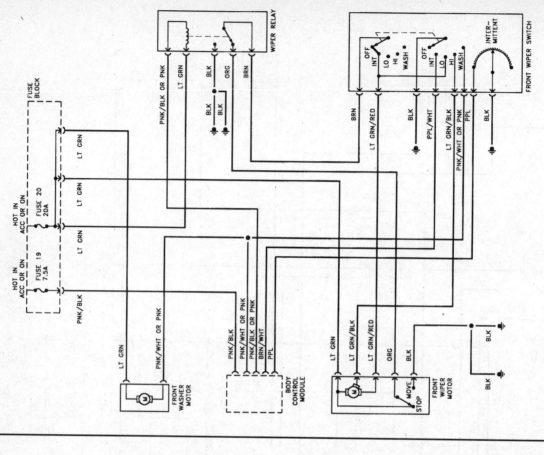

Windshield wiper and washer system - 1995 through 1999 models

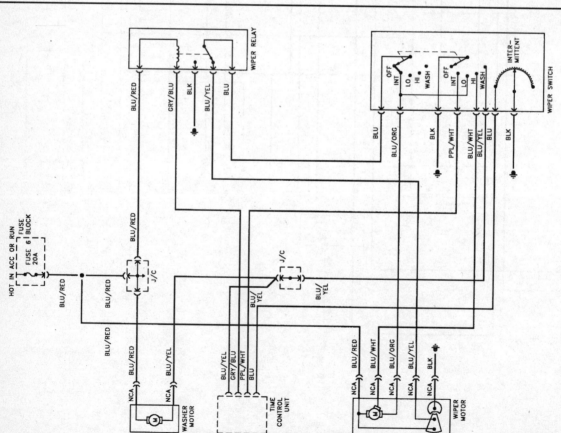

Windshield wiper and washer system - 1993 and 1994 models

Chapter 12 Chassis electrical system   12-63

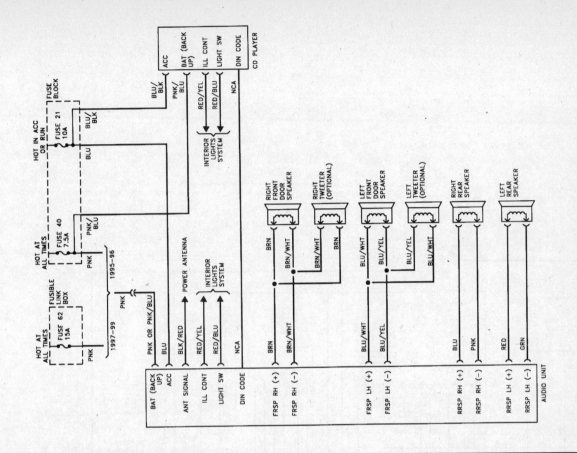

Audio system (base model) - 1995 through 1999 models

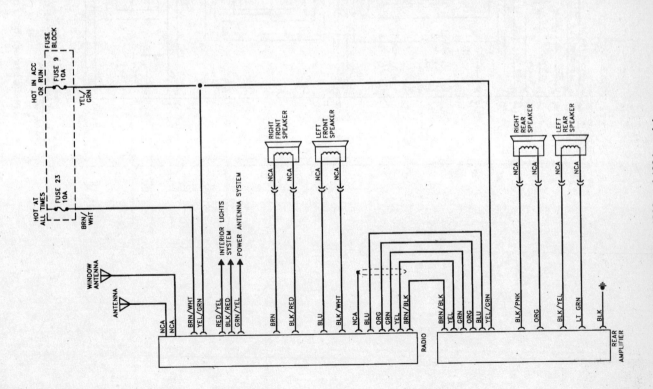

Audio system - 1993 and 1994 models

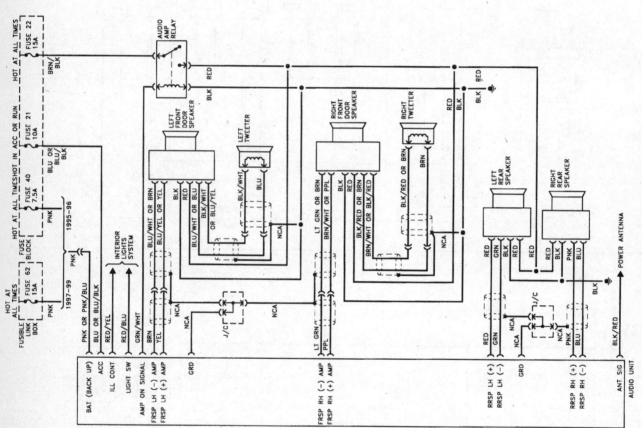

Index

A

Accelerator cable, removal, installation and adjustment, 4-8
Acknowledgements, 0-2
Air conditioning
    and heating system, check and maintenance, 3-11
    compressor, removal and installation, 3-14
    condenser, removal and installation, 3-15
    receiver/drier, removal and installation, 3-14
Air filter replacement, 1-20
Airbag system, general information, 12-18
Alignment, general information, 10-17
Alternator, removal and installation, 5-11
Antenna, removal and installation, 12-10
Antifreeze, general information, 3-2
Anti-lock Brake System (ABS), general information and trouble codes, 9-2
Automatic transaxle, 7B-1 through 7B-10
    diagnosis and trouble codes, 7B-2
    fluid change, 1-25
    fluid level check, 1-11
    Park Neutral Position switch, check, adjustment and replacement, 7B-7
    removal and installation, 7B-7
    shift cable, check, adjustment and replacement, 7B-4
    shift interlock system, description, check and component replacement, 7B-6
    shift lever, removal and installation, 7B-5
Automotive chemicals and lubricants, 0-18
Axle beam (rear), lateral link and control rod (1995 and later models), removal and installation, 10-11

B

Back-up light and neutral position switch, check and replacement, 7A-4
Balljoints, replacement, 10-9

Battery
    cables, check and replacement, 5-4
    check and replacement, 5-3
    check, maintenance and charging, 1-13
    emergency jump starting, 0-16
Blower motor circuit, check, 3-7
Blower motor, removal and installation, 3-8
Body, 11-1 through 11-18
    bumpers, removal and installation, 11-6
    center console, removal and installation, 11-14
    cowl cover, removal and installation, 11-9
    dashboard trim panels, removal and installation, 11-14
    door latch, lock cylinder and handle, removal and installation, 11-11
    door trim panel, removal and installation, 11-10
    door window glass
        regulator, removal and installation, 11-12
        removal and installation, 11-12
    door, removal, installation and adjustment, 11-10
    front fender, removal and installation, 11-8
    hinges and locks, maintenance, 11-5
    hood and hood support struts, removal, installation and adjustment, 11-5
    hood latch and release cable, removal and installation, 11-6
    instrument panel, removal and installation, 11-16
    maintenance, 11-1
    mirrors, removal and installation, 11-12
    radiator grille, removal and installation, 11-5
    rear package shelf, removal and installation, 11-18
    repair
        major damage, 11-5
        minor damage, 11-4
    seats, removal and installation, 11-18
    steering column covers, removal and installation, 11-15
    sunroof, adjustment, 11-18
    trunk lid latch and lock cylinder, removal and installation, 11-13
    trunk lid, removal, installation and adjustment, 11-13

Index

trunk release and fuel door cables, removal and installation, 11-14
upholstery and carpets, maintenance, 11-1
vinyl trim, maintenance, 11-1
windshield and fixed glass, replacement, 11-5

Booster battery (jump) starting, 0-16
Brake check, 1-18
Brake disc, inspection, removal and installation, 9-10
Brake light switch, check and replacement, 9-19
Brakes, 9-1 through 9-20

Anti-lock Brake System (ABS), general information and trouble codes, 9-2
brake disc, inspection, removal and installation, 9-10
brake light switch, check and replacement, 9-19
brake pedal, adjustment, 9-18
disc brake caliper, removal and installation, 9-9
disc brake pads, replacement, 9-5
drum brake shoes, replacement, 9-11
hoses and lines, inspection and replacement, 9-15
hydraulic system, bleeding, 9-16
master cylinder, removal and installation, 9-14
parking brake
  adjustment, 9-17
  cables, replacement, 9-17
power brake booster, check, replacement and adjustment, 9-16
proportioning valve, replacement, 9-15
wheel cylinder, removal and installation, 9-14

Bulb replacement, 12-12
Bumpers, removal and installation, 11-6
Buying parts, 0-8

C

Camshaft position sensor, check and replacement, 6-13
Camshafts and lifters, removal and installation, Dual Overhead Cam (DOHC) engine, 2B-10
Camshafts, lifters and bearings, inspection, 2C-21
Camshafts, lifters and seals, removal and installation, Single Overhead Cam (SOHC) engine, 2A-9
Catalytic converter, 6-22
Center console, removal and installation, 11-14
Charging system

check, 5-10
general information and precautions, 5-10

Chassis electrical system, 12-1 through 12-64

airbag system, general information, 12-18
antenna, removal and installation, 12-10
bulb replacement, 12-12
circuit breakers, general information, 12-3
cruise control system, description and check, 12-16
Daytime Running Lights (DRL), general information, 12-15
electric side view mirrors, description and check, 12-17
electric sunroof, description and check, 12-17
electrical troubleshooting, general information, 12-1
fuses and fusible links, general information, 12-3
headlight
  bulb, replacement, 12-11
  housing, replacement, 12-12
headlights, adjustment, 12-11
horn, check and replacement, 12-15
ignition switch and key lock cylinder, check and replacement, 12-7
instrument cluster, removal and installation, 12-9
instrument panel
  gauges, check, 12-9
  switches, check and replacement, 12-8
In-Vehicle Multiplexing System, description, 12-18
power door lock system, description and check, 12-17
power seats, description and check, 12-18
power window system, description and check, 12-16
radio and speakers, removal and installation, 12-9
rear window defogger, check and repair, 12-15
relays, general information and testing, 12-3
steering column switches, check and replacement, 12-5
turn signal/hazard flashers, check and replacement, 12-5
wiper motor, removal and installation, 12-14
wiring diagrams, general information, 12-19

Circuit breakers, general information, 12-3
Clutch

components, removal, inspection and installation, 8-3
description and check, 8-2
hydraulic system, bleeding, 8-3
master cylinder, removal and installation, 8-2
pedal, adjustment, 8-6
release bearing and lever, removal, inspection and installation, 8-5
release cylinder, removal and installation, 8-3
start switch, check and replacement, 8-7

Clutch and driveaxles, 8-1 through 8-14
Control arm (front), removal, inspection and installation, 10-7
Conversion factors, 0-19
Coolant temperature sending unit, check and replacement, 3-7
Cooling system

check, 1-17
servicing (draining, flushing and refilling), 1-24

Cooling, heating and air conditioning systems, 3-1 through 3-16

air conditioning and heating system, check and maintenance, 3-11
antifreeze, general information, 3-2
blower motor circuit, check, 3-7
blower motor, removal and installation, 3-8
compressor, removal and installation, 3-14
condenser, removal and installation, 3-15
coolant temperature sending unit, check and replacement, 3-7
engine cooling fan and circuit, check and component replacement, 3-4
heater and air conditioning control assembly, removal and installation, 3-9
heater core, removal and installation, 3-9

Index

radiator and coolant reservoir, removal and installation, 3-5
receiver/drier, removal and installation, 3-14
thermostat, check and replacement, 3-3
water pump, check and replacement, 3-6

Cowl cover, removal and installation, 11-9

Crankshaft
inspection, 2C-19
installation and main bearing oil clearance check, 2C-23
removal, 2C-15

Crankshaft front oil seal, replacement
Dual Overhead Cam (DOHC) engine, 2B-16
Single Overhead Cam (SOHC) engine, 2A-14

Crankshaft position sensor, check and replacement, 6-12

Crankshaft pulley, removal and installation
Dual Overhead Cam (DOHC) engine, 2B-15
Single Overhead Cam (SOHC) engine, 2A-14

Cruise control system, description and check, 12-16

Cylinder compression check, 2C-7

Cylinder head
cleaning and inspection, 2C-12
disassembly, 2C-11
reassembly, 2C-14
removal and installation
Dual Overhead Cam (DOHC) engine, 2B-14
Single Overhead Cam (SOHC) engine, 2A-13

Cylinder honing, 2C-17

D

Dashboard trim panels, removal and installation, 11-14

Daytime Running Lights (DRL), general information, 12-15

Disc brake
caliper, removal and installation, 9-9
pads, replacement, 9-5

Distributor (1993 and 1994 models), removal and installation, 5-9

Door latch, lock cylinder and handle, removal and installation, 11-11

Door lock system, power, description and check, 12-17

Door trim panel, removal and installation, 11-10

Door window glass
regulator, removal and installation, 11-12
removal and installation, 11-12

Door, removal, installation and adjustment, 11-10

Driveaxle boot, replacement, 8-9

Driveaxles
general information and inspection, 8-7
removal and installation, 8-7

Drivebelt check, adjustment and replacement, 1-15

Drum brake shoes, replacement, 9-11

Dual Overhead Cam (DOHC) engine, 2B-1 through 2B-22
camshafts and lifters, removal and installation, 2B-10
crankshaft front oil seal, replacement, 2B-16
crankshaft pulley, removal and installation, 2B-15
cylinder head, removal and installation, 2B-14
exhaust manifold, removal and installation, 2B-13
flywheel/driveplate, removal and installation, 2B-19
intake manifold, removal and installation, 2B-11
oil pan, removal and installation, 2B-16
oil pump, removal, inspection and installation, 2B-18
powertrain mounts, check and replacement, 2B-20
rear main oil seal, replacement, 2B-20
repair operations possible with the engine in the vehicle, 2B-3
timing chain and sprockets, removal, inspection and installation, 2B-7
Top Dead Center (TDC) for number one piston, locating, 2B-3
valve clearance check and adjustment, 2B-5
valve cover, removal and installation, 2B-4
valve springs, retainers and seals, replacement, 2B-6

E

Electric side view mirrors, description and check, 12-17

Electric sunroof, description and check, 12-17

Electrical troubleshooting, general information, 12-1

Emissions and engine control systems, 6-1 through 6-22
camshaft position sensor, check and replacement, 6-13
catalytic converter, 6-22
crankshaft position sensor, check and replacement, 6-12
engine coolant temperature sensor, check and replacement, 6-11
evaporative emissions control system, 6-21
exhaust gas recirculation system, 6-19
fuel temperature sensor, check and replacement, 6-16
idle air control system, 6-17
information sensors, 6-3
intake air temperature sensor, check and replacement, 6-11
knock sensor, check and replacement, 6-15
manifold absolute pressure sensor and solenoid valve, check and replacement, 6-10
mass airflow sensor, check and replacement, 6-9
obtaining diagnostic system trouble codes, 6-4
On-Board Diagnosis (OBD) system and trouble codes, 6-3
oxygen sensor, check and replacement, 6-14
positive crankcase ventilation system, 6-19
power steering pressure switch, check and replacement, 6-13
power valve control system, 6-18
Powertrain Control Module, removal and installation, 6-8
throttle position sensor, check, replacement and adjustment, 6-8
vehicle speed sensor, check and replacement, 6-16

Engine coolant temperature sensor, check and replacement, 6-11

Engine cooling fan and circuit, check and component replacement, 3-4
Engine electrical systems, 5-1 through 5-14
  alternator, removal and installation, 5-11
  battery
    cables, check and replacement, 5-4
    check and replacement, 5-3
    emergency jump starting, 0-16
  charging system
    check, 5-10
    general information and precautions, 5-10
  distributor (1993 and 1994 models), removal and installation, 5-9
  ignition coil, check and replacement, 5-7
  ignition system
    check, 5-5
    general information, 5-5
  ignition timing, check and adjustment, 5-10
  power transistor (1993 and 1994 models), check and replacement, 5-8
  starter motor and circuit, in-vehicle check, 5-13
  starter motor, removal and installation, 5-13
  starter solenoid, replacement, 5-13
  starting system, general information and precautions, 5-12
Engine idle speed, check and adjustment, 4-12
Engine oil and filter change, 1-11
Engines
  Single Overhead Cam (SOHC) engine, 2A-1 through 2A-20
    camshafts, lifters and seals, removal and installation, 2A-9
    crankshaft front oil seal, replacement, 2A-14
    crankshaft pulley, removal and installation, 2A-14
    cylinder heads, removal and installation, 2A-13
    exhaust manifolds, removal and installation, 2A-12
    flywheel/driveplate, removal and installation, 2A-17
    intake manifold, removal and installation, 2A-11
    oil pan, removal and installation, 2A-15
    oil pump, removal, inspection and installation, 2A-16
    powertrain mounts, check and replacement, 2A-18
    repair operations possible with the engine in the vehicle, 2A-2
    rear main oil seal, replacement, 2A-18
    rocker arm assembly, removal, inspection and installation, 2A-4
    timing belt and sprockets, removal and installation, 2A-6
    Top Dead Center (TDC) for number one piston, locating, 2A-3
    valve covers, removal and installation, 2A-4
    valve springs, retainers and seals, replacement, 2A-5
  Dual Overhead Cam (DOHC) engine, 2B-1 through 2B-22
    camshafts and lifters, removal and installation, 2B-10
    crankshaft front oil seal, replacement, 2B-16
    crankshaft pulley, removal and installation, 2B-15
    cylinder head, removal and installation, 2B-14
    exhaust manifold, removal and installation, 2B-13
    flywheel/driveplate, removal and installation, 2B-19
    intake manifold, removal and installation, 2B-11
    oil pan, removal and installation, 2B-16
    oil pump, removal, inspection and installation, 2B-18
    powertrain mounts, check and replacement, 2B-20
    rear main oil seal, replacement, 2B-20
    repair operations possible with the engine in the vehicle, 2B-3
    timing chain and sprockets, removal, inspection and installation, 2B-7
    Top Dead Center (TDC) for number one piston, locating, 2B-3
    valve clearance check and adjustment, 2B-5
    valve cover, removal and installation, 2B-4
    valve springs, retainers and seals, replacement, 2B-6
  General engine overhaul procedures, 2C-1 through 2C-28
    camshafts, lifters and bearings, inspection, 2C-21
    crankshaft
      inspection, 2C-19
      installation and main bearing oil clearance check, 2C-23
      removal, 2C-15
    cylinder compression check, 2C-7
    cylinder head
      cleaning and inspection, 2C-12
      disassembly, 2C-11
      reassembly, 2C-14
    cylinder honing, 2C-17
    engine block
      cleaning, 2C-16
      inspection, 2C-16
    engine overhaul
      disassembly sequence, 2C-11
      reassembly sequence, 2C-22
    engine rebuilding alternatives, 2C-8
    engine removal, methods and precautions, 2C-9
    engine, removal and installation, 2C-9
    initial start-up and break-in after overhaul, 2C-27
    main and connecting rod bearings, inspection and main bearing selection, 2C-20
    oil pressure check, 2C-6
    piston rings, installation, 2C-22
    pistons/connecting rods
      inspection, 2C-18
      installation and rod bearing oil clearance check, 2C-25
      removal, 2C-14
    rear main oil seal installation, 2C-25
    vacuum gauge diagnostic checks, 2C-7
    valves, servicing, 2C-13
Evaporative emissions control system, 6-21
Exhaust Gas Recirculation (EGR) valve check, 1-27
Exhaust gas recirculation system, 6-19
Exhaust manifold, removal and installation
  Dual Overhead Cam (DOHC) engine, 2B-13
  Single Overhead Cam (SOHC) engine, 2A-12
Exhaust system
  check, 1-19
  servicing, general information, 4-12

Index

F

Fender, front, removal and installation, 11-8
Fluid level checks, 1-7
   automatic transmission, 1-11
   battery electrolyte, 1-8
   brake and clutch fluid, 1-8
   engine coolant, 1-8
   engine oil, 1-7
   windshield washer fluid, 1-8
Flywheel/driveplate, removal and installation
   Dual Overhead Cam (DOHC) engine, 2B-19
   Single Overhead Cam (SOHC) engine, 2A-17
Fuel and exhaust systems, 4-1 through 4-14
   accelerator cable, removal, installation and adjustment, 4-8
   air filter assembly, removal and installation, 4-7
   engine idle speed, check and adjustment, 4-12
   exhaust system servicing, general information, 4-12
   fuel injection system
      check, 4-9
      general information, 4-8
   fuel level sending unit, check and replacement, 4-6
   fuel pressure relief procedure, 4-3
   fuel pump, removal and installation, 4-4
   fuel pump/fuel pressure, check, 4-3
   lines and fittings, replacement, 4-4
   pressure regulator, removal and installation, 4-10
   rail and injectors, removal and installation, 4-10
   tank
      cleaning and repair, 4-7
      removal and installation, 4-6
   throttle body, removal and installation, 4-10
Fuel filter replacement, 1-20
Fuel injection system
   check, 4-9
   general information, 4-8
Fuel system check, 1-18
Fuel temperature sensor, check and replacement, 6-16
Fuses and fusible links, general information, 12-3

G

Gear, steering, removal and installation, 10-14
General engine overhaul procedures, 2C-1 through 2C-28
   camshafts, lifters and bearings, inspection, 2C-21
   crankshaft
      inspection, 2C-19
      installation and main bearing oil clearance check, 2C-23
      removal, 2C-15
   cylinder compression check, 2C-7
   cylinder head
      cleaning and inspection, 2C-12
      disassembly, 2C-11
      reassembly, 2C-14
   cylinder honing, 2C-17
   engine block
      cleaning, 2C-16
      inspection, 2C-16
   engine overhaul
      disassembly sequence, 2C-11
      reassembly sequence, 2C-22
   engine rebuilding alternatives, 2C-8
   engine removal, methods and precautions, 2C-9
   engine, removal and installation, 2C-9
   initial start-up and break-in after overhaul, 2C-27
   main and connecting rod bearings, inspection and main bearing selection, 2C-20
   oil pressure check, 2C-6
   piston rings, installation, 2C-22
   pistons/connecting rods
      inspection, 2C-18
      installation and rod bearing oil clearance check, 2C-25
      removal, 2C-14
   rear main oil seal installation, 2C-25
   vacuum gauge diagnostic checks, 2C-7
   valves, servicing, 2C-13

H

Headlight
   bulb, replacement, 12-11
   housing, replacement, 12-12
Headlights, adjustment, 12-11
Heater and air conditioning control assembly, removal and installation, 3-9
Heater core, removal and installation, 3-9
Hinges and locks, maintenance, 11-5
Hood and hood support struts, removal, installation and adjustment, 11-5
Hood latch and release cable, removal and installation, 11-6
Horn, check and replacement, 12-15
Hub and wheel bearing assembly, removal and installation
   front, 10-7
   rear, 10-10

I

Idle air control system, 6-17
Ignition
   coil, check and replacement, 5-7
   switch and key lock cylinder, check and replacement, 12-7
   system
      check, 5-5
      general information, 5-5
   timing, check and adjustment, 5-10
Information sensors, 6-3
Initial start-up and break-in after overhaul, 2C-27
Instrument cluster, removal and installation, 12-9

Instrument panel
  gauges, check, 12-9
  removal and installation, 11-16
  switches, check and replacement, 12-8
Intake air temperature sensor, check and replacement, 6-11
Intake manifold, removal and installation
  Dual Overhead Cam (DOHC) engine, 2B-11
  Single Overhead Cam (SOHC) engine, 2A-11
Introduction to the Nissan Maxima, 0-5
In-Vehicle Multiplexing System, description, 12-18

J

Jacking and towing, 0-17

K

Knock sensor, check and replacement, 6-15

M

Main and connecting rod bearings, inspection and main bearing selection, 2C-20
Maintenance schedule, 1-6
Manifold absolute pressure sensor and solenoid valve, check and replacement, 6-10
Manual transaxle, 7A-1 through 7A-6
  back-up light and neutral position switch, check and replacement, 7A-4
  lubricant change, 1-26
  lubricant level check, 1-20
  oil seal replacement, 7A-2
  overhaul, general information, 7A-6
  removal and installation, 7A-5
  shift linkage and lever, removal and installation, 7A-2
Mass airflow sensor, check and replacement, 6-9
Master cylinder, removal and installation, 9-14
Mirrors, removal and installation, 11-12
Mounts, powertrain, check and replacement
  Dual Overhead Cam (DOHC) engine, 2B-20
  Single Overhead Cam (SOHC) engine, 2A-18

N

Nissan Maxima Maintenance schedule, 1-6

O

Obtaining diagnostic system trouble codes, 6-4
Oil pan, removal and installation
  Dual Overhead Cam (DOHC) engine, 2B-16
  Single Overhead Cam (SOHC) engine, 2A-15
Oil pressure check, 2C-6
Oil pump, removal, inspection and installation
  Dual Overhead Cam (DOHC) engine, 2B-18
  Single Overhead Cam (SOHC) engine, 2A-16
On-Board Diagnosis (OBD) system and trouble codes, 6-3
  information sensors, 6-3
  obtaining diagnostic system trouble codes, 6-4
Oxygen sensor, check and replacement, 6-14

P

Package shelf, removal and installation, 11-18
Park Neutral Position switch, check, adjustment and replacement, 7B-7
Parking brake
  adjustment, 9-17
  cables, replacement, 9-17
Piston rings, installation, 2C-22
Pistons/connecting rods
  inspection, 2C-18
  installation and rod bearing oil clearance check, 2C-25
  removal, 2C-14
Positive Crankcase Ventilation (PCV) valve and hose check and replacement, 1-23
Positive Crankcase Ventilation system, 6-19
Power brake booster, check, replacement and adjustment, 9-16
Power seats, description and check, 12-18
Power steering
  fluid level check, 1-10
  pressure switch, check and replacement, 6-13
  pump, removal and installation, 10-16
  system, bleeding, 10-16
Powertrain Control Module, removal and installation, 6-8
Powertrain mounts, check and replacement
  Dual Overhead Cam (DOHC) engine, 2B-20
  Single Overhead Cam (SOHC) engine, 2A-18
Proportioning valve, replacement, 9-15

R

Radiator and coolant reservoir, removal and installation, 3-5
Radiator grille, removal and installation, 11-5
Radio and speakers, removal and installation, 12-9
Rear main oil seal installation, 2C-25
Rear main oil seal, replacement, 2A-18
Rear suspension arms (1993 and 1994 models), removal and installation, 10-11
Relays, general information and testing, 12-3
Repair operations possible with the engine in the vehicle
  Dual Overhead Cam (DOHC) engine, 2B-3
  Single Overhead Cam (SOHC) engine, 2A-2

Index

Rocker arm assembly, removal, inspection and installation, Single Overhead Cam (SOHC) engine, 2A-4

S

Safety first!, 0-20
Seat belt check, 1-12
Seats, removal and installation, 11-18
Shift
  cable, check, adjustment and replacement, 7B-4
  interlock system, description, check and component replacement, 7B-6
  lever, removal and installation, 7B-5
  linkage and lever, removal and installation, 7A-2
Single Overhead Cam (SOHC) engine, 2A-1 through 2A-20
  camshafts, lifters and seals, removal and installation, 2A-9
  crankshaft front oil seal, replacement, 2A-14
  crankshaft pulley, removal and installation, 2A-14
  cylinder heads, removal and installation, 2A-13
  exhaust manifolds, removal and installation, 2A-12
  flywheel/driveplate, removal and installation, 2A-17
  intake manifold, removal and installation, 2A-11
  oil pan, removal and installation, 2A-15
  oil pump, removal, inspection and installation, 2A-16
  powertrain mounts, check and replacement, 2A-18
  rear main oil seal, replacement, 2A-18
  repair operations possible with the engine in the vehicle, 2A-2
  rocker arm assembly, removal, inspection and installation, 2A-4
  timing belt and sprockets, removal and installation, 2A-6
  Top Dead Center (TDC) for number one piston, locating, 2A-3
  valve covers, removal and installation, 2A-4
  valve springs, retainers and seals, replacement, 2A-5
Spark plug
  check and replacement, 1-21
  wire, distributor cap and rotor check and replacement, 1993 and 1994 models, 1-22
Stabilizer bar, removal and installation
  front, 10-7
  rear (1993 and 1994 models), 10-10
Starter motor
  and circuit, in-vehicle check, 5-13
  removal and installation, 5-13
Starter solenoid, replacement, 5-13
Starting system, general information and precautions, 5-12
Steering column
  covers, removal and installation, 11-15
  switches, check and replacement, 12-5
Steering knuckle and hub, removal and installation, 10-6

Steering system
  alignment, general information, 10-17
  power steering
    pump, removal and installation, 10-16
    system, bleeding, 10-16
  steering gear boots, replacement, 10-14
  steering gear, removal and installation, 10-14
  steering wheel, removal and installation, 10-12
  tie-rod ends, removal and installation, 10-13
  wheel studs, replacement, 10-16
  wheels and tires, general information, 10-17
Strut or coil spring, replacement, 10-5
Strut or shock absorber/coil spring assembly (rear), removal, inspection and installation, 10-9
Strut/coil spring assembly (front), removal, inspection and installation, 10-5
Sunroof, adjustment, 11-18
Support bearing assembly, check and replacement, 8-9
Suspension and steering systems, 10-1 through 10-18
Suspension system
  balljoints, replacement, 10-9
  control arm (front), removal, inspection and installation, 10-7
  hub and wheel bearing assembly, removal and installation
    front, 10-7
    rear, 10-10
  rear axle beam, lateral link and control rod (1995 and later models), removal and installation, 10-11
  rear suspension arms (1993 and 1994 models), removal and installation, 10-11
  stabilizer bar, removal and installation
    front, 10-7
    rear (1993 and 1994 models), 10-10
  steering knuckle and hub, removal and installation, 10-6
  strut or coil spring, replacement, 10-5
  strut or shock absorber/coil spring assembly (rear), removal, inspection and installation, 10-9
  strut/coil spring assembly (front), removal, inspection and installation, 10-5
Suspension, steering and driveaxle boot check, 1-26

T

Thermostat, check and replacement, 3-3
Throttle body, removal and installation, 4-10
Throttle position sensor, check, replacement and adjustment, 6-8
Tie-rod ends, removal and installation, 10-13
Timing belt and sprockets, removal and installation, Single Overhead Cam (SOHC) engine, 2A-6
Timing chain and sprockets, removal, inspection and installation, Dual Overhead Cam (DOHC) engine, 2B-7
Tire and tire pressure checks, 1-9
Tire rotation, 1-17
Tools, 0-11

Top Dead Center (TDC) for number one piston, locating
   Dual Overhead Cam (DOHC) engine, 2B-3
   Single Overhead Cam (SOHC) engine, 2A-3
Transaxle, automatic, 7B-1 through 7B-10
   diagnosis and trouble codes, 7B-2
   fluid change, 1-25
   fluid level check, 1-11
   Park Neutral Position switch, check, adjustment and replacement, 7B-7
   removal and installation, 7B-7
   shift cable, check, adjustment and replacement, 7B-4
   shift interlock system, description, check and component replacement, 7B-6
   shift lever, removal and installation, 7B-5
Transaxle, manual, 7A-1 through 7A-6
   back-up light and neutral position switch, check and replacement, 7A-4
   lubricant change, 1-26
   lubricant level check, 1-20
   oil seal replacement, 7A-2
   overhaul, general information, 7A-6
   removal and installation, 7A-5
   shift linkage and lever, removal and installation, 7A-2
Transistor, power (1993 and 1994 models), check and replacement, 5-8
Troubleshooting, 0-21
Trunk lid
   latch and lock cylinder, removal and installation, 11-13
   removal, installation and adjustment, 11-13
Trunk release and fuel door cables, removal and installation, 11-14
Tune-up and routine maintenance, 1-1 through 1-28
Tune-up general information, 1-7
Turn signal/hazard flashers, check and replacement, 12-5

U

Underhood hose check and replacement, 1-16
Upholstery and carpets, maintenance, 11-1

V

Vacuum gauge diagnostic checks, 2C-7
Valve clearance check and adjustment, Dual Overhead Cam (DOHC) engine, 2B-5
Valve control system, power, 6-18
Valve cover, removal and installation
   Dual Overhead Cam (DOHC) engine, 2B-4
   Single Overhead Cam (SOHC) engine, 2A-4
Valve springs, retainers and seals, replacement
   Dual Overhead Cam (DOHC) engine, 2B-6
   Single Overhead Cam (SOHC) engine, 2A-5
Valves, servicing, 2C-13
Vehicle identification numbers, 0-6
Vehicle speed sensor, check and replacement, 6-16
Vinyl trim, maintenance, 11-1

W

Water pump, check and replacement, 3-6
Wheel cylinder, removal and installation, 9-14
Wheel studs, replacement, 10-16
Wheels and tires, general information, 10-17
Window defogger, rear, check and repair, 12-15
Window system, power, description and check, 12-16
Windshield and fixed glass, replacement, 11-5
Wiper blade inspection and replacement, 1-12
Wiper motor, removal and installation, 12-14
Wiring diagrams, general information, 12-19
Working facilities, 0-15

Haynes Automotive Manuals

NOTE: New manuals are added to this list on a periodic basis. If you do not see a listing for your vehicle, consult your local Haynes dealer for the latest product information.

ACURA
- *12020 Integra '86 thru '89 & Legend '86 thru '90

AMC
- Jeep CJ - see JEEP (50020)
- 14020 Mid-size models, Concord, Hornet, Gremlin & Spirit '70 thru '83
- 14025 (Renault) Alliance & Encore '83 thru '87

AUDI
- 15020 4000 all models '80 thru '87
- 15025 5000 all models '77 thru '83
- 15026 5000 all models '84 thru '88

AUSTIN-HEALEY
- Sprite - see MG Midget (66015)

BMW
- *18020 3/5 Series not including diesel or all-wheel drive models '82 thru '92
- *18021 3 Series except 325iX models '92 thru '97
- 18025 320i all 4 cyl models '75 thru '83
- 18035 528i & 530i all models '75 thru '80
- 18050 1500 thru 2002 except Turbo '59 thru '77

BUICK
- Century (front wheel drive) - see GM (829)
- *19020 Buick, Oldsmobile & Pontiac Full-size (Front wheel drive) all models '85 thru '98
  Buick Electra, LeSabre and Park Avenue;
  Oldsmobile Delta 88 Royale, Ninety Eight and Regency; Pontiac Bonneville
- 19025 Buick Oldsmobile & Pontiac Full-size (Rear wheel drive)
  Buick Estate '70 thru '90, Electra '70 thru '84, LeSabre '70 thru '85, Limited '74 thru '79
  Oldsmobile Custom Cruiser '70 thru '90, Delta 88 '70 thru '85,Ninety-eight '70 thru '84
  Pontiac Bonneville '70 thru '81, Catalina '70 thru '81, Grandville '70 thru '75, Parisienne '83 thru '86
- 19030 Mid-size Regal & Century all rear-drive models with V6, V8 and Turbo '74 thru '87
  Regal - see GENERAL MOTORS (38010)
  Riviera - see GENERAL MOTORS (38030)
  Roadmaster - see CHEVROLET (24046)
  Skyhawk - see GENERAL MOTORS (38015)
  Skylark '80 thru '85 - see GM (38020)
  Skylark '86 on - see GM (38025)
  Somerset - see GENERAL MOTORS (38025)

CADILLAC
- *21030 Cadillac Rear Wheel Drive all gasoline models '70 thru '93
  Cimarron - see GENERAL MOTORS (38015)
  Eldorado - see GENERAL MOTORS (38030)
  Seville '80 thru '85 - see GM (38030)

CHEVROLET
- *24010 Astro & GMC Safari Mini-vans '85 thru '93
- 24015 Camaro V8 all models '70 thru '81
- 24016 Camaro all models '82 thru '92
  Cavalier - see GENERAL MOTORS (38015)
  Celebrity - see GENERAL MOTORS (38005)
- 24017 Camaro & Firebird '93 thru '97
- 24020 Chevelle, Malibu & El Camino '69 thru '87
- 24024 Chevette & Pontiac T1000 '76 thru '87
  Citation - see GENERAL MOTORS (38020)
- *24032 Corsica/Beretta all models '87 thru '96
- 24040 Corvette all V8 models '68 thru '82
- *24041 Corvette all models '84 thru '96
- 10305 Chevrolet Engine Overhaul Manual
- 24045 Full-size Sedans Caprice, Impala, Biscayne, Bel Air & Wagons '69 thru '90
- 24046 Impala SS & Caprice and Buick Roadmaster '91 thru '96
  Lumina - see GENERAL MOTORS (38010)
- 24048 Lumina & Monte Carlo '95 thru '98
  Lumina APV - see GM (38035)
- 24050 Luv Pick-up all 2WD & 4WD '72 thru '82
- *24055 Monte Carlo all models '70 thru '88
  Monte Carlo '95 thru '98 - see LUMINA (24048)
- 24059 Nova all V8 models '69 thru '79
- *24060 Nova and Geo Prizm '85 thru '92
- 24064 Pick-ups '67 thru '87 - Chevrolet & GMC, all V8 & in-line 6 cyl '67 thru '87; Suburbans, Blazers & Jimmys '67 thru '91
- *24065 Pick-ups '88 thru '98 - Chevrolet & GMC, all full-size pick-ups, '88 thru '98; Blazer & Jimmy '92 thru '94; Suburban '92 thru '98; Tahoe & Yukon '98
- 24070 S-10 & S-15 Pick-ups '82 thru '93, Blazer & Jimmy '83 thru '94,
- *24071 S-10 & S-15 Pick-ups '94 thru '96 Blazer & Jimmy '95 thru '96
- *24075 Sprint & Geo Metro '85 thru '94
- *24080 Vans - Chevrolet & GMC, V8 & in-line 6 cylinder models '68 thru '96

CHRYSLER
- 25015 Chrysler Cirrus, Dodge Stratus, Plymouth Breeze '95 thru '98
- 25025 Chrysler Concorde, New Yorker & LHS, Dodge Intrepid, Eagle Vision, '93 thru '97
- 10310 Chrysler Engine Overhaul Manual
- *25020 Full-size Front-Wheel Drive '88 thru '93
  K-Cars - see DODGE Aries (30008)
  Laser - see DODGE Daytona (30030)
- *25030 Chrysler & Plymouth Mid-size front wheel drive '82 thru '95
  Rear-wheel Drive - see Dodge (30050)

DATSUN
- 28005 200SX all models '80 thru '83
- 28007 B-210 all models '73 thru '78
- 28009 210 all models '79 thru '82
- 28012 240Z, 260Z & 280Z Coupe '70 thru '78
- 28014 280ZX Coupe & 2+2 '79 thru '83
  300ZX - see NISSAN (72010)
- 28016 310 all models '78 thru '82
- 28018 510 & PL521 Pick-up '68 thru '73
- 28020 510 all models '78 thru '81
- 28022 620 Series Pick-up all models '73 thru '79
  720 Series Pick-up - see NISSAN (72030)
- 28025 810/Maxima all gasoline models, '77 thru '84

DODGE
- 400 & 600 - see CHRYSLER (25030)
- *30008 Aries & Plymouth Reliant '81 thru '89
- 30010 Caravan & Plymouth Voyager Mini-Vans all models '84 thru '95
- *30011 Caravan & Plymouth Voyager Mini-Vans all models '96 thru '98
- 30012 Challenger/Plymouth Saporro '78 thru '83
- 30016 Colt & Plymouth Champ (front wheel drive) all models '78 thru '87
- *30020 Dakota Pick-ups all models '87 thru '96
- 30025 Dart, Demon, Plymouth Barracuda, Duster & Valiant 6 cyl models '67 thru '76
- *30030 Daytona & Chrysler Laser '84 thru '89
  Intrepid - see CHRYSLER (25025)
- *30034 Neon all models '95 thru '97
- *30035 Omni & Plymouth Horizon '78 thru '90
- *30040 Pick-ups all full-size models '74 thru '93
- *30041 Pick-ups all full-size models '94 thru '96
- *30045 Ram 50/D50 Pick-ups & Raider and Plymouth Arrow Pick-ups '79 thru '93
- 30050 Dodge/Plymouth/Chrysler rear wheel drive '71 thru '89
- *30055 Shadow & Plymouth Sundance '87 thru '94
- *30060 Spirit & Plymouth Acclaim '89 thru '95
- *30065 Vans - Dodge & Plymouth '71 thru '96

EAGLE
- Talon - see Mitsubishi Eclipse (68030)
- Vision - see CHRYSLER (25025)

FIAT
- 34010 124 Sport Coupe & Spider '68 thru '78
- 34025 X1/9 all models '74 thru '80

FORD
- 10355 Ford Automatic Transmission Overhaul
- *36004 Aerostar Mini-vans all models '86 thru '96
- *36006 Contour & Mercury Mystique '95 thru '98
- 36008 Courier Pick-up all models '72 thru '82
- 36012 Crown Victoria & Mercury Grand Marquis '88 thru '96
- 10320 Ford Engine Overhaul Manual
- 36016 Escort/Mercury Lynx all models '81 thru '90
- *36020 Escort/Mercury Tracer '91 thru '96
- *36024 Explorer & Mazda Navajo '91 thru '95
- 36028 Fairmont & Mercury Zephyr '78 thru '83
- 36030 Festiva & Aspire '88 thru '97
- 36032 Fiesta all models '77 thru '80
- 36036 Ford & Mercury Full-size,
  Ford LTD & Mercury Marquis ('75 thru '82);
  Ford Custom 500,Country Squire, Crown Victoria & Mercury Colony Park ('75 thru '87);
  Ford LTD Crown Victoria & Mercury Gran Marquis ('83 thru '87)
- 36040 Granada & Mercury Monarch '75 thru '80
- 36044 Ford & Mercury Mid-size,
  Ford Thunderbird & Mercury Cougar ('75 thru '82);
  Ford LTD & Mercury Marquis ('83 thru '86);
  Ford Torino,Gran Torino, Elite, Ranchero pick-up, LTD II, Mercury Montego, Comet, XR-7 & Lincoln Versailles ('75 thru '86)
- 36048 Mustang V8 all models '64-1/2 thru '73
- 36049 Mustang II 4 cyl, V6 & V8 models '74 thru '78
- 36050 Mustang & Mercury Capri all models Mustang, '79 thru '93; Capri, '79 thru '86
- *36051 Mustang all models '94 thru '97
- 36054 Pick-ups & Bronco '73 thru '79
- 36058 Pick-ups & Bronco '80 thru '96
- 36059 Pick-ups, Expedition & Mercury Navigator '97 thru '98
- 36062 Pinto & Mercury Bobcat '75 thru '80
- 36066 Probe all models '89 thru '92
- 36070 Ranger/Bronco II gasoline models '83 thru '92
- *36071 Ranger '93 thru '97 & Mazda Pick-ups '94 thru '97
- 36074 Taurus & Mercury Sable '86 thru '95
- *36075 Taurus & Mercury Sable '96 thru '98
- *36078 Tempo & Mercury Topaz '84 thru '94
- *36082 Thunderbird/Mercury Cougar '83 thru '88
- *36086 Thunderbird/Mercury Cougar '89 and '97
- 36090 Vans all V8 Econoline models '69 thru '91
- *36094 Vans full size '92-'95
- *36097 Windstar Mini-van '95-'98

GENERAL MOTORS
- *10360 GM Automatic Transmission Overhaul
- *38005 Buick Century, Chevrolet Celebrity, Oldsmobile Cutlass Ciera & Pontiac 6000 all models '82 thru '96
- *38010 Buick Regal, Chevrolet Lumina, Oldsmobile Cutlass Supreme & Pontiac Grand Prix front-wheel drive models '88 thru '95
- *38015 Buick Skyhawk, Cadillac Cimarron, Chevrolet Cavalier, Oldsmobile Firenza & Pontiac J-2000 & Sunbird '82 thru '94
- *38016 Chevrolet Cavalier & Pontiac Sunfire '95 thru '98
- 38020 Buick Skylark, Chevrolet Citation, Olds Omega, Pontiac Phoenix '80 thru '85
- 38025 Buick Skylark & Somerset, Oldsmobile Achieva & Calais and Pontiac Grand Am all models '85 thru '95
- 38030 Cadillac Eldorado '71 thru '85, Seville '80 thru '85,
  Oldsmobile Toronado '71 thru '85 & Buick Riviera '79 thru '85
- *38035 Chevrolet Lumina APV, Olds Silhouette & Pontiac Trans Sport all models '90 thru '95
  General Motors Full-size Rear-wheel Drive - see BUICK (19025)

(Continued on other side)

---

* Listings shown with an asterisk (*) indicate model coverage as of this printing. These titles will be periodically updated to include later model years - consult your Haynes dealer for more information.

Haynes North America, Inc., 861 Lawrence Drive, Newbury Park, CA 91320-1514 • (805) 498-6703

Haynes Automotive Manuals (continued)

NOTE: New manuals are added to this list on a periodic basis. If you do not see a listing for your vehicle, consult your local Haynes dealer for the latest product information.

GEO
- Metro - see CHEVROLET Sprint (24075)
- Prizm - '85 thru '92 see CHEVY (24060), '93 thru '96 see TOYOTA Corolla (92036)
- *40030 Storm all models '90 thru '93
- Tracker - see SUZUKI Samurai (90010)

GMC
- Safari - see CHEVROLET ASTRO (24010)
- Vans & Pick-ups - see CHEVROLET

HONDA
- 42010 Accord CVCC all models '76 thru '83
- 42011 Accord all models '84 thru '89
- 42012 Accord all models '90 thru '93
- 42013 Accord all models '94 thru '95
- 42020 Civic 1200 all models '73 thru '79
- 42021 Civic 1300 & 1500 CVCC '80 thru '83
- 42022 Civic 1500 CVCC all models '75 thru '79
- 42023 Civic all models '84 thru '91
- *42024 Civic & del Sol '92 thru '95
- *42040 Prelude CVCC all models '79 thru '89

HYUNDAI
- *43015 Excel all models '86 thru '94

ISUZU
- Hombre - see CHEVROLET S-10 (24071)
- *47017 Rodeo '91 thru '97; Amigo '89 thru '94; Honda Passport '95 thru '97
- *47020 Trooper & Pick-up, all gasoline models Pick-up, '81 thru '93; Trooper, '84 thru '91

JAGUAR
- *49010 XJ6 all 6 cyl models '68 thru '86
- *49011 XJ6 all models '88 thru '94
- *49015 XJ12 & XJS all 12 cyl models '72 thru '85

JEEP
- *50010 Cherokee, Comanche & Wagoneer Limited all models '84 thru '96
- 50020 CJ all models '49 thru '86
- *50025 Grand Cherokee all models '93 thru '98
- 50029 Grand Wagoneer & Pick-up '72 thru '91 Grand Wagoneer '84 thru '91, Cherokee & Wagoneer '72 thru '83, Pick-up '72 thru '88
- *50030 Wrangler all models '87 thru '95

LINCOLN
- Navigator - see FORD Pick-up (36059)
- 59010 Rear Wheel Drive all models '70 thru '96

MAZDA
- 61010 GLC Hatchback (rear wheel drive) '77 thru '83
- 61011 GLC (front wheel drive) '81 thru '85
- *61015 323 & Protegé '90 thru '97
- *61016 MX-5 Miata '90 thru '97
- *61020 MPV all models '89 thru '94
- Navajo - see Ford Explorer (36024)
- 61030 Pick-ups '72 thru '93
- Pick-ups '94 thru '96 - see Ford Ranger (36071)
- 61035 RX-7 all models '79 thru '85
- *61036 RX-7 all models '86 thru '91
- 61040 626 (rear wheel drive) all models '79 thru '82
- *61041 626/MX-6 (front wheel drive) '83 thru '91

MERCEDES-BENZ
- 63012 123 Series Diesel '76 thru '85
- *63015 190 Series four-cyl gas models, '84 thru '88
- 63020 230/250/280 6 cyl sohc models '68 thru '72
- 63025 280 123 Series gasoline models '77 thru '81
- 63030 350 & 450 all models '71 thru '80

MERCURY
- See FORD Listing.

MG
- 66010 MGB Roadster & GT Coupe '62 thru '80
- 66015 MG Midget, Austin Healey Sprite '58 thru '80

MITSUBISHI
- *68020 Cordia, Tredia, Galant, Precis & Mirage '83 thru '93
- *68030 Eclipse, Eagle Talon & Ply. Laser '90 thru '94
- *68040 Pick-up '83 thru '96 & Montero '83 thru '93

NISSAN
- 72010 300ZX all models including Turbo '84 thru '89
- *72015 Altima all models '93 thru '97
- *72020 Maxima all models '85 thru '91
- *72030 Pick-ups '80 thru '96 Pathfinder '87 thru '95
- 72040 Pulsar all models '83 thru '86
- *72050 Sentra all models '82 thru '94
- *72051 Sentra & 200SX all models '95 thru '98
- *72060 Stanza all models '82 thru '90

OLDSMOBILE
- *73015 Cutlass V6 & V8 gas models '74 thru '88
- For other OLDSMOBILE titles, see BUICK, CHEVROLET or GENERAL MOTORS listing.

PLYMOUTH
- For PLYMOUTH titles, see DODGE listing.

PONTIAC
- 79008 Fiero all models '84 thru '88
- 79018 Firebird V8 models except Turbo '70 thru '81
- 79019 Firebird all models '82 thru '92
- For other PONTIAC titles, see BUICK, CHEVROLET or GENERAL MOTORS listing.

PORSCHE
- *80020 911 except Turbo & Carrera 4 '65 thru '89
- 80025 914 all 4 cyl models '69 thru '76
- 80030 924 all models including Turbo '76 thru '82
- *80035 944 all models including Turbo '83 thru '89

RENAULT
- Alliance & Encore - see AMC (14020)

SAAB
- *84010 900 all models including Turbo '79 thru '88

SATURN
- 87010 Saturn all models '91 thru '96

SUBARU
- 89002 1100, 1300, 1400 & 1600 '71 thru '79
- *89003 1600 & 1800 2WD & 4WD '80 thru '94

SUZUKI
- *90010 Samurai/Sidekick & Geo Tracker '86 thru '96

TOYOTA
- 92005 Camry all models '83 thru '91
- 92006 Camry all models '92 thru '96
- 92015 Celica Rear Wheel Drive '71 thru '85
- *92020 Celica Front Wheel Drive '86 thru '93
- 92025 Celica Supra all models '79 thru '92
- 92030 Corolla all models '75 thru '79
- 92032 Corolla all rear wheel drive models '80 thru '87
- 92035 Corolla all front wheel drive models '84 thru '92
- *92036 Corolla & Geo Prizm '93 thru '97
- 92040 Corolla Tercel all models '80 thru '82
- 92045 Corona all models '74 thru '82
- 92050 Cressida all models '78 thru '82
- 92055 Land Cruiser FJ40, 43, 45, 55 '68 thru '82
- 92056 Land Cruiser FJ60, 62, 80, FZJ80 '80 thru '96
- *92065 MR2 all models '85 thru '87
- 92070 Pick-up all models '69 thru '78
- *92075 Pick-up all models '79 thru '95
- *92076 Tacoma '95 thru '98, 4Runner '96 thru '98, & T100 '93 thru '98
- *92080 Previa all models '91 thru '95
- 92085 Tercel all models '87 thru '94

TRIUMPH
- 94007 Spitfire all models '62 thru '81
- 94010 TR7 all models '75 thru '81

VW
- 96008 Beetle & Karmann Ghia '54 thru '79
- 96012 Dasher all gasoline models '74 thru '81
- *96016 Rabbit, Jetta, Scirocco, & Pick-up gas models '74 thru '91 & Convertible '80 thru '92
- 96017 Golf & Jetta all models '93 thru '97
- 96020 Rabbit, Jetta & Pick-up diesel '77 thru '84
- 96030 Transporter 1600 all models '68 thru '79
- 96035 Transporter 1700, 1800 & 2000 '72 thru '79
- 96040 Type 3 1500 & 1600 all models '63 thru '73
- 96045 Vanagon all air-cooled models '80 thru '83

VOLVO
- 97010 120, 130 Series & 1800 Sports '61 thru '73
- 97015 140 Series all models '66 thru '74
- *97020 240 Series all models '76 thru '93
- 97025 260 Series all models '75 thru '82
- *97040 740 & 760 Series all models '82 thru '88

TECHBOOK MANUALS
- 10205 Automotive Computer Codes
- 10210 Automotive Emissions Control Manual
- 10215 Fuel Injection Manual, 1978 thru 1985
- 10220 Fuel Injection Manual, 1986 thru 1996
- 10225 Holley Carburetor Manual
- 10230 Rochester Carburetor Manual
- 10240 Weber/Zenith/Stromberg/SU Carburetors
- 10305 Chevrolet Engine Overhaul Manual
- 10310 Chrysler Engine Overhaul Manual
- 10320 Ford Engine Overhaul Manual
- 10330 GM and Ford Diesel Engine Repair Manual
- 10340 Small Engine Repair Manual
- 10345 Suspension, Steering & Driveline Manual
- 10355 Ford Automatic Transmission Overhaul
- 10360 GM Automatic Transmission Overhaul
- 10405 Automotive Body Repair & Painting
- 10410 Automotive Brake Manual
- 10415 Automotive Detailing Manual
- 10420 Automotive Eelectrical Manual
- 10425 Automotive Heating & Air Conditioning
- 10430 Automotive Reference Manual & Dictionary
- 10435 Automotive Tools Manual
- 10440 Used Car Buying Guide
- 10445 Welding Manual
- 10450 ATV Basics

SPANISH MANUALS
- 98903 Reparación de Carrocería & Pintura
- 98905 Códigos Automotrices de la Computadora
- 98910 Frenos Automotriz
- 98915 Inyección de Combustible 1986 al 1994
- 99040 Chevrolet & GMC Camionetas '67 al '87 Incluye Suburban, Blazer & Jimmy '67 al '91
- 99041 Chevrolet & GMC Camionetas '88 al '95 Incluye Suburban '92 al '95, Blazer & Jimmy '92 al '94, Tahoe y Yukon '95
- 99042 Chevrolet & GMC Camionetas Cerradas '68 al '95
- 99055 Dodge Caravan & Plymouth Voyager '84 al '95
- 99075 Ford Camionetas y Bronco '80 al '94
- 99077 Ford Camionetas Cerradas '69 al '91
- 99083 Ford Modelos de Tamaño Grande '75 al '87
- 99088 Ford Modelos de Tamaño Mediano '75 al '86
- 99091 Ford Taurus & Mercury Sable '86 al '95
- 99095 GM Modelos de Tamaño Grande '70 al '90
- 99100 GM Modelos de Tamaño Mediano '70 al '88
- 99110 Nissan Camionetas '80 al '96, Pathfinder '87 al '95
- 99118 Nissan Sentra '82 al '94
- 99125 Toyota Camionetas y 4Runner '79 al '95

Over 100 Haynes motorcycle manuals also available

* Listings shown with an asterisk (*) indicate model coverage as of this printing. These titles will be periodically updated to include later model years - consult your Haynes dealer for more information.

Haynes North America, Inc., 861 Lawrence Drive, Newbury Park, CA 91320-1514 • (805) 498-6703